ESSENTIALS OF
COLLEGE ALGEBRA

ESSENTIALS OF
COLLEGE ALGEBRA

Richard N. Aufmann

Richard D. Nation

Palomar College

Houghton Mifflin Company

Boston   New York

Publisher: Jack Shira
Senior Sponsoring Editor: Lynn Cox
Associate Editor: Jennifer King
Assistant Editor: Melissa Parkin
Senior Project Editor: Tamela Ambush
Editorial Assistant: Sage Anderson
Manufacturing Manager: Karen Banks
Senior Marketing Manager: Danielle Potvin
Marketing Coordinator: Nicole Mollica

Cover photograph: Getty Images, Inc./Taxi

PHOTO CREDITS:

Chapter P: *p. 1* NASA, N. Benitez (JHU), T. Broadhurst (The Hebrew University), H. Ford (JHU), M. Clampin (STScI), G. Hartig (STScI), G. Illingworth (UCO/Link Observatory), the ACS Science Team and ESA; *p. 2* AP/Wide World Photos; *p. 58* NASA, ESA, and the Hubble Heritage Team (STScI/AURA). **Chapter 1:** *p. 75* Bettmann /CORBIS; *p. 75* Spencer Grant / PhotoEdit, Inc.; *p. 87* Mona Lisa by Leonardo da Vinci. © Gianni Dagli Orti/CORBIS; *p. 92* © Heritage / The Image Works; *p. 102* The Granger Collection; *p. 106* Jeff Greenberg / PhotoEdit, Inc.; *p. 133* AP/Wide World Photos; **Chapter 2:** *p. 145* Charles O'Rear / CORBIS; *p. 208* CORBIS. **Chapter 3:** *p. 247* Sonda Dawes / The Image Works; *p. 248* David Young-Wolff / PhotoEdit, Inc.; *p. 258* Syndicated Features Limited / The Image Works; *p. 279* The Granger Collection; *p. 289* Bettmann/CORBIS; *p. 291* Bettmann / CORBIS; *p. 312* Richard T. Nowitz / CORBIS. **Chapter 4:** *p. 323* Chris McLaughlin / CORBIS; *p. 343* Bettmann / CORBIS; *p. 349* Charles O'Rear / CORBIS; *p. 349* David James / Getty Images; *p. 353* Bettmann / CORBIS; *p. 398* Tom Brakefield / CORBIS; *p. 405* Bettmann / CORBIS. **Chapter 5:** n/a.

Printed in the U.S.A.

Library of Congress Control Number: 2004109016

ISBNs:
Student's Edition: 0-618-48096-X
Instructor's Annotated Edition: 0-618-48097-8

1 2 3 4 5 6 7 8 9-VH-09 08 07 06 05

CONTENTS

5   SYSTEMS OF EQUATIONS   *415*

PREFACE

Essentials of College Algebra provides students with material that focuses on selected key concepts of college algebra and how those concepts can be applied to a variety of problems. To help students master these concepts, we have tried to maintain a balance among theory, application, modeling, and drill. Carefully developed mathematics is complemented by applications that are both contemporary and representative of a wide range of disciplines. Many application exercises are accompanied by a diagram that helps the student visualize the mathematics of the application.

Technology is introduced naturally to support and advance better understanding of a concept. The optional *Integrating Technology* boxes and graphing calculator exercises are designed to promote an appreciation of both the power and the limitations of technology.

Features

Interactive Presentation   *Essentials of College Algebra* is written in a style that encourages the student to interact with the textbook. At various places throughout the text, we pose a question to the student about the material being presented. This question encourages the reader to pause and think about the current discussion and to answer the question. To ensure that the student does not miss important information, the answer to the question is provided as a footnote on the same page.

Each section contains a variety of worked examples. Each example is given a title so that the student can see at a glance the type of problem being illustrated. Most examples are accompanied by annotations that assist the student in moving from step to step. Following the worked example is a suggested exercise for the student to work. The *complete solution* to that exercise can be found in an appendix in the text. This feature allows students to self-assess their progress and to get immediate feedback by means of not just an answer, but a complete solution.

Focus on Problem Solving   Each chapter begins with a *Focus on Problem Solving* that demonstrates various strategies that are used by successful problem solvers. At the completion of the *Focus on Problem Solving*, the student is directed to an exercise in the text that can be solved using the problem-solving strategy that was just discussed.

Mathematics and Technology   Technology is introduced in the text to illustrate or enhance a concept. We attempt to foster the idea that technology, combined with analytical thinking, can lead to deeper understanding of a concept. The optional *Integrating Technology* boxes and graphing calculator exercises are designed to develop an awareness of technology's capabilities and limitations.

Topics for Discussion are found at the end of each section of the text. These topics can form the basis for a group discussion or serve as writing assignments.

Extensive Exercise Sets   The exercise sets of *Essentials of College Algebra* are carefully developed to provide the student with a variety of exercises. The exercises range from drill and practice to interesting challenges and were chosen to illustrate the many facets of the topics discussed in the text. Each exercise set emphasizes concept building, skill building and maintenance, and, as appropriate, applications.

Projects are included at the end of each exercise set and are designed to encourage students to research and write about mathematics and its applications. These projects encourage critical thinking beyond the scope of the regular problem sets. Responses to the projects are given in the *Instructor's Solutions Manual.*

Prepare for Next Section Exercises are found at the end of each exercise set, except for the last section of a chapter. These exercises concentrate on topics from previous sections of the text that are particularly relevant to the next section of the text. By completing these exercises, the student reviews some of the concepts and skills necessary for success in the next section.

Chapter Review Exercises allow the student to review concepts and skills presented in the chapter. Answers to all chapter review exercises are included in the student answer section. If a student incorrectly answers an exercise, there is a section reference next to each answer that directs the student to the section from which that exercise was taken. Using this reference, the student can review the concepts that are required to correctly solve the exercise.

Chapter Tests provide students with an opportunity to self-assess their understanding of the concepts presented in the chapter. The answers to all exercises in the chapter tests are included in the student answer section. As with the answers to the chapter review exercises, there is a section reference next to each answer that directs the student to the section from which that exercise was taken.

Cumulative Review Exercises end every chapter after Chapter P. These exercises allow students to refresh their knowledge of previously studied skills and concepts and help them maintain skills that promote success in college algebra. Answers to all cumulative review exercises are included in the student answer section. As with the answers to the chapter review exercises, there is a section reference next to each answer that directs the student to the section from which that exercise was taken.

CHAPTER OPENER FEATURES

CHAPTER OPENER

Each chapter begins with a **Chapter Opener** that illustrates a specific application of a concept from the chapter. There is a reference to a particular exercise within the chapter that asks the student to solve a problem related to the chapter opener topic.

The icons   at the bottom of the page let students know of additional resources available on CD, video/DVD, in the *Student Study Guide*, and online at math.college.hmco.com/students.

Page 323

CHAPTER **4**

EXPONENTIAL AND LOGARITHMIC FUNCTIONS

4.1 INVERSE FUNCTIONS
4.2 EXPONENTIAL FUNCTIONS AND THEIR APPLICATIONS
4.3 LOGARITHMIC FUNCTIONS AND THEIR APPLICATIONS
4.4 LOGARITHMS AND LOGARITHMIC SCALES
4.5 EXPONENTIAL AND LOGARITHMIC EQUATIONS
4.6 EXPONENTIAL GROWTH AND DECAY

Modeling Data with an Exponential Function

The following table shows the time, in hours, before the body of a scuba diver, wearing a 5-millimeter-thick wet suit, reaches hypothermia (95°F) for various water temperatures.

Water Temperature, °F	Time, hours
36	1.5
41	1.8
46	2.6
50	3.1
55	4.9

Source: Data extracted from the *American Journal of Physics*, vol. 71, no. 4 (April 2003), Fig. 3, p. 336.

The following function, which is an example of an exponential function, closely models the data in the table:

$$T(F) = 0.1509(1.0639)^F$$

In this function F represents the Fahrenheit temperature of the water, and T represents the time in hours. A diver can use the function to determine the time it takes to reach hypothermia for water temperatures that are not included in the table.

Exponential functions can be used to model many other situations. Exercise 43 on page 348 uses an exponential function to estimate the growth of broadband Internet connections.

Page 324

 ON PROBLEM SOLVING

Use Two Methods to Solve and Compare Results

Sometimes it is possible to solve a problem in two or more ways. In such situations it is recommended that you use at least two methods to solve the problem, and compare your results. Here is an example of an application that can be solved in more than one way.

Example

In a league of eight basketball teams, each team plays every other team in the league exactly once. How many league games will take place?

Solution

Method 1: Use an analytic approach. Each of the eight teams must play the other seven teams. Using this information, you might be tempted to conclude that there will be $8 \cdot 7 = 56$ games, but this result is too large because it counts each game between two individual teams as two different games. Thus the number of league games will be

$$\frac{8 \cdot 7}{2} = \frac{56}{2} = 28$$

Method 2: Make an organized list. Use the letters A, B, C, D, E, F, G, and H to represent the eight teams. Use the notation AB to represent the game between team A and team B. Do not include BA in your list because it represents the same game between team A and team B.

AB AC AD AE AF AG AH
　　BC BD BE BF BG BH
　　　　CD CE CF CG CH
　　　　　　DE DF DG DH
　　　　　　　　EF EG EH
　　　　　　　　　　FG FH
　　　　　　　　　　　　GH

The list shows that there will be 28 league games.

The procedure of using two different solution methods and comparing results is employed often in this chapter. For instance, see Example 2, page 379. In this example, a solution is found by applying algebraic procedures and also by graphing. Notice that both methods produce the same result.

43.   **INTERNET CONNECTIONS** Data from Forrester Research suggest that the number of broadband [cable and digital subscriber line (DSL)] connections to the Internet can be modeled by $f(x) = 1.353(1.9025)^x$, where x is the number of years after January 1, 1998, and $f(x)$ is the number of connections in millions.

a. How many broadband Internet connections, to the nearest million, does this model predict will exist on January 1, 2005?

b. According to the model, in what year will the number of broadband connections first reach 300 million? [*Hint:* Use the intersect feature of a graphing utility to determine the *x*-coordinate of the point of intersection of the graphs of $f(x)$ and $y = 300$.]

Page 348

FOCUS ON PROBLEM SOLVING

A **Focus on Problem Solving** follows the Chapter Opener. This feature highlights and demonstrates a problem-solving strategy that may be used to successfully solve some of the problems presented in the chapter.

AUFMANN INTERACTIVE METHOD (AIM)

Page 330

INTERACTIVE PRESENTATION

Essentials of College Algebra is written in a style that encourages the student to interact with the textbook.

EXAMPLES

Each section contains a variety of worked examples. Examples are **titled** so that the student can see at a glance the type of problem being illustrated, often accompanied by **annotations** that assist the student in moving from step to step; and offers the **final answer in color** so that it is readily identifiable.

TRY EXERCISES

Following every example is a suggested **Try Exercise** from that section's exercise set for the student to work. The exercises are color coded by number in the exercise set and the *complete solution* to that exercise can be found in an appendix to the text.

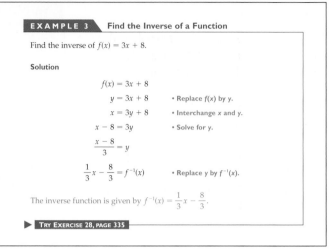

> **EXAMPLE 3**   Find the Inverse of a Function
>
> Find the inverse of $f(x) = 3x + 8$.
>
> **Solution**
>
> $$f(x) = 3x + 8$$
> $$y = 3x + 8 \qquad \text{• Replace } f(x) \text{ by } y.$$
> $$x = 3y + 8 \qquad \text{• Interchange } x \text{ and } y.$$
> $$x - 8 = 3y \qquad \text{• Solve for } y.$$
> $$\frac{x - 8}{3} = y$$
> $$\frac{1}{3}x - \frac{8}{3} = f^{-1}(x) \qquad \text{• Replace } y \text{ by } f^{-1}(x).$$
>
> The inverse function is given by $f^{-1}(x) = \frac{1}{3}x - \frac{8}{3}$.
>
> ▶ **TRY EXERCISE 28, PAGE 335**

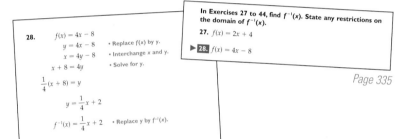

28.
$$f(x) = 4x - 8$$
$$y = 4x - 8 \qquad \text{• Replace } f(x) \text{ by } y.$$
$$x = 4y - 8 \qquad \text{• Interchange } x \text{ and } y.$$
$$x + 8 = 4y \qquad \text{• Solve for } y.$$
$$\frac{1}{4}(x + 8) = y$$
$$y = \frac{1}{4}x + 2$$
$$f^{-1}(x) = \frac{1}{4}x + 2 \qquad \text{• Replace } y \text{ by } f^{-1}(x).$$

In Exercises 27 to 44, find $f^{-1}(x)$. State any restrictions on the domain of $f^{-1}(x)$.

27. $f(x) = 2x + 4$

▶ **28.** $f(x) = 4x - 8$

Page 335

Page S15

SECTION 4.4   LOGARITHMS AND LOGARITHMIC SCALES

- PROPERTIES OF LOGARITHMS
- CHANGE-OF-BASE FORMULA
- LOGARITHMIC SCALES

● PROPERTIES OF LOGARITHMS

In Section 4.3 we introduced the following basic properties of logarithms.

$$\log_b b = 1 \quad \text{and} \quad \log_b 1 = 0$$

Also, because exponential functions and logarithmic functions are inverses of each other, we observed the relationships

$$\log_b(b^x) = x \quad \text{and} \quad b^{\log_b x} = x$$

We can use the properties of exponents to establish the following additional logarithmic properties.

Properties of Logarithms

In the following properties, b, M, and N are positive real numbers ($b \neq 1$).

Product property	$\log_b(MN) = \log_b M + \log_b N$
Quotient property	$\log_b \frac{M}{N} = \log_b M - \log_b N$
Power property	$\log_b(M^p) = p \log_b M$
Logarithm-of-each-side property	$M = N$ implies $\log_b M = \log_b N$
One-to-one property	$\log_b M = \log_b N$ implies $M = N$

? QUESTION Is it true that $\ln 5 + \ln 10 = \ln 50$?

The above properties of logarithms are often used to rewrite logarithmic expressions in an equivalent form.

take note

Pay close attention to these properties. Note that
$$\log_b(MN) \neq \log_b M \cdot \log_b N$$
and
$$\log_b \frac{M}{N} \neq \frac{\log_b M}{\log_b N}$$
Also,
$$\log_b(M + N) \neq \log_b M + \log_b N$$
In fact, the expression $\log_b(M + N)$ cannot be expanded at all.

? ANSWER Yes. By the product property, $\ln 5 + \ln 10 = \ln(5 \cdot 10)$.

QUESTION/ANSWER

In every section, we pose at least one **Question** to the student about the material being presented. This question encourages the reader to pause and think about the current discussion and to answer the question. To make sure that the student does not miss important information, the **Answer** to the question is provided as a footnote on the same page.

Page 365

REAL DATA AND APPLICATIONS

APPLICATIONS

One way to motivate an interest in mathematics is through applications. Applications require the student to use problem-solving strategies, along with the skills covered in a section, to solve practical problems. This careful integration of applications generates student awareness of the value of algebra as a real-life tool.

Applications are taken from many disciplines including agriculture, business, chemistry, construction, Earth science, education, economics, manufacturing, nutrition, real estate, and sociology.

---

398   Chapter 4   Exponential and Logarithmic Functions

In the following example we determine a logistic growth model for a coyote population.

EXAMPLE 8   **Find and Use a Logistic Model**

At the beginning of 2002, the coyote population in a wilderness area was estimated at 200. By the beginning of 2004, the coyote population had increased to 250. A park ranger estimates that the carrying capacity of the wilderness area is 500 coyotes.

a. Use the given data to determine the growth rate constant for the logistic model of this coyote population.

b. Use the logistic model determined in part **a.** to predict the year in which the coyote population will first reach 400.

Solution

a. If we represent the beginning of the year 2002 by $t = 0$, then the beginning of the year 2004 will be represented by $t = 2$. In the logistic model, make the following substitutions: $P(2) = 250$, $c = 500$, and

$$a = \frac{c - P_0}{P_0} = \frac{500 - 200}{200} = 1.5.$$

$$P(t) = \frac{c}{1 + ae^{-bt}}$$

$$P(2) = \frac{500}{1 + 1.5e^{-b \cdot 2}} \qquad \text{• Substitute the given values.}$$

$$250 = \frac{500}{1 + 1.5e^{-b \cdot 2}}$$

$$250(1 + 1.5e^{-b \cdot 2}) = 500 \qquad \text{• Solve for the growth rate constant } b.$$

$$1 + 1.5e^{-b \cdot 2} = \frac{500}{250}$$

$$1.5e^{-b \cdot 2} = 2 - 1$$

$$e^{-b \cdot 2} = \frac{1}{1.5}$$

$$-2b = \ln\left(\frac{1}{1.5}\right)$$

$$b = -\frac{1}{2}\ln\left(\frac{1}{1.5}\right)$$

$$b \approx 0.20273255$$

Page 398

Page 135

1.5   Inequalities   135

51. PERSONAL FINANCE A bank offers two checking account plans. The monthly fee and charge per check for each plan are shown below. Under what conditions is it less expensive to use the LowCharge plan?

Account Plan	Monthly Fee	Charge per Check
LowCharge	$5.00	$.01
FeeSaver	$1.00	$.08

▶ **52. PERSONAL FINANCE** You can rent a car for the day from Company A for $29.00 plus $0.12 a mile. Company B charges $22.00 plus $0.21 a mile. Find the number of miles m (to the nearest mile) per day for which it is cheaper to rent from Company A.

53.  **SHIPPING REQUIREMENTS** United Parcel Service (UPS) will only ship packages for which the length is less than or equal to 108 inches and the length plus the girth is less than or equal to 130 inches. The length of a package is defined as the length of the longest side. The girth is defined as twice the width plus twice the height of the package. If a box has a length of 34 inches and a width of 22 inches, determine the possible range of heights h for this package if you wish to ship it by UPS. (*Source:* http://www.iship.com.)

▶ **54.** **MOVIE TICKET PRICES** The average U.S. movie ticket price P, in dollars, can be modeled by

$$P = 0.218t + 4.02, \quad t \ge 0$$

where $t = 0$ represents the year 1994. According to this model, in what year will the average price of a movie ticket first exceed $6.50? (*Source:* National Association of Theatre Owners, http://www.natoonline.org/satistics-tickets.htm.)

Movie Ticket Prices

55. PERSONAL FINANCE A sales clerk has a choice between two payment plans. Plan A pays $100.00 a week plus $8.00 a sale. Plan B pays $250.00 a week plus $3.50 a sale. How many sales per week must be made for plan A to yield the greater paycheck?

56. PERSONAL FINANCE A video store offers two rental plans. The yearly membership fee and the daily charge per video for each plan are shown below. How many one-night rentals can be made per year if the No-fee plan is to be the less expensive of the plans?

THE VIDEO STORE

Rental Plan	Yearly Fee	Daily Charge per Video
Low-rate	$15.00	$1.49
No-fee	None	$1.99

57.  **AVERAGE TEMPERATURES** The average daily minimum-to-maximum temperature range for the city of Palm Springs during the month of September is 68 to 104 degrees Fahrenheit. What is the corresponding temperature range measured on the Celsius temperature scale? (*Hint:* Let F be the average daily temperature. Then $68 \le F \le 104$. Now substitute $\frac{9}{5}C + 32$ for F and solve the resulting inequality for C.)

▶ **58.** **AVERAGE TEMPERATURES** The average daily minimum-to-maximum temperature range for the city of Palm Springs during the month of January is 41 to 68 degrees Fahrenheit. What is the corresponding temperature range measured on the Celsius temperature scale? (*Hint:* See Exercise 57.)

59. CONSECUTIVE EVEN INTEGERS The sum of three consecutive even integers is between 36 and 54. Find all possible sets of integers that satisfy these conditions.

60. CONSECUTIVE ODD INTEGERS The sum of three consecutive odd integers is between 63 and 81. Find all possible sets of integers that satisfy these conditions.

61. **FORENSIC SCIENCE** Forensic specialists can estimate the height of a deceased person from the lengths of the person's bones. These lengths are substituted into mathematical inequalities. For instance, an inequality that relates the height h, in centimeters, of an adult female and the length f, in centimeters, of her femur is

$$|h - (2.47f + 54.10)| \le 3.72$$

REAL DATA

Real data examples and exercises, identified by 🌑, ask students to analyze and create mathematical models from actual situations. Students are often required to work with tables, graphs, and charts drawn from a variety of disciplines.

TECHNOLOGY

INTEGRATING TECHNOLOGY

The **Integrating Technology** feature contains optional discussions that can be used to further explore a concept using technology. Some introduce technology as an alternative way to solve certain problems and others provide suggestions for using a calculator to solve certain problems and applications. Additionally, optional graphing calculator examples and exercises (identified by) are presented throughout the text.

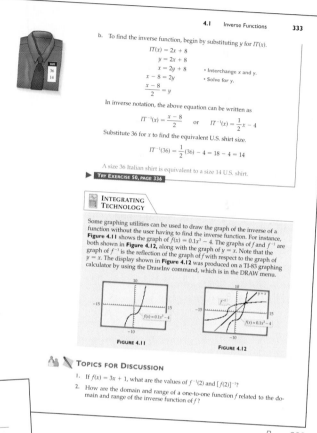

Page 333

Excerpt from page 333:

4.1   Inverse Functions   333

b. To find the inverse function, begin by substituting y for $IT(x)$.

$$IT(x) = 2x + 8$$
$$y = 2x + 8$$
$$x = 2y + 8 \quad \text{• Interchange } x \text{ and } y.$$
$$x - 8 = 2y \quad \text{• Solve for } y.$$
$$\frac{x - 8}{2} = y$$

In inverse notation, the above equation can be written as

$$IT^{-1}(x) = \frac{x - 8}{2} \quad \text{or} \quad IT^{-1}(x) = \frac{1}{2}x - 4$$

Substitute 36 for x to find the equivalent U.S. shirt size.

$$IT^{-1}(36) = \frac{1}{2}(36) - 4 = 18 - 4 = 14$$

A size 36 Italian shirt is equivalent to a size 14 U.S. shirt.

▶ **TRY EXERCISE 50, PAGE 336**

INTEGRATING TECHNOLOGY

Some graphing utilities can be used to draw the graph of the inverse of a function without the user having to find the inverse function. For instance, **Figure 4.11** shows the graph of $f(x) = 0.1x^3 - 4$. The graphs of f and f^{-1} are both shown in **Figure 4.12**, along with the graph of $y = x$. Note that the graph of f^{-1} is the reflection of the graph of f with respect to the graph of $y = x$. The display shown in **Figure 4.12** was produced on a TI-83 graphing calculator by using the DrawInv command, which is in the DRAW menu.

FIGURE 4.11   **FIGURE 4.12**

TOPICS FOR DISCUSSION

1. If $f(x) = 3x + 1$, what are the values of $f^{-1}(2)$ and $[f(2)]^{-1}$?

2. How are the domain and range of a one-to-one function f related to the domain and range of the inverse function of f?

Page 406

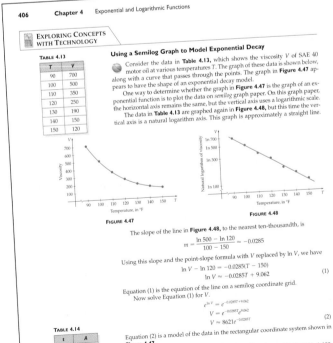

Excerpt from page 406:

406   Chapter 4   Exponential and Logarithmic Functions

EXPLORING CONCEPTS WITH TECHNOLOGY

Using a Semilog Graph to Model Exponential Decay

Consider the data in **Table 4.13**, which shows the viscosity V of SAE 40 motor oil at various temperatures T. The graph of these data is shown below, along with a curve that passes through the points. The graph in **Figure 4.47** appears to have the shape of an exponential decay model.

One way to determine whether the graph in **Figure 4.47** is the graph of an exponential function is to plot the data on *semilog* graph paper. On this graph paper, the horizontal axis remains the same, but the vertical axis uses a logarithmic scale.

The data in **Table 4.13** are graphed again in **Figure 4.48**, but this time the vertical axis is a natural logarithm axis. This graph is approximately a straight line.

TABLE 4.13

T	V
90	700
100	500
110	350
120	250
130	190
140	150
150	120

FIGURE 4.47   **FIGURE 4.48**

The slope of the line in **Figure 4.48**, to the nearest ten-thousandth, is

$$m = \frac{\ln 500 - \ln 120}{100 - 150} \approx -0.0285$$

Using this slope and the point-slope formula with V replaced by $\ln V$, we have

$$\ln V - \ln 120 = -0.0285(T - 150)$$
$$\ln V = -0.0285T + 9.062 \quad (1)$$

Equation (1) is the equation of the line on a semilog coordinate grid.
Now solve Equation (1) for V.

$$e^{\ln V} = e^{-0.0285T + 9.062}$$
$$V = e^{-0.0285T}e^{9.062}$$
$$V \approx 8621e^{-0.0285T} \quad (2)$$

Equation (2) is a model of the data in the rectangular coordinate system shown in **Figure 4.47.**

1. A chemist wishes to determine the decay characteristics of iodine-131. A 100-mg sample of iodine-131 is observed over a 30-day period. **Table 4.14** shows the amount A (in milligrams) of iodine-131 remaining after t days.

   a. Graph the ordered pairs (t, A) on semilog paper. (*Note:* Semilog paper comes in different varieties. Our calculations are based on semilog paper that has a natural logarithm scale on the vertical axis.)

TABLE 4.14

t	A
1	91.77
4	70.92
8	50.30
15	27.57
20	17.95
30	7.60

EXPLORING CONCEPTS WITH TECHNOLOGY

A special end-of-chapter feature, **Exploring Concepts with Technology**, extends ideas introduced in the text by using technology (graphing calculator, CAS, etc.) to investigate extended applications or mathematical topics. These explorations can serve as group projects, class discussions, or extra-credit assignments.

STUDENT PEDAGOGY

TOPIC LIST

At the beginning of each section is a list of the major topics covered in the section.

KEY TERMS AND CONCEPTS

Key terms, in bold, emphasize important terms. Key concepts are presented in blue boxes in order to highlight these important concepts and to provide for easy reference.

MATH MATTERS

These margin notes contain interesting sidelights about mathematics, its history, or its application.

TAKE NOTE

These margin notes alert students to a point requiring special attention or are used to amplify the concept under discussion.

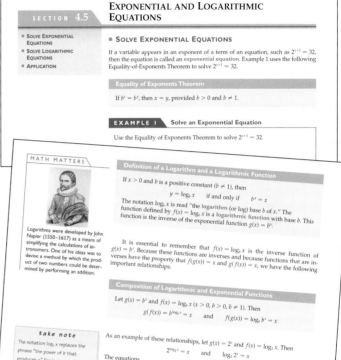

Page 378

SECTION 4.5

EXPONENTIAL AND LOGARITHMIC EQUATIONS

- SOLVE EXPONENTIAL EQUATIONS
- SOLVE LOGARITHMIC EQUATIONS
- APPLICATION

● SOLVE EXPONENTIAL EQUATIONS

If a variable appears in an exponent of a term of an equation, such as $2^{x+1} = 32$, then the equation is called an exponential equation. Example 1 uses the following Equality-of-Exponents Theorem to solve $2^{x+1} = 32$.

Equality of Exponents Theorem

If $b^x = b^y$, then $x = y$, provided $b > 0$ and $b \neq 1$.

EXAMPLE 1   Solve an Exponential Equation

Use the Equality of Exponents Theorem to solve $2^{x+1} = 32$.

MATH MATTERS

Logarithms were developed by John Napier (1550–1617) as a means of simplifying the calculations of astronomers. One of his ideas was to devise a method by which the product of two numbers could be determined by performing an addition.

Definition of a Logarithm and a Logarithmic Function

If $x > 0$ and b is a positive constant ($b \neq 1$), then

$$y = \log_b x \quad \text{if and only if} \quad b^y = x$$

The notation $\log_b x$ is read "the logarithm (or log) base b of x." The function defined by $f(x) = \log_b x$ is a logarithmic function with base b. This function is the inverse of the exponential function $g(x) = b^x$.

It is essential to remember that $f(x) = \log_b x$ is the inverse function of $g(x) = b^x$. Because these functions are inverses and because functions that are inverses have the property that $f(g(x)) = x$ and $g(f(x)) = x$, we have the following important relationships.

Composition of Logarithmic and Exponential Functions

Let $g(x) = b^x$ and $f(x) = \log_b x$ ($x > 0, b > 0, b \neq 1$). Then

$$g(f(x)) = b^{\log_b x} = x \quad \text{and} \quad f(g(x)) = \log_b b^x = x$$

take note

The notation $\log_b x$ replaces the phrase "the power of b that produces x." For instance, "3 is the power of 2 that produces 8" is abbreviated $3 = \log_2 8$. In your work with logarithms, remember that a logarithm is an exponent.

As an example of these relationships, let $g(x) = 2^x$ and $f(x) = \log_2 x$. Then

$$2^{\log_2 x} = x \quad \text{and} \quad \log_2 2^x = x$$

The equations

$$y = \log_b x \quad \text{and} \quad b^y = x$$

are different ways of expressing the same concept.

Exponential Form and Logarithmic Form

The exponential form of $y = \log_b x$ is $b^y = x$.

The logarithmic form of $b^y = x$ is $y = \log_b x$.

These concepts are illustrated in the next two examples.

Page 353

Page 157

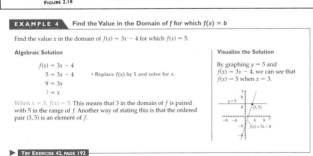

EXAMPLE 7   Find the Center and Radius of a Circle by Completing the Square

Find the center and the radius of the circle that is given by

$$x^2 + y^2 - 6x + 4y - 3 = 0$$

Solution

First rearrange and group the terms as shown.

$$(x^2 - 6x) + (y^2 + 4y) = 3$$

Now complete the squares of $(x^2 - 6x)$ and $(y^2 + 4y)$.

$$(x^2 - 6x + 9) + (y^2 + 4y + 4) = 3 + 9 + 4 \quad \text{• Add 9 and 4 to each side of the equation.}$$

$$(x - 3)^2 + (y + 2)^2 = 16$$
$$(x - 3)^2 + (y - (-2))^2 = 4^2$$

This equation is the standard form of the equation of a circle and indicates that the graph of the original equation is a circle centered at $(3, -2)$ with radius 4. See **Figure 2.18**.

▶ **TRY EXERCISE 66, PAGE 159**

To review COMPLETING THE SQUARE, see p. 99.

$x^2 + y^2 - 6x + 4y - 3 = 0$
FIGURE 2.18

EXAMPLE 4   Find the Value in the Domain of f for which $f(x) = b$

Find the value x in the domain of $f(x) = 3x - 4$ for which $f(x) = 5$.

Algebraic Solution

$$f(x) = 3x - 4$$
$$5 = 3x - 4 \quad \text{• Replace } f(x) \text{ by 5 and solve for } x.$$
$$9 = 3x$$
$$3 = x$$

When $x = 3$, $f(x) = 5$. This means that 3 in the domain of f is paired with 5 in the range of f. Another way of stating this is that the ordered pair $(3, 5)$ is an element of f.

Visualize the Solution

By graphing $y = 5$ and $f(x) = 3x - 4$, we can see that $f(x) = 5$ when $x = 3$.

$y = 5$
$(3, 5)$
$f(x) = 3x - 4$

▶ **TRY EXERCISE 42, PAGE 192**

REVIEW NOTES

A ⬅➡ directs the student to the place in the text where the student can review a concept that was previously discussed.

VISUALIZE THE SOLUTION

For appropriate examples within the text, we have provided both an algebraic solution and a graphical representation of the solution. This approach creates a link between the algebraic and visual components of a solution.

Page 186

EXERCISES

TOPICS FOR DISCUSSION

These special exercises provide questions related to key concepts in the section. Instructors can use these to initiate class discussions or to ask students to write about concepts presented.

EXERCISES

The exercise sets in *Essentials of College Algebra* were carefully developed to provide a wide variety of exercises. The exercises range from drill and practice to interesting challenges. They were chosen to illustrate the many facets of topics discussed in the text. Each exercise set emphasizes skill building, skill maintenance, and, as appropriate, applications. **Icons** identify appropriate writing , group , data analysis , web , and graphing calculator 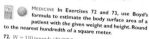 exercises.

Page 351

TOPICS FOR DISCUSSION

1. Explain how to use the graph of $f(x) = 2^x$ to produce the graph of $g(x) = 2^{(x-3)} + 4$.

2. At what point does the function $g(x) = e^{-x^2/2}$ take on its maximum value?

3. Without using a graphing utility, determine whether the revenue function $R(t) = 10 + e^{-0.05t}$ is an increasing function or a decreasing function.

4. Discuss the properties of the graph of $f(x) = b^x$ when $b > 1$.

5. What is the base of the natural exponential function? How is it calculated? What is its approximate value?

In Exercises 59 to 68, use a graphing utility to graph the function.

59. $f(x) = -2 \ln x$

60. $f(x) = -\log x$

61. $f(x) = |\ln x|$

62. $f(x) = \ln |x|$

63. $f(x) = \log \sqrt[3]{x}$

64. $f(x) = \ln \sqrt{x}$

65. $f(x) = \log(x + 10)$

66. $f(x) = \ln(x + 3)$

67. $f(x) = 3 \log |2x + 10|$

68. $f(x) = \frac{1}{2} \ln |x - 4|$

69. **MONEY MARKET RATES** The function
$$r(t) = 0.69607 + 0.60781 \ln t$$
gives the annual interest rate r, as a percent, a bank will pay on its money market accounts, where t is the term (the time the money is invested) in months.

a. What interest rate, to the nearest tenth of a percent, will the bank pay on a money market account with a term of 9 months?

b. What is the minimum number of complete months during which a person must invest to receive an interest rate of at least 3%?

▶ 70. **AVERAGE TYPING SPEED** The following function models the average typing speed S, in words per minute, of a student who has been typing for t months.
$$S(t) = 5 + 29 \ln(t + 1), \quad 0 \le t \le 16$$

... typing speed, to the ... the student first started ... average typing speed, ... after 3 months?

... ine how long, to the ... will take the student ... speed of 65 words

ADVERTISING COSTS AND SALES The function
$$N(x) = 2750 + 180 \ln\left(\frac{x}{1000} + 1\right)$$
models the relationship between the dollar amount x spent on advertising a product and the number of units N that a company can sell.

a. Find the number of units that will be sold with advertising expenditures of $20,000, $40,000, and $60,000.

b. How many units will be sold if the company does not pay to advertise the product?

In anesthesiology it is necessary to accurately estimate the body surface area of a patient. One formula for estimating body surface area (BSA) was developed by Edith Boyd (University of Minnesota Press, 1935). Her formula for the BSA (in square meters) of a patient of height H (in centimeters) and weight W (in grams) is
$$BSA = 0.0003207 \cdot H^{0.3} \cdot W^{(0.7285 - 0.0188 \log W)}$$

MEDICINE In Exercises 72 and 73, use Boyd's formula to estimate the body surface area of a patient with the given weight and height. Round to the nearest hundredth of a square meter.

72. $W = 110$ pounds (49,895.2 grams); $H = 5$ feet 4 inches (162.56 centimeters)

73. $W = 180$ pounds (81,646.6 grams); $H = 6$ feet 1 inch (185.42 centimeters)

74. **ASTRONOMY** Astronomers measure the apparent brightness of a star by a unit called the **apparent magnitude.** This unit was created in the second century B.C. when the Greek astronomer Hipparchus classified the relative brightness of several stars. In his list he assigned the number 1 to the stars that appeared to be the brightest (Sirius, Vega, and Deneb). They are first-magnitude stars. Hipparchus assigned the number 2 to all the stars in the Big Dipper. They are second-magnitude stars. The following table shows the

---

Page 362

Included in each exercise set are **Connecting Concepts** exercises. These exercises extend some of the concepts discussed in the section and require students to connect ideas studied earlier with new concepts.

EXERCISES TO PREPARE FOR THE NEXT SECTION

Every section's exercise set (except for the last section of a chapter) contains exercises that allow students to practice the previously-learned skills they will need to be successful in the next section. Next to each question, in brackets, is a reference to the section of the text that contains the concepts related to the question for students to easily review. All answers are provided in the Answer Appendix.

PROJECTS

Projects are provided at the end of each exercise set. They are designed to encourage students to do research and write about what they have learned. These Projects generally emphasize critical thinking skills and can be used as collaborative learning exercises or as extra-credit assignments.

4.2   Exponential Functions and Their Applications   **351**

CONNECTING CONCEPTS

60. Verify that the hyperbolic cosine function
$$\cosh(x) = \frac{e^x + e^{-x}}{2} \text{ is an even function.}$$

61. Verify that the hyperbolic sine function $\sinh(x) = \frac{e^x - e^{-x}}{2}$ is an odd function.

62. Graph $g(x) = 10^x$, and then sketch the graph of g reflected across the line given by $y = x$.

63. Graph $f(x) = e^x$, and then sketch the graph of f reflected across the line given by $y = x$.

In Exercises 64 to 67, determine the domain of the given function. Write the domain using interval notation.

64. $f(x) = \frac{e^x - e^{-x}}{e^x + e^{-x}}$

65. $f(x) = \frac{e^{|x|}}{1 + e^x}$

66. $f(x) = \sqrt{1 - e^x}$

67. $f(x) = \sqrt{e^x - e^{-x}}$

PREPARE FOR SECTION 4.3

68. If $2^x = 16$, determine the value of x. [4.2]

69. If $3^{-x} = \frac{1}{27}$, determine the value of x. [4.2]

70. If $x^4 = 625$, determine the value of x. [4.2]

71. Find the inverse of $f(x) = \frac{2x}{x + 3}$. [4.1]

72. State the domain of $g(x) = \sqrt{x - 2}$. [2.2]

73. If the range of $h(x)$ is the set of all positive real numbers, then what is the domain of $h^{-1}(x)$? [4.2]

PROJECTS

1. **THE SAINT LOUIS GATEWAY ARCH** The Gateway Arch in Saint Louis was designed in the shape of an inverted **catenary,** as shown by the red curve in the drawing at the right. The Gateway Arch is one of the largest optical illusions ever created. As you look at the arch (and its basic shape defined by the catenary curve), it appears to be much taller than it is wide. However, this is not the case. The height of the catenary is given by
$$h(x) = 693.8597 - 68.7672\left(\frac{e^{0.0100333x} + e^{-0.0100333x}}{2}\right)$$
where x and $h(x)$ are measured in feet and $x = 0$ represents the position at ground level that is directly below the highest point of the catenary.

a. Use a graphing utility to graph $h(x)$.

b. Use your graph to find the height of the catenary for $x = 0, 100, 200,$ and 299 feet. Round each result to the nearest tenth of a foot.

END OF CHAPTER

Page 316

CHAPTER SUMMARY

At the end of each chapter there is a Chapter Summary that provides a concise section-by-section review of the chapter topics.

TRUE/FALSE EXERCISES

Following each chapter summary are true/false exercises. These exercises are intended to help students understand concepts and can be used to initiate class discussions.

CHAPTER 3 SUMMARY

3.1 The Remainder Theorem and the Factor Theorem

- *The Remainder Theorem* If a polynomial function $P(x)$ is divided by $(x - c)$, then the remainder equals $P(c)$.

- *The Factor Theorem* A polynomial function $P(x)$ has a factor $(x - c)$ if and only if $P(c) = 0$.

3.2 Polynomial Functions of Higher Degree

- Characteristics and properties used in graphing polynomial functions include:

  ○ ... Polynomial functions are smooth continu-

has integer coefficients, and $\frac{p}{q}$ (where p and q have no common factors) is a rational zero of P, then p is a factor of a_0 and q is a factor of a_n.

- *Upper- and Lower-Bound Theorem*
Let $P(x)$ be a polynomial function with real coefficients. Use synthetic division to divide $P(x)$ by $x - b$, where b is a nonzero real number.

  Upper Bound
  a. If $b > 0$ and the leading coefficient of P is positive, then b is an upper bound for the real zeros of P provided none of the numbers in the bottom row of the synthetic division are negative.

  b. If $b > 0$ and the leading coefficient of P is negative, then

CHAPTER 3 TRUE/FALSE EXERCISES

In Exercises 1 to 12, answer true or false. If the statement is false, explain why the statement is false or give an example to show that the statement is false.

1. The complex zeros of a polynomial function with complex coefficients always occur in conjugate pairs.

2. Descartes' Rule of Signs indicates that the polynomial function $P(x) = x^3 - x^2 + x - 1$ must have three positive zeros.

3. The polynomial $2x^5 + x^4 - 7x^3 - 5x^2 + 4x + 10$ has two variations in sign.

4. If 4 is an upper bound of the zeros of the polynomial function P, then 5 is also an upper bound of the zeros of P.

5. The graph of every rational function has a vertical asymptote.

6. The graph of the rational function $F(x) = \dfrac{x^2 - 4x + 4}{x^2 - 5x + 6}$ has a vertical asymptote of $x = 2$.

7. If 7 is a zero of the polynomial function P, then $x - 7$ is a factor of P.

8. According to the Zero Location Theorem, the polynomial function $P(x) = x^3 + 6x - 2$ has a real zero between 0 and 1.

9. Every fourth-degree polynomial function with complex coefficients has exactly four complex zeros, provided each zero is counted according to its multiplicity.

10. The graph of a rational function can have at most one horizontal asymptote.

11. Descartes' Rule of Signs indicates that the polynomial function $P(x) = x^3 + 2x^2 + 4x - 7$ does have a positive zero.

12. Every polynomial function has at least one real zero.

Page 317

Page 318

CHAPTER 3 REVIEW EXERCISES

In Exercises 1 to 6, use synthetic division to divide the first polynomial by the second.

1. $4x^3 - 11x^2 + 5x - 2, x - 3$

2. $5x^3 - 18x + 2, x - 1$

3. $3x^3 - 5x + 1, x + 2$

4. $2x^3 + 7x^2 + 16x - 10, x - \frac{1}{2}$

5. $3x^3 - 10x^2 - 36x + 55, x - 5$

6. $x^4 + 9x^3 + 6x^2 - 65x - 63, x + 7$

In Exercises 21 to 26, use the Rational Zero Theorem to list all possible rational zeros for each polynomial function.

21. $P(x) = x^3 - 7x - 6$

22. $P(x) = 2x^3 + 3x^2 - 29x - 30$

23. $P(x) = 15x^3 - 91x^2 + 4x + 12$

24. $P(x) = x^4 - 12x^3 + 52x^2 - 96x + 64$

25. $P(x) = x^3 + x^2 - x - 1$

26. $P(x) = 6x^3 + 3x - 2$

CHAPTER REVIEW EXERCISES

Review exercises are found at the end of each chapter. These exercises are selected to help the student integrate all of the topics presented in the chapter.

CHAPTER 3 TEST

1. Use synthetic division to divide:
$(3x^3 + 5x^2 + 4x - 1) \div (x + 2)$

2. Use the Remainder Theorem to find $P(-2)$ if
$P(x) = -3x^3 + 7x^2 + 2x - 5$

3. Show that $x - 1$ is a factor of
$x^4 - 4x^3 + 7x^2 - 6x + 2$

4. Examine the leading term of the function given by the equation $P(x) = -3x^3 + 2x^2 - 5x + 2$ and determine the far-left and far-right behavior of the graph of P.

CHAPTER TEST

The Chapter Test exercises are designed to simulate a possible test of the material in the chapter.

Page 319

CUMULATIVE REVIEW EXERCISES

1. Write $\dfrac{3 + 4i}{1 - 2i}$ in $a + bi$ form.

2. Use the quadratic formula to solve $x^2 - x - 1 = 0$.

3. Solve: $\sqrt{2x + 5} - \sqrt{x - 1} = 2$

4. Solve: $|x - 3| \leq 11$

5. Find the distance between the points $(2, 5)$ and $(7, -11)$.

6. Explain how to use the graph of $y = x^2$ to produce the graph of $y = (x - 2)^2 + 4$.

7. Find the difference quotient for the function
$P(x) = x^2 - 2x - 3$.

8. Given $f(x) = 2x^2 + 5x - 3$ and $g(x) = 4x - 7$, find $(f \cdot g)(x)$.

9. Given $f(x) = x^2 - 2x + 7$ and $g(x) = x^2 - 3x - 4$, find $(f - g)(x)$.

10. Use synthetic division to divide $(4x^4 - 2x^2 - 4x - 5)$ by $(x + 2)$.

13. Determine the relative maximum of the polynomial function $P(x) = -3x^3 - x^2 + 4x - 1$. Round to the nearest ten thousandth.

14. Use the Rational Zero Theorem to list all possible rational zeros of $P(x) = 3x^4 - 4x^3 - 11x^2 + 16x - 4$.

15. Use Descartes' Rule of Signs to state the number of possible positive and negative real zeros of $P(x) = x^3 + x^2 + 2x + 4$.

16. Find all zeros of $P(x) = x^3 + x + 10$.

17. Find a polynomial function of smallest degree that has real coefficients and -2 and $3 + i$ as zeros.

18. Write $P(x) = x^3 - 2x^2 + 9x - 18$ as a product of linear factors.

19. Determine the vertical and horizontal asymptotes of the graph of $F(x) = \dfrac{4x^2}{x^2 + x - 6}$.

20. Find the equation of the slant asymptote for the graph of $F(x) = \dfrac{x^3 + 4x^2 + 1}{x^2 + 4}$.

CUMULATIVE REVIEW EXERCISES

Cumulative Review Exercises, which appear at the end of each chapter (except Chapter P), help students maintain skills learned in previous chapters.

The answers to all **Chapter Review Exercises**, all **Chapter Test Exercises**, and all **Cumulative Review Exercises** are given in the Answer Section. Along with the answer, there is a reference to the section that pertains to each exercise.

Instructor Resources

Essentials of College Algebra has a complete set of support materials for the instructor.

Instructor's Annotated Edition   This edition contains a replica of the student text with additional resources for the instructor. These include: *Instructor Notes, Alternative Example* notes, *PowerPoint* icons, *Suggested Assignments,* and answers to all exercises.

Instructor's Solutions Manual   The *Instructor's Solutions Manual* contains worked-out solutions for all exercises in the text.

Instructor's Resource Manual with Testing   This resource includes six ready-to-use printed *Chapter Tests* per chapter, and a *Printed Test Bank* providing a printout of one example of each of the algorithmic items on the *HM Testing* CD-ROM program.

HM ClassPrep with HM Testing CD-ROM   *HM ClassPrep* contains a multitude of text-specific resources for instructors to use to enhance the classroom experience. These resources can be easily accessed by chapter or resource type and also can link you to the text's website. *HM Testing* is our computerized test generator and contains a database of algorithmic test items, as well as providing **online testing** and **gradebook** functions.

Instructor Text-specific Website   The resources available on the *ClassPrep CD* are also available on the instructor website at math.college.hmco.com/instructors. Appropriate items are password protected. Instructors also have access to the student part of the text's website.

Student Resources

Student Study Guide   The *Student Study Guide* contains complete solutions to all odd-numbered exercises in the text, as well as study tips and a practice test for each chapter.

Math Study Skills Workbook *by Paul D. Nolting*   This workbook is designed to reinforce skills and minimize frustration for students in any math class, lab, or study skills course. It offers a wealth of study tips and sound advice on note taking, time management, and reducing math anxiety. In addition, numerous opportunities for self assessment enable students to track their own progress.

HM Eduspace® Online Learning Environment   *Eduspace* is a text-specific, web-based learning environment that combines an algorithmic tutorial program with homework capabilities. Specific content is available 24 hours a day to help you further understand your textbook.

HM mathSpace® Tutorial CD-ROM   This tutorial CD-ROM allows students to practice skills and review concepts as many times as necessary by providing algorithmically-generated exercises and step-by-step solutions for practice.

SMARTHINKING™ Live, Online Tutoring   Houghton Mifflin has partnered with SMARTHINKING to provide an easy-to-use and effective online tutorial service. **Whiteboard Simulations** and **Practice Area** promote real-time visual interaction.

Three levels of service are offered.

- **Text-specific Tutoring** provides real-time, one-on-one instruction with a specially qualified 'e-structor.'
- **Questions Any Time** allows students to submit questions to the tutor outside the scheduled hours and receive a reply within 24 hours.
- **Independent Study Resources** connect students with around-the-clock access to additional educational services, including interactive websites, diagnostic tests, and Frequently Asked Questions posed to SMARTHINKING e-structors.

Houghton Mifflin Instructional Videos and DVDs Text-specific videos and DVDs, hosted by Dana Mosely, cover all sections of the text and provide a valuable resource for further instruction and review.

Student Text-specific Website Online student resources can be found at this text's website at math.college.hmco.com/students.

Acknowledgments

The authors would like to thank the people who have reviewed this manuscript and provided many valuable suggestions.

Ioannis K. Argyros, *Cameron University, OK*
Peter Arvanites, *Rockland Community College, NY*
Linda Berg, *University of Great Falls, MT*
Paul Bialek, *Trinity College, IL*
Zhixiong Cai, *Barton College, NC*
Cheryl F. Cavaliero, *Butler County Community College, PA*
Jennie Cox, *Central Georgia Technical College, GA*
Marilyn Danchanko, *Cambria County Area Community College, PA*
Laura Davis, *Garland County Community College, AR*
Sylvia Dorminey, *Savannah Technical College, GA*
Gay Grubbs, *Griffin Technical College, GA*
Kathryn Hodge, *Midland College, TX*
Clement S. Lam, *Mission College, CA*
Dr. Helen Medley
Carla Monticelli, *Camden County College, NJ*
J. Reid Mowrer, *University of New Mexico–Valencia Campus, NM*
Sue Neal, *Wichita State University, KS*
Georgie O'Leary, *Warner Southern College, FL*
Suzanne Pauly, *Delaware Technical and Community College, DE*
Lauri Semarne
Mike Shirazi, *Germanna Community College, VA*
Anthony Tongen, *Trinity International University, IL*
Hanson Umoh, *Delaware State University, DE*
Rebecca Wells, *Henderson Community College, KY*

Special thanks to Christi Verity for her diligent preparation of the solutions manuals.

PRELIMINARY CONCEPTS

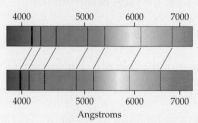

4000   5000   6000   7000
Angstroms

Source: Original art at
http://www.astro.ucla.edu/~wright/
doppler.htm.

The Red Shift

You may have noticed that the sound of an approaching siren has a higher pitch (frequency) than that of a receding siren. Like sound, when a light source moves toward or away from us, the frequency of the light changes. Astronomers use this fact to measure the distance of a galaxy from us.

The photograph at the left was taken by the Hubble Space Telescope. Many of the bright objects in this photograph are galaxies. When a galaxy is moving away from us, the characteristic frequencies of certain light waves, the black lines in the spectrum below the photo, are shifted to the red side of the spectrum. The amount of this *red shift*, as astronomers call it, can be used to determine the speed at which the galaxy is receding from Earth. The formula used to compute the speed involves a rational expression, which is one of the topics of this chapter. **Exercise 67 on page 58** is an example of this formula.

VIDEO & DVD   SSG   WWW

Polya's Four-Step Process

Your success in mathematics and your success in the workplace are heavily dependent on your ability to solve problems. George Polya (1887–1985) was one of the foremost mathematicians to study problem solving. The basic structure that Polya advocated for problem solving has four steps, as outlined below.

1. Understand the problem.
   - Can you restate the problem in your own words?
   - Can you determine what is known about this type of problem?
   - Is there missing information that you need in order to solve the problem?
   - Is there information given that is not needed?
   - What is the goal?

2. Devise a plan.
   - Make a list of the known information.
   - Make a list of information that is needed to solve the problem.
   - Make a table or draw a diagram.
   - Work backwards.
   - Try to solve a similar but simpler problem.
   - Research the problem to determine whether there are known techniques for solving problems of its kind.
   - Try to determine whether some pattern exists.
   - Write an equation.

3. Carry out the plan.
   - Work carefully.
   - Keep an accurate and neat record of all your attempts.
   - Realize that some of your initial plans will not work and that you may have to return to step 2 and devise another plan or modify your existing plan.

4. Review your solution.
   - Make sure that the solution is consistent with the facts of the problem.
   - Interpret the solution in the context of the problem.
   - Ask yourself whether there are generalizations of the solution that could apply to other problems.
   - Determine the strengths and weaknesses of your solution. For instance, is your solution only an approximation to the actual solution?
   - Consider the possibility of alternative solutions.

As you go through this course, make a conscious effort to develop good problem-solving skills. One way to do this is to create problems and then solve them. Do not focus only on math problems. Think about your major or strategy games and create problems in those areas and then solve them. Here is one to get you started.

> Three containers A, B, and C can hold, respectively, 8, 5, and 3 gallons of water. Initially container A is filled completely, and the other two containers are empty. Without using any measuring devices other than these containers, divide the water into two equal parts by pouring from one container to another.

THE REAL NUMBER SYSTEM

● SETS

Human beings share the desire to organize and classify. Ancient astronomers classified stars into groups called *constellations*. Modern astronomers continue to classify stars by such characteristics as color, mass, size, temperature, and distance from Earth. In mathematics it is useful to place numbers with similar characteristics into **sets**. The following sets of numbers are used extensively in the study of algebra:

Integers	$\{\ldots, -3, -2, -1, 0, 1, 2, 3, \ldots\}$
Rational numbers	{all terminating or repeating decimals}
Irrational numbers	{all nonterminating, nonrepeating decimals}
Real numbers	{all rational or irrational numbers}

If a number in decimal form terminates or repeats a block of digits, then the number is a rational number. Here are two examples of rational numbers.

0.75 is a terminating decimal.

$0.2\overline{45}$ is a repeating decimal. The bar over the 45 means that the digits 45 repeat without end. That is, $0.2\overline{45} = 0.24545454\ldots$.

Rational numbers also can be written in the form $\dfrac{p}{q}$, where p and q are integers and $q \neq 0$. Examples of rational numbers written in this form are

$$\frac{3}{4} \qquad \frac{27}{110} \qquad -\frac{5}{2} \qquad \frac{7}{1} \qquad \frac{-4}{3}$$

Note that $\dfrac{7}{1} = 7$, and in general, $\dfrac{n}{1} = n$ for any integer n. Therefore, all integers are rational numbers.

When a rational number is written in the form $\dfrac{p}{q}$, the decimal form of the rational number can be found by dividing the numerator by the denominator.

$$\frac{3}{4} = 0.75 \qquad \frac{27}{110} = 0.2\overline{45}$$

In its decimal form, an irrational number neither terminates nor repeats. For example, $0.272272227\ldots$ is a nonterminating, nonrepeating decimal and thus is an irrational number. One of the best-known irrational numbers is pi, denoted by the Greek symbol π. The number π is defined as the ratio of the circumference of a circle to its diameter. Often in applications the rational number 3.14 or the rational number $\dfrac{22}{7}$ is used as an approximation of the irrational number π.

Every real number is either a rational number or an irrational number. If a real number is written in decimal form, it is a terminating decimal, a repeating decimal, or a nonterminating and nonrepeating decimal.

MATH MATTERS

Archimedes (c. 287–212 B.C.) was the first to calculate π with any degree of precision. He was able to show that

$$3\frac{10}{71} < \pi < 3\frac{1}{7}$$

from which we get the approximation

$$3\frac{1}{7} = \frac{22}{7} \approx \pi.$$

The use of the symbol π for this quantity was introduced by Leonhard Euler (1707–1783) in 1739, approximately 2000 years after Archimedes.

The relationship between the various sets of numbers is shown in **Figure P.1**.

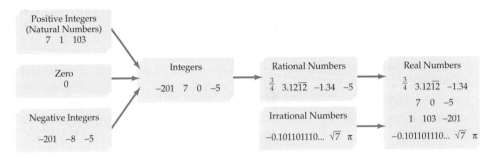

FIGURE P.1

Prime numbers and *composite numbers* play an important role in almost every branch of mathematics. A **prime number** is a positive integer other than 1 that has no positive-integer factors[1] other than itself and 1. The 10 smallest prime numbers are 2, 3, 5, 7, 11, 13, 17, 19, 23, and 29. Each of these numbers has only itself and 1 as factors.

A **composite number** is a positive integer greater than 1 that is not a prime number. For example, 10 is a composite number because 10 has both 2 and 5 as factors. The 10 smallest composite numbers are 4, 6, 8, 9, 10, 12, 14, 15, 16, and 18.

EXAMPLE 1   **Classify Real Numbers**

Determine which of the following numbers are

a. integers   b. rational numbers   c. irrational numbers
d. real numbers   e. prime numbers   f. composite numbers

$$-0.2, \quad 0, \quad 0.\overline{3}, \quad \pi, \quad 6, \quad 7, \quad 41, \quad 51, \quad 0.71771777177771\ldots$$

Solution

a. Integers: $0, 6, 7, 41, 51$

b. Rational numbers: $-0.2, 0, 0.\overline{3}, 6, 7, 41, 51$

c. Irrational numbers: $0.71771777177771\ldots, \pi$

d. Real numbers: $-0.2, 0, 0.\overline{3}, \pi, 6, 7, 41, 51, 0.71771777177771\ldots$

e. Prime numbers: $7, 41$

f. Composite numbers: $6, 51$

▶ **TRY EXERCISE 2, PAGE 15**

---

[1] Recall that a factor of a number divides the number evenly. For instance, 3 and 7 are factors of 21; 5 is not a factor of 21.

Each member of a set is called an **element** of the set. For instance, if $C = \{2, 3, 5\}$, then the elements of C are 2, 3, and 5. The notation $2 \in C$ is read "2 is an element of C." Set A is a **subset** of set B if every element of A is also an element of B, and we write $A \subseteq B$. For instance, the set of **negative integers** $\{-1, -2, -3, -4, \ldots\}$ is a subset of the set of integers. The set of **positive integers** $\{1, 2, 3, 4, \ldots\}$ (also known as the set of **natural numbers**) is also a subset of the set of integers.

❓ QUESTION Are the integers a subset of the rational numbers?

> ### take note
>
> The order of the elements of a set is not important. For instance, the set of natural numbers less than 6 given at the right could have been written $\{3, 5, 2, 1, 4\}$. It is customary, however, to list elements of a set in numerical order.

The **empty set**, or **null set**, is the set that contains no elements. The symbol $\varnothing$ is used to represent the empty set. The set of people who have run a two-minute mile is the empty set.

The set of natural numbers less than 6 is $\{1, 2, 3, 4, 5\}$. This is an example of a **finite set**; all the elements of the set can be listed. The set of all natural numbers is an example of an **infinite set**. There is no largest natural number, so all the elements of the set of natural numbers cannot be listed.

Sets are often written using **set-builder notation**. Set-builder notation can be used to describe almost any set, but it is especially useful when writing infinite sets. For instance, the set

$$\{2n \mid n \in \text{natural numbers}\}$$

is read as "the set of elements $2n$ such that n is a natural number." By replacing n by each of the natural numbers, this is the set of positive even integers: $\{2, 4, 6, 8, \ldots\}$.

The set of real numbers greater than 2 is written:

$$\{x \mid x > 2, x \in \text{real numbers}\}$$

and is read "the set of x such that x is greater than 2 and x is an element of the real numbers."

Much of the work we do in this text uses the real numbers. With this in mind, we will frequently write, for instance, $\{x \mid x > 2, x \in \text{real numbers}\}$ in a shortened form as $\{x \mid x > 2\}$, where we assume that x is a real number.

> ### MATH MATTERS
>
> A **fuzzy set** is one in which each element is given a "degree" of membership. The concepts behind fuzzy sets are used in a wide variety of applications such as traffic lights, washing machines, and computer speech recognition programs.

EXAMPLE 2   Use Set-Builder Notation

List the four smallest elements in $\{n^3 \mid n \in \text{natural numbers}\}$.

Solution
Because we want the four *smallest* elements, we choose the four smallest natural numbers. Thus $n = 1, 2, 3,$ and 4. Therefore, the four smallest elements of $\{n^3 \mid n \in \text{natural numbers}\}$ are 1, 8, 27, and 64.

▶ **TRY EXERCISE 6, PAGE 15**

❓ ANSWER Yes.

● **UNION AND INTERSECTION OF SETS**

Just as operations such as addition and multiplication are performed on real numbers, operations are performed on sets. Two operations performed on sets are union and intersection. The union of two sets A and B is the set of elements that belong to A or to B or to both A and B.

Union of Two Sets

The **union** of two sets, written $A \cup B$, is the set of all elements that belong to either A or B. In set-builder notation, this is written

$$A \cup B = \{x \mid x \in A \text{ or } x \in B\}$$

For instance, given $A = \{2, 3, 4\}$ and $B = \{0, 1, 2, 3\}$, then $A \cup B = \{0, 1, 2, 3, 4\}$. Note that an element that belongs to both sets is listed only once.

The intersection of the two sets A and B is the set of elements that belong to both A and B.

Intersection of Two Sets

The **intersection** of two sets, written $A \cap B$, is the set of all elements that are common to both A and B. In set-builder notation, this is written

$$A \cap B = \{x \mid x \in A \text{ and } x \in B\}$$

For instance, given $A = \{2, 3, 4\}$ and $B = \{0, 1, 2, 3\}$, $A \cap B = \{2, 3\}$.

If the intersection of two sets is the empty set, the two sets are said to be **disjoint**. For example, if $A = \{2, 3, 4\}$ and $B = \{7, 8\}$, then $A \cap B = \varnothing$ and A and B are disjoint sets.

EXAMPLE 3   **Find the Union and Intersection of Sets**

Find each intersection or union given $A = \{0, 2, 4, 6, 10, 12\}$, $B = \{0, 3, 6, 12, 15\}$, and $C = \{1, 2, 3, 4, 5, 6, 7\}$.

a. $A \cup C$ **b.** $B \cap C$
c. $A \cap (B \cup C)$ **d.** $B \cup (A \cap C)$

Solution

a. $A \cup C = \{0, 1, 2, 3, 4, 5, 6, 7, 10, 12\}$ • The elements that belong to A or C

b. $B \cap C = \{3, 6\}$ • The elements that belong to B and C

c. First determine $B \cup C = \{0, 1, 2, 3, 4, 5, 6, 7, 12, 15\}$. Then

$$A \cap (B \cup C) = \{0, 2, 4, 6, 12\}$$

• The elements that belong to A and $(B \cup C)$

d. First determine $A \cap C = \{2, 4, 6\}$. Then

$$B \cup (A \cap C) = \{0, 2, 3, 4, 6, 12, 15\}$$

• The elements that belong to B or $(A \cap C)$

▶ **TRY EXERCISE 16, PAGE 15**

● ABSOLUTE VALUE AND DISTANCE

FIGURE P.2

The real numbers can be represented geometrically by a **coordinate axis** called a **real number line.** **Figure P.2** shows a portion of a real number line. The number associated with a point on a real number line is called the **coordinate** of the point. The point corresponding to zero is called the **origin.** Every real number corresponds to a point on the number line, and every point on the number line corresponds to a real number.

The *absolute value* of a real number a, denoted $|a|$, is the distance between a and 0 on the number line. For instance, $|3| = 3$ and $|-3| = 3$ because both 3 and -3 are 3 units from zero. See **Figure P.3.**

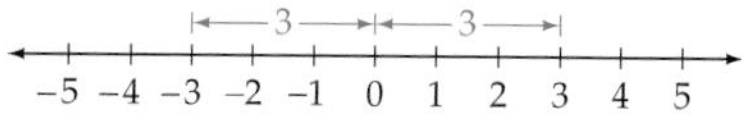

FIGURE P.3

In general, if $a \geq 0$, then $|a| = a$; however, if $a < 0$, then $|a| = -a$ because $-a$ is positive when $a < 0$. This leads to the following definition.

take note

The second part of the definition of absolute value states that if $a < 0$, then $|a| = -a$. For instance, if $a = -4$, then

$$|a| = |-4| = -(-4) = 4$$

Definition of Absolute Value

The **absolute value** of the real number a is defined by

$$|a| = \begin{cases} a & \text{if } a \geq 0 \\ -a & \text{if } a < 0 \end{cases}$$

The definition of *distance* between two points on a real number line makes use of absolute value.

Distance Between Points on a Real Number Line

If a and b are the coordinates of two points on a number line, the **distance** between the graph of a and the graph of b, denoted by $d(a, b)$, is given by $d(a, b) = |a - b|$.

As an example of this definition, the distance between the point whose coordinate is -2 and the point whose coordinate is 5 is given by

$$d(-2, 5) = |-2 - 5| = |-7| = 7$$

FIGURE P.4

Note from **Figure P.4** that there are 7 units between -2 and 5 on the number line. Also note that the order of the coordinates does not matter.

$$d(5, -2) = |5 - (-2)| = |7| = 7$$

EXAMPLE 4   **Use Absolute Value to Express the Distance Between Two Points**

Express the distance between a and -3 on the number line using absolute value.

Solution

$$d(a, -3) = |a - (-3)| = |a + 3|$$

▶ **TRY EXERCISE 48, PAGE 16**

● **INTERVAL NOTATION**

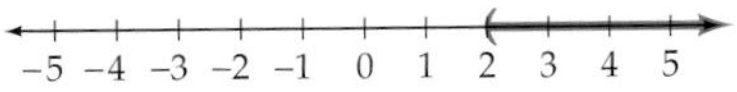

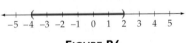

FIGURE P.5

The graph of $\{x \,|\, x > 2\}$ is shown in **Figure P.5**. The set is the real numbers greater than 2. The parenthesis at 2 indicates that 2 is not included in the set. Rather than write this set of real numbers using set-builder notation, we frequently write the set in **interval notation** as $(2, \infty)$.

In general, the interval notation

FIGURE P.6

(a, b) represents all real numbers between a and b, not including a and b. This is an **open interval**. In set-builder notation, we write $\{x \,|\, a < x < b\}$. For instance, the graph of $(-4, 2)$ is shown in **Figure P.6**.

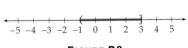

FIGURE P.7

$[a, b]$ represents all real numbers between a and b, including a and b. This is a **closed interval**. In set-builder notation, we write $\{x \,|\, a \le x \le b\}$. For instance, the graph of $[0, 4]$ is shown in **Figure P.7**. The brackets at 0 and 4 indicate that those numbers are included in the graph.

FIGURE P.8

$(a, b]$ represents all real numbers between a and b, not including a but including b. This is a **half-open interval**. In set-builder notation, we write $\{x \,|\, a < x \le b\}$. For instance, the graph of $(-1, 3]$ is shown in **Figure P.8**.

FIGURE P.9

$[a, b)$ represents all real numbers between a and b, including a but not including b. This is a **half-open interval**. In set-builder notation, we write $\{x \,|\, a \le x < b\}$. For instance, the graph of $[-4, -1)$ is shown in **Figure P.9**.

Subsets of the real numbers whose graphs extend forever in one or both directions can be represented by interval notation using the **infinity symbol** ∞ or the **negative infinity symbol** −∞.

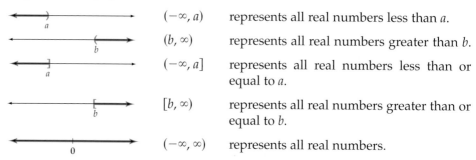

$(-\infty, a)$ represents all real numbers less than a.

(b, ∞) represents all real numbers greater than b.

$(-\infty, a]$ represents all real numbers less than or equal to a.

$[b, \infty)$ represents all real numbers greater than or equal to b.

$(-\infty, \infty)$ represents all real numbers.

EXAMPLE 5 Graph a Set in Interval Notation

Graph $(-\infty, 3]$. Write the interval in set-builder notation.

Solution

The set is the real numbers less than or equal to 3. In set-builder notation, this is the set $\{x \mid x \le 3\}$. Draw a right bracket at 3, and darken the number line to the left of 3, as shown in **Figure P.10.**

FIGURE P.10

▶ TRY EXERCISE 54, PAGE 16

TRY EXERCISE 54, PAGE 16

take note

It is *never* correct to use a bracket when using the infinity symbol. For instance, $[-\infty, 3]$ is not correct. Nor is $[2, \infty]$ correct. Neither negative infinity nor positive infinity is a real number and therefore cannot be contained in an interval.

The set $\{x \mid x \le -2\} \cup \{x \mid x > 3\}$ is the set of real numbers that are either less than or equal to −2 or greater than 3. We also could write this in interval notation as $(-\infty, -2] \cup (3, \infty)$. The graph of the set is shown in **Figure P.11.**

The set $\{x \mid x > -4\} \cap \{x \mid x < 1\}$ is the set of real numbers that are greater than −4 *and* less than 1. Note from **Figure P.12** that this set is the interval $(-4, 1)$, which can be written in set-builder notation as $\{x \mid -4 < x < 1\}$.

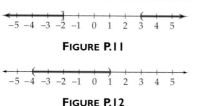

FIGURE P.11

FIGURE P.12

EXAMPLE 6 Graph Intervals

Graph the following. Write **a.** and **b.** using interval notation. Write **c.** and **d.** using set-builder notation.

a. $\{x \mid x \le -1\} \cup \{x \mid x \ge 2\}$ 　　b. $\{x \mid x \ge -1\} \cap \{x \mid x < 5\}$
c. $(-\infty, 0) \cup [1, 3]$ 　　d. $[-1, 3] \cap (1, 5)$

Solution

a. $(-\infty, -1] \cup [2, \infty)$

b. $[-1, 5)$

Continued ▶

c.  $\{x\,|\,x < 0\} \cup \{x\,|\,1 \le x \le 3\}$

d. The graphs of $[-1, 3]$, in red, and $(1, 5)$, in blue, are shown below.

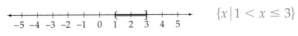

Note that the intersection of the sets occurs where the graphs intersect. Although $1 \in [-1, 3]$, $1 \notin (1, 5)$. Therefore, 1 does not belong to the intersection of the sets. On the other hand, $3 \in [-1, 3]$ and $3 \in (1, 5)$. Therefore, 3 belongs to the intersection of the sets. Thus we have the following.

$\{x\,|\,1 < x \le 3\}$

▶ **TRY EXERCISE 64, PAGE 16**

● ORDER OF OPERATIONS AGREEMENT

The approximate pressure p, in pounds per square inch, on a scuba diver x feet below the water's surface is given by

$$p = 15 + 0.5x$$

The pressure on the diver at various depths is given below.

10 feet	$15 + 0.5(10) = 15 + 5 = 20$ pounds
20 feet	$15 + 0.5(20) = 15 + 10 = 25$ pounds
40 feet	$15 + 0.5(40) = 15 + 20 = 35$ pounds
70 feet	$15 + 0.5(70) = 15 + 35 = 50$ pounds

Note that the expression $15 + 0.5(70)$ has two operations, addition and multiplication. When an expression contains more than one operation, the operations must be performed in a specified order, as listed below in the Order of Operations Agreement.

The Order of Operations Agreement

If grouping symbols are present, evaluate by performing the operations within the grouping symbols, innermost grouping symbols first, while observing the order given in steps 1 to 3.

Step 1 Evaluate exponential expressions.
Step 2 Do multiplication and division as they occur from left to right.
Step 3 Do addition and subtraction as they occur from left to right.

Therefore, the expression $15 + 0.5(70)$ is simplified by first performing the multiplication and then performing the addition, as we did above.

Pg 70 #1-74 review Poblems

take note

Recall that subtraction can be rewritten as addition of the opposite. Therefore,

$3x^2 - 4xy + 5x - y - 7$
$= 3x^2 + (-4)xy + 5x + (-y) + (-7)$

In this form, we can see that the terms (addends) are $3x^2$, $-4xy$, $5x$, $-y$, and -7.

One of the ways the Order of Operations Agreement is used is to evaluate variable expressions. The addends of a variable expression are called **terms**. The terms for the expression at the ... re $3x^2$, $-4xy$, $5x$, ...

$$3x^2 - 4xy + 5x - y - 7$$

... e terms. The term -7 is a **constant** ... ficient and a **variable part**. The nu- ... erical coefficient for the term $-4xy$ ... is 5; and the numerical coefficient ... ficient is 1 or -1 (as in x and $-x$),

... e variables by their given values ... t to simplify the result.

... pression

... 3.

b. ... when $x = 3$, $y = -2$, and $z = -4$.

Solution

a. $\dfrac{x^3 - y^3}{x^2 + xy + y^2}$

$$\dfrac{2^3 - (-3)^3}{2^2 + 2(-3) + (-3)^2} = \dfrac{8 - (-27)}{4 - 6 + 9} = \dfrac{35}{7} = 5$$

b. $(x + 2y)^2 - 4z$

$$(3 + 2(-2))^2 - 4(-4) = (3 + (-4))^2 - 4(-4)$$
$$= (-1)^2 - 4(-4) = 1 - 4(-4)$$
$$= 1 + 16 = 17$$

▶ **TRY EXERCISE 74, PAGE 16**

● SIMPLIFYING VARIABLE EXPRESSIONS

Addition, multiplication, subtraction, and *division* are the operations of arithmetic. **Addition** of the two real numbers a and b is designated by $a + b$. If $a + b = c$, then c is the **sum** and the real numbers a and b are called **terms**.

Multiplication of the real numbers a and b is designated by ab or $a \cdot b$. If $ab = c$, then c is the **product** and the real numbers a and b are called **factors** of c.

The number $-b$ is referred to as the **additive inverse** of b. **Subtraction** of the real numbers a and b is designated by $a - b$ and is defined as the sum of a and the additive inverse of b. That is,

$$a - b = a + (-b)$$

If $a - b = c$, then c is called the **difference** of a and b.

The **multiplicative inverse** or **reciprocal** of the nonzero number b is $1/b$. The **division** of a and b, designated by $a \div b$ with $b \neq 0$, is defined as the product of a and the reciprocal of b. That is,

$$a \div b = a\left(\frac{1}{b}\right) \qquad \text{provided that } b \neq 0$$

If $a \div b = c$, then c is called the **quotient** of a and b.

The notation $a \div b$ is often represented by the fractional notation a/b or $\dfrac{a}{b}$.

The real number a is the **numerator,** and the nonzero real number b is the **denominator** of the fraction.

Properties of Real Numbers

Let a, b, and c be real numbers.

	Addition Properties	**Multiplication Properties**
Closure	$a + b$ is a unique real number.	ab is a unique real number.
Commutative	$a + b = b + a$	$ab = ba$
Associative	$(a + b) + c = a + (b + c)$	$(ab)c = a(bc)$
Identity	There exists a unique real number 0 such that $a + 0 = 0 + a = a.$	There exists a unique real number 1 such that $a \cdot 1 = 1 \cdot a = a.$
Inverse	For each real number a, there is a unique real number $-a$ such that $a + (-a) = (-a) + a = 0.$	For each *nonzero* real number a, there is a unique real number $1/a$ such that $a \cdot \dfrac{1}{a} = \dfrac{1}{a} \cdot a = 1.$
Distributive		$a(b + c) = ab + ac$

EXAMPLE 8   Identify Properties of Real Numbers

Identify the property of real numbers illustrated in each statement.

a. $(2a)b = 2(ab)$

b. $\left(\dfrac{1}{5}\right)11$ is a real number.

c. $4(x + 3) = 4x + 12$

d. $(a + 5b) + 7c = (5b + a) + 7c$

e. $\left(\dfrac{1}{2} \cdot 2\right)a = 1 \cdot a$

f. $1 \cdot a = a$

Solution

a.   Associative property of multiplication

b.   Closure property of multiplication of real numbers

c.   Distributive property

d.   Commutative property of addition

e.   Inverse property of multiplication

f.   Identity property of multiplication

▶ **TRY EXERCISE 86, PAGE 16**

We can identify which property of real numbers has been used to rewrite expressions by closely comparing the expressions and noting any changes. For instance, to simplify $(6x)2$, both the commutative property and associative property of multiplication are used.

$$(6x)2 = 2(6x)$$ • **Commutative property of multiplication**
$$= (2 \cdot 6)x$$ • **Associative property of multiplication**
$$= 12x$$

To simplify $3(4p + 5)$, use the distributive property.

$$3(4p + 5) = 3(4p) + 3(5)$$ • **Distributive property**
$$= 12p + 15$$

Terms that have the same variable part are called **like terms**. The distributive property is also used to simplify an expression with like terms such as $3x^2 + 9x^2$.

$$3x^2 + 9x^2 = (3 + 9)x^2$$ • **Distributive property**
$$= 12x^2$$

Note from this example that like terms are combined by adding the coefficients of the like terms.

> **take note**
>
> Normally, we will not show, as we did at the right, all the steps in the simplification of a variable expression. For instance, we will just write $(6x)2 = 12x$, $3(4p + 5) = 12p + 15$, and $3x^2 + 9x^2 = 12x^2$. It is important to know, however, that every step in the simplification depends on one of the properties of real numbers.

❓ **QUESTION** Are the terms $2x^2$ and $3x$ like terms?

EXAMPLE 9   **Simplify Variable Expressions**

a.   Simplify $5 + 3(2x - 6)$.

b.   Simplify $4x - 2[7 - 5(2x - 3)]$.

Continued ▶

❓ **ANSWER** No. The variable parts are not the same. The variable part of $2x^2$ is $x \cdot x$. The variable part of $3x$ is x.

Solution

a. $5 + 3(2x - 6) = 5 + 6x - 18$ • Use the distributive property.

$\qquad\qquad\quad = 6x - 13$ • Add the constant terms.

b. $4x - 2[7 - 5(2x - 3)]$

$\qquad = 4x - 2[7 - 10x + 15]$ • Use the distributive property to remove the inner parentheses.

$\qquad = 4x - 2[-10x + 22]$ • Simplify.

$\qquad = 4x + 20x - 44$ • Use the distributive property to remove the brackets.

$\qquad = 24x - 44$ • Simplify.

▶ **TRY EXERCISE 106, PAGE 17**

An **equation** is a statement of equality between two numbers or two expressions. There are four basic properties of equality that relate to equations.

Properties of Equality

Let a, b, and c be real numbers.

Reflexive	$a = a$
Symmetric	If $a = b$, then $b = a$.
Transitive	If $a = b$ and $b = c$, then $a = c$.
Substitution	If $a = b$, then a may be replaced by b in any expression that involves a.

EXAMPLE 10   **Identify Properties of Equality**

Identify the property of equality illustrated in each statement.

a.   If $3a + b = c$, then $c = 3a + b$.

b. $5(x + y) = 5(x + y)$

c.   If $4a - 1 = 7b$ and $7b = 5c + 2$, then $4a - 1 = 5c + 2$.

d.   If $a = 5$ and $b(a + c) = 72$, then $b(5 + c) = 72$.

Solution

a.   Symmetric   b.   Reflexive   c.   Transitive   d.   Substitution

▶ **TRY EXERCISE 90, PAGE 16**

TOPICS FOR DISCUSSION

1. Archimedes determined that $\dfrac{223}{71} < \pi < \dfrac{22}{7}$. Is it possible to find an exact representation for π of the form $\dfrac{a}{b}$, where a and b are integers?

2. If $I = \{\text{irrational numbers}\}$ and $Q = \{\text{rational numbers}\}$, name the sets $I \cup Q$ and $I \cap Q$.

3. If the proposed simplification shown at the right is correct, so state. If it is incorrect, show a correct simplification.
$$2 \cdot 3^2 = 6^2 = 36$$

4. Are there any even prime numbers? If so, name them.

5. Does every real number have an additive inverse? Does every real number have a multiplicative inverse?

6. What is the difference between an open interval and a closed interval?

EXERCISE SET P.1

In Exercises 1 and 2, determine whether each number is an integer, a rational number, an irrational number, a prime number, or a real number.

1. $-\dfrac{1}{5}, 0, -44, \pi, 3.14, 5.05005000500005\ldots, \sqrt{81}, 53$

▶ 2. $\dfrac{5}{\sqrt{7}}, \dfrac{5}{7}, 31, -2\dfrac{1}{2}, 4.235653907493, 51, 0.888\ldots$

In Exercises 3 to 8, list the four smallest elements of each set.

3. $\{2x \mid x \in \text{positive integers}\}$

4. $\{|x| \mid x \in \text{integers}\}$

5. $\{y \mid y = 2x + 1, x \in \text{natural numbers}\}$

▶ 6. $\{y \mid y = x^2 - 1, x \in \text{integers}\}$

7. $\{z \mid z = |x|, x \in \text{integers}\}$

8. $\{z \mid z = |x| - x, x \in \text{negative integers}\}$

In Exercises 9 to 18, perform the operations given $A = \{-3, -2, -1, 0, 1, 2, 3\}$, $B = \{-2, 0, 2, 4, 6\}$, $C = \{0, 1, 2, 3, 4, 5, 6\}$, and $D = \{-3, -1, 1, 3\}$.

9. $A \cup B$

10. $C \cup D$

11. $A \cap C$

12. $C \cap D$

13. $B \cap D$

14. $B \cup (A \cap C)$

15. $D \cap (B \cup C)$

▶ 16. $(A \cap B) \cup (A \cap C)$

17. $(B \cup C) \cap (B \cup D)$

18. $(A \cap C) \cup (B \cap D)$

In Exercises 19 to 30, graph each set. Write sets given in interval notation in set-builder notation and write sets given in set-builder notation in interval notation.

19. $(-2, 3)$

20. $[1, 5]$

21. $[-5, -1]$

22. $(-3, 3)$

23. $[2, \infty)$

24. $(-\infty, 4)$

25. $\{x \mid 3 < x < 5\}$

26. $\{x \mid x < -1\}$

27. $\{x \mid x \geq -2\}$

28. $\{x \mid -1 \leq x < 5\}$

29. $\{x \mid 0 \leq x \leq 1\}$

30. $\{x \mid -4 < x \leq 5\}$

In Exercises 31 to 40, write each expression without absolute value symbols.

31. $-|-5|$

32. $-|-4|^2$

33. $|3| \cdot |-4|$

34. $|3| - |-7|$

35. $|\pi^2 + 10|$

36. $|\pi^2 - 10|$

37. $|x - 4| + |x + 5|$, given $0 < x < 1$

38. $|x + 6| + |x - 2|$, given $2 < x < 3$

39. $|2x| - |x - 1|$, given $0 < x < 1$

40. $|x + 1| + |x - 3|$, given $x > 3$

In Exercises 41 to 50, use absolute value notation to describe the given situation.

41. Distance between x and 3

42. Distance between a and -2

43. The distance between x and -2 is 4.

44. The distance between z and 5 is 1.

45. $d(m, n)$

46. $d(p, 8)$

47. The distance between a and 4 is less than 5.

▶ **48.** The distance between z and 5 is greater than 7.

49. The distance between x and -2 is greater than 4.

50. The distance between y and -3 is greater than 6.

In Exercises 51 to 66, graph each set.

51. $(-\infty, 0) \cup [2, 4]$

52. $(-3, 1) \cup (3, 5)$

53. $(-4, 0) \cap [-2, 5]$

▶ **54.** $(-\infty, 3] \cap (2, 6)$

55. $(1, \infty) \cup (-2, \infty)$

56. $(-4, \infty) \cup (0, \infty)$

57. $(1, \infty) \cap (-2, \infty)$

58. $(-4, \infty) \cap (0, \infty)$

59. $[-2, 4] \cap [4, 5]$

60. $(-\infty, 1] \cap [1, \infty)$

61. $(-2, 4) \cap (4, 5)$

62. $(-\infty, 1) \cap (1, \infty)$

63. $\{x \mid x < -3\} \cup \{x \mid 1 < x < 2\}$

▶ **64.** $\{x \mid -3 \leq x < 0\} \cup \{x \mid x \geq 2\}$

65. $\{x \mid x < -3\} \cup \{x \mid x < 2\}$

66. $\{x \mid x < -3\} \cap \{x \mid x < 2\}$

In Exercises 67 to 78, evaluate the variable expression for $x = 3$, $y = -2$, and $z = -1$.

67. $-y^3$

68. $-y^2$

69. $2xyz$

70. $-3xz$

71. $-2x^2y^2$

72. $2y^3z^2$

73. $xy - z(x - y)^2$

▶ **74.** $(z - 2y)^2 - 3z^3$

75. $\dfrac{x^2 + y^2}{x + y}$

76. $\dfrac{2xy^2z^4}{(y - z)^4}$

77. $\dfrac{3y}{x} - \dfrac{2z}{y}$

78. $(x - z)^2(x + z)^2$

In Exercises 79 to 92, state the property of real numbers or the property of equality that is used.

79. $(ab^2)c = a(b^2c)$

80. $2x - 3y = -3y + 2x$

81. $4(2a - b) = 8a - 4b$

82. $6 + (7 + a) = 6 + (a + 7)$

83. $(3x)y = y(3x)$

84. $4ab + 0 = 4ab$

85. $1 \cdot (4x) = 4x$

▶ **86.** $7(a + b) = 7(b + a)$

87. $x^2 + 1 = x^2 + 1$

88. If $a + b = 2$, then $2 = a + b$.

89. If $2x + 1 = y$ and $3x - 2 = y$, then $2x + 1 = 3x - 2$.

▶ **90.** If $4x + 2y = 7$ and $x = 3$, then $4(3) + 2y = 7$.

91. $4 \cdot \dfrac{1}{4} = 1$

92. $ab + (-ab) = 0$

In Exercises 93 to 106, simplify the variable expression.

93. $3(2x)$

94. $-2(4y)$

95. $3(2 + x)$

96. $-2(4 + y)$

97. $\dfrac{2}{3}a + \dfrac{5}{6}a$

98. $\dfrac{3}{4}x - \dfrac{1}{2}x$

99. $2 + 3(2x - 5)$

100. $4 + 2(2a - 3)$

101. $5 - 3(4x - 2y)$

102. $7 - 2(5n - 8m)$

103. $3(2a - 4b) - 4(a - 3b)$

104. $5(4r - 7t) - 2(10r + 3t)$

105. $5a - 2[3 - 2(4a + 3)]$

▶ **106.** $6 + 3[2x - 4(3x - 2)]$

107. **AREA OF A TRIANGLE** The area of a triangle is given by area $= \dfrac{1}{2}bh$, where b is the base of the triangle and h is its height. Find the area of a triangle whose base is 3 inches and whose height is 4 inches.

108. **VOLUME OF A BOX** The volume of a rectangular box is given by

$$\text{volume} = lwh$$

where l is the length, w is the width, and h is the height of the box. Find the volume of a classroom that is 40 feet long, 30 feet wide, and 12 feet high.

109. **PROFIT FROM SALES** The profit, in dollars, a company earns from the sale of x bicycles is given by

$$\text{profit} = -0.5x^2 + 120x - 2000$$

Find the profit the company earns from selling 110 bicycles.

110. **MAGAZINE CIRCULATION** The circulation, in thousands of subscriptions, of a new magazine n months after its introduction can be approximated by

$$\text{circulation} = \sqrt{n^2 - n + 1}$$

Find, to the nearest hundred, the circulation of the magazine after 12 months.

111. **HEART RATE** The heart rate, in beats per minute, of a certain runner during a cool-down period can be approximated by

$$\text{Heart rate} = 65 + \frac{53}{4t + 1}$$

where t is the number of minutes after the start of cool-down. Find the runner's heart rate in 10 minutes. Round to the nearest whole number.

112. **BODY MASS INDEX** According to the National Institutes of Health, body mass index (BMI) is a measure of body fat based on height and weight that applies to both adult men and women, with values between 18.5 and 24.9 considered healthy. BMI is calculated as BMI $= \dfrac{705w}{h^2}$, where w is the weight of the person in pounds and h is the person's height in inches. Find the BMI for a person who weighs 160 pounds and is 5 feet 10 inches tall. Round to the nearest whole number.

113. **PHYSICS** The height, in feet, of a ball t seconds after it is thrown upward is given by height $= -16t^2 + 80t + 4$. Find the height of the ball 2 seconds after it has been released.

114. **CHEMISTRY** Salt is being added to water in such a way that the concentration, in grams per liter, is given by concentration $= \dfrac{50t}{t + 1}$, where t is the time in minutes after the introduction of the salt. Find the concentration of salt after 24 minutes.

CONNECTING CONCEPTS

In Exercises 115 to 118, perform the operation given A is any set.

115. $A \cup A$

116. $A \cap A$

117. $A \cap \varnothing$

118. $A \cup \varnothing$

119. If A and B are two sets and $A \cup B = A$, what can be said about B?

120. If A and B are two sets and $A \cap B = B$, what can be said about B?

121. Is division of real numbers an associative operation? Give a reason for your answer.

122. Is subtraction of real numbers a commutative operation? Give a reason for your answer.

123. Which of the properties of real numbers are satisfied by the integers?

124. Which of the properties of real numbers are satisfied by the rational numbers?

In Exercises 125 and 126, write each expression without absolute value symbols.

125. $\left| \dfrac{x + 7}{|x| + |x - 1|} \right|$, given $0 < x < 1$.

126. $\left| \dfrac{x + 3}{\left|x - \dfrac{1}{2}\right| + \left|x + \dfrac{1}{2}\right|} \right|$, given $0 < x < 0.2$.

In Exercises 127 to 132, use absolute value notation to describe the given statement.

127. x is closer to 2 than it is to 6.

128. x is closer to a than it is to b.

129. x is farther from 3 than it is from -7.

130. x is farther from 0 than it is from 5.

131. x is more than 2 units from 4 but less than 7 units from 4.

132. x is more than b units from a but less than c units from a.

PREPARE FOR SECTION P.2

133. Simplify: $2^2 \cdot 2^3$ [P.1]

134. Simplify: $\dfrac{4^3}{4^5}$ [P.1]

135. Simplify: $(2^3)^2$ [P.1]

136. Simplify: $3.14(10^5)$ [P.1]

137. True or false: $3^4 \cdot 3^2 = 9^6$ [P.1]

138. True or false: $(3 + 4)^2 = 3^2 + 4^2$ [P.1]

PROJECTS

1. NUMBER PUZZLE A number n has the following properties:

When n is divided by 6, the remainder is 5.
When n is divided by 5, the remainder is 4.
When n is divided by 4, the remainder is 3.
When n is divided by 3, the remainder is 2.
When n is divided by 2, the remainder is 1.

What is the smallest value of n?

2. OPERATIONS ON INTERVALS Besides finding unions and intersections of intervals, it is possible to apply other operations to intervals. For instance, $(-1, 2)^2$ is the interval that results from squaring every number in the interval $(-1, 2)$. This gives $[0, 4)$. Thus $(-1, 2)^2 = [0, 4)$.

a. Find $(-4, 2)^2$.

b. Find $ABS(-4, 5)$, the absolute value of every number in $(-4, 5)$.

c. Find $\sqrt{(0, 9)}$, the square root of every number in $(0, 9)$.

d. Find $\dfrac{1}{(0, 1)}$, the reciprocal of every number in $(0, 1)$.

3. FACTORS OF A NUMBER Explain why the square of a natural number always has an odd number of natural number factors.

INTEGER AND RATIONAL NUMBER EXPONENTS

• PROPERTIES OF EXPONENTS

A compact method of writing $5 \cdot 5 \cdot 5 \cdot 5$ is 5^4. The expression 5^4 is written in **exponential notation.** Similarly, we can write

$$\frac{2x}{3} \cdot \frac{2x}{3} \cdot \frac{2x}{3} \quad \text{as} \quad \left(\frac{2x}{3}\right)^3$$

Exponential notation can be used to express the product of any expression that is used repeatedly as a factor.

Definition of Natural Number Exponents

If b is any real number and n is a natural number, then

$$b^n = \overbrace{b \cdot b \cdot b \cdot \cdots \cdot b}^{b \text{ is a factor } n \text{ times}}$$

where b is the **base** and n is the **exponent.**

For instance,

$$5^4 = 5 \cdot 5 \cdot 5 \cdot 5 = 625$$
$$-5^4 = -(5 \cdot 5 \cdot 5 \cdot 5) = -625$$
$$(-5)^4 = (-5)(-5)(-5)(-5) = 625$$

Pay close attention to the difference between -5^4 (the base is 5) and $(-5)^4$ (the base is -5).

❓ QUESTION What is the value of **a.** -2^5 and **b.** $(-2)^5$?

We can extend the definition of an exponent to all the integers. We first deal with the case of zero as an exponent.

Definition of b^0

For any nonzero real number b, $b^0 = 1$.

Some examples of this definition are

$$3^0 = 1 \qquad \left(\frac{3}{4}\right)^0 = 1 \qquad (-7)^0 = 1 \qquad (a^2 + 1)^0 = 1$$

❓ ANSWER   **a.** $-2^5 = -(2 \cdot 2 \cdot 2 \cdot 2 \cdot 2) = -32$
  b. $(-2)^5 = (-2)(-2)(-2)(-2)(-2) = -32$

take note

Using the definition of b^{-n},

$$\frac{5^{-2}}{7^{-1}} = \frac{\dfrac{1}{5^2}}{\dfrac{1}{7}}$$

Using the rules for dividing fractions, we have

$$\frac{\dfrac{1}{5^2}}{\dfrac{1}{7}} = \frac{1}{5^2} \div \frac{1}{7} = \frac{1}{5^2} \cdot \frac{7}{1} = \frac{7}{5^2}$$

Now we extend the definition to include negative integers.

Definition of b^{-n}

If $b \neq 0$ and n is a natural number, then $b^{-n} = \dfrac{1}{b^n}$ and $\dfrac{1}{b^{-n}} = b^n$.

Here are some examples.

$$3^{-2} = \frac{1}{3^2} = \frac{1}{9} \qquad \frac{1}{4^{-3}} = 4^3 = 64 \qquad \frac{5^{-2}}{7^{-1}} = \frac{7}{5^2} = \frac{7}{25}$$

EXAMPLE 1   Evaluate an Exponential Expression

a. $(-2^4)(-3)^2$ b. $\dfrac{(-4)^{-3}}{(-2)^{-5}}$ c. $-\pi^0$

Solution

a. $(-2^4)(-3)^2 = -(2 \cdot 2 \cdot 2 \cdot 2)(-3)(-3) = -(16)(9) = -144$

b. $\dfrac{(-4)^{-3}}{(-2)^{-5}} = \dfrac{(-2)(-2)(-2)(-2)(-2)}{(-4)(-4)(-4)} = \dfrac{-32}{-64} = \dfrac{1}{2}$

c. $-\pi^0 = -(\pi^0) = -1$

▶ TRY EXERCISE 10, PAGE 31

take note

Part **c.** is similar to $-5^4 = -625$, which was discussed earlier.

When working with exponential expressions containing variables, we must ensure that a value of the variable does not result in an undefined expression. For instance, $x^{-2} = \dfrac{1}{x^2}$. Because division by zero is not allowed, for the expression x^{-2}, we must assume that $x \neq 0$. Therefore, to avoid problems with undefined expressions, we will use the following restriction agreement.

Restriction Agreement

The expressions 0^0, 0^n (where n is a negative integer), and $\dfrac{a}{0}$ are all undefined expressions. Therefore, all values of variables in this text are restricted to avoid any one of these expressions.

For instance, in the expression

$$\frac{x^0 y^{-3}}{z - 4}, \ x \neq 0, y \neq 0, \text{ and } z \neq 4.$$

For the expression

$$\frac{(a - 1)^0}{b + 2}, a \neq 1 \text{ and } b \neq -2.$$

Exponential expressions containing variables are simplified by using the following properties of exponents.

Properties of Exponents

If m, n, and p are integers and a and b are real numbers, then

Product $\qquad b^m \cdot b^n = b^{m+n}$

Quotient $\qquad \dfrac{b^m}{b^n} = b^{m-n}, \qquad b \neq 0$

Power $\qquad (b^m)^n = b^{mn} \qquad (a^m b^n)^p = a^{mp} b^{np}$

$\qquad\qquad \left(\dfrac{a^m}{b^n}\right)^p = \dfrac{a^{mp}}{b^{np}}, \qquad b \neq 0$

Here are some examples of these properties.

$a^4 \cdot a \cdot a^3 = a^{4+1+3} = a^8$ $\qquad$ • Recall that $a = a^1$.

$(x^4 y^3)(x y^5 z^2) = x^{4+1} y^{3+5} z^2 = x^5 y^8 z^2$ $\qquad$ • Add the exponents on the like bases.

$\dfrac{a^7 b}{a^2 b^5} = a^{7-2} b^{1-5} = a^5 b^{-4} = \dfrac{a^5}{b^4}$ $\qquad$ • Subtract the exponents on the like bases.

$(u v^3)^5 = u^{1 \cdot 5} v^{3 \cdot 5} = u^5 v^{15}$ $\qquad$ • Multiply the exponents.

? QUESTION Can the exponential expression $x^5 y^3$ be simplified using the properties of exponents?

INTEGRATING TECHNOLOGY

Exponential expressions such as a^{b^c} can be confusing. The generally accepted meaning of a^{b^c} is $a^{(b^c)}$. However, some graphing calculators do not evaluate exponential expressions in this way. Enter 2^3^4 in a graphing calculator. If the result is approximately 2.42×10^{24}, then the calculator evaluated $2^{(3^4)}$. If the result is 4096, then the calculator evaluated $(2^3)^4$. To ensure that you calculate the value you intend, we strongly urge you to use parentheses. For instance, entering 2^(3^4) will produce 2.42×10^{24} and entering (2^3)^4 will produce 4096.

? ANSWER No. The bases are not the same.

To simplify an expression involving exponents, write the expression in a form in which *each base appears at most once* and *no powers of powers or negative exponents appear.*

EXAMPLE 2   **Simplify Exponential Expressions**

Simplify.   **a.** $(5x^2y)(-4x^3y^5)$ **b.** $(3x^2yz^{-4})^3$ **c.** $\left(\dfrac{4p^2q}{6pq^4}\right)^{-2}$

Solution

a. $(5x^2y)(-4x^3y^5) = [5(-4)]x^{2+3}y^{1+5}$ • Multiply the coefficients. Multiply the variables by adding the exponents on the like bases.

$$= -20x^5y^6$$

b. $(3x^2yz^{-4})^3 = 3^{1\cdot3}x^{2\cdot3}y^{1\cdot3}z^{-4\cdot3}$ • Use the power property of exponents.

$$= 3^3x^6y^3z^{-12} = \frac{27x^6y^3}{z^{12}}$$

c. $\left(\dfrac{4p^2q}{6pq^4}\right)^{-2} = \left(\dfrac{2p^{2-1}q^{1-4}}{3}\right)^{-2} = \left(\dfrac{2pq^{-3}}{3}\right)^{-2}$ • Use the quotient property of exponents.

$$= \frac{2^{1(-2)}p^{1(-2)}q^{-3(-2)}}{3^{1(-2)}} = \frac{2^{-2}p^{-2}q^6}{3^{-2}}$$ • Use the power property of exponents.

$$= \frac{9q^6}{4p^2}$$ • Write the answer in simplest form.

▶ **TRY EXERCISE 30, PAGE 32**

● SCIENTIFIC NOTATION

The exponent theorems provide a compact method of writing very large or very small numbers. The method is called *scientific notation*. A number written in **scientific notation** has the form $a \cdot 10^n$, where n is an integer and $1 \leq a < 10$. The following procedure is used to change a number from its decimal form to scientific notation.

For numbers greater than 10, move the decimal point to the position to the right of the first digit. The exponent n will equal the number of places the decimal point has been moved. For example,

$$7{,}430{,}000 = 7.43 \times 10^6$$

6 places

For numbers less than 1, move the decimal point to the right of the first nonzero digit. The exponent n will be negative, and its absolute value will equal the number of places the decimal point has been moved. For example,

$$0.00000078 = 7.8 \times 10^{-7}$$

7 places

To change a number from scientific notation to its decimal form, reverse the procedure. That is, if the exponent is positive, move the decimal point to the right the same number of places as the exponent. For example,

$$3.5 \times 10^5 = 350{,}000$$

5 places

If the exponent is negative, move the decimal point to the left the same number of places as the absolute value of the exponent. For example,

$$2.51 \times 10^{-8} = 0.0000000251$$

8 places

Most scientific calculators display very large and very small numbers in scientific notation. The number $450{,}000^2$ is displayed as $\boxed{2.025 \quad E\ 11}$. This means $450{,}000^2 = 2.025 \times 10^{11}$.

EXAMPLE 3 — Simplify an Expression Using Scientific Notation

The Andromeda galaxy is approximately 1.4×10^{19} miles from Earth. If a spacecraft could travel 2.8×10^{12} miles in 1 year (about one-half the speed of light), how many years would it take to reach the Andromeda galaxy?

Solution

To find the time, divide the distance by the speed.

$$t = \frac{1.4 \times 10^{19}}{2.8 \times 10^{12}} = \frac{1.4}{2.8} \times 10^{19-12} = 0.5 \times 10^7 = 5.0 \times 10^6$$

It would take 5.0×10^6 (or 5,000,000) years to reach the Andromeda galaxy.

▶ **TRY EXERCISE 46, PAGE 32**

● RATIONAL EXPONENTS AND RADICALS

To this point, the expression b^n has been defined for real numbers b and integers n. Now we wish to extend the definition of exponents to include rational numbers so that expressions such as $2^{1/2}$ will be meaningful. Not just any definition will do. We want a definition of rational exponents for which the properties of integer exponents are true. The following example shows the direction we can take to accomplish our goal.

If the product property for exponential expressions is to hold for rational exponents, then for rational numbers p and q, $b^p b^q = b^{p+q}$. For example,

$$9^{1/2} \cdot 9^{1/2} \quad \text{must equal} \quad 9^{1/2+1/2} = 9^1 = 9$$

Thus $9^{1/2}$ must be a square root of 9. That is, $9^{1/2} = 3$.

The example suggests that $b^{1/n}$ can be defined in terms of roots according to the following definition.

Definition of $b^{1/n}$

If n is an even positive integer and $b \geq 0$, then $b^{1/n}$ is the nonnegative real number such that $(b^{1/n})^n = b$.

If n is an odd positive integer, then $b^{1/n}$ is the real number such that $(b^{1/n})^n = b$.

As examples,

- $25^{1/2} = 5$ because $5^2 = 25$.
- $(-64)^{1/3} = -4$ because $(-4)^3 = -64$.
- $16^{1/2} = 4$ because $4^2 = 16$.
- $-16^{1/2} = -(16^{1/2}) = -4$.
- $(-16)^{1/2}$ is not a real number.
- $(-32)^{1/5} = -2$ because $(-2)^5 = -32$.

If n is an even positive integer and $b < 0$, then $b^{1/n}$ is a *complex number*. Complex numbers are discussed in Section P.5.

To define expressions such as $8^{2/3}$, we will extend our definition of exponents even further. Because we want the power property $(b^p)^q = b^{pq}$ to be true for rational exponents also, we must have $(b^{1/n})^m = b^{m/n}$. With this in mind, we make the following definition.

Definition of $b^{m/n}$

For all positive integers m and n such that m/n is in simplest form, and for all real numbers b for which $b^{1/n}$ is a real number,

$$b^{m/n} = (b^{1/n})^m = (b^m)^{1/n}$$

Because $b^{m/n}$ is defined as $(b^{1/n})^m$ and also as $(b^m)^{1/n}$, we can evaluate expressions such as $8^{4/3}$ in more than one way. For example, because $8^{1/3}$ is a real number, $8^{4/3}$ can be evaluated in either of the following ways:

$$8^{4/3} = (8^{1/3})^4 = 2^4 = 16$$
$$8^{4/3} = (8^4)^{1/3} = 4096^{1/3} = 16$$

Of the two methods, the $b^{m/n} = (b^{1/n})^m$ method is usually easier to apply, provided you can evaluate $b^{1/n}$.

Here are some additional examples.

$$64^{2/3} = (64^{1/3})^2 = 4^2 = 16$$

$$32^{-6/5} = \frac{1}{32^{6/5}} = \frac{1}{(32^{1/5})^6} = \frac{1}{2^6} = \frac{1}{64}$$

$$81^{0.75} = 81^{3/4} = (81^{1/4})^3 = 3^3 = 27$$

INTEGRATING TECHNOLOGY

For the examples on page 24, the base of the exponential expression was an integer power of the denominator of the exponent.

$$64 = 4^3 \qquad 32 = 2^5 \qquad 81 = 3^4$$

If the base of the exponential expression is not a power of the denominator of the exponent, a calculator is used to evaluate the expression. Here are some examples.

```
16^(3/8)
             2.828427125
25^(-1/5)
              .5253055609
42^(.14)
              1.687543205
```

In each of these examples, the value of the exponential expression is an irrational number. The decimal display is only an approximation of the actual result.

The following exponent properties were stated earlier, but they are restated here to remind you that they have now been extended to apply to rational exponents.

Properties of Rational Exponents

If p, q, and r represent rational numbers and a and b are positive real numbers, then

Product $b^p \cdot b^q = b^{p+q}$

Quotient $\dfrac{b^p}{b^q} = b^{p-q}$

Power $(b^p)^q = b^{pq} \qquad (a^p b^q)^r = a^{pr} b^{qr}$

            $\left(\dfrac{a^p}{b^q}\right)^r = \dfrac{a^{pr}}{b^{qr}} \qquad b^{-p} = \dfrac{1}{b^p}$

Recall that an exponential expression is in simplest form when no powers of powers or negative exponents appear and each base occurs at most once.

EXAMPLE 4 Simplify Exponential Expressions

Simplify: $\left(\dfrac{x^2y^3}{x^{-3}y^5}\right)^{1/2}$ (Assume $x > 0$, $y > 0$.)

Solution

$$\left(\frac{x^2y^3}{x^{-3}y^5}\right)^{1/2} = (x^{2-(-3)}y^{3-5})^{1/2} = (x^5y^{-2})^{1/2} = x^{5/2}y^{-1} = \frac{x^{5/2}}{y}$$

▶ **TRY EXERCISE 62, PAGE 32**

TRY EXERCISE 62, PAGE 32

● SIMPLIFY RADICAL EXPRESSIONS

MATH MATTERS

The formula for kinetic energy (energy of motion) that is used in Einstein's Theory of Relativity involves a radical.

$$\text{K.E.}_r = mc^2\left(\frac{1}{\sqrt{1 - \dfrac{v^2}{c^2}}} - 1\right)$$

where m is the mass of the object at rest, v is the speed of the object, and c is the speed of light.

Radicals, expressed by the notation $\sqrt[n]{b}$, are also used to denote roots. The number b is the **radicand,** and the positive integer n is the **index** of the radical.

Definition of $\sqrt[n]{b}$

If n is a positive integer and b is a real number such that $b^{1/n}$ is a real number, then $\sqrt[n]{b} = b^{1/n}$.

If the index n equals 2, then the radical $\sqrt[2]{b}$ is written as simply $\sqrt{b}$, and it is referred to as the **principal square root of b** or simply the **square root of b.**

The symbol $\sqrt{b}$ is reserved to represent the nonnegative square root of b. To represent the negative square root of b, write $-\sqrt{b}$. For example, $\sqrt{25} = 5$, whereas $-\sqrt{25} = -5$.

Definition of $(\sqrt[n]{b})^m$

For all positive integers n, all integers m, and all real numbers b such that $\sqrt[n]{b}$ is a real number, $(\sqrt[n]{b})^m = \sqrt[n]{b^m} = b^{m/n}$.

When $\sqrt[n]{b}$ is a real number, the equations

$$b^{m/n} = \sqrt[n]{b^m} \qquad \text{and} \qquad b^{m/n} = (\sqrt[n]{b})^m$$

can be used to write exponential expressions such as $b^{m/n}$ in radical form. Use the denominator n as the index of the radical and the numerator m as the power of the radicand or as the power of the radical. For example,

$$(5xy)^{2/3} = (\sqrt[3]{5xy})^2 = \sqrt[3]{25x^2y^2}$$

• Use the denominator 3 as the index of the radical and the numerator 2 as the power of the radical.

The equations

$$b^{m/n} = \sqrt[n]{b^m} \qquad \text{and} \qquad b^{m/n} = (\sqrt[n]{b})^m$$

also can be used to write radical expressions in exponential form. For example,

$$\sqrt{(2ab)^3} = (2ab)^{3/2}$$ • **Use the index 2 as the denominator of the power and the exponent 3 as the numerator of the power.**

The definition of $\left(\sqrt[n]{b}\right)^m$ often can be used to evaluate radical expressions. For instance,

$$(\sqrt[3]{8})^4 = 8^{4/3} = (8^{1/3})^4 = 2^4 = 16$$

Care must be exercised when simplifying even roots (square roots, fourth roots, sixth roots,…) of variable expressions. Consider $\sqrt{x^2}$ when $x = 5$ and when $x = -5$.

Case 1   If $x = 5$, then $\sqrt{x^2} = \sqrt{5^2} = \sqrt{25} = 5 = x$.

Case 2   If $x = -5$, then $\sqrt{x^2} = \sqrt{(-5)^2} = \sqrt{25} = 5 = -x$.

These two cases suggest that

To review **ABSOLUTE VALUE,** *see p. 7.*

$$\sqrt{x^2} = \begin{cases} x, & \text{if } x \geq 0 \\ -x, & \text{if } x < 0 \end{cases}$$

Recalling the definition of absolute value, we can write this more compactly as $\sqrt{x^2} = |x|$.

Simplifying odd roots of a variable expression does not require using the absolute value symbol. Consider $\sqrt[3]{x^3}$ when $x = 5$ and when $x = -5$.

Case 1   If $x = 5$, then $\sqrt[3]{x^3} = \sqrt[3]{5^3} = \sqrt[3]{125} = 5 = x$.

Case 2   If $x = -5$, then $\sqrt[3]{x^3} = \sqrt[3]{(-5)^3} = \sqrt[3]{-125} = -5 = x$.

Thus $\sqrt[3]{x^3} = x$.

Although we have illustrated this principle only for square roots and cube roots, the same reasoning can be applied to other cases. The general result is given below.

Definition of $\sqrt[n]{b^n}$

If n is an even natural number and b is a real number, then

$$\sqrt[n]{b^n} = |b|$$

If n is an odd natural number and b is a real number, then

$$\sqrt[n]{b^n} = b$$

Here are some examples of these properties.

$$\sqrt[4]{16z^4} = 2|z| \qquad \sqrt[5]{32a^5} = 2a$$

Because radicals are defined in terms of rational powers, the properties of radicals are similar to those of exponential expressions.

Properties of Radicals

If m and n are natural numbers and a and b are nonnegative real numbers, then

Product $\quad \sqrt[n]{a} \cdot \sqrt[n]{b} = \sqrt[n]{ab}$

Quotient $\quad \dfrac{\sqrt[n]{a}}{\sqrt[n]{b}} = \sqrt[n]{\dfrac{a}{b}}$

Index $\quad \sqrt[m]{\sqrt[n]{a}} = \sqrt[mn]{a}$

A radical is in **simplest form** if it meets all of the following criteria.

1. The radicand contains only powers less than the index. ($\sqrt{x^5}$ does not satisfy this requirement because 5, the exponent, is greater than 2, the index.)

2. The index of the radical is as small as possible. ($\sqrt[9]{x^3}$ does not satisfy this requirement because $\sqrt[9]{x^3} = x^{3/9} = x^{1/3} = \sqrt[3]{x}$.)

3. The denominator has been rationalized. That is, no radicals appear in the denominator. ($1/\sqrt{2}$ does not satisfy this requirement.)

4. No fractions appear under the radical sign. ($\sqrt[4]{2/x^3}$ does not satisfy this requirement.)

Radical expressions are simplified by using the properties of radicals. Here are some examples.

EXAMPLE 5 **Simplify Radical Expressions**

Simplify.

a. $\sqrt[4]{32x^3y^4}$ b. $\sqrt[3]{162x^4y^6}$

Solution

a. $\sqrt[4]{32x^3y^4} = \sqrt[4]{2^5x^3y^4} = \sqrt[4]{(2^4y^4) \cdot (2x^3)}$

 • **Factor and group factors that can be written as a power of the index.**

$\quad = \sqrt[4]{2^4y^4} \cdot \sqrt[4]{2x^3}$

 • **Use the product property of radicals.**

$\quad = 2|y|\sqrt[4]{2x^3}$

 • **Recall that for n even, $\sqrt[n]{b^n} = |b|$.**

b. $\sqrt[3]{162x^4y^6} = \sqrt[3]{(2 \cdot 3^4)x^4y^6}$

 • **Factor and group factors that can be written as a power of the index.**

$\quad = \sqrt[3]{(3xy^2)^3 \cdot (2 \cdot 3x)}$

$\quad = \sqrt[3]{(3xy^2)^3} \cdot \sqrt[3]{6x}$

 • **Use the product property of radicals.**

$\quad = 3xy^2\sqrt[3]{6x}$

 • **Recall that for n odd, $\sqrt[n]{b^n} = b$.**

▶ **TRY EXERCISE 78, PAGE 32**

Like radicals have the same radicand and the same index. For instance,

$$3\sqrt[3]{5xy^2} \quad \text{and} \quad -4\sqrt[3]{5xy^2}$$

are like radicals. Addition and subtraction of like radicals are accomplished by using the distributive property. For example,

$$4\sqrt{3x} - 9\sqrt{3x} = (4-9)\sqrt{3x} = -5\sqrt{3x}$$
$$2\sqrt[3]{y^2} + 4\sqrt[3]{y^2} - \sqrt[3]{y^2} = (2+4-1)\sqrt[3]{y^2} = 5\sqrt[3]{y^2}$$

The sum $2\sqrt{3} + 6\sqrt{5}$ cannot be simplified further because the radicands are not the same. The sum $3\sqrt[3]{x} + 5\sqrt[4]{x}$ cannot be simplified because the indices are not the same.

Sometimes it is possible to simplify radical expressions that do not appear to be like radicals by simplifying each radical expression.

EXAMPLE 6   **Combine Radical Expressions**

Simplify: $5x\sqrt[3]{16x^4} - \sqrt[3]{128x^7}$

Solution

$5x\sqrt[3]{16x^4} - \sqrt[3]{128x^7}$

$\quad = 5x\sqrt[3]{2^4x^4} - \sqrt[3]{2^7x^7}$ • Factor.

$\quad = 5x\sqrt[3]{2^3x^3} \cdot \sqrt[3]{2x} - \sqrt[3]{2^6x^6} \cdot \sqrt[3]{2x}$ • Group factors that can be written as a power of the index.

$\quad = 5x(2x\sqrt[3]{2x}) - 2^2x^2 \cdot \sqrt[3]{2x}$ • Use the product property of radicals.

$\quad = 10x^2\sqrt[3]{2x} - 4x^2\sqrt[3]{2x}$ • Simplify.

$\quad = 6x^2\sqrt[3]{2x}$

▶ **TRY EXERCISE 86, PAGE 32**

Multiplication of radical expressions is accomplished by using the distributive property. For instance,

$$\sqrt{5}(\sqrt{20} - 3\sqrt{15}) = \sqrt{5}(\sqrt{20}) - \sqrt{5}(3\sqrt{15})$$ • Use the distributive property.

$$= \sqrt{100} - 3\sqrt{75}$$ • Multiply the radicals.

$$= 10 - 3 \cdot 5\sqrt{3}$$ • Simplify.

$$= 10 - 15\sqrt{3}$$

The product of more complicated radical expressions may require repeated use of the distributive property.

EXAMPLE 7   **Multiply Radical Expressions**

Perform the indicated operation:

$$(\sqrt{3} + 5)(\sqrt{3} - 2)$$

Solution

$$(\sqrt{3} + 5)(\sqrt{3} - 2)$$
$$= (\sqrt{3} + 5)\sqrt{3} - (\sqrt{3} + 5)2 \qquad \text{• Use the distributive property.}$$
$$= (\sqrt{3}\sqrt{3} + 5\sqrt{3}) - (2\sqrt{3} + 2 \cdot 5) \qquad \text{• Use the distributive property.}$$
$$= 3 + 5\sqrt{3} - 2\sqrt{3} - 10$$
$$= -7 + 3\sqrt{3}$$

▶ **TRY EXERCISE 96, PAGE 33**

To **rationalize the denominator** of a fraction means to write it in an equivalent form that does not involve any radicals in its denominator.

EXAMPLE 8   **Rationalize the Denominator**

Rationalize the denominator.   **a.** $\dfrac{5}{\sqrt[3]{a}}$ **b.** $\sqrt{\dfrac{3}{32y}}$

Solution

a. $\dfrac{5}{\sqrt[3]{a}} = \dfrac{5}{\sqrt[3]{a}} \cdot \dfrac{\sqrt[3]{a^2}}{\sqrt[3]{a^2}} = \dfrac{5\sqrt[3]{a^2}}{\sqrt[3]{a^3}} = \dfrac{5\sqrt[3]{a^2}}{a}$ • Use $\sqrt[3]{a} \cdot \sqrt[3]{a^2} = \sqrt[3]{a^3} = a$.

b. $\sqrt{\dfrac{3}{32y}} = \dfrac{\sqrt{3}}{\sqrt{32y}} = \dfrac{\sqrt{3}}{4\sqrt{2y}} = \dfrac{\sqrt{3}}{4\sqrt{2y}} \cdot \dfrac{\sqrt{2y}}{\sqrt{2y}} = \dfrac{\sqrt{6y}}{8y}$

▶ **TRY EXERCISE 106, PAGE 33**

take note

In Example 8b, y must be positive or the original expression does not represent a real number. Therefore, an absolute value symbol is unnecessary in the final answer.

To rationalize the denominator of a fractional expression such as

$$\frac{1}{\sqrt{m} + \sqrt{n}}$$

we make use of the conjugate of $\sqrt{m} + \sqrt{n}$, which is $\sqrt{m} - \sqrt{n}$. The product of these conjugate pairs does not involve a radical.

$$(\sqrt{m} + \sqrt{n})(\sqrt{m} - \sqrt{n}) = m - n$$

In Example 9 we use the conjugate of the denominator to rationalize the denominator.

EXAMPLE 9   **Rationalize the Denominator**

Rationalize the denominator.

a. $\dfrac{2}{\sqrt{3} + \sqrt{2}}$ **b.** $\dfrac{a + \sqrt{5}}{a - \sqrt{5}}$

Solution

a. $\dfrac{2}{\sqrt{3} + \sqrt{2}} = \dfrac{2}{\sqrt{3} + \sqrt{2}} \cdot \dfrac{\sqrt{3} - \sqrt{2}}{\sqrt{3} - \sqrt{2}} = \dfrac{2\sqrt{3} - 2\sqrt{2}}{3 - 2} = 2\sqrt{3} - 2\sqrt{2}$

b. $\dfrac{a + \sqrt{5}}{a - \sqrt{5}} = \dfrac{a + \sqrt{5}}{a - \sqrt{5}} \cdot \dfrac{a + \sqrt{5}}{a + \sqrt{5}} = \dfrac{a^2 + 2a\sqrt{5} + 5}{a^2 - 5}$

▶ **TRY EXERCISE 110, PAGE 33**

  TOPICS FOR DISCUSSION

1. Given that a is a real number, discuss when the expression $a^{p/q}$ represents a real number.

2. The expressions $-a^n$ and $(-a)^n$ do not always represent the same number. Discuss the situations in which the two expressions are equal and those in which they are not equal.

3. The following calculator screen shows the value of the quotient of two radical expressions. Is the answer correct? Explain what happened.

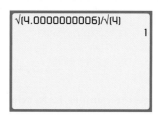

4. If you enter the expression for $\sqrt{5}$ on your calculator, the calculator will respond with 2.236067977 or some number close to that. Is this the exact value of $\sqrt{5}$? Is it possible to find the exact decimal value of $\sqrt{5}$ with a calculator? with a computer?

EXERCISE SET P.2

In Exercises 1 to 12, evaluate each expression.

1. -5^3

2. $(-5)^3$

3. $\left(\dfrac{2}{3}\right)^0$

4. -6^0

5. 4^{-2}

6. 3^{-4}

7. $\dfrac{1}{2^{-5}}$

8. $\dfrac{1}{3^{-3}}$

9. $\dfrac{2^{-3}}{6^{-3}}$

▶ **10.** $\dfrac{4^{-2}}{2^{-3}}$

11. $-2x^0$

12. $\dfrac{x^0}{4}$

In Exercises 13 to 32, write the exponential expression in simplest form.

13. $2x^{-4}$

14. $3y^{-2}$

15. $(-2ab^4)(-3a^2b^4)$

16. $(9xy^2)(-2x^2y^2)$

17. $\dfrac{6a^4}{8a^8}$

18. $\dfrac{12x^3}{16x^4}$

19. $\dfrac{12x^3y^4}{18x^5y^2}$

20. $\dfrac{5v^4w^{-3}}{10v^8}$

21. $\dfrac{36a^{-2}b^3}{3ab^4}$

22. $\dfrac{-48ab^{10}}{-32a^4b^3}$

23. $(-2m^3n^2)(-3mn^2)^2$

24. $(2a^3b^2)^3(-4a^4b^2)$

25. $(x^{-2}y)^2(xy)^{-2}$

26. $(x^{-1}y^2)^{-3}(x^2y^{-4})^{-3}$

27. $\left(\dfrac{3a^2b^3}{6a^4b^4}\right)^2$

28. $\left(\dfrac{2ab^2c^3}{5ab^2}\right)^3$

29. $\dfrac{(-4x^2y^3)^2}{(2xy^2)^3}$

▶ 30. $\dfrac{(-3a^2b^3)^2}{(-2ab^4)^3}$

31. $\left(\dfrac{a^{-2}b}{a^3b^{-4}}\right)^2$

32. $\left(\dfrac{x^{-3}y^{-4}}{x^{-2}y}\right)^{-2}$

In Exercises 33 to 36, write the number in scientific notation.

33. 2,011,000,000,000

34. 49,100,000,000

35. 0.000000000562

36. 0.000000402

In Exercises 37 to 40, change the number from scientific notation to decimal notation.

37. 3.14×10^7

38. 4.03×10^9

39. -2.3×10^{-6}

40. 6.14×10^{-8}

In Exercises 41 to 48, perform the indicated operation and write the answer in scientific notation.

41. $(3 \times 10^{12})(9 \times 10^{-5})$

42. $(8.9 \times 10^{-5})(3.4 \times 10^{-6})$

43. $\dfrac{9 \times 10^{-3}}{6 \times 10^8}$

44. $\dfrac{2.5 \times 10^8}{5 \times 10^{10}}$

45. $\dfrac{(3.2 \times 10^{-11})(2.7 \times 10^{18})}{1.2 \times 10^{-5}}$

▶ 46. $\dfrac{(6.9 \times 10^{27})(8.2 \times 10^{-13})}{4.1 \times 10^{15}}$

47. $\dfrac{(4.0 \times 10^{-9})(8.4 \times 10^5)}{(3.0 \times 10^{-6})(1.4 \times 10^{18})}$

48. $\dfrac{(7.2 \times 10^8)(3.9 \times 10^{-7})}{(2.6 \times 10^{-10})(1.8 \times 10^{-8})}$

In Exercises 49 to 70, simplify each exponential expression.

49. $4^{3/2}$

50. $-16^{3/2}$

51. $-64^{2/3}$

52. $125^{4/3}$

53. $9^{-3/2}$

54. $32^{-3/5}$

55. $\left(\dfrac{4}{9}\right)^{1/2}$

56. $\left(\dfrac{16}{25}\right)^{3/2}$

57. $\left(\dfrac{1}{8}\right)^{-4/3}$

58. $\left(\dfrac{8}{27}\right)^{-2/3}$

59. $(4a^{2/3}b^{1/2})(2a^{1/3}b^{3/2})$

60. $(6a^{3/5}b^{1/4})(-3a^{1/5}b^{3/4})$

61. $(-3x^{2/3})(4x^{1/4})$

▶ 62. $(-5x^{1/3})(-4x^{1/2})$

63. $(81x^8y^{12})^{1/4}$

64. $(27x^3y^6)^{2/3}$

65. $\dfrac{16z^{3/5}}{12z^{1/5}}$

66. $\dfrac{6a^{2/3}}{9a^{1/3}}$

67. $(2x^{2/3}y^{1/2})(3x^{1/6}y^{1/3})$

68. $\dfrac{x^{1/3}y^{5/6}}{x^{2/3}y^{1/6}}$

69. $\dfrac{9a^{3/4}b}{3a^{2/3}b^2}$

70. $\dfrac{12x^{1/6}y^{1/4}}{16x^{3/4}y^{1/2}}$

In Exercises 71 to 80, simplify each radical expression.

71. $\sqrt{45}$

72. $\sqrt{75}$

73. $\sqrt[3]{24}$

74. $\sqrt[3]{135}$

75. $\sqrt[3]{-135}$

76. $\sqrt[3]{-250}$

77. $\sqrt{24x^2y^3}$

▶ 78. $\sqrt{18x^2y^5}$

79. $\sqrt[3]{16a^3y^7}$

80. $\sqrt[3]{54m^2n^7}$

In Exercises 81 to 88, simplify each radical and then combine like radicals.

81. $2\sqrt{32} - 3\sqrt{98}$

82. $5\sqrt[3]{32} + 2\sqrt[3]{108}$

83. $-8\sqrt[4]{48} + 2\sqrt[4]{243}$

84. $2\sqrt[3]{40} - 3\sqrt[3]{135}$

85. $4\sqrt[3]{32y^4} + 3y\sqrt[3]{108y}$

▶ 86. $-3x\sqrt[3]{54x^4} + 2\sqrt[3]{16x^7}$

87. $x\sqrt[3]{8x^3y^4} - 4y\sqrt[3]{64x^6y}$

88. $4\sqrt{a^5b} - a^2\sqrt{ab}$

In Exercises 89 to 98, find the indicated products and express each result in simplest form.

89. $(\sqrt{5} + 3)(\sqrt{5} + 4)$

90. $(\sqrt{7} + 2)(\sqrt{7} - 5)$

91. $(\sqrt{2} - 3)(\sqrt{2} + 3)$

92. $(2\sqrt{7} + 3)(2\sqrt{7} - 3)$

93. $(3\sqrt{z} - 2)(4\sqrt{z} + 3)$

94. $(4\sqrt{a} - \sqrt{b})(3\sqrt{a} + 2\sqrt{b})$

95. $(\sqrt{x} + 2)^2$

▶ **96.** $(3\sqrt{5y} - 4)^2$

97. $(\sqrt{x - 3} + 2)^2$

98. $(\sqrt{2x + 1} - 3)^2$

In Exercises 99 to 112, simplify each expression by rationalizing the denominator. Write the result in simplest form.

99. $\dfrac{2}{\sqrt{2}}$

100. $\dfrac{3x}{\sqrt{3}}$

101. $\sqrt{\dfrac{5}{18}}$

102. $\sqrt{\dfrac{7}{40}}$

103. $\dfrac{3}{\sqrt[3]{2}}$

104. $\dfrac{2}{\sqrt[3]{4}}$

105. $\dfrac{4}{\sqrt[3]{8x^2}}$

▶ **106.** $\dfrac{2}{\sqrt[4]{4y}}$

107. $\dfrac{3}{\sqrt{3} + 4}$

108. $\dfrac{2}{\sqrt{5} - 2}$

109. $\dfrac{6}{2\sqrt{5} + 2}$

▶ **110.** $\dfrac{-7}{3\sqrt{2} - 5}$

111. $\dfrac{3}{\sqrt{5} + \sqrt{x}}$

112. $\dfrac{5}{\sqrt{y} - \sqrt{3}}$

113. NATIONAL DEBT In February of 2003, the U.S. national debt was approximately 6.4×10^{12} dollars. At that time, the population of the U.S. was 2.89×10^8 people. In February of 2003, what was the U.S. debt per person?

114. COLOR MONITORS A color monitor for a computer can display 2^{32} colors. A physiologist estimates that the human eye can detect approximately 36,000 different colors. How many colors, to the nearest thousand, would go undetected by a human eye using this monitor?

115. WEIGHT OF AN ORCHID SEED An orchid seed weighs approximately 3.2×10^{-8} ounce. If a package of seeds contains 1 ounce of orchid seeds, how many seeds are in the package?

116. LASER WAVELENGTH The wavelength of a certain helium-neon laser is 800 nanometers. (1 nanometer is 1×10^{-9} meter.) The frequency, in cycles per second, of this wave is $\dfrac{1}{\text{wavelength}}$. What is the frequency of this laser?

117. DOPPLER EFFECT Astronomers can approximate the distance to a galaxy by measuring its *red shift*, which is a shift in the wavelength of light due to the velocity of the galaxy. This is similar to the way the sound of a siren coming toward you seems to have a higher pitch than when the siren is moving away from you. A formula for red shift is red shift $= \dfrac{\lambda_r - \lambda_s}{\lambda_s}$, where λ_r and λ_s are wavelengths of a certain frequency of light. Calculate the red shift for a galaxy for which $\lambda_r = 5.13 \times 10^{-7}$ meter and $\lambda_s = 5.06 \times 10^{-7}$ meter.

118. ASTRONOMICAL UNIT Earth's mean distance from the sun is 9.3×10^7 miles. This distance is called the *astronomical unit* (AU). Jupiter is 5.2 AU from the sun. Find the distance in miles from Jupiter to the sun.

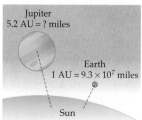

Jupiter
5.2 AU = ? miles

Earth
1 AU = 9.3×10^7 miles

Sun

119. ASTRONOMY The sun is approximately 1.44×10^{11} meters from Earth. If light travels 3×10^8 meters per second, how many minutes does it take light from the sun to reach Earth?

120. MASS OF AN ATOM One gram of hydrogen contains 6.023×10^{23} atoms. Find the mass of one hydrogen atom.

121. CELLULAR PHONE PRODUCTION An electronics firm estimates that the revenue R it will receive from the sale of x cell phones (in thousands) can be approximated by $R = 1250x(2^{-0.007x})$. What is the estimated revenue when the company sells 20,000 cell phones? Round to the nearest dollar.

122. DRUG POTENCY The amount A (in milligrams) of digoxin, a drug taken by cardiac patients, remaining in the blood t hours after a patient takes a 2-milligram dose is given by $A = 2(10^{-0.0078t})$.

a. How much digoxin remains in the blood of a patient 4 hours after taking a 2-milligram dose?

b. Suppose that a patient takes a 2-milligram dose of digoxin at 1:00 P.M. and another 2-milligram dose at 5:00 P.M. How much digoxin remains in the patient's blood at 6:00 P.M.?

123. WORLD POPULATION An estimate of the world's future population P is given by $P = 5.9(2^{0.0119n})$, where n is the number of years after 2000 and P is in billions. Using this estimate, what will the world's population be in 2050?

124. LEARNING THEORY In a psychology experiment, students were given a nine-digit number to memorize. The percent P of students who remembered the number t minutes after it was read to them can be given by $P = 90 - 3t^{2/3}$. What percent of the students remembered the number after 1 hour?

125. OCEANOGRAPHY The percent P of light that will pass to a depth d, in meters, at a certain place in the ocean is given by $P = 10^{2-(d/40)}$. Find, to the nearest percent, the amount of light that will pass to a depth of **a.** 10 meters and **b.** 25 meters below the surface of the ocean.

CONNECTING CONCEPTS

126. If $2^x = y$, then find 2^{x-4} in terms of y.

127. If a and b are nonzero numbers and $a < b$, is the statement $a^{-1} < b^{-1}$ a true statement? Give a reason for your answer.

128. How many digits are in the product $4^{50} \cdot 5^{100}$?

In Exercises 129 to 132, find the value of p for which the statement is true.

129. $a^{2/5}a^p = a^2$

130. $b^{-3/4}b^{2p} = b^3$

131. $\dfrac{x^{-3/4}}{x^{3p}} = x^4$

132. $(x^4 x^{2p})^{1/2} = x$

In Exercises 133 to 136, rationalize the numerator.

133. $\dfrac{\sqrt{4+h}-2}{h}$

134. $\dfrac{\sqrt{9+h}-3}{h}$

135. $\sqrt{n^2+1} - n$ $\left(Hint: \sqrt{n^2+1} - n = \dfrac{\sqrt{n^2+1}-n}{1} \right)$

136. $\sqrt{n^2+n} - n$ $\left(Hint: \sqrt{n^2+n} - n = \dfrac{\sqrt{n^2+n}-n}{1} \right)$

137. Evaluate: $\left(\sqrt{2^{\sqrt{2}}} \right)^{\sqrt{2}}$

PREPARE FOR SECTION P.3

138. Simplify: $-3(2a - 4b)$ [P.1]

139. Simplify: $5 - 2(2x - 7)$ [P.1]

140. Simplify: $2x^2 + 3x - 5 + x^2 - 6x - 1$ [P.1]

141. Simplify: $4x^2 - 6x - 1 - 5x^2 + x$ [P.1]

142. True or false: $4 - 3x - 2x^2 = 2x^2 - 3x + 4$ [P.1]

143. True or false: $\dfrac{12 + 15}{4} = \dfrac{\overset{3}{\cancel{12}} + 15}{\cancel{4}} = 18$ [P.1]

PROJECTS

1. RELATIVITY THEORY A moving object has energy, called *kinetic energy*, by virtue of its motion. As mentioned earlier in this chapter, the Theory of Relativity uses the formula at the right for kinetic energy.

$$K.E_r = mc^2 \left(\dfrac{1}{\sqrt{1 - \dfrac{v^2}{c^2}}} - 1 \right)$$

When the speed of an object is much less than the speed of light (3.0×10^8 meters per second), the formula

$$\text{K.E}_n = \frac{1}{2}mv^2$$

is used. In each formula, v is the velocity of the object in meters per second, m is its rest mass in kilograms, and c is the speed of light given above. Calculate the percent error (in **a.** through **e.**) for each of the given velocities. The formula for percent error is

$$\% \text{ error} = \frac{|\text{K.E}_r - \text{K.E}_n|}{\text{K.E}_r} \times 100$$

a. $v = 30$ meters per second (speeding car on an expressway)

b. $v = 240$ meters per second (speed of a commercial jet)

c. $v = 3.0 \times 10^7$ meters per second (10% of the speed of light)

d. $v = 1.5 \times 10^8$ meters per second (50% of the speed of light)

e. $v = 2.7 \times 10^8$ meters per second (90% of the speed of light)

f. Use your answers from **a.** through **e.** to give a reason why the formula for kinetic energy given by K.E_n is adequate for most of our common experiences involving motion (walking, running, bicycle, car, plane).

g. According to relativity theory, the mass, m, of an object changes as its velocity according to

$$m = \frac{m_0}{\sqrt{1 - \dfrac{v^2}{c_2}}}$$

where m_0 is the rest mass of the object. The approximate rest mass of an electron is 9.11×10^{-31} kilogram. What is the percent change, from its rest mass, in the mass of an electron that is traveling at $0.99c$ (99% of the speed of light)?

h. According to the Theory of Relativity, a particle (such as an electron or a spacecraft) cannot exceed the speed of light. Explain why the equation for K.E_r suggests such a conclusion.

SECTION P.3

POLYNOMIALS

OPERATIONS ON POLYNOMIALS

A **monomial** is a constant, a variable, or a product of a constant and one or more variables, with the variables having only nonnegative integer exponents. The constant is called the **numerical coefficient** or simply the **coefficient** of the monomial. The **degree of a monomial** is the sum of the exponents of the variables. For example, $-5xy^2$ is a monomial with coefficient -5 and degree 3.

The algebraic expression $3x^{-2}$ is not a monomial because it cannot be written as a product of a constant and a variable with a *nonnegative* integer exponent.

A sum of a finite number of monomials is called a **polynomial.** Each monomial is called a **term** of the polynomial. The **degree of a polynomial** is the largest degree of the terms in the polynomial.

Terms that have exactly the same variables raised to the same powers are called **like terms.** For example, $14x^2$ and $-31x^2$ are like terms; however, $2x^3y$ and $7xy$ are not like terms because x^3y and xy are not identical.

A polynomial is said to be simplified if all its like terms have been combined. For example, the simplified form of $4x^2 + 3x + 5x$ is $4x^2 + 8x$. A simplified polynomial that has two terms is a **binomial,** and a simplified polynomial that has three terms is a **trinomial.** For example, $4x + 7$ is a binomial, and $2x^3 - 7x^2 + 11$ is a trinomial.

A nonzero constant, such as 5, is called a **constant polynomial**. It has degree zero because $5 = 5x^0$. The number 0 is defined to be a polynomial with no degree.

Standard Form of a Polynomial

The **standard form of a polynomial** of degree n in the variable x is

$$a_n x^n + a_{n-1} x^{n-1} + \cdots + a_2 x^2 + a_1 x + a_0$$

where $a_n \neq 0$ and n is a nonnegative integer. The coefficient a_n is the **leading coefficient**, and a_0 is the **constant term**.

If a polynomial in the variable x is written with decreasing powers of x, then it is in **standard form**. For example, the polynomial

$$3x^2 - 4x^3 + 7x^4 - 1$$

is written in standard form as

$$7x^4 - 4x^3 + 3x^2 - 1$$

The following table shows the leading coefficient, degree, terms, and coefficients of the given polynomials.

Polynomial	Leading Coefficient	Degree	Terms	Coefficients
$9x^2 - x + 5$	9	2	$9x^2, -x, 5$	$9, -1, 5$
$11 - 2x$	-2	1	$-2x, 11$	$-2, 11$
$x^3 + 5x - 3$	1	3	$x^3, 5x, -3$	$1, 5, -3$

To add polynomials, add the coefficients of the like terms.

EXAMPLE 1   Add Polynomials

Add: $(3x^3 - 2x^2 - 6) + (4x^2 - 6x - 7)$

Solution

$(3x^3 - 2x^2 - 6) + (4x^2 - 6x - 7)$
$= 3x^3 + (-2x^2 + 4x^2) + (-6x) + [(-6) + (-7)]$
$= 3x^3 + 2x^2 - 6x - 13$

▶ **TRY EXERCISE 22, PAGE 46**

The **additive inverse** of the polynomial $3x - 7$ is

$$-(3x - 7) = -3x + 7$$

? QUESTION What is the additive inverse of $3x^2 - 8x + 7$?

To subtract a polynomial, we add its additive inverse. For example,

$$(2x - 5) - (3x - 7) = (2x - 5) + (-3x + 7)$$
$$= [2x + (-3x)] + [(-5) + 7]$$
$$= -x + 2$$

The distributive property is used to find the product of polynomials. For instance, to find the product of $(3x - 4)$ and $(2x^2 + 5x + 1)$, we treat $3x - 4$ as a *single* quantity and *distribute it* over the trinomial $2x^2 + 5x + 1$, as shown in Example 2.

EXAMPLE 2   **Multiply Polynomials**

Simplify: $(3x - 4)(2x^2 + 5x + 1)$

Solution

$(3x - 4)(2x^2 + 5x + 1)$
$$= (3x - 4)(2x^2) + (3x - 4)(5x) + (3x - 4)(1)$$
$$= (3x)(2x^2) - 4(2x^2) + (3x)(5x) - 4(5x) + (3x)(1) - 4(1)$$
$$= 6x^3 - 8x^2 + 15x^2 - 20x + 3x - 4$$
$$= 6x^3 + 7x^2 - 17x - 4$$

▶ **TRY EXERCISE 26, PAGE 46**

A vertical format also can be used to find the product of $(2x^2 + 5x + 1)$ and $(3x - 4)$ in Example 2. Note that like terms are arranged in the same vertical column.

$$
\begin{array}{r}
2x^2 + 5x + 1 \\
3x - 4 \\
\hline
-8x^2 - 20x - 4 = -4(2x^2 + 5x + 1) \\
6x^3 + 15x^2 + 3x = 3x(2x^2 + 5x + 1) \\
\hline
6x^3 + 7x^2 - 17x - 4
\end{array}
$$

This is the same answer as we obtained in Example 2.

If the terms of the binomials $(a + b)$ and $(c + d)$ are labeled as shown below, then the product of the two binomials can be computed mentally by the **FOIL** method.

$$(a + b) \cdot (c + d) = \underset{\text{First}}{ac} + \underset{\text{Outer}}{ad} + \underset{\text{Inner}}{bc} + \underset{\text{Last}}{bd}$$

First   Last   Inner   Outer

? ANSWER The additive inverse is $-3x^2 + 8x - 7$.

In the following illustration, we find the product of $(7x - 2)$ and $(5x + 4)$ by the FOIL method.

$$
\begin{array}{cccc}
& \text{First} & \text{Outer} & \text{Inner} & \text{Last}
\end{array}
$$

$$(7x - 2)(5x + 4) = (7x)(5x) + (7x)(4) + (-2)(5x) + (-2)(4)$$
$$= \quad 35x^2 \quad + \quad 28x \quad - \quad 10x \quad - \quad 8$$
$$= \quad 35x^2 + 18x - 8$$

Certain products occur so frequently in algebra that they deserve special attention.

Special Product Formulas

Special Form	Formula(s)
(Sum)(Difference)	$(x + y)(x - y) = x^2 - y^2$
(Binomial)^2	$(x + y)^2 = x^2 + 2xy + y^2$ $(x - y)^2 = x^2 - 2xy + y^2$

The variables x and y in these special product formulas can be replaced by other algebraic expressions, as shown in Example 3.

EXAMPLE 3   **Use the Special Product Formulas**

Find each special product.   **a.** $(7x + 10)(7x - 10)$ **b.** $(2y^2 + 11z)^2$

Solution

a. $(7x + 10)(7x - 10) = (7x)^2 - (10)^2 = 49x^2 - 100$

b. $(2y^2 + 11z)^2 = (2y^2)^2 + 2[(2y^2)(11z)] + (11z)^2 = 4y^4 + 44y^2z + 121z^2$

▶ **TRY EXERCISE 46, PAGE 47**

● **GREATEST COMMON FACTOR**

Writing a polynomial as a product of polynomials of lower degree is called **factoring**. Factoring is an important procedure that is often used to simplify fractional expressions and to solve equations.

In this section we consider only the factorization of polynomials that have integer coefficients. Also, we are concerned only with **factoring over the integers.** That is, we search only for polynomial factors that have integer coefficients.

The first step in any factorization of a polynomial is to use the distributive property to factor out the **greatest common factor (GCF)** of the terms of the polynomial. Given two or more exponential expressions with the same prime number base or the same variable base, the GCF is the exponential expression with the smallest exponent. For example,

2^3 is the GCF of 2^3, 2^5, and 2^8 and a is the GCF of a^4 and a

The GCF of two or more monomials is the product of the GCFs of all the *common* bases. For example, to find the GCF of $27a^3b^4$ and $18b^3c$, factor the coefficients into prime factors and then write each common base with its smallest exponent.

$$27a^3b^4 = 3^3 \cdot a^3 \cdot b^4 \qquad 18b^3c = 2 \cdot 3^2 \cdot b^3 \cdot c$$

The only common bases are 3 and b. The product of these common bases with their smallest exponents is 3^2b^3. The GCF of $27a^3b^4$ and $18b^3c$ is $9b^3$.

The expressions $3x(2x + 5)$ and $4(2x + 5)$ have a common *binomial* factor that is $2x + 5$. Thus the GCF of $3x(2x + 5)$ and $4(2x + 5)$ is $2x + 5$.

EXAMPLE 4   **Factor Out the Greatest Common Factor**

Factor out the GCF.

a. $10x^3 + 6x$ **b.** $15x^{2n} + 9x^{n+1} - 3x^n$ (where n is a positive integer)

c. $(m + 5)(x + 3) + (m + 5)(x - 10)$

Solution

a. $10x^3 + 6x = (2x)(5x^2) + (2x)(3)$ • The GCF is 2x.

$= 2x(5x^2 + 3)$ • Factor out the GCF.

b. $15x^{2n} + 9x^{n+1} - 3x^n$

$= (3x^n)(5x^n) + (3x^n)(3x) - (3x^n)(1)$ • The GCF is 3x^n.

$= 3x^n(5x^n + 3x - 1)$ • Factor out the GCF.

c. $(m + 5)(x + 3) + (m + 5)(x - 10)$ • Use the distributive property

$= (m + 5)[(x + 3) + (x - 10)]$ to factor out (m + 5).

$= (m + 5)(2x - 7)$ • Simplify.

▶ **TRY EXERCISE 56, PAGE 47**

● **FACTORING TRINOMIALS**

Some trinomials of the form $x^2 + bx + c$ can be factored by a trial procedure. This method makes use of the FOIL method in reverse. For example, consider the following products:

$(x + 3)(x + 5) = x^2 + 5x + 3x + (3)(5) \quad = x^2 + 8x + 15$

$(x - 2)(x - 7) = x^2 - 7x - 2x + (-2)(-7) = x^2 - 9x + 14$

$(x + 4)(x - 9) = x^2 - 9x + 4x + (4)(-9) \quad = x^2 - 5x - 36$

The coefficient of x is the sum of the constant terms of the binomials.

The constant term of the trinomial is the product of the constant terms of the binomials.

? **QUESTION** Is $(x - 2)(x + 7)$ the correct factorization of $x^2 - 5x - 14$?

Points to Remember to Factor $x^2 + bx + c$

1. The constant term c of the trinomial is the product of the constant terms of the binomials.

2. The coefficient b in the trinomial is the sum of the constant terms of the binomials.

3. If the constant term c of the trinomial is positive, the constant terms of the binomials have the same sign as the coefficient b of the trinomial.

4. If the constant term c of the trinomial is negative, the constant terms of the binomials have opposite signs.

EXAMPLE 5 Factor a Trinomial of the Form $x^2 + bx + c$

Factor: $x^2 + 7x - 18$

Solution

We must find two binomials whose first terms have a product of x^2 and whose last terms have a product of -18; also, the sum of the product of the outer terms and the product of the inner terms must be $7x$. Begin by listing the possible integer factorizations of -18 and the sum of those factors.

Factors of -18	Sum of the Factors
$1 \cdot (-18)$	$1 + (-18) = -17$
$(-1) \cdot 18$	$(-1) + 18 = 17$
$2 \cdot (-9)$	$2 + (-9) = -7$
$(-2) \cdot 9$	$(-2) + 9 = 7$

• Stop. This is the desired sum.

Thus -2 and 9 are the numbers whose sum is 7 and whose product is -18. Therefore,

$$x^2 + 7x - 18 = (x - 2)(x + 9)$$

The FOIL method can be used to verify that the factorization is correct.

▶ **TRY EXERCISE 62, PAGE 47**

The trial method sometimes can be used to factor trinomials of the form $ax^2 + bx + c$, which do not have a leading coefficient of 1. We use the factors of a

---

? **ANSWER** No. $(x - 2)(x + 7) = x^2 + 5x - 14$.

and c to form trial binomial factors. Factoring trinomials of this type may require testing many factors. To reduce the number of trial factors, make use of the following points.

Points to Remember to Factor $ax^2 + bx + c, a > 0$

1. If the constant term of the trinomial is positive, the constant terms of the binomials have the same sign as the coefficient b in the trinomial.

2. If the constant term of the trinomial is negative, the constant terms of the binomials have opposite signs.

3. If the terms of the trinomial do not have a common factor, then neither binomial will have a common factor.

EXAMPLE 6   **Factor a Trinomial of the Form $ax^2 + bx + c$**

Factor: $6x^2 - 11x + 4$

Solution

Because the constant term of the trinomial is positive and the coefficient of the x term is negative, the constant terms of the binomials will both be negative. This time we find factors of the first term as well as factors of the constant term.

Factors of $6x^2$	Factors of 4 (both negative)
$x, 6x$	$-1, -4$
$2x, 3x$	$-2, -2$

Use these factors to write trial factors. Use the FOIL method to see whether any of the trial factors produce the correct middle term. If the terms of a trinomial do not have a common factor, then a binomial factor cannot have a common factor (point 3). Such trial factors need not be checked.

Trial Factors	Middle Term	
$(x - 1)(6x - 4)$	Common factor	• 6x and 4 have a common factor.
$(x - 4)(6x - 1)$	$-1x - 24x = -25x$	
$(x - 2)(6x - 2)$	Common factor	• 6x and 2 have a common factor.
$(2x - 1)(3x - 4)$	$-8x - 3x = -11x$	• This is the correct middle term.

Thus $6x^2 - 11x + 4 = (2x - 1)(3x - 4)$.

▶ **TRY EXERCISE 66, PAGE 47**

Sometimes it is impossible to factor a polynomial into the product of two polynomials having integer coefficients. Such polynomials are said to be **nonfactorable over the integers.** For example, $x^2 + 3x + 7$ is nonfactorable over the integers because there are no integers whose product is 7 and whose sum or difference is 3.

Certain trinomials can be expressed as quadratic trinomials by making suitable variable substitutions. A trinomial is **quadratic in form** if it can be written as

$$au^2 + bu + c$$

If we let $x^2 = u$, the trinomial $x^4 + 5x^2 + 6$ can be written as shown at the right.

$$x^4 + 5x^2 + 6$$
$$= (x^2)^2 + 5(x^2) + 6$$
$$= u^2 + 5u + 6$$

The trinomial is quadratic in form.

If we let $xy = u$, the trinomial $2x^2y^2 + 3xy - 9$ can be written as shown at the right.

$$2x^2y^2 + 3xy - 9$$
$$= 2(xy)^2 + 3(xy) - 9$$
$$= 2u^2 + 3u - 9$$

The trinomial is quadratic in form.

When a trinomial that is quadratic in form is factored, the variable part of the first term in each binomial factor will be u. For example, because $x^4 + 5x^2 + 6$ is quadratic in form when $x^2 = u$, the first term in each binomial factor will be x^2.

$$x^4 + 5x^2 + 6 = (x^2)^2 + 5(x^2) + 6$$
$$= (x^2 + 2)(x^2 + 3)$$

The trinomial $x^2y^2 - 2xy - 15$ is quadratic in form when $xy = u$. The first term in each binomial factor will be xy.

$$x^2y^2 - 2xy - 15 = (xy)^2 - 2(xy) - 15$$
$$= (xy + 3)(xy - 5)$$

EXAMPLE 7

Factor: **a.** $6x^2y^2 - xy - 12$ **b.** $2x^4 + 5x^2 - 12$

Solution

a. $6x^2y^2 - xy - 12$

$= (3xy + 4)(2xy - 3)$

• The trinomial is quadratic in form when $xy = u$.

b. $2x^4 + 5x^2 - 12$

$= (x^2 + 4)(2x^2 - 3)$

• The trinomial is quadratic in form when $x^2 = u$.

▶ **TRY EXERCISE 78, PAGE 47**

● SPECIAL FACTORING

The product of a term and itself is called a **perfect square.** The exponents on variables of perfect squares are always even numbers. The **square root of a perfect square** is one of the two equal factors of the perfect square. To find the square root of a perfect square variable term, divide the exponent by 2. For the examples that follow, assume the variables represent positive numbers.

Term		Perfect Square	Square Root
7	$7 \cdot 7 =$	49	$\sqrt{49} = 7$
y	$y \cdot y =$	y^2	$\sqrt{y^2} = y$
$2x^3$	$2x^3 \cdot 2x^3 =$	$4x^6$	$\sqrt{4x^6} = 2x^3$
x^n	$x^n \cdot x^n =$	x^{2n}	$\sqrt{x^{2n}} = x^n$

take note

The **sum** of two squares does not factor over the integers. For instance, $49x^2 + 144$ does not factor over the integers.

The factors of the difference of two perfect squares are the sum and difference of the square roots of the perfect squares.

Factors of the Difference of Two Perfect Squares

$$a^2 - b^2 = (a + b)(a - b)$$

EXAMPLE 8　Factor the Difference of Squares

Factor: $49x^2 - 144$

Solution

$49x^2 - 144 = (7x)^2 - (12)^2$　• Recognize the difference-of-squares form.

$= (7x + 12)(7x - 12)$　• The binomial factors are the sum and the difference of the square roots of the squares.

▶ **TRY EXERCISE 82, PAGE 47**

A **perfect-square trinomial** is a trinomial that is the square of a binomial. For example, $x^2 + 6x + 9$ is a perfect-square trinomial because

$$(x + 3)^2 = x^2 + 6x + 9$$

Every perfect-square trinomial can be factored by the trial method, but it generally is faster to factor perfect-square trinomials by using the following factoring formulas.

Factors of a Perfect-Square Trinomial

$$a^2 + 2ab + b^2 = (a + b)^2$$
$$a^2 - 2ab + b^2 = (a - b)^2$$

EXAMPLE 9　Factor a Perfect-Square Trinomial

Factor: $16m^2 - 40mn + 25n^2$

Continued ▶

Solution

Because $16m^2 = (4m)^2$ and $25n^2 = (5n)^2$, try factoring $16m^2 - 40mn + 25n^2$ as the square of a binomial.

$$16m^2 - 40mn + 25n^2 \overset{?}{=} (4m - 5n)^2$$

Check:

$$(4m - 5n)^2 = (4m - 5n)(4m - 5n)$$
$$= 16m^2 - 20mn - 20mn + 25n^2$$
$$= 16m^2 - 40mn + 25n^2$$

The factorization checks. Therefore, $16m^2 - 40mn + 25n^2 = (4m - 5n)^2$.

▶ **TRY EXERCISE 88, PAGE 47**

The product of the same three terms is called a **perfect cube**. The exponents on variables of perfect cubes are always divisible by 3. The **cube root of a perfect cube** is one of the three equal factors of the perfect cube. To find the cube root of a perfect cube variable term, divide the exponent by 3.

Term		Perfect Cube	Cube Root
5	$5 \cdot 5 \cdot 5 =$	125	$\sqrt[3]{125} = 5$
z	$z \cdot z \cdot z =$	z^3	$\sqrt[3]{z^3} = z$
$3x^2$	$3x^2 \cdot 3x^2 \cdot 3x^2 =$	$27x^6$	$\sqrt[3]{27x^6} = 3x^2$
x^n	$x^n \cdot x^n \cdot x^n =$	x^{3n}	$\sqrt[3]{x^{3n}} = x^n$

The following factoring formulas are used to factor the sum or difference of two perfect cubes.

Factors of the Sum or Difference of Two Perfect Cubes

$$a^3 + b^3 = (a + b)(a^2 - ab + b^2)$$
$$a^3 - b^3 = (a - b)(a^2 + ab + b^2)$$

EXAMPLE 10   **Factor the Sum or Difference of Cubes**

Factor: **a.** $8a^3 + b^3$ **b.** $a^3 - 64$

Solution

a. $8a^3 + b^3 = (2a)^3 + b^3$ • Recognize the sum-of-cubes form.

$= (2a + b)(4a^2 - 2ab + b^2)$ • Factor.

b. $a^3 - 64 = a^3 - 4^3$ • Recognize the difference-of-cubes form.

$= (a - 4)(a^2 + 4a + 16)$ • Factor.

▶ **TRY EXERCISE 94, PAGE 47**

● FACTOR BY GROUPING

<table>
<tr><td>

take note

$-a + b = -(a - b)$. Thus,

$-4y + 14 = -(4y - 14)$.

</td><td>

Some polynomials can be factored by grouping. Pairs of terms that have a common factor are first grouped together. The process makes repeated use of the distributive property, as shown in the following factorization of $6y^3 - 21y^2 - 4y + 14$.

</td></tr>
</table>

$$6y^3 - 21y^2 - 4y + 14$$

$$= (6y^3 - 21y^2) - (4y - 14) \qquad \text{• Group the first two terms and the last two terms.}$$

$$= 3y^2(2y - 7) - 2(2y - 7) \qquad \text{• Factor out the GCF from each of the groups.}$$

$$= (2y - 7)(3y^2 - 2) \qquad \text{• Factor out the common binomial factor.}$$

When you factor by grouping, some experimentation may be necessary to find a grouping that is of the form of one of the special factoring formulas.

EXAMPLE 11   **Factor by Grouping**

Factor by grouping.   **a.** $a^2 + 10ab + 25b^2 - c^2$ **b.** $p^2 + p - q - q^2$

Solution

a. $a^2 + 10ab + 25b^2 - c^2$

$$= (a^2 + 10ab + 25b^2) - c^2 \qquad \text{• Group the terms of the perfect-square trinomial.}$$

$$= (a + 5b)^2 - c^2 \qquad \text{• Factor the trinomial.}$$

$$= [(a + 5b) + c][(a + 5b) - c] \qquad \text{• Factor the difference of squares.}$$

$$= (a + 5b + c)(a + 5b - c) \qquad \text{• Simplify.}$$

b. $p^2 + p - q - q^2$

$$= p^2 - q^2 + p - q \qquad \text{• Rearrange the terms.}$$

$$= (p^2 - q^2) + (p - q) \qquad \text{• Regroup.}$$

$$= (p + q)(p - q) + (p - q) \qquad \text{• Factor the difference of squares.}$$

$$= (p - q)(p + q + 1) \qquad \text{• Factor out the common factor } (p - q).$$

▶ **TRY EXERCISE 102, PAGE 47**

TOPICS FOR DISCUSSION

1. Discuss the definition of the term *polynomial*. Give some examples of expressions that are polynomials and some examples of expressions that are not polynomials.

2. Suppose that P and Q are both polynomials of degree n. Discuss the degrees of $P + Q$, $P - Q$, PQ, $P + P$, and $P - P$.

3. Discuss the meaning of the phrase *nonfactorable over the integers*.

4. You know that if $ab = 0$, then $a = 0$ or $b = 0$. Suppose a polynomial is written in factored form and then set equal to zero. For instance, suppose

$$x^2 - 2x - 15 = (x - 5)(x + 3) = 0$$

Discuss what implications this has for the values of x. Do not answer this question only for the polynomial above, but also for any polynomial written as a product of linear factors and then set equal to zero.

5. Let P be a polynomial of degree n. Discuss the number of possible distinct linear polynomials that can be a factor of P.

EXERCISE SET P.3

In Exercises 1 to 10, match the descriptions, labeled A, B, C,…, J, with the appropriate examples.

A. $x^3y + xy$

B. $7x^2 + 5x - 11$

C. $\dfrac{1}{2}x^2 + xy + y^2$

D. $4xy$

E. $8x^3 - 1$

F. $3 - 4x^2$

G. 8

H. $3x^5 - 4x^2 + 7x - 11$

I. $8x^4 - \sqrt{5}x^3 + 7$

J. 0

1. A monomial of degree 2.

2. A binomial of degree 3.

3. A polynomial of degree 5.

4. A binomial with leading coefficient of -4.

5. A zero-degree polynomial.

6. A fourth-degree polynomial that has a third-degree term.

7. A trinomial with integer coefficients.

8. A trinomial in x and y.

9. A polynomial with no degree.

10. A fourth-degree binomial.

In Exercises 11 to 16, for each polynomial determine its a. standard form, b. degree, c. coefficients, d. leading coefficient, e. terms.

11. $2x + x^2 - 7$

12. $-3x^2 - 11 - 12x^4$

13. $x^3 - 1$

14. $4x^2 - 2x + 7$

15. $2x^4 + 3x^3 + 5 + 4x^2$

16. $3x^2 - 5x^3 + 7x - 1$

In Exercises 17 to 20, determine the degree of the given polynomial.

17. $3xy^2 - 2xy + 7x$

18. $x^3 + 3x^2y + 3xy^2 + y^3$

19. $4x^2y^2 - 5x^3y^2 + 17xy^3$

20. $-9x^5y + 10xy^4 - 11x^2y^2$

In Exercises 21 to 28, perform the indicated operations and simplify if possible by combining like terms. Write the result in standard form.

21. $(3x^2 + 4x + 5) + (2x^2 + 7x - 2)$

▶ 22. $(5y^2 - 7y + 3) + (2y^2 + 8y + 1)$

23. $(r^2 - 2r - 5) - (3r^2 - 5r + 7)$

24. $(7s^2 - 4s + 11) - (-2s^2 + 11s - 9)$

25. $(4x - 5)(2x^2 + 7x - 8)$

▶ 26. $(5x - 7)(3x^2 - 8x - 5)$

27. $(3x^2 - 2x + 5)(2x^2 - 5x + 2)$

28. $(2y^3 - 3y + 4)(2y^2 - 5y + 7)$

In Exercises 29 to 38, use the FOIL method to find the indicated product.

29. $(y + 2)(y + 1)$

30. $(y + 5)(y + 3)$

31. $(4z - 3)(z - 4)$

32. $(5z - 6)(z - 1)$

33. $(a + 6)(a - 3)$

34. $(a - 10)(a + 4)$

35. $(9x + 5y)(2x + 5y)$

36. $(3x - 7z)(5x - 7z)$

37. $(3p + 5q)(2p - 7q)$

38. $(2r - 11s)(5r + 8s)$

In Exercises 39 to 44, perform the indicated operations and simplify.

39. $(4d - 1)^2 - (2d - 3)^2$

40. $(5c - 8)^2 - (2c - 5)^2$

41. $(r + s)(r^2 - rs + s^2)$

42. $(r - s)(r^2 + rs + s^2)$

43. $(3c - 2)(4c + 1)(5c - 2)$

44. $(4d - 5)(2d - 1)(3d - 4)$

In Exercises 45 to 50, use the special product formulas to perform the indicated operation.

45. $(3x + 5)(3x - 5)$

▶ **46.** $(4x^2 - 3y)(4x^2 + 3y)$

47. $(3x^2 - y)^2$

48. $(6x + 7y)^2$

49. $(4w + z)^2$

50. $(3x - 5y^2)^2$

In Exercises 51 to 58, factor out the GCF from each polynomial.

51. $5x + 20$

52. $8x^2 + 12x - 40$

53. $-15x^2 - 12x$

54. $-6y^2 - 54y$

55. $10x^2y + 6xy - 14xy^2$

▶ **56.** $6a^3b^2 - 12a^2b + 72ab^3$

57. $(x - 3)(a + b) + (x - 3)(a + 2b)$

58. $(x - 4)(2a - b) + (x + 4)(2a - b)$

In Exercises 59 to 70, factor each trinomial over the integers.

59. $x^2 + 7x + 12$

60. $x^2 + 9x + 20$

61. $a^2 - 10a - 24$

▶ **62.** $b^2 + 12b - 28$

63. $6x^2 + 25x + 4$

64. $8a^2 - 26a + 15$

65. $51x^2 - 5x - 4$

▶ **66.** $57y^2 + y - 6$

67. $x^4 + 6x^2 + 5$

68. $x^4 + 11x^2 + 18$

69. $6x^4 + 23x^2 + 15$

70. $9x^4 + 10x^2 + 1$

In Exercises 71 to 78, factor over the integers.

71. $x^4 - x^2 - 6$

72. $x^4 + 3x^2 + 2$

73. $x^2y^2 - 2xy - 8$

74. $2x^2y^2 + xy - 1$

75. $3x^4 + 11x^2 - 4$

76. $2x^4 + 3x^2 - 9$

77. $3x^6 + 2x^3 - 8$

▶ **78.** $8x^6 - 10x^3 - 3$

In Exercises 79 to 86, factor each difference of squares over the integers.

79. $x^2 - 9$

80. $x^2 - 64$

81. $4a^2 - 49$

▶ **82.** $81b^2 - 16c^2$

83. $1 - 100x^2$

84. $1 - 121y^2$

85. $x^4 - 9$

86. $y^4 - 196$

In Exercises 87 to 92, factor each perfect-square trinomial.

87. $a^2 - 14a + 49$

▶ **88.** $b^2 - 24b + 144$

89. $4x^2 + 12x + 9$

90. $25y^2 + 40y + 16$

91. $z^4 + 4z^2w^2 + 4w^4$

92. $9x^4 - 30x^2y^2 + 25y^4$

In Exercises 93 to 98, factor each sum or difference of cubes over the integers.

93. $x^3 - 8$

▶ **94.** $b^3 + 64$

95. $8x^3 - 27y^3$

96. $64u^3 - 27v^3$

97. $8 - x^6$

98. $1 + y^{12}$

In Exercises 99 to 104, factor (over the integers) by grouping in pairs.

99. $3x^3 + x^2 + 6x + 2$

100. $18w^3 + 15w^2 + 12w + 10$

101. $ax^2 - ax + bx - b$

▶ **102.** $a^2y^2 - ay^3 + ac - cy$

103. $6w^3 + 4w^2 - 15w - 10$

104. $10z^3 - 15z^2 - 4z + 6$

In Exercises 105 to 118, use the general factoring strategy to completely factor each polynomial. If the polynomial does not factor, then state that it is nonfactorable over the integers.

105. $18x^2 - 2$

106. $4bx^3 + 32b$

107. $16x^4 - 1$

108. $81y^4 - 16$

113. $(w - 5)^3 + 8$

114. $5xy + 20y - 15x - 60$

109. $3bx^3 + 4bx^2 - 3bx - 4b$

110. $2x^6 - 2$

115. $x^2 + 6xy + 9y^2 - 1$

116. $4y^2 - 4yz + z^2 - 9$

111. $72bx^2 + 24bxy + 2by^2$

112. $64y^3 - 16y^2z + yz^2$

117. $8x^2 + 3x - 4$

118. $16x^2 + 81$

CONNECTING CONCEPTS

In Exercises 119 and 120, find all positive values of k such that the trinomial is a perfect-square trinomial.

119. $x^2 + kx + 16$

120. $36x^2 + kxy + 100y^2$

In Exercises 121 and 122, find k such that the trinomial is a perfect-square trinomial.

121. $x^2 + 16x + k$

122. $x^2 - 14xy + ky^2$

In Exercises 123 to 126, write, in its factored form, the area of the shaded portion of each geometric figure.

123.

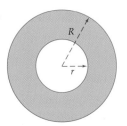

124.

125.

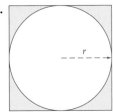

126.

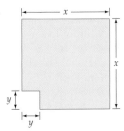

PREPARE FOR SECTION P.4

127. Simplify: $1 + \dfrac{1}{2 - \dfrac{1}{3}}$ [P.1]

128. Simplify: $\left(\dfrac{w}{x}\right)^{-1}\left(\dfrac{y}{z}\right)^{-1}$ [P.2]

129. What is the common binomial factor of $x^2 + 2x - 3$ and $x^2 + 7x + 12$? [P.3]

In Exercises 130 to 132, factor completely over the integers.

130. $(2x - 3)(3x + 2) - (2x - 3)(x + 2)$ [P.3]

131. $x^2 - 5x - 6$ [P.3]

132. $x^3 - 64$ [P.3]

—— *PROJECTS* ——

1. GEOMETRY The ancient Greeks used geometric figures and the concept of area to illustrate many algebraic concepts. The factoring formula $x^2 - y^2 = (x + y)(x - y)$ can be illustrated by the figure below.

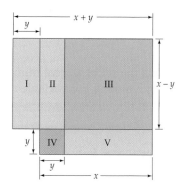

a. Which regions are represented by $(x + y)(x - y)$?

b. Which regions are represented by $x^2 - y^2$?

c. Explain why the area of the regions listed in **a.** must equal the area of the regions listed in **b.**

2. GEOMETRY What algebraic formula does the geometric figure below illustrate?

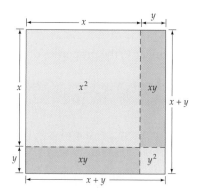

3. GEOMETRY Show how the figure below can be used to illustrate the factoring formula for the difference of two cubes.

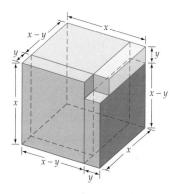

RATIONAL EXPRESSIONS

A **rational expression** is a fraction in which the numerator and denominator are polynomials. For example,

$$\frac{3}{x+1} \quad \text{and} \quad \frac{x^2 - 4x - 21}{x^2 - 9}$$

are rational expressions.

The **domain of a rational expression** is the set of all real numbers that can be used as replacements for the variable. Any value of the variable that causes division by zero is excluded from the domain of the rational expression. For example, the domain of

$$\frac{x+3}{x^2 - 5x}, \quad x \neq 0, x \neq 5$$

is the set of all real numbers except 0 and 5. Both 0 and 5 are excluded values because the denominator $x^2 - 5x$ equals zero when $x = 0$ and also when $x = 5$. Sometimes the excluded values are specified to the right of a rational expression, as shown here. However, a rational expression is meaningful only for those real numbers that are not excluded values, regardless of whether the excluded values are specifically stated.

❓ **QUESTION** What values of x must be excluded from the domain of $\frac{3x}{x^2 - x - 2}$?

Rational expressions have properties similar to the properties of rational numbers.

Properties of Rational Expressions

For all rational expressions $\dfrac{P}{Q}$ and $\dfrac{R}{S}$ where $Q \neq 0$ and $S \neq 0$,

Equality $\qquad \dfrac{P}{Q} = \dfrac{R}{S}$ if and only if $PS = QR$

Equivalent expressions $\qquad \dfrac{P}{Q} = \dfrac{PR}{QR}, \quad R \neq 0$

Sign $\qquad -\dfrac{P}{Q} = \dfrac{-P}{Q} = \dfrac{P}{-Q}$

MATH MATTERS

Evidence from work left by early Egyptians more than 3600 years ago shows that they used, with one exception, unit fractions—that is, fractions whose numerators are 1. The one exception was 2/3. A unit fraction was represented by placing an oval over the symbol for the number in the denominator. For instance, $1/4 = $ ⬭ .

❓ **ANSWER** When $x = -1$ and $x = 2$, $x^2 - x - 2 = 0$. Therefore, -1 and 2 must be excluded from the domain.

• SIMPLIFY A RATIONAL EXPRESSION

To **simplify a rational expression,** factor the numerator and the denominator. Then use the equivalent expressions property to eliminate factors common to both the numerator and the denominator. A rational expression is *simplified* when 1 is the only common factor of both the numerator and the denominator.

EXAMPLE 1   Simplify a Rational Expression

Simplify: $\dfrac{7 + 20x - 3x^2}{2x^2 - 11x - 21}$

Solution

$$\frac{7 + 20x - 3x^2}{2x^2 - 11x - 21} = \frac{(7 - x)(1 + 3x)}{(x - 7)(2x + 3)}$$ • Factor.

$$= \frac{-(x - 7)(1 + 3x)}{(x - 7)(2x + 3)}$$ • Use $(7 - x) = -(x - 7)$.

$$= \frac{-\cancel{(x - 7)}(1 + 3x)}{\cancel{(x - 7)}(2x + 3)}$$

$$= \frac{-(1 + 3x)}{2x + 3} = -\frac{3x + 1}{2x + 3}, \quad x \neq 7, x \neq -\frac{3}{2}$$

▶ **TRY EXERCISE 2, PAGE 56**

take note

A rational expression like $(x + 3)/3$ does not simplify to $x + 1$ because

$$\frac{x + 3}{3} = \frac{x}{3} + \frac{3}{3} = \frac{x}{3} + 1$$

Rational expressions can be simplified by dividing nonzero factors common to the numerator and the denominator, but not terms.

• OPERATIONS ON RATIONAL EXPRESSIONS

Arithmetic operations are defined on rational expressions just as they are on rational numbers.

Arithmetic Operations Defined on Rational Expressions

For all rational expressions $\dfrac{P}{Q}, \dfrac{R}{Q}$, and $\dfrac{R}{S}$ where $Q \neq 0$ and $S \neq 0$,

Addition $\qquad \dfrac{P}{Q} + \dfrac{R}{Q} = \dfrac{P + R}{Q}$

Subtraction $\qquad \dfrac{P}{Q} - \dfrac{R}{Q} = \dfrac{P - R}{Q}$

Multiplication $\qquad \dfrac{P}{Q} \cdot \dfrac{R}{S} = \dfrac{PR}{QS}$

Division $\qquad \dfrac{P}{Q} \div \dfrac{R}{S} = \dfrac{P}{Q} \cdot \dfrac{S}{R} = \dfrac{PS}{QR}, \quad R \neq 0$

Factoring and the equivalent expressions property of rational expressions are used in the multiplication and division of rational expressions.

EXAMPLE 2   **Divide a Rational Expression**

Simplify: $\dfrac{x^2 + 6x + 9}{x^3 + 27} \div \dfrac{x^2 + 7x + 12}{x^3 - 3x^2 + 9x}$

Solution

$\dfrac{x^2 + 6x + 9}{x^3 + 27} \div \dfrac{x^2 + 7x + 12}{x^3 - 3x^2 + 9x}$

$= \dfrac{(x + 3)^2}{(x + 3)(x^2 - 3x + 9)} \div \dfrac{(x + 4)(x + 3)}{x(x^2 - 3x + 9)}$ • Factor.

$= \dfrac{(x + 3)^2}{(x + 3)(x^2 - 3x + 9)} \cdot \dfrac{x(x^2 - 3x + 9)}{(x + 4)(x + 3)}$ • Multiply by the reciprocal.

$= \dfrac{\cancel{(x + 3)^2}x\cancel{(x^2 - 3x + 9)}}{\cancel{(x + 3)}\cancel{(x^2 - 3x + 9)}(x + 4)\cancel{(x + 3)}}$ • Simplify.

$= \dfrac{x}{x + 4}$

To review **FACTORING THE SUM OF TWO CUBES**, *see p. 44.*

▶ **TRY EXERCISE 16, PAGE 57**

Addition of rational expressions with a **common denominator** is accomplished by writing the sum of the numerators over the common denominator. For example,

$$\frac{5x}{18} + \frac{x}{18} = \frac{5x + x}{18} = \frac{6x}{18} = \frac{x}{3}$$

If the rational expressions do not have a common denominator, then they can be written as equivalent rational expressions that have a common denominator by multiplying the numerator and denominator of each of the rational expressions by the required polynomials. The following procedure can be used to determine the least common denominator (LCD) of rational expressions. It is similar to the process used to find the LCD of rational numbers.

● **DETERMINING THE LCD OF RATIONAL EXPRESSIONS**

1. Factor each denominator completely and express repeated factors using exponential notation.

2. Identify the largest power of each factor in any single factorization. The LCD is the product of each factor raised to its largest power.

For example,

$$\frac{1}{x + 3} \qquad \text{and} \qquad \frac{5}{2x - 1}$$

have an LCD of $(x + 3)(2x - 1)$. The rational expressions

$$\frac{5x}{(x + 5)(x - 7)^3} \quad \text{and} \quad \frac{7}{x(x + 5)^2(x - 7)}$$

have an LCD of $x(x + 5)^2(x - 7)^3$.

EXAMPLE 3   **Add and Subtract Rational Expressions**

Perform the indicated operation and then simplify if possible.

a. $\dfrac{5x}{48} + \dfrac{x}{15}$ b. $\dfrac{x}{x^2 - 4} - \dfrac{2x - 1}{x^2 - 3x - 10}$

Solution

a. Determine the prime factorization of the denominators.

$$48 = 2^4 \cdot 3 \quad \text{and} \quad 15 = 3 \cdot 5$$

The desired common denominator is the product of each of the prime factors raised to its largest power. Thus the common denominator is $2^4 \cdot 3 \cdot 5 = 240$. Write each rational expression as an equivalent rational expression with a denominator of 240.

$$\frac{5x}{48} + \frac{x}{15} = \frac{5x \cdot 5}{48 \cdot 5} + \frac{x \cdot 16}{15 \cdot 16} = \frac{25x}{240} + \frac{16x}{240} = \frac{41x}{240}$$

b. Factor each denominator to determine the LCD of the rational expressions.

$$x^2 - 4 = (x + 2)(x - 2)$$
$$x^2 - 3x - 10 = (x + 2)(x - 5)$$

The LCD is $(x + 2)(x - 2)(x - 5)$. Forming equivalent rational expressions that have the LCD, we have

$$\frac{x}{x^2 - 4} - \frac{2x - 1}{x^2 - 3x - 10}$$

$$= \frac{x(x - 5)}{(x + 2)(x - 2)(x - 5)} - \frac{(2x - 1)(x - 2)}{(x + 2)(x - 5)(x - 2)}$$

$$= \frac{x^2 - 5x - (2x^2 - 5x + 2)}{(x + 2)(x - 2)(x - 5)} = \frac{x^2 - 5x - 2x^2 + 5x - 2}{(x + 2)(x - 2)(x - 5)}$$

$$= \frac{-x^2 - 2}{(x + 2)(x - 2)(x - 5)} = -\frac{x^2 + 2}{(x + 2)(x - 2)(x - 5)}$$

▶ **TRY EXERCISE 30, PAGE 57**

● COMPLEX FRACTIONS

A **complex fraction** is a fraction whose numerator or denominator contains one or more fractions. Complex fractions can be simplified by using one of the following two methods.

Methods for Simplifying Complex Fractions

Method 1: Multiply by 1 in the form $\dfrac{LCD}{LCD}$.

1. Determine the LCD of all the fractions in the complex fraction.

2. Multiply both the numerator and the denominator of the complex fraction by the LCD.

3. If possible, simplify the resulting rational expression.

Method 2: Multiply the numerator by the reciprocal of the denominator.

1. Simplify the numerator to a single fraction and the denominator to a single fraction.

2. Using the definition for dividing fractions, multiply the numerator by the reciprocal of the denominator.

3. If possible, simplify the resulting rational expression.

EXAMPLE 4   **Simplify Complex Fractions**

Simplify: $\dfrac{\dfrac{2}{x-2}+\dfrac{1}{x}}{\dfrac{3x}{x-5}-\dfrac{2}{x-5}}$

Solution

First simplify the numerator to a single fraction and then simplify the denominator to a single fraction.

$$\dfrac{\dfrac{2}{x-2}+\dfrac{1}{x}}{\dfrac{3x}{x-5}-\dfrac{2}{x-5}}=\dfrac{\dfrac{2\cdot x}{(x-2)\cdot x}+\dfrac{1\cdot(x-2)}{x\cdot(x-2)}}{\dfrac{3x-2}{x-5}}$$ • Simplify numerator and denominator.

$$=\dfrac{\dfrac{2x+(x-2)}{x(x-2)}}{\dfrac{3x-2}{x-5}}=\dfrac{\dfrac{3x-2}{x(x-2)}}{\dfrac{3x-2}{x-5}}$$

$$=\dfrac{3x-2}{x(x-2)}\cdot\dfrac{x-5}{3x-2}$$ • Multiply the numerator by the reciprocal of the denominator.

$$=\dfrac{x-5}{x(x-2)}$$

▶ **TRY EXERCISE 42, PAGE 57**

EXAMPLE 5 Simplify a Fraction

Simplify the fraction $\dfrac{c^{-1}}{a^{-1} + b^{-1}}$.

Solution

The fraction written without negative exponents becomes

$$\frac{c^{-1}}{a^{-1} + b^{-1}} = \frac{\dfrac{1}{c}}{\dfrac{1}{a} + \dfrac{1}{b}} \qquad \text{• Using } x^{-n} = \frac{1}{x^n}$$

$$= \frac{\dfrac{1}{c} \cdot abc}{\left(\dfrac{1}{a} + \dfrac{1}{b}\right)abc} \qquad \text{• Multiply the numerator and the denominator by } abc, \text{ which is the LCD of the fraction in the numerator and the fraction in the denominator.}$$

$$= \frac{ab}{bc + ac}$$

take note

It is a mistake to write

$$\frac{c^{-1}}{a^{-1} + b^{-1}} \qquad as \qquad \frac{a + b}{c}$$

because a^{-1} and b^{-1} are terms and cannot be treated as factors.

▶ **TRY EXERCISE 60, PAGE 58**

● **APPLICATION OF RATIONAL EXPRESSIONS**

EXAMPLE 6 Solve an Application

The *average speed* for a round trip is given by the complex fraction

$$\frac{2}{\dfrac{1}{v_1} + \dfrac{1}{v_2}}$$

where v_1 is the average speed on the way to your destination and v_2 is the average speed on your return trip. Find the average speed for a round trip if $v_1 = 50$ mph and $v_2 = 40$ mph.

Solution

Evaluate the complex fraction with $v_1 = 50$ and $v_2 = 40$.

$$\frac{2}{\dfrac{1}{v_1} + \dfrac{1}{v_2}} = \frac{2}{\dfrac{1}{50} + \dfrac{1}{40}} = \frac{2}{\dfrac{1 \cdot 4}{50 \cdot 4} + \dfrac{1 \cdot 5}{40 \cdot 5}} \qquad \text{• Substitute and simplify the denominator.}$$

$$= \frac{2}{\dfrac{4}{200} + \dfrac{5}{200}} = \frac{2}{\dfrac{9}{200}}$$

$$= 2 \cdot \frac{200}{9} = \frac{400}{9} = 44\frac{4}{9}$$

The average speed for the round trip is $44\dfrac{4}{9}$ mph.

▶ **TRY EXERCISE 64, PAGE 58**

? QUESTION In Example 6, why is the speed for the round trip *not* the average of v_1 and v_2?

 TOPICS FOR DISCUSSION

1. Discuss the meaning of the phrase *rational expression*. Is a rational expression the same as a fraction? If not, give some examples of fractions that are not rational expressions.

2. What is the domain of a rational expression?

3. Explain why the following is *not* correct.

$$\frac{2x^2 + 5}{x^2} = 2 + 5 = 7$$

4. Consider the rational expression $\dfrac{x^2 - 3x - 10}{x^2 + x - 30}$. By simplifying this expression, we have

$$\frac{x^2 - 3x - 10}{x^2 + x - 30} = \frac{(x - 5)(x + 2)}{(x - 5)(x + 6)} = \frac{x + 2}{x + 6}$$

Does this really mean that $\dfrac{x^2 - 3x - 10}{x^2 + x - 30} = \dfrac{x + 2}{x + 6}$ for every value of x? If not, for what values of x are the two expressions equal?

? ANSWER Because you were traveling slower on the return trip, the return trip took longer than the time spent going to your destination. More time was spent traveling at the slower speed. Thus the average speed is less than the average of v_1 and v_2.

EXERCISE SET P.4

In Exercises 1 to 10, simplify each rational expression.

1. $\dfrac{x^2 - x - 20}{3x - 15}$

▶ 2. $\dfrac{2x^2 - 5x - 12}{2x^2 + 5x + 3}$

3. $\dfrac{x^3 - 9x}{x^3 + x^2 - 6x}$

4. $\dfrac{x^3 + 125}{2x^3 - 50x}$

5. $\dfrac{a^3 + 8}{a^2 - 4}$

6. $\dfrac{y^3 - 27}{-y^2 + 11y - 24}$

7. $\dfrac{x^2 + 3x - 40}{-x^2 + 3x + 10}$

8. $\dfrac{2x^3 - 6x^2 + 5x - 15}{9 - x^2}$

9. $\dfrac{4y^3 - 8y^2 + 7y - 14}{-y^2 - 5y + 14}$

10. $\dfrac{x^3 - x^2 + x}{x^3 + 1}$

In Exercises 11 to 40, simplify each expression.

11. $\left(-\dfrac{4a}{3b^2}\right)\left(\dfrac{6b}{a^4}\right)$

12. $\left(\dfrac{12x^2y}{5z^4}\right)\left(-\dfrac{25x^2z^3}{15y^2}\right)$

13. $\left(\dfrac{6p^2}{5q^2}\right)^{-1}\left(\dfrac{2p}{3q^2}\right)^2$

14. $\left(\dfrac{4r^2s}{3t^3}\right)^{-1}\left(\dfrac{6rs^3}{5t^2}\right)$

15. $\dfrac{x^2 + x}{2x + 3} \cdot \dfrac{3x^2 + 19x + 28}{x^2 + 5x + 4}$

▶ **16.** $\dfrac{x^2 - 16}{x^2 + 7x + 12} \cdot \dfrac{x^2 - 4x - 21}{x^2 - 4x}$

17. $\dfrac{3x - 15}{2x^2 - 50} \cdot \dfrac{2x^2 + 16x + 30}{6x + 9}$

18. $\dfrac{y^3 - 8}{y^2 + y - 6} \cdot \dfrac{y^2 + 3y}{y^3 + 2y^2 + 4y}$

19. $\dfrac{12y^2 + 28y + 15}{6y^2 + 35y + 25} \div \dfrac{2y^2 - y - 3}{3y^2 + 11y - 20}$

20. $\dfrac{z^2 - 81}{z^2 - 16} \div \dfrac{z^2 - z - 20}{z^2 + 5z - 36}$

21. $\dfrac{a^2 + 9}{a^2 - 64} \div \dfrac{a^3 - 3a^2 + 9a - 27}{a^2 + 5a - 24}$

22. $\dfrac{6x^2 + 13xy + 6y^2}{4x^2 - 9y^2} \div \dfrac{3x^2 - xy - 2y^2}{2x^2 + xy - 3y^2}$

23. $\dfrac{p + 5}{r} + \dfrac{2p - 7}{r}$

24. $\dfrac{2s + 5t}{4t} + \dfrac{-2s + 3t}{4t}$

25. $\dfrac{x}{x - 5} + \dfrac{7x}{x + 3}$

26. $\dfrac{2x}{3x + 1} + \dfrac{5x}{x - 7}$

27. $\dfrac{5y - 7}{y + 4} - \dfrac{2y - 3}{y + 4}$

28. $\dfrac{6x - 5}{x - 3} - \dfrac{3x - 8}{x - 3}$

29. $\dfrac{4z}{2z - 3} + \dfrac{5z}{z - 5}$

▶ **30.** $\dfrac{3y - 1}{3y + 1} - \dfrac{2y - 5}{y - 3}$

31. $\dfrac{x}{x^2 - 9} - \dfrac{3x - 1}{x^2 + 7x + 12}$

32. $\dfrac{m - n}{m^2 - mn - 6n^2} + \dfrac{3m - 5n}{m^2 + mn - 2n^2}$

33. $\dfrac{1}{x} + \dfrac{2}{3x - 1} \cdot \dfrac{3x^2 + 11x - 4}{x - 5}$

34. $\dfrac{2}{y} - \dfrac{3}{y + 1} \cdot \dfrac{y^2 - 1}{y + 4}$

35. $\dfrac{q + 1}{q - 3} - \dfrac{2q}{q - 3} \div \dfrac{q + 5}{q - 3}$

36. $\dfrac{p}{p + 5} + \dfrac{p}{p - 4} \div \dfrac{p + 2}{p^2 - p - 12}$

37. $\dfrac{1}{x^2 + 7x + 12} + \dfrac{1}{x^2 - 9} + \dfrac{1}{x^2 - 16}$

38. $\dfrac{2}{a^2 - 3a + 2} + \dfrac{3}{a^2 - 1} - \dfrac{5}{a^2 + 3a - 10}$

39. $\left(1 + \dfrac{2}{x}\right)\left(3 - \dfrac{1}{x}\right)$ **40.** $\left(4 - \dfrac{1}{z}\right)\left(4 + \dfrac{2}{z}\right)$

In Exercises 41 to 58, simplify each complex fraction.

41. $\dfrac{4 + \dfrac{1}{x}}{1 - \dfrac{1}{x}}$ ▶ **42.** $\dfrac{3 - \dfrac{2}{a}}{5 + \dfrac{3}{a}}$ **43.** $\dfrac{\dfrac{x}{y} - 2}{y - x}$

44. $\dfrac{3 + \dfrac{2}{x - 3}}{4 + \dfrac{1}{2 + \dfrac{1}{x}}}$ **45.** $\dfrac{5 - \dfrac{1}{x + 2}}{1 + \dfrac{3}{1 + \dfrac{3}{x}}}$ **46.** $\dfrac{\dfrac{1}{(x + h)^2} - 1}{h}$

47. $\dfrac{1 + \dfrac{1}{b - 2}}{1 - \dfrac{1}{b + 3}}$ **48.** $r - \dfrac{r}{r + \dfrac{1}{3}}$

49. $\dfrac{1 - \dfrac{1}{x^2}}{1 + \dfrac{1}{x}}$ **50.** $\dfrac{1}{\dfrac{1}{a} + \dfrac{1}{b}}$

51. $2 - \dfrac{m}{1 - \dfrac{1 - m}{-m}}$ **52.** $\dfrac{\dfrac{x + h + 1}{x + h} - \dfrac{x}{x + 1}}{h}$

53. $\dfrac{\dfrac{1}{x} - \dfrac{x - 4}{x + 1}}{\dfrac{x}{x + 1}}$ **54.** $\dfrac{\dfrac{2}{y} - \dfrac{3y - 2}{y - 1}}{\dfrac{y}{y - 1}}$

55. $\dfrac{\dfrac{1}{x + 3} - \dfrac{2}{x - 1}}{\dfrac{x}{x - 1} + \dfrac{3}{x + 3}}$ **56.** $\dfrac{\dfrac{x + 2}{x^2 - 1} + \dfrac{1}{x + 1}}{\dfrac{x}{2x^2 - x - 1} + \dfrac{1}{x - 1}}$

57. $\dfrac{\dfrac{x^2 + 3x - 10}{x^2 + x - 6}}{\dfrac{x^2 - x - 30}{2x^2 - 15x + 18}}$ **58.** $\dfrac{\dfrac{2y^2 + 11y + 15}{y^2 - 4y - 21}}{\dfrac{6y^2 + 11y - 10}{3y^2 - 23y + 14}}$

In Exercises 59 to 62, simplify each algebraic fraction. Write all answers with positive exponents.

59. $\dfrac{a^{-1} + b^{-1}}{a - b}$

▶ **60.** $\dfrac{e^{-2} - f^{-1}}{ef}$

61. $\dfrac{a^{-1}b - ab^{-1}}{a^2 + b^2}$

62. $(a + b^{-2})^{-1}$

63. AVERAGE SPEED According to Example 6, the average speed for a round trip in which the average speed on the way to your destination is v_1 and the average speed on your return is v_2 is given by the complex fraction

$$\dfrac{2}{\dfrac{1}{v_1} + \dfrac{1}{v_2}}$$

a. Find the average speed for a round trip by helicopter with $v_1 = 180$ mph and $v_2 = 110$ mph.

b. Simplify the complex fraction.

▶ **64. RELATIVITY THEORY** Using Einstein's Theory of Relativity, the "sum" of the two speeds v_1 and v_2 is given by the complex fraction

$$\dfrac{v_1 + v_2}{1 + \dfrac{v_1 v_2}{c^2}}$$

where c is the speed of light.

a. Evaluate this expression with $v_1 = 1.2 \times 10^8$ mph, $v_2 = 2.4 \times 10^8$ mph, and $c = 6.7 \times 10^8$ mph.

b. Simplify the complex fraction.

65.  **DOPPLER SHIFT** The photograph at the right shows the central region of a galaxy named NGC 1705, which is approximately 17 million light-years from Earth. The distance to this galaxy can be determined by measuring its red shift. (See the Chapter Opener

on page 1.) The amount of this red shift z is given by

$$z = \dfrac{\lambda_O - \lambda_S}{\lambda_S}$$

where λ_O is the wavelength of light from the receding galaxy and λ_S is the wavelength of the same light on Earth. Find the red shift of a galaxy for a wavelength that normally measures 375.4×10^{-9} meters on Earth but measures 390.5×10^{-9} meters coming from the receding galaxy. Round to the nearest thousandth.

66. DOPPLER SHIFT Calculate the red shift (see Exercise 65) for a wavelength that normally measures 401.5×10^{-9} meters on Earth but measures 412.3×10^{-9} meters coming from the receding galaxy. Round to the nearest thousandth.

67. SPEED OF A GALAXY The relative speed v, in kilometers per second, at which a galaxy is receding from Earth can be determined from

$$v = c\left[\dfrac{(z + 1)^2 - 1}{(z + 1)^2 + 1}\right]$$

where c is the speed of light (3×10^5 kilometers per second) and z is the red shift (see Exercise 65) of the galaxy. Find the relative speed of a receding galaxy whose red shift is 0.032. Round to the nearest kilometer per second.

68. SPEED OF A GALAXY Find the relative speed (see Exercise 67) of a receding galaxy whose red shift is 0.041. Round to the nearest kilometer per second.

69. Find the rational expression in simplest form that represents the sum of the reciprocals of the consecutive integers x and $x + 1$.

70. Find the rational expression in simplest form that represents the positive difference between the reciprocals of the consecutive even integers x and $x + 2$.

71. Find the rational expression in simplest form that represents the sum of the reciprocals of the consecutive even integers $x - 2$, x, and $x + 2$.

72. Find the rational expression in simplest form that represents the sum of the reciprocals of the squares of the consecutive even integers $x - 2$, x, and $x + 2$.

─── *CONNECTING CONCEPTS* ───

In Exercises 73 to 76, simplify each algebraic fraction.

73. $\dfrac{(x + 5) - x(x + 5)^{-1}}{x + 5}$

74. $\dfrac{(y + 2) + y^2(y + 2)^{-1}}{y + 2}$

75. $\dfrac{x^{-1} - 4y}{(x^{-1} - 2y)(x^{-1} + 2y)}$

76. $\dfrac{x + y}{x - y} \cdot \dfrac{x^{-1} - y^{-1}}{x^{-1} + y^{-1}}$

77. FINANCE The **present value** of an ordinary annuity is given by

$$R\left[\dfrac{1 - \dfrac{1}{(1 + i)^n}}{i}\right]$$

where n is the number of payments of R dollars invested at an interest rate of i per conversion period. Simplify the complex fraction.

78. ELECTRICITY The total resistance of the three resistances R_1, R_2, and R_3 in parallel is given by

$$\dfrac{1}{\dfrac{1}{R_1} + \dfrac{1}{R_2} + \dfrac{1}{R_3}}$$

Simplify the complex fraction.

PREPARE FOR SECTION P.5

In Exercises 79 to 84, simplify the expression.

79. $(2 - 3x)(4 - 5x)$ [P.3]

80. $(2 - 5x)^2$ [P.3]

81. $\sqrt{96}$ [P.2]

82. $\left(2 + 3\sqrt{5}\right)\left(3 - 4\sqrt{5}\right)$ [P.2]

83. $\dfrac{5 + \sqrt{2}}{3 - \sqrt{2}}$ [P.2]

84. Which of the following polynomials, if any, does not factor over the integers? [P.3]

  a. $81 - x^2$ **b.** $9 + z^2$

PROJECTS

1. CONTINUED FRACTIONS The complex fraction shown at the right is called a **continued fraction.** The three dots in $\dfrac{1}{1 + \cdots}$ indicate that the pattern continues in the same manner. A **convergent** of a complex fraction is an approximation of the continued fraction that is found by stopping the process at some point.

$$\cfrac{1}{1 + \cfrac{1}{1 + \cfrac{1}{1 + \cfrac{1}{1 + \cdots}}}}$$

a. Calculate the convergent $C_2 = \cfrac{1}{1 + \cfrac{1}{1 + 1}}$.

b. Calculate the convergent $C_3 = \cfrac{1}{1 + \cfrac{1}{1 + \cfrac{1}{1 + 1}}}$.

c. Calculate the convergent $C_5 = \cfrac{1}{1 + \cfrac{1}{1 + \cfrac{1}{1 + \cfrac{1}{1 + \cfrac{1}{1 + 1}}}}}$.

d. Show that $C_5 \approx \dfrac{-1 + \sqrt{5}}{2}$. Using some techniques from more advanced math courses, it can be shown that as n increases, C_n becomes closer and closer to $\dfrac{-1 + \sqrt{5}}{2}$.

2. REPRESENTATION OF π There are a few continued-fraction representations for π. Find two of these representations. Compute the value of π accurate to four decimal places using a convergent from each of the continued fractions you found.

COMPLEX NUMBERS

● INTRODUCTION TO COMPLEX NUMBERS

Recall that $\sqrt{9} = 3$ because $3^2 = 9$. Now consider the expression $\sqrt{-9}$. To find $\sqrt{-9}$, we need to find a number c such that $c^2 = -9$. However, the square of any real number c (except zero) is a *positive* number. Consequently, we must expand our concept of number to include numbers whose squares are negative numbers.

Around the 17th century, a new number, called an *imaginary number*, was defined so that a negative number would have a square root. The letter i was chosen to represent the number whose square is -1.

Definition of i

The number i, called the **imaginary unit**, is the number such that $i^2 = -1$.

The principal square root of a negative number is defined in terms of i.

Principal Square Root of a Negative Number

If a is a positive real number, then $\sqrt{-a} = i\sqrt{a}$. The number $i\sqrt{a}$ is called an **imaginary number.**

Here are some examples of imaginary numbers.

$$\sqrt{-36} = i\sqrt{36} = 6i \qquad \sqrt{-18} = i\sqrt{18} = 3i\sqrt{2}$$
$$\sqrt{-23} = i\sqrt{23} \qquad \sqrt{-1} = i\sqrt{1} = i$$

It is customary to write i in front of a radical sign, as we did for $i\sqrt{23}$, to avoid confusing $\sqrt{a}\,i$ with $\sqrt{ai}$.

Complex Numbers

A **complex number** is a number of the form $a + bi$, where a and b are real numbers and $i = \sqrt{-1}$. The number a is the **real part** of $a + bi$, and b is the **imaginary part.**

Here are some examples of complex numbers.

$-3 + 5i$	• Real part: -3; imaginary part: 5
$2 - 6i$	• Real part: 2; imaginary part: -6
5	• Real part: 5; imaginary part: 0
$7i$	• Real part: 0; imaginary part: 7

Note from these examples that a real number is a complex number whose imaginary part is zero and that an imaginary number is a complex number whose real part is zero.

MATH MATTERS

It may seem strange to just invent new numbers, but that is how mathematics evolves. For instance, negative numbers were not an accepted part of mathematics until well into the 13th century. In fact, these numbers often were referred to as "fictitious numbers."

In the 17th century, Rene Descartes called square roots of negative numbers "imaginary numbers," an unfortunate choice of words, and started using the letter i to denote these numbers. These numbers were subjected to the same skepticism as negative numbers.

It is important to understand that these numbers are not *imaginary* in the dictionary sense of the word. It is similar to the situation of negative numbers being called fictitious.

If you think of a number line, then the numbers to the right of zero are positive numbers and the numbers to the left of zero are negative numbers. One way to think of an imaginary number is to visualize it as *up* or *down* from zero. See the Project on page 67 for more information on this topic.

MATH MATTERS

The imaginary unit i is important in the field of electrical engineering. However, because the letter i is used by engineers as the symbol for electric current, these engineers use j for the complex unit.

? QUESTION What are the real part and the imaginary part of $3 - 5i$?

Note from the following diagram that the real numbers are a subset of the complex numbers, and the imaginary numbers are a subset of the complex numbers. The real numbers and imaginary numbers are disjoint sets.

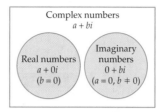

Example 1 illustrates writing a complex number in the standard form $a + bi$.

EXAMPLE I   **Write a Complex Number in Standard Form**

Write $7 + \sqrt{-45}$ in the form $a + bi$.

Solution

$$7 + \sqrt{-45} = 7 + i\sqrt{45}$$
$$= 7 + i\sqrt{9} \cdot \sqrt{5}$$
$$= 7 + 3i\sqrt{5}$$

▶ **TRY EXERCISE 8, PAGE 65**

● **ADDITION AND SUBTRACTION OF COMPLEX NUMBERS**

All the standard arithmetic operations that are applied to real numbers can be applied to complex numbers.

Definition of Addition and Subtraction of Complex Numbers

If $a + bi$ and $c + di$ are complex numbers, then

Addition $(a + bi) + (c + di) = (a + c) + (b + d)i$

Subtraction $(a + bi) - (c + di) = (a - c) + (b - d)i$

Basically, these rules say that to add two complex numbers, add the real parts and add the imaginary parts. To subtract two complex numbers, subtract the real parts and subtract the imaginary parts.

? ANSWER Real part: 3; imaginary part: -5

| EXAMPLE 2 | Add or Subtract Complex Numbers |

Simplify.

   a. $(7 - 2i) + (-2 + 4i)$ **b.** $(-9 + 4i) - (2 - 6i)$

Solution

   a. $(7 - 2i) + (-2 + 4i) = (7 + (-2)) + (-2 + 4)i = 5 + 2i$

   b. $(-9 + 4i) - (2 - 6i) = (-9 - 2) + (4 - (-6))i = -11 + 10i$

▶ **TRY EXERCISE 18, PAGE 66**

● **MULTIPLICATION OF COMPLEX NUMBERS**

When multiplying complex numbers, the term i^2 is frequently a part of the product. Recall that $i^2 = -1$. Therefore,

$$3i(5i) = 15i^2 = 15(-1) = -15$$
$$-2i(6i) = -12i^2 = -12(-1) = 12$$
$$4i(3 - 2i) = 12i - 8i^2 = 12i - 8(-1) = 8 + 12i$$

When multiplying square roots of negative numbers, first rewrite the radical expressions using i. For instance,

$$\sqrt{-6} \cdot \sqrt{-24} = i\sqrt{6} \cdot i\sqrt{24} \qquad \bullet\ \sqrt{-6} = i\sqrt{6},\ \sqrt{-24} = i\sqrt{24}$$
$$= i^2\sqrt{144} = -1 \cdot 12$$
$$= -12$$

> *take note*
>
> Recall that the definition of the product of radical expressions required that the radicand be a positive number. Therefore, when multiplying expressions containing negative radicands, we must first rewrite the expression using i and a positive radicand.

Note from this example that it would have been incorrect to multiply the radicands of the two radical expressions. To illustrate:

$$\sqrt{-6} \cdot \sqrt{-24} \neq \sqrt{(-6)(-24)}$$

❓ **QUESTION** What is the product of $\sqrt{-2}$ and $\sqrt{-8}$?

To multiply two complex numbers, we use the following definition.

Definition of Multiplication of Complex Numbers

If $a + bi$ and $c + di$ are complex numbers, then

$$(a + bi)(c + di) = (ac - bd) + (ad + bc)i$$

Because every complex number can be written as a sum of two terms, it is natural to perform multiplication on complex numbers in a manner consistent

---

❓ **ANSWER** $\sqrt{-2} \cdot \sqrt{-8} = i\sqrt{2} \cdot i\sqrt{8} = i^2\sqrt{16} = -1 \cdot 4 = -4$

with the operation defined on binomials and the definition $i^2 = -1$. By using this analogy, you can multiply complex numbers without memorizing the definition.

EXAMPLE 3 **Multiply Complex Numbers**

Simplify. **a.** $(3 - 4i)(2 + 5i)$ **b.** $\left(2 + \sqrt{-3}\right)\left(4 - 5\sqrt{-3}\right)$

Solution

a. $(3 - 4i)(2 + 5i) = 6 + 15i - 8i - 20i^2$
$\qquad\qquad\qquad\quad = 6 + 15i - 8i - 20(-1)$ • Replace i^2 by -1.
$\qquad\qquad\qquad\quad = 6 + 15i - 8i + 20$ • Simplify.
$\qquad\qquad\qquad\quad = 26 + 7i$

b. $\left(2 + \sqrt{-3}\right)\left(4 - 5\sqrt{-3}\right) = \left(2 + i\sqrt{3}\right)\left(4 - 5i\sqrt{3}\right)$
$\qquad\qquad\qquad\qquad\qquad = 8 - 10i\sqrt{3} + 4i\sqrt{3} - 5i^2\sqrt{9}$
$\qquad\qquad\qquad\qquad\qquad = 8 - 10i\sqrt{3} + 4i\sqrt{3} - 5(-1)(3)$
$\qquad\qquad\qquad\qquad\qquad = 8 - 10i\sqrt{3} + 4i\sqrt{3} + 15 = 23 - 6i\sqrt{3}$

▶ **TRY EXERCISE 34, PAGE 66**

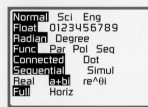
● **DIVISION OF COMPLEX NUMBERS**

Recall that the number $\dfrac{3}{\sqrt{2}}$ is not in simplest form because there is a radical expression in the denominator. Similarly, $\dfrac{3}{i}$ is not in simplest form because $i = \sqrt{-1}$. To write this expression in simplest form, multiply the numerator and denominator by i.

$$\frac{3}{i} \cdot \frac{i}{i} = \frac{3i}{i^2} = \frac{3i}{-1} = -3i$$

Here is another example.

$$\frac{3 - 6i}{2i} = \frac{3 - 6i}{2i} \cdot \frac{i}{i} = \frac{3i - 6i^2}{2i^2} = \frac{3i - 6(-1)}{2(-1)}$$

$$= \frac{3i + 6}{-2} = -3 - \frac{3}{2}i$$

Recall that to simplify the quotient $\dfrac{2 + \sqrt{3}}{5 + 2\sqrt{3}}$, we multiply the numerator and denominator by the conjugate of $5 + 2\sqrt{3}$, which is $5 - 2\sqrt{3}$. In a similar manner, to find the quotient of two complex numbers, we multiply the numerator and denominator by the conjugate of the denominator.

The complex numbers $a + bi$ and $a - bi$ are called **complex conjugates** or **conjugates** of each other. The conjugate of the complex number z is denoted by $\bar{z}$. For instance,

$$\overline{2 + 5i} = 2 - 5i \qquad \text{and} \qquad \overline{3 - 4i} = 3 + 4i$$

Consider the product of a complex number and its conjugate. For instance,

$$(2 + 5i)(2 - 5i) = 4 - 10i + 10i - 25i^2$$
$$= 4 - 25(-1) = 4 + 25$$
$$= 29$$

Note that the product is a *real* number. This is always true.

Product of Complex Conjugates

The product of a complex number and its conjugate is a real number. That is, $(a + bi)(a - bi) = a^2 + b^2$.

For instance, $(5 + 3i)(5 - 3i) = 5^2 + 3^2 = 25 + 9 = 34$.

The next example shows how the quotient of two complex numbers is determined by using conjugates.

EXAMPLE 4   Divide Complex Numbers

Simplify: $\dfrac{16 - 11i}{5 + 2i}$

Solution

$$\frac{16 - 11i}{5 + 2i} = \frac{16 - 11i}{5 + 2i} \cdot \frac{5 - 2i}{5 - 2i}$$

• Multiply numerator and denominator by the conjugate of the denominator.

$$= \frac{80 - 32i - 55i + 22i^2}{5^2 + 2^2}$$

$$= \frac{80 - 32i - 55i + 22(-1)}{25 + 4}$$

$$= \frac{80 - 87i - 22}{29}$$

$$= \frac{58 - 87i}{29}$$

$$= \frac{29(2 - 3i)}{29} = 2 - 3i$$

▶ **TRY EXERCISE 48, PAGE 66**

● POWERS OF i

The following powers of i illustrate a pattern:

$$i^1 = i \qquad\qquad\qquad i^5 = i^4 \cdot i = 1 \cdot i = i$$
$$i^2 = -1 \qquad\qquad\qquad i^6 = i^4 \cdot i^2 = 1(-1) = -1$$
$$i^3 = i^2 \cdot i = (-1)i = -i \qquad i^7 = i^4 \cdot i^3 = 1(-i) = -i$$
$$i^4 = i^2 \cdot i^2 = (-1)(-1) = 1 \qquad i^8 = (i^4)^2 = 1^2 = 1$$

Because $i^4 = 1$, $(i^4)^n = 1^n = 1$ for any integer n. Thus it is possible to evaluate powers of i by factoring out powers of i^4, as shown in the following:

$$i^{27} = (i^4)^6 \cdot i^3 = 1^6 \cdot i^3 = 1 \cdot (-i) = -i$$

The following theorem can be used to evaluate powers of i.

Powers of i

If n is a positive integer, then $i^n = i^r$, where r is the remainder of the division of n by 4.

EXAMPLE 5   Evaluate a Power of i

Evaluate: i^{153}

Solution

Use the powers of i theorem.

$$i^{153} = i^1 = i \qquad \bullet \text{ Remainder of } 153 \div 4 \text{ is } 1.$$

▶ TRY EXERCISE 60, PAGE 66

TOPICS FOR DISCUSSION

1. What is an imaginary number? What is a complex number?

2. How are the real numbers related to the complex numbers?

3. Is zero a complex number?

4. What is the conjugate of a complex number?

5. If a and b are real numbers and $ab = 0$, then $a = 0$ or $b = 0$. Is the same true for complex numbers? That is, if u and v are complex numbers and $uv = 0$, must one of the numbers be zero?

EXERCISE SET P.5

In Exercises 1 to 10, write the complex number in standard form.

1. $\sqrt{-81}$

2. $\sqrt{-64}$

3. $\sqrt{-98}$

4. $\sqrt{-27}$

5. $\sqrt{16} + \sqrt{-81}$

6. $\sqrt{25} + \sqrt{-9}$

7. $5 + \sqrt{-49}$

▶ **8.** $6 - \sqrt{-1}$

9. $8 - \sqrt{-18}$

10. $11 + \sqrt{-48}$

In Exercises 11 to 36, simplify and write the complex number in standard form.

11. $(5 + 2i) + (6 - 7i)$

12. $(4 - 8i) + (5 + 3i)$

13. $(-2 - 4i) - (5 - 8i)$ **14.** $(3 - 5i) - (8 - 2i)$

15. $(1 - 3i) + (7 - 2i)$ **16.** $(2 - 6i) + (4 - 7i)$

17. $(-3 - 5i) - (7 - 5i)$ ▶ **18.** $(5 - 3i) - (2 + 9i)$

19. $8i - (2 - 8i)$ **20.** $3 - (4 - 5i)$

21. $5i \cdot 8i$ **22.** $(-3i)(2i)$

23. $\sqrt{-50} \cdot \sqrt{-2}$ **24.** $\sqrt{-12} \cdot \sqrt{-27}$

25. $3(2 + 5i) - 2(3 - 2i)$ **26.** $3i(2 + 5i) + 2i(3 - 4i)$

27. $(4 + 2i)(3 - 4i)$ **28.** $(6 + 5i)(2 - 5i)$

29. $(-3 - 4i)(2 + 7i)$ **30.** $(-5 - i)(2 + 3i)$

31. $(4 - 5i)(4 + 5i)$ **32.** $(3 + 7i)(3 - 7i)$

33. $\left(3 + \sqrt{-4}\right)\left(2 - \sqrt{-9}\right)$

▶ **34.** $\left(5 + 2\sqrt{-16}\right)\left(1 - \sqrt{-25}\right)$

35. $\left(3 + 2\sqrt{-18}\right)\left(2 + 2\sqrt{-50}\right)$

36. $\left(5 - 3\sqrt{-48}\right)\left(2 - 4\sqrt{-27}\right)$

In Exercises 37 to 54, write each expression as a complex number in standard form.

37. $\dfrac{6}{i}$ **38.** $\dfrac{-8}{2i}$

39. $\dfrac{6 + 3i}{i}$ **40.** $\dfrac{4 - 8i}{4i}$

41. $\dfrac{1}{7 + 2i}$ **42.** $\dfrac{5}{3 + 4i}$

43. $\dfrac{2i}{1 + i}$ **44.** $\dfrac{5i}{2 - 3i}$

45. $\dfrac{5 - i}{4 + 5i}$ **46.** $\dfrac{4 + i}{3 + 5i}$

47. $\dfrac{3 + 2i}{3 - 2i}$ ▶ **48.** $\dfrac{8 - i}{2 + 3i}$

49. $\dfrac{-7 + 26i}{4 + 3i}$ **50.** $\dfrac{-4 - 39i}{5 - 2i}$

51. $(3 - 5i)^2$ **52.** $(2 + 4i)^2$

53. $(1 + 2i)^3$ **54.** $(2 - i)^3$

In Exercises 55 to 62, evaluate the power of i.

55. i^{15} **56.** i^{66} **57.** $-i^{40}$ **58.** $-i^{51}$

59. $\dfrac{1}{i^{25}}$ ▶ **60.** $\dfrac{1}{i^{83}}$ **61.** i^{-34} **62.** i^{-52}

In Exercises 63 to 68, evaluate $\dfrac{-b + \sqrt{b^2 - 4ac}}{2a}$ for the given values of a, b, and c. Write your answer as a complex number in standard form.

63. $a = 3, b = -3, c = 3$ **64.** $a = 2, b = 4, c = 4$

65. $a = 2, b = 6, c = 6$ **66.** $a = 2, b = 1, c = 3$

67. $a = 4, b = -4, c = 2$ **68.** $a = 3, b = -2, c = 4$

CONNECTING CONCEPTS

The property that the product of conjugates of the form $(a + bi)(a - bi)$ is equal to $a^2 + b^2$ can be used to factor the sum of two perfect squares over the set of complex numbers. For example, $x^2 + y^2 = (x + yi)(x - yi)$. In Exercises 69 to 74, factor the binomial over the set of complex numbers.

69. $x^2 + 16$ **70.** $x^2 + 9$

71. $z^2 + 25$ **72.** $z^2 + 64$

73. $4x^2 + 81$ **74.** $9x^2 + 1$

75. Show that if $x = 1 + 2i$, then $x^2 - 2x + 5 = 0$.

76. Show that if $x = 1 - 2i$, then $x^2 - 2x + 5 = 0$.

77. When we think of the cube root of 8, $\sqrt[3]{8}$, we normally mean the *real* cube root of 8 and write $\sqrt[3]{8} = 2$. However, there are two other cube roots of 8 that are complex numbers. Verify that $-1 + i\sqrt{3}$ and $-1 - i\sqrt{3}$ are cube roots of 8 by showing that $\left(-1 + i\sqrt{3}\right)^3 = 8$ and $\left(-1 - i\sqrt{3}\right)^3 = 8$.

78. It is possible to find the square root of a complex number.

Verify that $\sqrt{i} = \dfrac{\sqrt{2}}{2}(1 + i)$ by showing that

$$\left[\frac{\sqrt{2}}{2}(1+i)\right]^2 = i.$$

79. Simplify $i + i^2 + i^3 + i^4 + \cdots + i^{28}$.

80. Simplify $i + i^2 + i^3 + i^4 + \cdots + i^{100}$.

PROJECTS

ARGAND DIAGRAM Just as we can graph a real number on a real number line, we can graph a complex number. This is accomplished by using one number line for the real part of the complex number and one number line for the imaginary part of the complex number. These two number lines are drawn perpendicular to each other and pass through their respective origins, as shown below.

The result is called the *complex plane* or an *Argand diagram* after Jean-Robert Argand (1768–1822), an accountant and amateur mathematician. Although he is given credit for this representation of complex numbers, Caspar Wessel (1745–1818) actually conceived the idea before Argand.

To graph the complex number $3 + 4i$, start at 3 on the real axis. Now move 4 units up (for positive numbers move up, for negative numbers move down) and place a dot at that point, as shown in the diagram. Graphs of several other complex numbers are also shown.

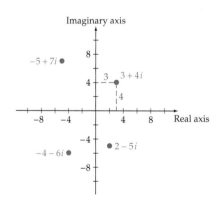

In Exercises 1 to 8, graph the complex number.

1. $2 + 5i$ **2.** $4 - 3i$ **3.** $-2 + 6i$ **4.** $-3 - 5i$

5. 4 **6.** $-2i$ **7.** $3i$ **8.** -5

The absolute value of a complex number is given by $|a + bi| = \sqrt{a^2 + b^2}$. In Exercises 9 to 12, find the absolute value of the complex number.

9. $2 + 5i$ **10.** $4 - 3i$ **11.** $-2 + 6i$ **12.** $-3 - 5i$

13. The additive inverse of $a + bi$ is $-a - bi$. Show that the absolute value of a complex number and the absolute value of its additive inverse are equal.

14. A *real* number and its additive inverse are the same distance from zero but on opposite sides of zero on a real number line. Describe the relationship between the graphs of a complex number and its additive inverse.

EXPLORING CONCEPTS
WITH TECHNOLOGY

Can You Trust Your Calculator?

You may think that your calculator always produces correct results in a *predictable* manner. However, the following experiment may change your opinion.

First note that the algebraic expression

$$p + 3p(1 - p)$$

is equal to the expression

$$4p - 3p^2$$

Use a graphing calculator to evaluate both expressions with $p = 0.05$. You should find that both expressions equal 0.1925. So far we do not observe any unexpected results. Now replace p in each expression with the current value of that expression (0.1925 in this case). This is called *feedback* because we are feeding our outputs back into each expression as inputs. Each new evaluation is referred to as an *iteration*. This time each expression takes on the value 0.65883125. Still no surprises. Continue the feedback process. That is, replace p in each expression *with the current value* of that expression. Now each expression takes on the value 1.33314915207, as shown in the following table. The iterations were performed on a *TI-85* calculator.

INTEGRATING TECHNOLOGY

To perform these iterations with a *TI* graphing calculator, first store 0.05 in p and then store $p + 3p(1 - p)$ in p as shown below.

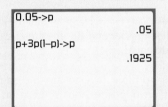

Each time you press ENTER, the expression $p + 3p(1 - p)$ will be evaluated with p equal to the previous result.

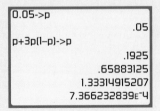

Iteration	$p + 3p(1 - p)$	$4p - 3p^2$
1	0.1925	0.1925
2	0.65883125	0.65883125
3	1.33314915207	1.33314915207

The following table shows that if we continue this feedback process on a calculator, the expressions $p + 3p(1 - p)$ and $4p - 3p^2$ will start to take on different values starting with the fourth iteration. By the 37th iteration, the values do not even agree to two decimal places.

Iteration	$p + 3p(1 - p)$	$4p - 3p^2$
4	7.366232839E-4	7.366232838E-4
5	0.002944865294	0.002944865294
6	0.011753444481	0.0117534448
7	0.046599347553	0.046599347547
20	1.12135618652	1.12135608405
30	0.947163304835	0.947033128433
37	0.285727963839	0.300943417861

1. Use a calculator to find the first 20 iterations of $p + 3p(1 - p)$ and $4p - 3p^2$, with the initial value of $p = 0.5$.

2. Write a report on chaos and fractals. Include information on the "butterfly effect." An excellent source is *Chaos and Fractals, New Frontiers of Science* by Heinz-Otto Peitgen, Hartmut Jurgens, and Dietmar Saupe (New York: Springer-Verlag, 1992).

3. Equations of the form $p_{n+1} = p_n + rp_n(1 - p_n)$ are called Verhulst population models. Write a report on Verhulst population models.

CHAPTER P SUMMARY

P.1 The Real Number System

- The following sets of numbers are used extensively in algebra:

Integers	$\{\ldots, -3, -2, -1, 0, 1, 2, 3, \ldots\}$
Rational numbers	{all terminating or repeating decimals}
Irrational numbers	{all nonterminating, nonrepeating decimals}
Real numbers	{all rational and irrational numbers}

- Set-builder notation is a method of writing sets and has the form {variable | condition on the variable}.

- Union and intersection are two operations on sets.

- The absolute value of a real number a is given by

$$|a| = \begin{cases} a, & a \geq 0 \\ -a, & a < 0 \end{cases}$$

- For any two real numbers a and b, the distance between the graph of a and the graph of b is given by $d(a, b) = |a - b|$.

- The Order of Operations Agreement is used to simplify expressions.

P.2 Integer and Rational Number Exponents

- If b is any real number and n is any natural number, then

$$b^n = \underbrace{b \cdot b \cdot b \cdot \cdots \cdot b}_{n \text{ factors of } b}$$

- For any nonzero real number b, $b^0 = 1$.

- If $b \neq 0$ and n is any natural number, then $b^{-n} = \dfrac{1}{b^n}$ and $\dfrac{1}{b^{-n}} = b^n$.

- **Properties of Rational Exponents**
  If p, q, and r represent rational numbers, and a and b are positive real numbers, then

Product	$b^p \cdot b^q = b^{p+q}$
Quotient	$\dfrac{b^p}{b^q} = b^{p-q}$
Power	$(b^p)^q = b^{pq} \qquad (a^p b^q)^r = a^{pr} b^{qr}$
	$\left(\dfrac{a^p}{b^q}\right)^r = \dfrac{a^{pr}}{b^{qr}} \qquad b^{-p} = \dfrac{1}{b^p}$

- **Properties of Radicals**
  If m and n are natural numbers, and a and b are positive real numbers, then

Product	$\sqrt[n]{a} \cdot \sqrt[n]{b} = \sqrt[n]{ab}$
Quotient	$\dfrac{\sqrt[n]{a}}{\sqrt[n]{b}} = \sqrt[n]{\dfrac{a}{b}}$
Index	$\sqrt[m]{\sqrt[n]{b}} = \sqrt[mn]{b}$

P.3 Polynomials

- A polynomial is an expression of the form

$$a_n x^n + a_{n-1} x^{n-1} + \cdots + a_2 x^2 + a_1 x + a_0$$

- Special product formulas are as follows:

Special Form	Formula(s)
(Sum)(Difference)	$(x + y)(x - y) = x^2 - y^2$
(Binomial)^2	$(x + y)^2 = x^2 + 2xy + y^2$ $(x - y)^2 = x^2 - 2xy + y^2$

- Factoring formulas are as follows:

Special Form	Formula(s)
Difference of two squares	$x^2 - y^2 = (x + y)(x - y)$
Perfect-square trinomials	$x^2 + 2xy + y^2 = (x + y)^2$ $x^2 - 2xy + y^2 = (x - y)^2$
Sum of cubes	$x^3 + y^3 = (x + y)(x^2 - xy + y^2)$
Difference of cubes	$x^3 - y^3 = (x - y)(x^2 + xy + y^2)$

P.4 Rational Expressions

- A rational expression is a fraction in which the numerator and denominator are polynomials. The properties of rational expressions are used to simplify a rational expression and to find the sum, difference, product, and quotient of two rational expressions.

- Complex fractions can be simplified in either of the following ways:

  Method 1: Multiply both the numerator and the denominator by the LCD of all the fractions in the complex fraction.

  Method 2: Simplify the numerator to a single fraction and the denominator to a single fraction. Multiply the numerator by the reciprocal of the denominator.

P.5 Complex Numbers

- The number i, called the *imaginary unit,* is the number such that $i^2 = -1$.

- If a is a positive real number, then $\sqrt{-a} = i\sqrt{a}$. The number $i\sqrt{a}$ is called an *imaginary number.*

- A *complex number* is a number of the form $a + bi$, where a and b are real numbers and $i = \sqrt{-1}$. The number a is the *real part* of $a + bi$, and b is the *imaginary part.*

- The complex numbers $a + bi$ and $a - bi$ are called *complex conjugates* or *conjugates* of each other.

- **Operations on Complex Numbers**

  $$(a + bi) + (c + di) = (a + c) + (b + d)i$$

  $$(a + bi) - (c + di) = (a - c) + (b - d)i$$

  $$(a + bi)(c + di) = (ac - bd) + (ad + bc)i$$

  $$\frac{a + bi}{c + di} = \frac{a + bi}{c + di} \cdot \frac{c - di}{c - di}$$

  - **Multiply numerator and denominator by the conjugate of the denominator.**

CHAPTER P TRUE/FALSE EXERCISES

In Exercises 1 to 10, answer true or false. If the statement is false, give an example or a reason to show that the statement is false.

1. If a and b are real numbers, then $|a - b| = |b - a|$.

2. If a is a real number, then $a^2 \geq a$.

3. The set of rational numbers is closed under the operation of addition.

4. The set of irrational numbers is closed under the operation of addition.

5. Let $x \oplus y$ denote the average of the two real numbers x and y. That is,

$$x \oplus y = \frac{x + y}{2}$$

The operation $\oplus$ is an associative operation because $(x \oplus y) \oplus z = x \oplus (y \oplus z)$ for all real numbers x, y, and z.

6. Using interval notation, we write the inequality $x > a$ as $[a, \infty)$.

7. If n is a real number, then $\sqrt{n^2} = n$.

8. $(a + b)^2 = a^2 + b^2$

9. $\sqrt[3]{a^3 + b^3} = a + b$

10. $\sqrt{-2}\sqrt{-8} = 4$

CHAPTER P REVIEW EXERCISES

In Exercises 1 to 4, classify each number as one or more of the following: integer, rational number, irrational number, real number, prime number, composite number.

1. 3   2. $\sqrt{7}$ 3. $-\dfrac{1}{2}$ 4. $0.\overline{5}$

In Exercises 5 and 6, use $A = \{1, 5, 7\}$ and $B = \{2, 3, 5, 11\}$ to find the indicated intersection or union.

5. $A \cup B$ 6. $A \cap B$

In Exercises 7 to 14, identify the real number property or the property of equality that is illustrated.

7. $5(x + 3) = 5x + 15$

8. $a(3 + b) = a(b + 3)$

9. $(6c)d = 6(cd)$

10. $\sqrt{2} + 3$ is a real number.

11. $7 + 0 = 7$

12. $1x = x$

13. If $7 = x$, then $x = 7$.

14. If $3x + 4 = y$, and $y = 5z$, then $3x + 4 = 5z$.

In Exercises 15 and 16, graph each inequality and write the inequality using interval notation.

15. $-4 < x \le 2$

16. $x \le -1$ or $x > 3$

In Exercises 17 and 18, graph each interval and write each interval as an inequality.

17. $[-3, 2)$

18. $(-1, \infty)$

In Exercises 19 to 22, write each real number without absolute value symbols.

19. $|7|$ **20.** $|2 - \pi|$ **21.** $|4 - \pi|$ **22.** $|-11|$

In Exercises 23 and 24, find the distance on the real number line between the points whose coordinates are given.

23. $-3, 14$

24. $\sqrt{5}, -\sqrt{2}$

In Exercises 25 and 26, evaluate each expression.

25. $-5^2 + (-11)$

26. $\dfrac{(2^2 \cdot 3^{-2})^2}{3^{-1} \cdot 2^3}$

In Exercises 27 and 28, simplify each expression.

27. $(3x^2y)(2x^3y)^2$

28. $\left(\dfrac{2a^2b^3c^{-2}}{3ab^{-1}}\right)^2$

In Exercises 29 and 30, evaluate each exponential expression.

29. $25^{1/2}$

30. $-27^{2/3}$

In Exercises 31 to 34, simplify each expression.

31. $x^{2/3} \cdot x^{3/4}$

32. $\left(\dfrac{8x^{5/4}}{x^{1/2}}\right)^{2/3}$

33. $\left(\dfrac{x^2y}{x^{1/2}y^{-3}}\right)^{1/2}$

34. $(x^{1/2} - y^{1/2})(x^{1/2} + y^{1/2})$

In Exercises 35 to 44, simplify each radical expression. Assume the variables are positive real numbers.

35. $\sqrt{48a^2b^7}$

36. $\sqrt{12a^3b}$

37. $\sqrt{72x^2y}$

38. $\sqrt{18x^3y^5}$

39. $\sqrt{\dfrac{54xy^3}{10x}}$

40. $-\sqrt{\dfrac{24xyz^3}{15z^6}}$

41. $\dfrac{7x}{\sqrt[3]{2x^2}}$

42. $\dfrac{5y}{\sqrt[3]{9y}}$

43. $\sqrt[3]{-135x^2y^7}$

44. $\sqrt[3]{-250xy^6}$

In Exercises 45 and 46, write each number in scientific notation.

45. $620,000$

46. 0.0000017

In Exercises 47 and 48, change each number from scientific notation to decimal form.

47. 3.5×10^4

48. 4.31×10^{-7}

In Exercises 49 to 52, perform the indicated operation and express each result as a polynomial in standard form.

49. $(2a^2 + 3a - 7) + (-3a^2 - 5a + 6)$

50. $(5b^2 - 11) - (3b^2 - 8b - 3)$

51. $(2x^2 + 3x - 5)(3x^2 - 2x + 4)$

52. $(3y - 5)^3$

In Exercises 53 to 56, completely factor each polynomial over the integers.

53. $3x^2 + 30x + 75$

54. $25x^2 - 30xy + 9y^2$

55. $20a^2 - 4b^2$

56. $16a^3 + 250$

In Exercises 57 and 58, simplify each rational expression.

57. $\dfrac{6x^2 - 19x + 10}{2x^2 + 3x - 20}$

58. $\dfrac{4x^3 - 25x}{8x^4 + 125x}$

In Exercises 59 to 62, perform the indicated operation and simplify if possible.

59. $\dfrac{10x^2 + 13x - 3}{6x^2 - 13x - 5} \cdot \dfrac{6x^2 + 5x + 1}{10x^2 + 3x - 1}$

60. $\dfrac{15x^2 + 11x - 12}{25x^2 - 9} \div \dfrac{3x^2 + 13x + 12}{10x^2 + 11x + 3}$

61. $\dfrac{x}{x^2 - 9} + \dfrac{2x}{x^2 + x - 12}$

62. $\dfrac{3x}{x^2 + 7x + 12} - \dfrac{x}{2x^2 + 5x - 3}$

In Exercises 63 and 64, simplify each complex fraction.

63. $\dfrac{2 + \dfrac{1}{x - 5}}{3 - \dfrac{2}{x - 5}}$

64. $\dfrac{1}{2 + \dfrac{3}{1 + \dfrac{4}{x}}}$

In Exercises 65 and 66, write the complex number in standard form.

65. $5 + \sqrt{-64}$

66. $2 - \sqrt{-18}$

In Exercises 67 to 74, perform the indicated operation and write the answer in simplest form.

67. $(2 - 3i) + 4 + 2i$

68. $(4 + 7i) - (6 - 3i)$

69. $2i(3 - 4i)$

70. $(4 - 3i)(2 + 7i)$

71. $(3 + i)^2$

72. i^{345}

73. $\dfrac{4 - 6i}{2i}$

74. $\dfrac{2 - 5i}{3 + 4i}$

CHAPTER P TEST

1. For real numbers a, b, and c, identify the property that is illustrated by $(a + b)c = ac + bc$.

2. Given $A = \{0, 2, 4, 6, 8\}$ and $B = \{1, 3, 5, 7, 9\}$, find $A \cup B$.

3. Find the distance between the points -12 and -5 on the number line.

4. Simplify: $(-2x^0y^{-2})^2(-3x^2y^{-1})^{-2}$

5. Simplify: $\dfrac{(2a^{-1}bc^{-2})^2}{(3^{-1}b)(2^{-1}ac^{-2})^3}$

6. Write 0.00137 in scientific notation.

7. Simplify: $\dfrac{x^{1/3}y^{-3/4}}{x^{-1/2}y^{3/2}}$

8. Simplify: $3x\sqrt[3]{81xy^4} - 2y\sqrt[3]{3x^4y}$

9. Simplify: $\dfrac{x}{\sqrt[4]{2x^3}}$

10. Simplify: $\dfrac{3}{\sqrt{x} + 2}$

11. Simplify: $(x - 2y)(x^2 - 2x + y)$

12. Factor: $81x^2 - 25$

13. Factor: $7x^2 + 34x - 5$

14. Factor: $3ax - 12bx - 2a + 8b$

15. Factor: $16x^4 - 2xy^3$

16. Simplify: $\dfrac{x^2 - 2x - 15}{25 - x^2}$

17. Simplify: $\dfrac{x}{x^2 + x - 6} - \dfrac{2}{x^2 - 5x + 6}$

18. Simplify: $\dfrac{2x^2 + 3x - 2}{x^2 - 3x} \div \dfrac{2x^2 - 7x + 3}{x^3 - 3x^2}$

19. Simplify: $\dfrac{3}{a+b} \cdot \dfrac{a^2 - b^2}{2a - b} - \dfrac{5}{a}$

20. Simplify: $x - \dfrac{x}{x + \dfrac{1}{2}}$

21. Write $7 + \sqrt{-20}$ in standard form.

In Exercises 22 to 25, write the complex number in simplest form.

22. $(4 - 3i) - (2 - 5i)$

23. $(2 + 5i)(1 - 4i)$

24. $\dfrac{3 + 4i}{5 - i}$

25. i^{97}

EQUATIONS AND INEQUALITIES

Quarterback Ratings and the Reading Level of Written Material

This chapter concerns equations and inequalities. Equations that express a relationship between two or more variables are called *formulas*. Formulas are used in a wide variety of applications. For instance, the National Football League (NFL) uses a formula to rate quarterbacks based on their performance in the passing aspect of the game. According to this formula, the top five all-time leading passers in the NFL are as follows:

Player	Rating
1. Steve Young	96.8
2. Joe Montana	92.3
3. Brett Favre	86.8
4. Dan Marino	86.4
5. Peyton Manning	85.1

Source: Time Almanac 2003, p. 936.

The NFL quarterback rating formula is given in **Example 2 on page 86.** In **Exercises 11 and 12, page 92,** you will use the quarterback rating formula to rate quarterbacks.

Formulas are also used to estimate the reading level of written material. In **Exercises 13 and 14, page 92,** you will use the SMOG (Simplified Measure of Gobbledygook) readability formula to estimate the reading level of two novels.

Verifying Results

One important aspect of problem solving involves the process of checking to see if your results satisfy the conditions of the original problem. This process will be especially important in this chapter when you solve an equation or an inequality.

Here is an example that illustrates the importance of checking your results. The problem seems easy, but many students fail to get the correct answer on their first attempt.

Two volumes of the series *Mathematics: Its Content, Methods, and Meaning* are on a shelf, with no space between the volumes. Each volume is 1 inch thick without its covers. Each cover is $\frac{1}{8}$ inch thick. A bookworm bores horizontally from the first page of Volume I to the last page of Volume II. How far does the bookworm travel?

Once you have obtained your solution, try to check it by closely examining two books placed as shown above. Check to make sure you have the proper starting and ending positions. The correct answer is $\frac{1}{4}$ inch.

SECTION 1.1 LINEAR AND ABSOLUTE VALUE EQUATIONS

● LINEAR EQUATIONS

An **equation** is a statement about the equality of two expressions. If either of the expressions contains a variable, the equation may be a true statement for some values of the variable and a false statement for other values. For example, the equation $2x + 1 = 7$ is a true statement for $x = 3$, but it is false for any number except 3. The number 3 is said to **satisfy** the equation $2x + 1 = 7$ because substituting 3 for x produces $2(3) + 1 = 7$, which is a true statement.

To **solve** an equation means to find all values of the variable that satisfy the equation. The values that satisfy an equation are called **solutions** or **roots** of the equation. For instance, 2 is a solution of $x + 3 = 5$.

Equivalent equations are equations that have exactly the same solution(s). The process of solving an equation involving the variable x is often accomplished by producing a sequence of equivalent equations until we produce an equation or equations of the form

$$x = \text{a constant}$$

To produce these equivalent equations that lead us to the solution(s), we often perform one or more of the following procedures.

Procedures That Produce Equivalent Equations

1. Simplification of an expression on either side of the equation by such procedures as (i) combining like terms and (ii) applying the properties explained in Chapter P, such as the commutative, associative, and distributive properties.

$2x + 3 + 5x = -11$ and $7x + 3 = -11$ are equivalent equations.

2. Addition or subtraction of the same quantity on both sides of an equation.

$3x - 7 = 2$ and $3x = 9$ are equivalent equations.

3. Multiplication or division by the same nonzero quantity on both sides of an equation.

$\dfrac{5}{6}x = 10$ and $x = 12$ are equivalent equations.

Many applications can be modeled by *linear equations*.

Definition of a Linear Equation

A **linear equation** in the single variable x is an equation that can be written in the form

$$ax + b = 0$$

where a and b are real numbers, with $a \neq 0$.

Linear equations generally are solved by applying the procedures that produce equivalent equations.

EXAMPLE 1   **Solve a Linear Equation**

Solve: $\dfrac{3}{4}x - 6 = 0$

Solution

$$\dfrac{3}{4}x - 6 = 0$$

• Use the procedures on page 83 to rewrite the equation in the form x = a constant.

$$\dfrac{3}{4}x - 6 + 6 = 0 + 6$$

• Add **6** to each side.

$$\dfrac{3}{4}x = 6$$

• Simplify.

$$\left(\dfrac{4}{3}\right)\left(\dfrac{3}{4}x\right) = \left(\dfrac{4}{3}\right)(6)$$

• Multiply each side by $\dfrac{4}{3}$.

$$x = 8$$

Because 8 satisfies the original equation (see the *Take Note*), 8 is the solution.

▶ **TRY EXERCISE 2, PAGE 82**

---

take note

Check the proposed solution by substituting 8 for x in the original equation.

$$\dfrac{3}{4}x - 6 = 0$$

$$\dfrac{3}{4}(8) - 6 \stackrel{?}{=} 0$$

$$0 = 0 \qquad \text{True}$$

If an equation involves fractions, it is helpful to multiply each side of the equation by the LCD (least common denominator) of all the denominators to produce an equivalent equation that does not contain fractions.

EXAMPLE 2   **Solve by Clearing Fractions**

Solve: $\dfrac{2}{3}x + 10 - \dfrac{x}{5} = \dfrac{36}{5}$

Solution

$$\dfrac{2}{3}x + 10 - \dfrac{x}{5} = \dfrac{36}{5}$$

$$15\left(\dfrac{2}{3}x + 10 - \dfrac{x}{5}\right) = 15\left(\dfrac{36}{5}\right)$$

• Multiply each side of the equation by **15**, the LCD of the denominators.

$$10x + 150 - 3x = 108$$

• Simplify.

$$7x + 150 = 108$$

$$7x + 150 - 150 = 108 - 150$$

• Subtract **150** from each side.

$$7x = -42$$

$$\dfrac{7x}{7} = \dfrac{-42}{7}$$

• Divide each side by **7**.

$$x = -6$$

• Check as before.

▶ **TRY EXERCISE 12, PAGE 82**

EXAMPLE 3 Solve an Equation by Applying Properties

Solve: $(x + 2)(5x + 1) = 5x(x + 1)$

Solution

$$(x + 2)(5x + 1) = 5x(x + 1)$$
$$5x^2 + 11x + 2 = 5x^2 + 5x \qquad \text{• Simplify each product.}$$
$$11x + 2 = 5x \qquad \text{• Subtract } 5x^2 \text{ from each side.}$$
$$6x + 2 = 0 \qquad \text{• Subtract } 5x \text{ from each side.}$$
$$6x = -2 \qquad \text{• Subtract 2 from each side.}$$
$$x = -\frac{1}{3} \qquad \text{• Divide each side of the equation by 6.}$$

▶ **TRY EXERCISE 18, PAGE 82**

● CONTRADICTIONS, CONDITIONAL EQUATIONS, AND IDENTITIES

An equation that has no solutions is called a **contradiction.** The equation $x = x + 1$ is a contradiction. No number is equal to itself increased by 1.

An equation that is true for some values of the variable but not true for other values of the variable is called a **conditional equation.** For example, $x + 2 = 8$ is a conditional equation because it is true for $x = 6$ and false for any number not equal to 6.

An **identity** is an equation that is true for all values of the variable for which all terms of the equation are defined. Examples of identities include the equations $x + x = 2x$ and $4(x + 3) - 1 = 4x + 11$.

EXAMPLE 4 Classify Equations

Classify each equation as a contradiction, a conditional equation, or an identity.

a. $x + 1 = x + 4$ **b.** $4x + 3 = x - 9$

c. $5(3x - 2) - 7(x - 4) = 8x + 18$

Solution

a. Subtract x from both sides of $x + 1 = x + 4$ to produce the equivalent equation $1 = 4$. Because $1 = 4$ is a false statement, the original equation $x + 1 = x + 4$ has no solutions. It is a contradiction.

b. Solve using the procedures that produce equivalent equations.

$$4x + 3 = x - 9$$
$$3x + 3 = -9 \qquad \text{• Subtract } x \text{ from each side.}$$
$$3x = -12 \qquad \text{• Subtract 3 from each side.}$$
$$x = -4 \qquad \text{• Divide each side by 3.}$$

Continued ▶

Check to confirm that -4 is a solution. The equation $4x + 3 = x - 9$ is true for $x = -4$, but it is not true for any other values of x. Thus $4x + 3 = x - 9$ is a conditional equation.

c. Simplify the left side of the equation to show that it is *identical* to the right side.

$$5(3x - 2) - 7(x - 4) = 8x + 18$$
$$15x - 10 - 7x + 28 = 8x + 18$$
$$8x + 18 = 8x + 18$$

The original equation $5(3x - 2) - 7(x - 4) = 8x + 18$ is true for all real numbers x. The equation is an identity.

▶ **TRY EXERCISE 24, PAGE 82**

❓ **QUESTION**   Dividing each side of $x = 4x$ by x produces $1 = 4$. Are the equations $x = 4x$ and $1 = 4$ equivalent equations?

● ABSOLUTE VALUE EQUATIONS

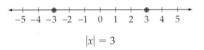

$|x| = 3$

FIGURE 1.1

The absolute value of a real number x is the distance between the number x and 0 on the real number line. Thus the solutions of $|x| = 3$ are all real numbers that are 3 units from 0. Therefore, the solutions of $|x| = 3$ are $x = 3$ or $x = -3$. See **Figure 1.1.**

The following property is used to solve absolute value equations.

A Property of Absolute Value Equations

For any variable expression E and any nonnegative real number k,

$$|E| = k \qquad \text{if and only if} \qquad E = k \quad \text{or} \quad E = -k$$

EXAMPLE 5   **Solve an Absolute Value Equation**

Solve: $|2x - 5| = 21$

Solution

$|2x - 5| = 21$ implies $2x - 5 = 21$ or $2x - 5 = -21$. Solving each of these linear equations produces

$$\begin{array}{ccc} 2x - 5 = 21 & \text{or} & 2x - 5 = -21 \\ 2x = 26 & & 2x = -16 \\ x = 13 & & x = -8 \end{array}$$

The solutions of $|2x - 5| = 21$ are -8 and 13.

▶ **TRY EXERCISE 38, PAGE 82**

take note

Some absolute value equations have no solutions. For example, $|x + 2| = -5$ is false for all values of x. Because an absolute value is always nonnegative, the equation is never true.

❓ **ANSWER**   No. The real number 0 is a solution of $x = 4x$, but 0 is not a solution of $1 = 4$.

● APPLICATIONS

Linear equations often can be used to model real-world data.

EXAMPLE 6 **Movie Theater Ticket Prices**

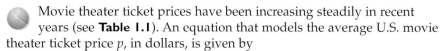

Movie theater ticket prices have been increasing steadily in recent years (see **Table 1.1**). An equation that models the average U.S. movie theater ticket price p, in dollars, is given by

$$p = 0.281t + 4.756$$

where t is the number of years after 1998. (This means that $t = 0$ corresponds to the year 1998.) Use this equation to predict in what year the average U.S. movie theater ticket price will reach $7.50.

Solution

$$p = 0.281t + 4.756$$
$$7.50 = 0.281t + 4.756 \qquad \bullet \textbf{ Substitute 7.50 for } p.$$
$$2.744 = 0.281t \qquad \bullet \textbf{ Solve for } t.$$
$$t \approx 9.8$$

Our equation predicts that the average U.S. movie theater ticket price will reach $7.50 about 9.8 years after 1998, which is the year 2007.

▶ **TRY EXERCISE 50, PAGE 83**

TABLE 1.1   Average U.S. Movie Theater Ticket Price

Year	Price (in dollars)
1998	4.69
1999	5.06
2000	5.39
2001	5.65
2002	5.80

Source: National Association of Theatre Owners, http://www.na-toonline.org/statisticstickets.htm.

EXAMPLE 7 **Driving Time**

Alicia is driving along a highway that passes through Centerville (see **Figure 1.2**). Her distance d, in miles, from Centerville is given by the equation

$$d = |135 - 60t|$$

where t is the time in hours since the start of her trip and $0 \le t \le 5$. Determine when Alicia will be exactly 15 miles from Centerville.

Solution

Substitute 15 for d.

$$d = |135 - 60t|$$
$$15 = |135 - 60t|$$

$$15 = 135 - 60t \qquad \text{or} \qquad -15 = 135 - 60t \qquad \bullet \textbf{ Solve for } t.$$
$$-120 = -60t \qquad\qquad\qquad -150 = -60t$$
$$2 = t \qquad\qquad\qquad\qquad \frac{5}{2} = t$$

Alicia will be exactly 15 miles from Centerville after she has driven for 2 hours and also after she has driven for $2\frac{1}{2}$ hours.

▶ **TRY EXERCISE 52, PAGE 83**

Highway

Starting point ● ————————— ● Centerville

FIGURE 1.2

TOPICS FOR DISCUSSION

1. A student multiplies each side of the equation $\frac{1}{2}x + 3 = 4$ by 2 to produce the equation $x + 3 = 8$. Has the student produced an equivalent equation? Explain.

2. If $P = Q$, is it also true that $Q = P$?

3. Consider the equation $|x + y| = |x| + |y|$. Is this equation true for all values of x and y, true for some values of x and y, or never true?

EXERCISE SET 1.1

In Exercises 1 to 28, solve each equation and check your solution.

1. $2x + 10 = 40$

▶ **2.** $-3y + 20 = 2$

3. $5x + 2 = 2x - 10$

4. $4x - 11 = 7x + 20$

5. $2(x - 3) - 5 = 4(x - 5)$

6. $5(x - 4) - 7 = -2(x - 3)$

7. $4(2r - 17) + 5(3r - 8) = 0$

8. $6(5s - 11) - 12(2s + 5) = 0$

9. $\frac{3}{4}x + \frac{1}{2} = \frac{2}{3}$

10. $\frac{x}{4} - 5 = \frac{1}{2}$

11. $\frac{2}{3}x - 5 = \frac{1}{2}x - 3$

▶ **12.** $\frac{1}{2}x + 7 - \frac{1}{4}x = \frac{19}{2}$

13. $0.2x + 0.4 = 3.6$

14. $0.04x - 0.2 = 0.07$

15. $x + 0.08(60) = 0.20(60 + x)$

16. $6(t + 1.5) = 12t$

17. $3(x + 5)(x - 1) = (3x + 4)(x - 2)$

▶ **18.** $5(x + 4)(x - 4) = (x - 3)(5x + 4)$

19. $5[x - (4x - 5)] = 3 - 2x$

20. $6[3y - 2(y - 1)] - 2 + 7y = 0$

21. $\frac{40 - 3x}{5} = \frac{6x + 7}{8}$

22. $\frac{12 + x}{-4} = \frac{5x - 7}{3} + 2$

In Exercises 23 to 32, classify each equation as a contradiction, a conditional equation, or an identity.

23. $-3(x - 5) = -3x + 15$

▶ **24.** $2x + \frac{1}{3} = \frac{6x + 1}{3}$

25. $2x + 7 = 3(x - 1)$

26. $4[2x - 5(x - 3)] = 6$

27. $\frac{4x + 8}{4} = x + 8$

28. $3[x - (4x - 1)] = -3(2x - 5)$

29. $3[x - 2(x - 5)] - 1 = -3x + 29$

30. $4[3(x - 5) + 7] = 12x - 32$

31. $2x - 8 = -x + 9$

32. $|3(x - 4) + 7| = |3x - 5|$

In Exercises 33 to 48, solve each absolute value equation for x.

33. $|x| = 4$

34. $|x| = 7$

35. $|x - 5| = 2$

36. $|x - 8| = 3$

37. $|2x - 5| = 11$

▶ **38.** $|2x - 3| = 21$

39. $|2x + 6| = 10$

40. $|2x + 14| = 60$

41. $\left|\frac{x - 4}{2}\right| = 8$

42. $\left|\frac{x + 3}{4}\right| = 6$

43. $|2x + 5| = -8$

44. $|4x - 1| = -17$

45. $2|x + 3| + 4 = 34$

46. $3|x - 5| - 16 = 2$

47. $|2x - a| = b$ $(b > 0)$

48. $3|x - d| = c$ $(c > 0)$

49. **RECREATION** The revenues of all the amusement and theme parks in the United States have been increasing since 1990. An equation that approximates the total revenues of all parks is given by

Revenues (in billions of dollars) $= 0.35x + 5.7$

where x is the number of years after 1990. Use this equation to predict the year in which the revenues for all amusement and theme parks in the U.S. will first reach $12 billion. (*Source: Amusement Business* magazine as reported in the San Diego *Union-Tribune*, March 19, 2000)

▶ **50.** **PATENTS** Data from the U.S. Patent and Trademark Office suggest that the number of patents P, in thousands, that have been issued each year in the U.S. since 1993 can be approximated by the equation

$$P = 5.4x + 110$$

where x is the number of years after 1993. Use this equation to predict in what year the number of patents will first exceed 175,000 patents.

51. **TRAVEL** Ruben is driving along a highway that passes through Barstow. His distance d, in miles, from Barstow is given by the equation $d = |210 - 50t|$, where t is the time, in hours, since the start of his trip and $0 \le t \le 6$. When will Ruben be exactly 60 miles from Barstow?

▶ **52.** **AUTOMOBILE GAS MILEAGE** The gas mileage m, in miles per gallon, obtained during a long trip is given by

$$m = -\frac{1}{2}|s - 55| + 25$$

where s is the speed of Kate's automobile in miles per hour and $40 \le s \le 70$. At what constant speed can Kate drive to obtain a gas mileage of exactly 22 miles per hour?

53. **OFFICE CARPETING** The cost to install new carpet in an office is determined by a $550 fixed fee plus a fee of $45 per square yard of floor space to be covered. How many square yards of floor space can be carpeted at a cost of $3800? Round to the nearest square yard.

54. **WHOLESALE PRICE** A retailer determines the retail price of a coat by first computing 175% of the wholesale price of the coat and then adding an additional markup of $8.00. What is the wholesale price of a coat that has a retail price of $156.75?

55. **COMPUTER SCIENCE** The percent of a file that remains to be downloaded using a dialup Internet connection for a certain modem is given by the equation

Percent remaining $= 100 - \dfrac{42,000}{N}t$

where N is the size of the file in bytes and t is the number of seconds since the download began. In how many minutes will 25% of a 500,000-byte file remain to be downloaded? Round to the nearest minute.

56. **AVIATION** The number of miles that remain to be flown by a commercial jet traveling from Boston to Los Angeles can be approximated by the equation

Miles remaining $= 2650 - 475t$

where t is the number of hours since leaving Boston. In how many hours will the plane be 1000 miles from Los Angeles? Round to the nearest tenth of an hour.

To benefit from an aerobic exercise program, many experts recommend that you exercise three to five times a week for 20 minutes to an hour. It is also important that your heart rate be in the *training zone,* **which is defined by the following linear equations, where** *a* **is your age in years and the heart rate is in beats per minute.**[1]

Maximum exercise heart rate $= 0.85(220 - a)$

Minimum exercise heart rate $= 0.65(220 - a)$

57. **MAXIMUM EXERCISE HEART RATE** Find the maximum exercise heart rate and the minimum exercise heart rate for a person who is 25 years of age. (Round to the nearest beat per minute.)

58. **MAXIMUM EXERCISE HEART RATE** How old is a person who has a maximum exercise heart rate of 153 beats per minute?

[1] "The Heart of the Matter," *American Health*, September 1995.

CONNECTING CONCEPTS

59. Let a, b, and c be real constants. Show that an equation of the form $ax + b = c$ has $x = \dfrac{c - b}{a}$ ($a \neq 0$) as its solution.

60. Let a, b, c, and d be real constants. Show that an equation of the form $ax + b = cx + d$ has $x = \dfrac{d - b}{a - c}$ ($a - c \neq 0$) as its solution.

In Exercises 61 to 66, solve each equation.

61. $|x + 4| = x + 4$

62. $|x - 1| = x - 1$

63. $|x + 7| = -(x + 7)$

64. $|x - 3| = -(x - 3)$

65. $|2x + 7| = 2x + 7$

66. $|3x - 11| = -3x + 11$

PREPARE FOR SECTION 1.2

67. The sum of two numbers is 32. If one of the numbers is represented by x, then the expression $32 - x$ represents the other number. Evaluate $32 - x$ for $x = 8\dfrac{1}{2}$. [P.1]

68. Evaluate $\dfrac{1}{2} bh$ for $b = \dfrac{2}{3}$ and $h = \dfrac{4}{5}$. [P.1]

69. What property has been applied to rewrite $2l + 2w$ as $2(l + w)$? [P.1]

70. What property has been applied to rewrite $\left(\dfrac{1}{2} b\right) h$ as $\dfrac{1}{2}(bh)$? [P.1]

71. Add: $\dfrac{2}{5} x + \dfrac{1}{3} x$ [P.1]

72. Simplify: $\dfrac{1}{\dfrac{1}{a} + \dfrac{1}{b}}$ [P.4]

PROJECTS

1. **PERFECT GAMES** In baseball, a **perfect game** is a game in which one of the teams gives up no hits, no walks, and no errors. Statistics show that a batter will get on base roughly 30% of the time. Thus the probability that a pitcher will retire the batter is 70%, or 0.7 as a decimal. The probability that a pitcher will retire two batters in a row is $0.7^2 = 0.49$. The probability is 0.7^{27} that a pitcher will retire 27 batters in succession and thus pitch a perfect game.[2]

a. Explain why the linear equation

$$p = 2(0.7^{27})x$$

provides a good estimate of the number of perfect games p we can expect after x games are completed.

b. Check a major league baseball almanac to determine how many perfect games have been played in the last 40 years and how many games have been played in the last 40 years.

c. Use the linear equation in **a.** to estimate how many perfect games we should expect to have been pitched over the last 40 years of major league baseball. How does this result compare with the actual result found in **b.**?

---

[2] *A Mathematician Reads the Newspaper*, by John Allen Paulos (New York: BasicBooks, A Division of HarperCollins Publishers, Inc., 1995).

SECTION 1.2

FORMULAS AND APPLICATIONS

- FORMULAS
- APPLICATIONS

● FORMULAS

A **formula** is an equation that expresses known relationships between two or more variables. **Table 1.2** lists several formulas from geometry that are used in this text. The variable P represents perimeter, C represents circumference of a circle, A represents area, S represents surface area of an enclosed solid, and V represents volume.

TABLE 1.2   Formulas from Geometry

Rectangle	Square	Triangle	Circle	Parallelogram
$P = 2l + 2w$	$P = 4s$	$P = a + b + c$	$C = \pi d = 2\pi r$	$P = 2b + 2s$
$A = lw$ 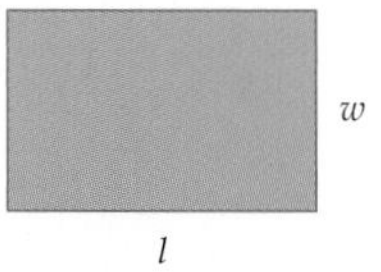	$A = s^2$ 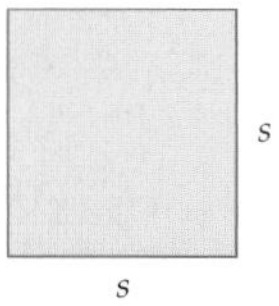	$A = \dfrac{1}{2}bh$ 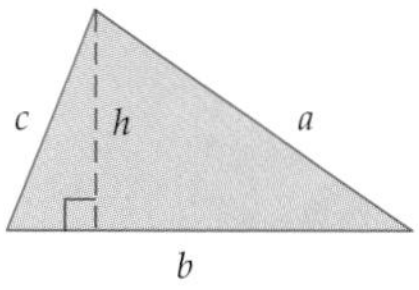	$A = \pi r^2$ 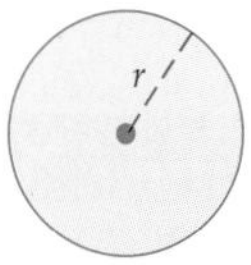	$A = bh$ 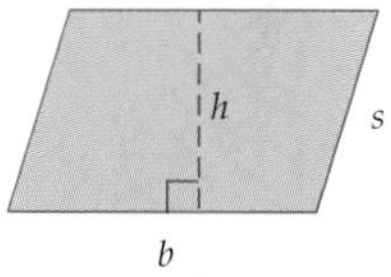

Rectangular Solid	Right Circular Cone	Sphere	Right Circular Cylinder	Frustum of a Cone
$S = 2(wh + lw + hl)$	$S = \pi r\sqrt{r^2 + h^2} + \pi r^2$	$S = 4\pi r^2$	$S = 2\pi rh + 2\pi r^2$	$S = \pi(R + r)\sqrt{h^2 + (R - r)^2} + \pi r^2 + \pi R^2$
$V = lwh$ 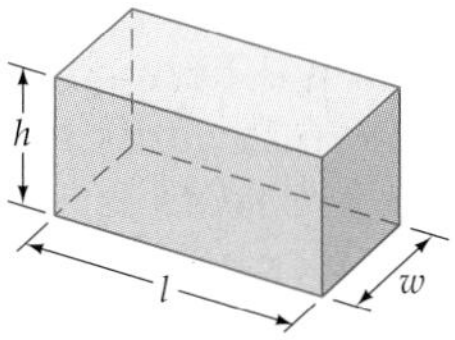	$V = \dfrac{1}{3}\pi r^2 h$ 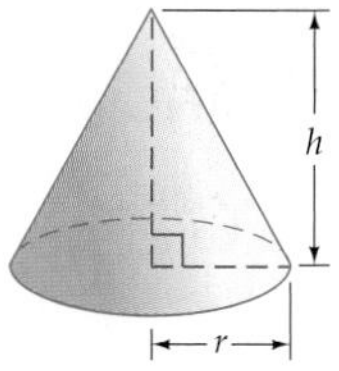	$V = \dfrac{4}{3}\pi r^3$ 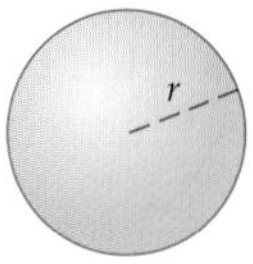	$V = \pi r^2 h$ 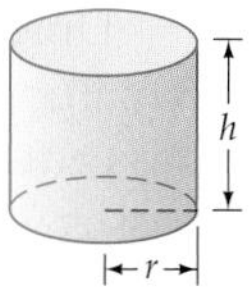	$V = \dfrac{1}{3}\pi h(r^2 + rR + R^2)$ 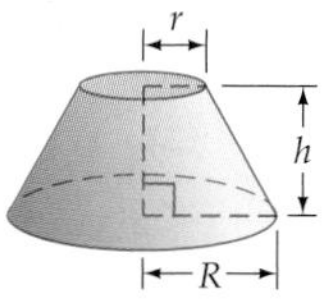

It is often necessary to solve a formula for a specified variable. Begin the process by isolating all terms that contain the specified variable on one side of the equation and all terms that do not contain the specified variable on the other side.

EXAMPLE 1   Solve a Formula for a Specified Variable

a. Solve $2l + 2w = P$ for l.   **b.** Solve $S = 2(wh + lw + hl)$ for h.

Solution

a.
$$2l + 2w = P$$
$$2l = P - 2w$$
- Subtract $2w$ from each side to isolate the $2l$ term.

$$l = \frac{P - 2w}{2}$$
- Divide each side by 2.

Continued ▶

b.
$$S = 2(wh + lw + hl)$$
$$S = 2wh + 2lw + 2hl$$

$$S - 2lw = 2wh + 2hl$$
• Isolate the terms that involve the variable *h* on the right side.

$$S - 2lw = 2h(w + l)$$
• Factor 2*h* from the right side.

$$\dfrac{S - 2lw}{2(w + l)} = h$$
• Divide each side by 2(*w* + *l*).

▶ **TRY EXERCISE 4, PAGE 92**

Formulas are often used to compare the performances of athletes. Here is an example of a formula that is used in professional football.

EXAMPLE 2   Calculate a Quarterback Rating

  The National Football League uses the following formula to rate quarterbacks:

$$\text{qb rating} = \dfrac{100}{6}[0.05(C - 30) + 0.25(Y - 3) + 0.2T + (2.375 - 0.25I)]$$

In this formula, *C* is the percentage of pass completions, *Y* is the average number of yards gained per pass attempt, *T* is the percentage of touchdown passes, and *I* is the percentage of interceptions.

During the 2002 season, Brad Johnson, the quarterback of the Tampa Bay Buccaneers, completed 62.31% of his passes. He averaged 6.76 yards per pass attempt, 4.88% of his passes were for touchdowns, and 1.33% of his passes were intercepted. Determine Brad Johnson's quarterback rating for the 2002 season.

Solution

Because *C* is defined as a percentage, *C* = 62.31. We are also given *Y* = 6.76, *T* = 4.88, and *I* = 1.33.

Substitute 62.31 for *C*, 6.76 for *Y*, 4.88 for *T*, and 1.33 for *I* in the rating formula.

qb rating

$$= \dfrac{100}{6}[0.05(62.31 - 30) + 0.25(6.76 - 3) + 0.2(4.88) + (2.375 - 0.25(1.33))]$$

$$= 92.9$$

Brad Johnson's quarterback rating for the 2002 season was 92.9.

▶ **TRY EXERCISE 14, PAGE 92**

❷ **QUESTION** If $ax + b = c$, does $x = \dfrac{c}{a} - b$?

● **APPLICATIONS**

Linear equations emerge in a variety of application problems. In solving such problems, it generally helps to apply specific techniques in a series of small steps. The following general strategies should prove to be helpful in the remaining portion of this section.

Strategies for Solving Application Problems

1. Read the problem carefully. If necessary, reread the problem several times.

2. When appropriate, draw a sketch and label parts of the drawing with the specific information given in the problem.

3. Determine the unknown quantities, and label them with variables. Write down any equation that relates the variables.

4. Use the information from step 3, along with a known formula or some additional information given in the problem, to write an equation.

5. Solve the equation obtained in step 4, and check to see whether these results satisfy all the conditions of the original problem.

EXAMPLE 3   **Dimensions of a Painting**

One of the best known paintings is *Mona Lisa* by Leonardo da Vinci. It is on display at the Musee du Louvre, in Paris. The length (or height) of this rectangular-shaped painting is 24 centimeters more than its width. The perimeter of the painting is 260 centimeters. Find the width and the length of the painting.

Solution

1. Read the problem carefully.

2. Draw a rectangle. See **Figure 1.3**.

3. Label the rectangle. We have used w for its width and l for its length. The problem states that the length is 24 centimeters more than the width. Thus l and w are related by the equation

$$l = w + 24$$

Continued ▶

FIGURE 1.3

❷ **ANSWER**   No. $x = \dfrac{c - b}{a}$, provided $a \neq 0$.

4. The perimeter of a rectangle is given by the formula $P = 2l + 2w$. To produce an equation that involves only constants and a single variable (say, w), substitute 260 for P and $w + 24$ for l.

$$P = 2l + 2w$$
$$260 = 2(w + 24) + 2w$$

5. Solve for w.

$$260 = 2w + 48 + 2w$$
$$260 = 4w + 48$$
$$212 = 4w$$
$$w = 53$$

The length is 24 centimeters more than the width. Thus $l = 53 + 24 = 77$.

A check verifies that 77 is 24 more than 53 and that twice the length (77) plus twice the width (53) gives the perimeter (260). The width of the painting is 53 centimeters and its length is 77 centimeters.

▶ **TRY EXERCISE 20, PAGE 93**

Many *uniform motion* problems can be solved by using the formula $d = rt$, where d is the distance traveled, r is the rate of speed, and t is the time.

EXAMPLE 4   **A Uniform Motion Problem**

A runner runs a course at a constant speed of 6 mph. One hour after the runner begins, a cyclist starts on the same course at a constant speed of 15 mph. How long after the runner starts does the cyclist overtake the runner?

Solution

If we represent the time the runner has spent on the course by t, then the time the cyclist takes to overtake the runner is $t - 1$. The following table organizes the information and helps us determine how to write the distance each person travels.

	rate r	·	time t	=	distance d
Runner	6	·	t	=	$6t$
Cyclist	15	·	$t - 1$	=	$15(t - 1)$

Figure 1.4 indicates that the runner and the cyclist cover the same distance. Thus

$$6t = 15(t - 1)$$
$$6t = 15t - 15$$
$$-9t = -15$$
$$t = 1\frac{2}{3}$$

The cyclist overtakes the runner $1\frac{2}{3}$ hours after the runner starts.

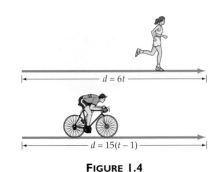

$d = 6t$

$d = 15(t - 1)$

FIGURE 1.4

▶ **TRY EXERCISE 24, PAGE 93**

Many business applications can be solved by using the equation

$$\text{Profit} = \text{revenue} - \text{cost}$$

EXAMPLE 5 A Business Application

It costs a tennis shoe manufacturer $26.55 to produce a pair of tennis shoes that sells for $49.95. How many pairs of tennis shoes must the manufacturer sell to make a profit of $14,274.00?

Solution

The *profit* is equal to the *revenue* minus the *cost*. If x equals the number of pairs of tennis shoes to be sold, then the revenue will be $49.95x$ and the cost will be $26.55x$. Therefore,

$$\text{Profit} = \text{revenue} - \text{cost}$$
$$14{,}274.00 = 49.95x - 26.55x$$
$$14{,}274.00 = 23.40x$$
$$610 = x$$

The manufacturer must sell 610 pairs of tennis shoes to make the desired profit.

▶ **TRY EXERCISE 32, PAGE 94**

Simple interest problems can be solved by using the formula $I = Prt$, where I is the interest, P is the principal, r is the simple interest rate per period, and t is the number of periods.

EXAMPLE 6 An Investment Problem

An accountant invests part of a $6000 bonus in a 5% simple interest account and the remainder of the money is invested at 8.5% simple interest. Together the investments earn $370 per year. Find the amount invested at each rate.

Solution

Let x be the amount invested at 5%. The remainder of the money is $6000 - x$, which will be the amount invested at 8.5%. Using the simple interest formula, $I = Prt$, with $t = 1$ year yields

$$\text{Interest at 5\%} = x \cdot 0.05 = 0.05x$$
$$\text{Interest at 8.5\%} = (6000 - x) \cdot (0.085) = 510 - 0.085x$$

The interest earned on the two accounts equals $370.

$$0.05x + (510 - 0.085x) = 370$$
$$-0.035x + 510 = 370$$
$$-0.035x = -140$$
$$x = 4000$$

The accountant invested $4000 at 5% and the remaining $2000 at 8.5%. Check as before.

▶ **TRY EXERCISE 36, PAGE 94**

Percent mixture problems involve combining solutions or alloys that have different concentrations of a common substance. Percent mixture problems can be solved by using the formula $pA = Q$, where p is the percent of concentration, A is the amount of the solution or alloy, and Q is the quantity of a substance in the solution or alloy. For example, in 4 liters of a 25% acid solution, p is the percent of acid (25%), A is the amount of solution (4 liters), and Q is the amount of acid in the solution, which equals $(0.25) \cdot (4)$ liters $= 1$ liter.

EXAMPLE 7　A Percent Mixture Problem

A chemist mixes an 11% hydrochloric acid solution with a 6% hydrochloric acid solution. How many milliliters (ml) of each solution should the chemist use to make a 600-milliliter solution that is 8% hydrochloric acid?

Solution

Let x be the number of milliliters of the 11% solution. Because the final solution will have a total of 600 milliliters of fluid, $600 - x$ is the number of milliliters of the 6% solution. See **Figure 1.5**.

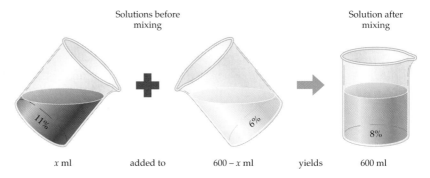

Solutions before mixing

Solution after mixing

x ml　　added to　　$600 - x$ ml　　yields　　600 ml

FIGURE 1.5

Because all the hydrochloric acid in the final solution comes from either the 11% solution or the 6% solution, the number of milliliters of hydrochloric acid in the 11% solution added to the number of milliliters of hydrochloric acid in the 6% solution must equal the number of milliliters of hydrochloric acid in the 8% solution.

$$\begin{pmatrix} \text{ml of acid in} \\ \text{11\% solution} \end{pmatrix} + \begin{pmatrix} \text{ml of acid in} \\ \text{6\% solution} \end{pmatrix} = \begin{pmatrix} \text{ml of acid in} \\ \text{8\% solution} \end{pmatrix}$$

$$0.11x \quad + \quad 0.06(600 - x) \quad = \quad 0.08(600)$$
$$0.11x + 36 - 0.06x = 48$$
$$0.05x + 36 = 48$$
$$0.05x = 12$$
$$x = 240$$

The chemist should use 240 milliliters of the 11% solution and 360 milliliters of the 6% solution to make a 600-milliliter solution that is 8% hydrochloric acid.

▶ **TRY EXERCISE 40, PAGE 94**

To solve a *work problem*, use the equation

Rate of work $\times$ time worked = part of task completed

For example, if a painter can paint a wall in 15 minutes, then the painter can paint 1/15 of the wall in 1 minute. The painter's *rate of work* is 1/15 of the wall each minute. In general, if a task can be completed in x minutes, then the rate of work is $1/x$ of the task each minute.

EXAMPLE 8   **A Work Problem**

Pump A can fill a pool in 6 hours, and pump B can fill the same pool in 3 hours. How long will it take to fill the pool if both pumps are used?

Solution

Because pump A fills the pool in 6 hours, $\dfrac{1}{6}$ represents the part of the pool filled by pump A in 1 hour. Because pump B fills the pool in 3 hours, $\dfrac{1}{3}$ represents the part of the pool filled by pump B in 1 hour.

Let t = the number of hours to fill the pool together. Then

$$t \cdot \frac{1}{6} = \frac{t}{6} \qquad \text{• Part of the pool filled by pump A}$$

$$t \cdot \frac{1}{3} = \frac{t}{3} \qquad \text{• Part of the pool filled by pump B}$$

$$\left(\begin{array}{c}\text{Part filled}\\\text{by pump A}\end{array}\right) + \left(\begin{array}{c}\text{Part filled}\\\text{by pump B}\end{array}\right) = \left(\begin{array}{c}\text{1 filled}\\\text{pool}\end{array}\right)$$

$$\frac{t}{6} \qquad + \qquad \frac{t}{3} \qquad = \qquad 1$$

Multiplying each side of the equation by 6 produces

$$t + 2t = 6$$
$$3t = 6$$
$$t = 2$$

Check: Pump A fills $\dfrac{2}{6}$, or $\dfrac{1}{3}$, of the pool in 2 hours and pump B fills $\dfrac{2}{3}$ of the pool in 2 hours, so 2 hours is the time required to fill the pool if both pumps are used.

▶ **TRY EXERCISE 50, PAGE 94**

TOPICS FOR DISCUSSION

1. A student solves the formula $A = P + Prt$ for the variable P. The student's answer is $P = A - Prt$. Is this a correct response? Explain.

2. A student takes the reciprocal of each term to write the formula

$$\frac{1}{f} = \frac{1}{d_0} + \frac{1}{d_i}$$

as $f = d_0 + d_i$. Did this technique produce a valid formula? Explain.

3. In the formula $S = \dfrac{a_1}{(1 - r)}$, what restrictions are placed on the variable r?

4. The formula $A = \dfrac{1}{2}bh$ can also be expressed as $A = \dfrac{bh}{2}$. Do you agree?

EXERCISE SET 1.2

In Exercises 1 to 10, solve the formula for the specified variable.

1. $V = \dfrac{1}{3}\pi r^2 h;$ h (geometry)

2. $P = S - Sdt;$ t (business)

3. $I = Prt;$ t (business)

▶ **4.** $A = P + Prt;$ P (business)

5. $F = \dfrac{Gm_1m_2}{d^2};$ m_1 (physics)

6. $A = \dfrac{1}{2}h(b_1 + b_2);$ b_1 (geometry)

7. $a_n = a_1 + (n - 1)d;$ d (mathematics)

8. $y - y_1 = m(x - x_1);$ x (mathematics)

9. $S = \dfrac{a_1}{1 - r};$ r (mathematics)

10. $\dfrac{P_1V_1}{T_1} = \dfrac{P_2V_2}{T_2};$ V_2 (chemistry)

11. **QUARTERBACK RATING** During the 2002 season, Peyton Manning, the quarterback of the Indianapolis Colts, completed 66.33% of his passes. He averaged 7.11 yards per pass attempt, 4.57% of his passes were for touchdowns, and 3.21% of his passes were intercepted. Determine Manning's quarterback rating for the 2002 season. Round to the nearest tenth. (*Hint:* See Example 2, page 86.)

12. **QUARTERBACK RATING** During the 2002 season, Drew Bledsoe, the quarterback of the Buffalo Bills, completed 61.48% of his passes. He averaged 7.15 yards per pass attempt, 3.93% of his passes were for touchdowns, and 2.46% of his passes were intercepted. Determine Bledsoe's quarterback rating for

the 2002 season. Round to the nearest tenth. (*Hint:* See Example 2, page 86.)

The SMOG (Simplified Measure of Gobbledygook) readability formula is often used to estimate the reading grade level required by a person if he or she is to *fully* understand the written material being assessed. The formula is given by

SMOG reading grade level = $\sqrt{w} + 3$

where w is the number of words that have three or more syllables in a sample of 30 sentences.

13. **ASSESSING A READING LEVEL** A sample of 30 sentences from *Alice's Adventures in Wonderland*, by Lewis Carroll, shows a total of 42 words that have three or more syllables. Use the SMOG reading grade level formula to estimate the reading grade level required to fully understand this novel. Round the reading grade level to the nearest tenth.

▶ **14.** **ASSESSING A READING LEVEL** A sample of 30 sentences from *A Tale of Two Cities*, by Charles Dickens, shows a total of 105 words that have three or more syllables. Use the SMOG reading grade level formula to estimate the reading grade level required to fully understand this novel. Round the reading grade level to the nearest tenth.

Another popular readability formula is the Gunning Fog Index. Here is the formula:

Gunning Fog Index = $0.4(A + W)$

where A is the average number of words per sentence and W is the percentage of words that have four or more syllables. The Gunning Fog Index is defined as the minimum grade level at which the writing is *easily* read.

15. **ASSESSING A READING LEVEL** In a sample of sentences from the novel *Bridget Jones's Diary*, by Helen Fielding, the average number of words per sentence is

23.0, and five words have four or more syllables. Use the Gunning Fog Index to estimate the reading grade level required to easily read this material.

16. **ASSESSING A READING LEVEL** In a sample of sentences from the book *Winnie-the-Pooh Meets Tigger*, by A. A. Milne, the average number of words per sentence is 11.47, and only one word has four or more syllables. Use the Gunning Fog Index to estimate the reading grade level required to easily read this material. Round the reading grade level to the nearest tenth.

In Exercises 17 to 52, solve by using the Strategies for Solving Application Problems (see page 87).

17. One-fifth of a number plus one-fourth of the number is 5 less than one-half the number. What is the number?

18. The numerator of a fraction is 4 less than the denominator. If the numerator is increased by 14 and the denominator is decreased by 10, the resulting number is 5. What is the original fraction?

19. **GEOMETRY** The length of a rectangle is 3 feet less than twice the width of the rectangle. If the perimeter of the rectangle is 174 feet, find the width and the length.

▶ 20. **GEOMETRY** The width of a rectangle is 1 meter more than half the length of the rectangle. If the perimeter of the rectangle is 110 meters, find the width and the length.

21. **GEOMETRY** A triangle has a perimeter of 84 centimeters. Each of the two longer sides of the triangle is three times as long as the shortest side. Find the length of each side of the triangle.

22. **GEOMETRY** A triangle has a perimeter of 161 miles. Each of the two smaller sides of the triangle is two-thirds the length of the longest side. Find the length of each side of the triangle.

23. **UNIFORM MOTION** Running at an average rate of 6 meters per second, a sprinter ran to the end of a track and then jogged back to the starting point at an average rate of 2 meters per second. The total time for the sprint and the jog back was 2 minutes 40 seconds. Find the length of the track.

▶ 24. **UNIFORM MOTION** A motorboat left a harbor and traveled to an island at an average rate of 15 knots. The average speed on the return trip was 10 knots. If the total trip took 7.5 hours, how far is the harbor from the island?

25. **UNIFORM MOTION** A plane leaves an airport traveling at an average speed of 240 kilometers per hour. How long will it take a second plane traveling the same route at an average speed of 600 kilometers per hour to catch up with the first plane if it leaves 3 hours later?

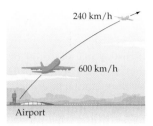

240 km/h

600 km/h

Airport

26. **UNIFORM MOTION** A plane leaves Chicago headed for Los Angeles at 540 mph. One hour later, a second plane leaves Los Angeles headed for Chicago at 660 mph. If the air route from Chicago to Los Angeles is 1800 miles, how long will it take for the first plane to pass the second plane? How far from Chicago will they be at that time?

1800 miles

Chicago

Los Angeles

27. **UNIFORM MOTION** Marlene rides her bicycle to her friend Jon's house and returns home by the same route. Marlene rides her bike at constant speeds of 6 mph on level ground, 4 mph when going uphill, and 12 mph when going downhill. If her total time riding was 1 hour, how far is it to Jon's house?

28. **UNIFORM MOTION** A car traveling at 80 kilometers per hour is passed by a second car going in the same direction at a constant speed. After 30 seconds, the two cars are 500 meters apart. Find the speed of the second car.

29. **FINDING AN AVERAGE** A student has test scores of 80, 82, 94, and 71. What score does the student need on the next test to produce an average score of 85?

What score do I need to get on the next test to have an average of 85?

Scores 80, 82, 94, 71

30. FINDING AN AVERAGE A student has test scores of 90, 74, 82, and 90. The next examination is the final examination, which will count as two tests. What score does the student need on the final examination to produce an average score of 85?

31. BUSINESS It costs a manufacturer of sunglasses $8.95 to produce sunglasses that sell for $29.99. How many sunglasses must the manufacturer sell to make a profit of $17,884?

▶ **32. BUSINESS** It costs a restaurant owner 18 cents per glass for orange juice, which is sold for 75 cents per glass. How many glasses of orange juice must the restaurant owner sell to make a profit of $2337?

33. BUSINESS The price of a computer fell 20% this year. If the computer now costs $750, how much did it cost last year?

34. BUSINESS The price of a magazine subscription rose 4% this year. If the subscription now costs $26, how much did it cost last year?

35. INVESTMENT An investment adviser invested $14,000 in two accounts. One investment earned 8% annual simple interest, and the other investment earned 6.5% annual simple interest. The amount of interest earned for 1 year was $1024. How much was invested in each account?

▶ **36. INVESTMENT** A total of $7500 is deposited into two simple interest accounts. On one account the annual simple interest rate is 5%, and on the second account the annual simple interest rate is 7%. The amount of interest earned for 1 year was $405. How much was invested in each account?

37. INVESTMENT An investment of $2500 is made at an annual simple interest rate of 5.5%. How much additional money must be invested at an annual simple interest rate of 8% so that the total interest earned is 7% of the total investment?

38. INVESTMENT An investment of $4600 is made at an annual simple interest rate of 6.8%. How much additional money must be invested at an annual simple interest rate of 9% so that the total interest earned is 8% of the total investment?

39. METALLURGY How many grams of pure silver must a silversmith mix with a 45% silver alloy to produce 200 grams of a 50% alloy?

▶ **40. CHEMISTRY** How many liters of a 40% sulfuric acid solution should be mixed with 4 liters of a 24% sulfuric acid solution to produce a 30% solution?

41. CHEMISTRY How many liters of water should be evaporated from 160 liters of a 12% saline solution so that the solution that remains is a 20% saline solution?

42. AUTOMOTIVE A radiator contains 6 liters of a 25% antifreeze solution. How much should be drained and replaced with pure antifreeze to produce a 33% antifreeze solution?

43. COMMERCE A ballet performance brought in $61,800 on the sale of 3000 tickets. If the tickets sold for $14 and $25, how many of each were sold?

44. COMMERCE A vending machine contains $41.25. The machine contains 255 coins, which consist only of nickels, dimes, and quarters. If the machine contains twice as many dimes as nickels, how many of each type of coin does the machine contain?

45. COMMERCE A coffee shop decides to blend a coffee that sells for $12 per pound with a coffee that sells for $9 per pound to produce a blend that will sell for $10 per pound. How much of each should be used to yield 20 pounds of the new blend?

46. DETERMINE NUMBER OF COINS A bag contains 42 coins, with a total weight of 246 grams. If the bag contains only gold coins that weigh 8 grams each and silver coins that weigh 5 grams each, how many gold and how many silver coins are in the bag?

47. METALLURGY How much pure gold should be melted with 15 grams of 14-karat gold to produce 18-karat gold? (*Hint:* A karat is a measure of the purity of gold in an alloy. Pure gold measures 24 karats. An alloy that measures x karats is $\dfrac{x}{24}$ gold. For example, 18-karat gold is $\dfrac{18}{24} = \dfrac{3}{4}$ gold.)

48. METALLURGY How much 14-karat gold should be melted with 4 ounces of pure gold to produce 18-karat gold? (*Hint:* See Exercise 47.)

49. INSTALL ELECTRICAL WIRES An electrician can install the electric wires in a house in 14 hours. A second electrician requires 18 hours. How long would it take both electricians, working together, to install the wires?

▶ **50. PRINT A REPORT** Printer A can print a report in 3 hours. Printer B can print the same report in 4 hours. How long would it take both printers, working together, to print the report?

51. DETERMINE INDIVIDUAL PRICES A book and a bookmark together sell for $10.10. If the price of the book is $10.00 more than the price of the bookmark, find the price of the book and the price of the bookmark.

52. SHARE AN EXPENSE Three people decide to share the cost of a yacht. By bringing in an additional partner, they can reduce the cost for each person by $4000. What is the total cost of the yacht?

CONNECTING CONCEPTS

The *Archimedean law of the lever* **states that for a lever to be in a state of balance with respect to a point called the** *fulcrum,* **the sum of the downward forces times their respective distances from the fulcrum on one side of the fulcrum must equal the sum of the downward forces times their respective distances from the fulcrum on the other side of the fulcrum. The accompanying figure shows this relationship.**

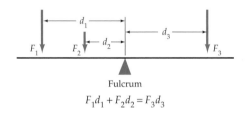

Fulcrum

$$F_1 d_1 + F_2 d_2 = F_3 d_3$$

53. LOCATE THE FULCRUM A 100-pound person 8 feet to the left of the fulcrum and a 40-pound person 5 feet to the left of the fulcrum balance with a 160-pound person on a teeter-totter. How far from the fulcrum is the 160-pound person?

54. LOCATE THE FULCRUM A lever 21 feet long has a force of 117 pounds applied to one end of the lever and a force of 156 pounds applied to the other end. Where should the fulcrum be located to produce a state of balance?

55. DETERMINE A FORCE How much force applied 5 feet from the fulcrum is needed to lift a 400-pound weight that is located on the other side, 0.5 foot from the fulcrum?

56. DETERMINE A FORCE Two workers need to lift a 1440-pound rock. They use a 6-foot steel bar with the fulcrum 1 foot from the rock, as the accompanying figure shows. One worker applies 180 pounds of force to the other end of the lever. How much force will the second worker need to apply 1 foot from that end to lift the rock?

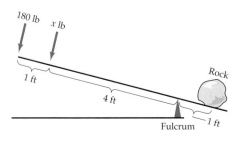

57. SPEED OF SOUND IN AIR Two seconds after firing a rifle at a target, the shooter hears the impact of the bullet. Sound travels at 1100 feet per second and the bullet at 1865 feet per second. Determine the distance to the target (to the nearest foot).

58. SPEED OF SOUND IN WATER Sound travels through sea water 4.62 times faster than through air. The sound of an exploding mine on the surface of the water and partially submerged reaches a ship through the water 4 seconds before it reaches the ship through the air. How far is the ship from the explosion (to the nearest foot)? Use 1100 feet per second as the speed of sound through the air.

59. AGE OF DIOPHANTUS The work of the ancient Greek mathematician Diophantus had great influence on later European number theorists. Nothing is known about his personal life except for the information given in the following epigram. "Diophantus passed $\frac{1}{6}$ of his life in childhood, $\frac{1}{12}$ in youth, and $\frac{1}{7}$ more as a bachelor. Five years after his marriage was born a son who died four years before his father, at $\frac{1}{2}$ his father's (final) age." How old was Diophantus when he died?

60. EQUIVALENT TEMPERATURES The relationship between the Fahrenheit temperature (F) and the Celsius temperature (C) is given by the formula

$$F = \frac{9}{5}C + 32$$

At what temperature will a Fahrenheit thermometer and a Celsius thermometer read the same?

PREPARE FOR SECTION 1.3

61. Factor: $x^2 - x - 42$ [P.3]

62. Factor: $6x^2 - x - 15$ [P.3]

63. Write $3 + \sqrt{-16}$ in $a + bi$ form. [P.5]

64. If $a = -3$, $b = -2$, and $c = 5$, evaluate

$$\frac{-b - \sqrt{b^2 - 4ac}}{2a} \quad \text{[P.1/P.2]}$$

65. If $a = 2$, $b = -3$, and $c = 1$, evaluate

$$\frac{-b + \sqrt{b^2 - 4ac}}{2a} \quad \text{[P.1/P.2]}$$

66. If $x = 3 - i$, evaluate $x^2 - 6x + 10$. [P.5]

PROJECTS

1. A Work Problem and Its Extensions If a pump can fill a pool in A hours, and a second pump can fill the same pool in B hours, then the total time T, in hours, needed to fill the pool with both pumps working together is given by

$$T = \frac{AB}{A + B}$$

a. Verify this formula.

b. Consider the case where a pool is to be filled by three pumps. One pump can fill the pool in A hours, a second in B hours, and a third in C hours. Derive a formula in terms of A, B, and C for the total time T needed to fill the pool.

c. Consider the case where a pool is to be filled by n pumps. One pump can fill the pool in A_1 hours, a second in A_2 hours, a third in A_3 hours,..., and the nth pump can fill the pool in A_n hours. Write a formula in terms of $A_1, A_2, A_3, \ldots, A_n$ for the total time T needed to fill the pool.

The following chart is called an *alignment chart* or a *nomogram*. If you know any two of the values A, B, and T, then you can use the alignment chart to determine the unknown value. For example, the straight line segment that connects 3 on the A-axis with 6 on the B-axis crosses the T-axis at 2. Thus the total time required for a pump that takes 3 hours to fill the pool and a pump that takes 6 hours to fill the pool is 2 hours when they work together.

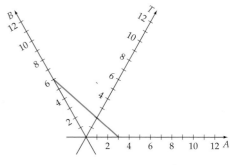

Alignment Chart for $T = \dfrac{AB}{A + B}$

d. Consider the case where a pool is to be filled by three pumps. One pump can fill the pool in $A = 6$ hours, a second in $B = 8$ hours, and a third in $C = 12$ hours. Write a few sentences explaining how you could make use of the alignment chart above to show that it takes about 2.7 hours for the three pumps to fill the pool when they work together.

2. Resistance of Parallel Circuits The alignment chart shown in Project 1 can be used to solve some problems in electronics that concern the total resistance of a *parallel* circuit. Read an electronics text, and write a paragraph or two that explain this problem and how it is related to the problem of filling a pool with two pumps.

SECTION 1.3

QUADRATIC EQUATIONS

● SOLVE QUADRATIC EQUATIONS BY FACTORING

In Section 1.1 you solved linear equations. In this section you will learn to solve a type of equation that is referred to as a *quadratic equation*.

Definition of a Quadratic Equation

A **quadratic equation** in x is an equation that can be written in the **standard quadratic form**

$$ax^2 + bx + c = 0$$

where a, b, and c are real numbers and $a \neq 0$.

MATH MATTERS

The term *quadratic* is derived from the Latin word *quadrare*, which means "to make square." Because the area of a square that measures x units on each side is x^2, we refer to equations that can be written in the form $ax^2 + bx + c = 0$ as equations that are quadratic in x.

Several methods can be used to solve a quadratic equation. For instance, if you can factor $ax^2 + bx + c$ into linear factors, then $ax^2 + bx + c = 0$ can be solved by applying the following property.

The Zero Product Principle

If A and B are algebraic expressions such that $AB = 0$, then $A = 0$ or $B = 0$.

The zero product principle states that if the product of two factors is zero, then at least one of the factors must be zero. In Example 1, the zero product principle is used to solve a quadratic equation.

EXAMPLE I Solve by Factoring

Solve each quadratic equation by factoring.

a. $x^2 + 2x - 15 = 0$ **b.** $2x^2 - 5x = 12$

Solution

a. $x^2 + 2x - 15 = 0$
$(x - 3)(x + 5) = 0$ • Factor.
$x - 3 = 0 \qquad x + 5 = 0$ • Set each factor equal to zero.
$x = 3 \qquad\quad x = -5$ • Solve each linear equation.

A check shows that 3 and -5 are both solutions of $x^2 + 2x - 15 = 0$.

Continued ▶

b. $2x^2 - 5x = 12$

$2x^2 - 5x - 12 = 0$ • **Write in standard quadratic form.**

$(x - 4)(2x + 3) = 0$ • **Factor.**

$x - 4 = 0 \qquad 2x + 3 = 0$ • **Set each factor equal to zero.**

$x = 4 \qquad\qquad 2x = -3$ • **Solve each linear equation.**

$$x = -\frac{3}{2}$$

A check shows that 4 and $-\dfrac{3}{2}$ are both solutions of $2x^2 - 5x = 12$.

▶ **TRY EXERCISE 6, PAGE 107**

Some quadratic equations have a solution that is called a *double root*. For instance, consider $x^2 - 8x + 16 = 0$. Solving this equation by factoring, we have

$x^2 - 8x + 16 = 0$

$(x - 4)(x - 4) = 0$ • **Factor.**

$x - 4 = 0 \qquad x - 4 = 0$ • **Set each factor equal to zero.**

$x = 4 \qquad\qquad x = 4$ • **Solve each linear equation.**

The only solution of $x^2 - 8x + 16 = 0$ is 4. In this situation, the single solution 4 is called a **double solution** or **double root** because it was produced by solving the two identical equations $x - 4 = 0$, both of which have 4 as a solution.

● SOLVE QUADRATIC EQUATIONS BY TAKING SQUARE ROOTS

In the following example we solve $x^2 = 25$ by factoring.

$x^2 = 25$

$x^2 - 25 = 0$

$(x - 5)(x + 5) = 0$ • **Factor.**

$x - 5 = 0 \qquad x + 5 = 0$ • **Set each factor equal to zero.**

$x = 5 \qquad\qquad x = -5$ • **Solve each linear equation.**

The solutions of $x^2 = 25$ also can be found by taking the square root of each side of the equation. In the following work a plus-or-minus sign is placed in front of the square root of 25 to produce both solutions. The notation $x = \pm 5$ means $x = 5$ or $x = -5$.

$x^2 = 25$

$x = \pm\sqrt{25}$ • **Take the square root of each side of the equation. Insert a plus-or-minus sign in front of the radical on the right.**

$x = \pm 5$

$x = 5 \quad\text{or}\quad x = -5$

We will refer to the preceding method of solving a quadratic equation as the *square root procedure*.

The Square Root Procedure

If $x^2 = c$, then $x = \sqrt{c}$ or $x = -\sqrt{c}$, which can also be written as $x = \pm\sqrt{c}$.

EXAMPLE 2   **Solve by Using the Square Root Procedure**

Use the square root procedure to solve each equation.

a. $3x^2 + 12 = 0$ b. $(x + 1)^2 = 49$

Solution

a. $3x^2 + 12 = 0$

$\qquad\qquad 3x^2 = -12$ • Solve for x^2.

$\qquad\qquad x^2 = -4$

$\qquad\qquad x = \pm\sqrt{-4}$ • Take the square root of each side of the equation and insert a plus-or-minus sign in front of the radical.

$\qquad\qquad x = \pm 2i$

b. $(x + 1)^2 = 49$

$\qquad\quad x + 1 = \pm\sqrt{49}$ • Take the square root of each side of the equation and insert a plus-or-minus sign in front of the radical.

$\qquad\qquad x = -1 \pm 7$ • Simplify.

$\qquad\qquad x = 6$ or -8

▶ **TRY EXERCISE 20, PAGE 107**

● **SOLVE QUADRATIC EQUATIONS BY COMPLETING THE SQUARE**

Consider the following binomial squares and their perfect-square trinomial products.

Square of a Binomial	=	Perfect-Square Trinomial
$(x + 5)^2$	=	$x^2 + 10x + 25$
$(x - 3)^2$	=	$x^2 - 6x + 9$

In each of the preceding perfect-square trinomials, the coefficient of x^2 is 1, and the constant term is the square of half the coefficient of the x term.

$$x^2 + 10x + 25, \qquad \left(\frac{1}{2} \cdot 10\right)^2 = 25$$

$$x^2 - 6x + 9, \qquad \left(\frac{1}{2} \cdot (-6)\right)^2 = 9$$

Adding to a binomial of the form $x^2 + bx$ the constant term that makes the binomial a perfect-square trinomial is called **completing the square**. For example, to complete the square of $x^2 + 8x$, add

$$\left(\frac{1}{2} \cdot 8\right)^2 = 16$$

to produce the perfect-square trinomial $x^2 + 8x + 16$.

Completing the square is a powerful procedure because it can be used to solve *any* quadratic equation. For instance, to solve $x^2 - 6x + 13 = 0$, begin by writing the variable terms on one side of the equation and the constant term on the other side.

$$x^2 - 6x = -13$$ • Subtract 13 from each side of the equation.

$$x^2 - 6x + 9 = -13 + 9$$ • Complete the square by adding $\left[\frac{1}{2}(-6)\right]^2 = 9$ to each side of the equation.

$$(x - 3)^2 = -4$$ • Factor and solve by the square root procedure.

$$x - 3 = \pm\sqrt{-4}$$

$$x - 3 = \pm 2i$$

$$x = 3 \pm 2i$$

The solutions of $x^2 - 6x + 13 = 0$ are $3 - 2i$ and $3 + 2i$. You can check these solutions by substituting each solution into the original equation. For instance, the following check shows that $3 - 2i$ does satisfy the original equation.

$$x^2 - 6x + 13 = 0$$

$$(3 - 2i)^2 - 6(3 - 2i) + 13 = 0$$ • Substitute $3 - 2i$ for x.

$$9 - 12i + 4i^2 - 18 + 12i + 13 = 0$$ • Simplify.

$$4 + 4(-1) = 0$$

$$0 = 0$$ • The left side equals the right side, so $3 - 2i$ checks.

EXAMPLE 3   **Solve by Completing the Square**

Solve $x^2 = 2x + 6$ by completing the square.

Solution

$$x^2 = 2x + 6$$

$$x^2 - 2x = 6$$ • Isolate the constant term.

$$x^2 - 2x + 1 = 6 + 1$$ • Complete the square.

$$(x - 1)^2 = 7$$ • Factor and simplify.

$$x - 1 = \pm\sqrt{7}$$ • Apply the square root procedure.

$$x = 1 \pm \sqrt{7}$$ • Solve for x.

The exact solutions of $x^2 = 2x + 6$ are $x = 1 - \sqrt{7}$ and $x = 1 + \sqrt{7}$. A calculator can be used to show that $1 - \sqrt{7} \approx -1.646$ and $1 + \sqrt{7} \approx 3.646$. The decimals -1.646 and 3.646 are approximate solutions of $x^2 = 2x + 6$.

▶ **TRY EXERCISE 26, PAGE 107**

Completing the square by adding the square of half the coefficient of the x term requires that the coefficient of the x^2 term be 1. If the coefficient of the x^2 term is not 1, then first multiply each term on each side of the equation by the reciprocal of the coefficient of x^2 to produce a coefficient of 1 for the x^2 term.

EXAMPLE 4   Solve by Completing the Square

Solve $2x^2 + 8x - 1 = 0$ by completing the square.

Solution

$$2x^2 + 8x - 1 = 0$$

$$2x^2 + 8x = 1 \qquad \text{• Isolate the constant term.}$$

$$\frac{1}{2}(2x^2 + 8x) = \frac{1}{2}(1) \qquad \text{• Multiply both sides of the equation by the}$$
$$\qquad\qquad\qquad\qquad \text{reciprocal of the coefficient of } x^2.$$

$$x^2 + 4x = \frac{1}{2}$$

$$x^2 + 4x + 4 = \frac{1}{2} + 4 \qquad \text{• Complete the square.}$$

$$(x + 2)^2 = \frac{9}{2} \qquad \text{• Factor and simplify.}$$

$$x + 2 = \pm\sqrt{\frac{9}{2}} \qquad \text{• Apply the square root procedure.}$$

$$x = -2 \pm 3\sqrt{\frac{1}{2}} \qquad \text{• Solve for } x.$$

$$x = -2 \pm 3\frac{\sqrt{2}}{2} \qquad \text{• Simplify.}$$

$$x = \frac{-4 \pm 3\sqrt{2}}{2}$$

The solutions of $2x^2 + 8x - 1 = 0$ are $x = \dfrac{-4 + 3\sqrt{2}}{2}$ and $x = \dfrac{-4 - 3\sqrt{2}}{2}$.

▶ **TRY EXERCISE 30, PAGE 107**

● SOLVE QUADRATIC EQUATIONS BY USING THE QUADRATIC FORMULA

Completing the square on $ax^2 + bx + c = 0$ $(a \neq 0)$ produces a formula for x in terms of the coefficients a, b, and c. The formula is known as the *quadratic formula*, and it can be used to solve *any* quadratic equation.

The Quadratic Formula

If $ax^2 + bx + c = 0$, $a \neq 0$, then

$$x = \frac{-b \pm \sqrt{b^2 - 4ac}}{2a}$$

MATH MATTERS

Ancient mathematicians thought of "completing the square" in a geometric manner. For instance, to complete the square of $x^2 + 8x$, draw a square that measures x units on each side, and add four rectangles that measure 1 unit by x units to the right side and the bottom of the square.

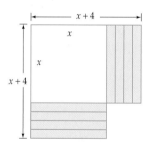

Each of the rectangles has an area of x square units, so the total area of the figure is $x^2 + 8x$. To make this figure a complete square, we must add 16 squares that measure 1 unit by 1 unit, as shown below.

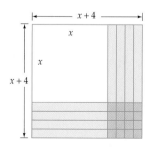

This figure is a *complete square* whose area is

$$(x + 4)^2 = x^2 + 8x + 16$$

Evariste Galois (1811–1832)

The quadratic formula provides the solutions to the general quadratic equation

$$ax^2 + bx + c = 0$$

and formulas have been developed to solve the general cubic

$$ax^3 + bx^2 + cx + d = 0$$

and the general quartic

$$ax^4 + bx^3 + cx^2 + dx + e = 0$$

However, the French mathematician Evariste Galois, shown above, was able to prove that there are no formulas that can be used to solve "by radicals" general equations of degree 5 or larger.

Shortly after completion of his remarkable proof, Galois was shot in a duel. It has been reported that as Galois lay dying, he asked his brother, Alfred, to "Take care of my work. Make it known. Important." When Alfred broke into tears, Evariste said, "Don't cry Alfred. I need all my courage to die at twenty." (*Source: Whom the Gods Love,* by Leopold Infeld, The National Council of Teachers of Mathematics, 1978, p. 299.)

Proof We assume a is a positive real number. If a were a negative real number, then we could multiply each side of the equation by -1 to make it positive.

$ax^2 + bx + c = 0 \quad (a \neq 0)$ • Given.

$ax^2 + bx = -c$ • Isolate the constant term.

$x^2 + \dfrac{b}{a}x = -\dfrac{c}{a}$ • Multiply each term on each side of the equation by $\dfrac{1}{a}$.

$x^2 + \dfrac{b}{a}x + \left(\dfrac{b}{2a}\right)^2 = \left(\dfrac{b}{2a}\right)^2 - \dfrac{c}{a}$ • Complete the square.

$\left(x + \dfrac{b}{2a}\right)^2 = \dfrac{b^2}{4a^2} - \dfrac{c}{a}$ • Factor the left side. Simplify the powers on the right side.

$\left(x + \dfrac{b}{2a}\right)^2 = \dfrac{b^2}{4a^2} - \dfrac{4a}{4a} \cdot \dfrac{c}{a}$ • Use a common denominator to simplify the right side.

$x + \dfrac{b}{2a} = \pm\sqrt{\dfrac{b^2 - 4ac}{4a^2}}$ • Apply the square root procedure.

$x + \dfrac{b}{2a} = \pm\dfrac{\sqrt{b^2 - 4ac}}{2a}$ • Because $a > 0$, $\sqrt{4a^2} = 2a$.

$x = -\dfrac{b}{2a} \pm \dfrac{\sqrt{b^2 - 4ac}}{2a}$ • Add $-\dfrac{b}{2a}$ to each side.

$x = \dfrac{-b \pm \sqrt{b^2 - 4ac}}{2a}$

◆

As a general rule, you should first try to solve quadratic equations by factoring. If the factoring process proves difficult, then solve by using the quadratic formula.

EXAMPLE 5   Solve by Using the Quadratic Formula

Use the quadratic formula to solve each of the following.

a. $4x^2 - 4x - 3 = 0$ **b.** $x^2 = 3x + 5$

Solution

a. For the equation $4x^2 - 4x - 3 = 0$, we have $a = 4$, $b = -4$, and $c = -3$. Substituting in the quadratic formula produces

$$x = \frac{-b \pm \sqrt{b^2 - 4ac}}{2a}$$

$$= \frac{-(-4) \pm \sqrt{(-4)^2 - 4(4)(-3)}}{2(4)}$$

$$= \frac{4 \pm \sqrt{64}}{8} = \frac{4 \pm 8}{8} = \frac{1 \pm 2}{2} = \frac{3}{2} \text{ or } -\frac{1}{2}$$

The solutions of $4x^2 - 4x - 3 = 0$ are $x = \dfrac{3}{2}$ and $x = -\dfrac{1}{2}$.

take note

Although the equation in Example **5a** can be solved by factoring, we have solved it by using the quadratic formula to illustrate the procedures involved in applying the quadratic formula.

b. The standard form of $x^2 = 3x + 5$ is $x^2 - 3x - 5 = 0$. Substituting $a = 1$, $b = -3$, and $c = -5$ in the quadratic formula produces

$$x = \frac{-(-3) \pm \sqrt{(-3)^2 - 4(1)(-5)}}{2(1)}$$

$$= \frac{3 \pm \sqrt{29}}{2}$$

The solutions of $x^2 = 3x + 5$ are $x = \dfrac{3 + \sqrt{29}}{2}$ and $x = \dfrac{3 - \sqrt{29}}{2}$.

▶ **TRY EXERCISE 38, PAGE 107**

❓ QUESTION Can the quadratic formula be used to solve any quadratic equation $ax^2 + bx + c = 0$ with real coefficients and $a \neq 0$?

● THE DISCRIMINANT OF A QUADRATIC EQUATION

The solutions of $ax^2 + bx + c = 0$, $a \neq 0$, are given by

$$x = \frac{-b \pm \sqrt{b^2 - 4ac}}{2a}$$

The expression under the radical, $b^2 - 4ac$, is called the **discriminant** of the equation $ax^2 + bx + c = 0$. If $b^2 - 4ac \geq 0$, then $\sqrt{b^2 - 4ac}$ is a real number. If $b^2 - 4ac < 0$, then $\sqrt{b^2 - 4ac}$ is not a real number. Thus the sign of the discriminant can be used to determine whether the solutions of a quadratic equation are real numbers.

To review **COMPLEX CONJUGATES,** *see p. 63.*

The Discriminant and the Solutions of a Quadratic Equation

The equation $ax^2 + bx + c = 0$, with real coefficients and $a \neq 0$, has as its discriminant $b^2 - 4ac$.

● If $b^2 - 4ac > 0$, then $ax^2 + bx + c = 0$ has *two distinct real solutions.*

● If $b^2 - 4ac = 0$, then $ax^2 + bx + c = 0$ has *one real solution.* The solution is a double solution.

● If $b^2 - 4ac < 0$, then $ax^2 + bx + c = 0$ has *two distinct nonreal complex solutions.* The solutions are conjugates of each other.

EXAMPLE 6   Use the Discriminant to Determine the Number of Real Solutions

For each equation, determine the discriminant and state the number of real solutions.

a. $2x^2 - 5x + 1 = 0$ **b.** $3x^2 + 6x + 7 = 0$ **c.** $x^2 + 6x + 9 = 0$

Continued ▶

❓ ANSWER   Yes. However, it is sometimes easier to find the solutions by factoring, by the square root procedure, or by completing the square.

Solution

a. The discriminant of $2x^2 - 5x + 1 = 0$ is $b^2 - 4ac = (-5)^2 - 4(2)(1) = 17$. Because the discriminant is positive, $2x^2 - 5x + 1 = 0$ has two distinct real solutions.

b. The discriminant of $3x^2 + 6x + 7 = 0$ is $b^2 - 4ac = 6^2 - 4(3)(7) = -48$. Because the discriminant is negative, $3x^2 + 6x + 7 = 0$ has no real solutions.

c. The discriminant of $x^2 + 6x + 9 = 0$ is $b^2 - 4ac = 6^2 - 4(1)(9) = 0$. Because the discriminant is 0, $x^2 + 6x + 9 = 0$ has one real solution.

▶ **TRY EXERCISE 48, PAGE 107**

• APPLICATIONS OF QUADRATIC EQUATIONS

A **right triangle** contains one 90° angle. The side opposite the 90° angle is called the **hypotenuse.** The other two sides are called **legs.** The lengths of the sides of a right triangle are related by a theorem known as the Pythagorean Theorem.

The Pythagorean Theorem

If a and b denote the lengths of the legs of a right triangle and c the length of the hypotenuse, then $c^2 = a^2 + b^2$.

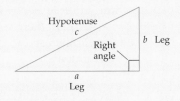

The Pythagorean Theorem states that the square of the length of the hypotenuse of a right triangle is equal to the sum of the squares of the lengths of the legs. This theorem is often used to solve applications that involve right triangles.

EXAMPLE 7   Determine the Dimensions of a Television Screen

A television screen measures 60 inches diagonally, and its *aspect ratio* is 16 to 9. This means that the ratio of the width of the screen to the height of the screen is 16 to 9. Find the width and height of the screen.

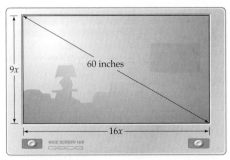

A 60-inch television screen with a 16:9 aspect ratio.

Solution

Let $16x$ represent the width of the screen and let $9x$ represent the height of the screen. Applying the Pythagorean Theorem gives us

$$(16x)^2 + (9x)^2 = 60^2$$

$$256x^2 + 81x^2 = 3600 \qquad \bullet \text{ Solve for } x.$$

$$337x^2 = 3600$$

$$x^2 = \frac{3600}{337} \qquad \bullet \text{ Apply the square root procedure.}$$

$$x = \sqrt{\frac{3600}{337}} \approx 3.268 \text{ inches} \qquad \bullet \text{ The plus-or-minus sign is not used in this application because we know } x \text{ is positive.}$$

The height of the screen is about $9(3.268) \approx 29.4$ inches and the width of the screen is about $16(3.268) \approx 52.3$ inches.

▶ **TRY EXERCISE 58, PAGE 108**

EXAMPLE 8 **Determine the Dimensions of a Candy Bar**

At the present time, a company makes rectangular solid candy bars that measure 5 inches by 2 inches by 0.5 inch. Due to difficult financial times, the company has decided to keep the price of the candy bar fixed and reduce the volume of the bar by 20%. What should be the dimensions of the new candy bar if it is decided to keep the height at 0.5 inch and to keep the length of the candy bar 3 inches longer than the width?

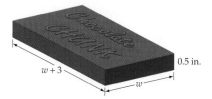

Solution

The volume of a rectangular solid is given by $V = lwh$. The original candy bar had a volume of $5 \cdot 2 \cdot 0.5 = 5$ cubic inches. The new candy bar will have a volume of $80\%(5) = 0.80(5) = 4$ cubic inches.

Let w represent the width and $w + 3$ represent the length of the new candy bar. For the new candy bar we have:

$$lwh = V$$

$$(w + 3)(w)(0.5) = 4 \qquad \bullet \text{ Substitute in the volume formula.}$$

$$(w + 3)(w) = 8 \qquad \bullet \text{ Multiply each side by 2.}$$

$$w^2 + 3w = 8$$

$$w^2 + 3w - 8 = 0 \qquad \bullet \text{ Write in } ax^2 + bx + c = 0 \text{ form.}$$

Continued ▶

INTEGRATING TECHNOLOGY

In many application problems it is helpful to use a calculator to estimate the solutions of a quadratic equation by applying the quadratic formula. For instance, the following figure shows the use of a graphing calculator to estimate the solutions of $w^2 + 3w - 8 = 0$.

```
(-3+√(3²-4*1*(-8)))/2
                1.701562119
(-3-√(3²-4*1*(-8)))/2
               -4.701562119
```

$$w = \frac{-(3) \pm \sqrt{(3)^2 - 4(1)(-8)}}{2(1)}$$ • Use the quadratic formula.

$$= \frac{-3 \pm \sqrt{41}}{2}$$

$$\approx 1.7 \quad \text{or} \quad -4.7$$

We can disregard the negative value because the width must be positive. The width of the new candy bar should be 1.7 inches, to the nearest tenth of an inch. The length should be 3 inches longer, which is 4.7 inches.

▶ **TRY EXERCISE 70, PAGE 109**

Quadratic equations are often used to determine the height (position) of an object that has been dropped or projected. For instance, the *position equation* $s = -16t^2 + v_0t + s_0$ can be used to estimate the height of a projected object near the surface of the earth at a given time t, in seconds. In this equation, v_0 is the initial velocity of the object in feet per second, and s_0 is the initial height of the object in feet.

EXAMPLE 9   **Determine the Time of Descent**

A ball is thrown downward with an initial velocity of 5 feet per second from the Golden Gate Bridge, which is 220 feet above the water. How long will it take for the ball to hit the water? Round your answer to the nearest hundreth of a second.

Solution

The distance s, in feet, of the ball above the water after t seconds is given by $s = -16t^2 - 5t + 220$. We have replaced v_0 with -5 because the ball is thrown downward. (If the ball had been thrown upward, we would use $v_0 = 5$). To determine the time it takes the ball to hit the water, substitute 0 for s in the equation $s = -16t^2 - 5t + 220$ and solve for t. In the following work, we have solved by using the quadratic formula.

$$0 = -16t^2 - 5t + 220$$

$$t = \frac{-(-5) \pm \sqrt{(-5)^2 - 4(-16)(220)}}{2(-16)}$$ • Use the quadratic formula.

$$= \frac{5 \pm \sqrt{14,105}}{-32}$$ • Use a calculator to estimate t.

$$\approx -3.87 \quad \text{or} \quad 3.56$$

Because the time must be positive, we disregard the negative value. The ball will hit the water in about 3.56 seconds.

▶ **TRY EXERCISE 72, PAGE 109**

 TOPICS FOR DISCUSSION

1. Name the four methods of solving a quadratic equation that have been discussed in this section. What are the advantages and disadvantages of each?

2. If x and y are real numbers and $xy = 0$, then $x = 0$ or $y = 0$. Do you agree with this statement? Explain.

3. If x and y are real numbers and $xy = 1$, then $x = 1$ or $y = 1$. Do you agree with this statement? Explain.

4. Explain how to complete the square on $x^2 + bx$.

5. If the discriminant of $ax^2 + bx + c = 0$ with real coefficients and $a \neq 0$ is negative, then what can be said concerning the solutions of the equation?

EXERCISE SET 1.3

In Exercises 1 to 10, solve each quadratic equation by factoring and applying the zero product principle.

1. $x^2 - 2x - 15 = 0$

2. $x^2 + 3x - 10 = 0$

3. $2x^2 - x = 1$

4. $2x^2 + 5x = 3$

5. $8x^2 + 189x - 72 = 0$

▶ **6.** $12x^2 - 41x + 24 = 0$

7. $3x^2 - 7x = 0$

8. $5x^2 = -8x$

9. $(x - 5)^2 - 9 = 0$

10. $(3x + 4)^2 - 16 = 0$

In Exercises 11 to 20, use the square root procedure to solve each quadratic equation.

11. $x^2 = 81$

12. $x^2 = 225$

13. $2x^2 = 48$

14. $3x^2 = 144$

15. $3x^2 + 12 = 0$

16. $4x^2 + 20 = 0$

17. $(x - 5)^2 = 36$

18. $(x + 4)^2 = 121$

19. $(x - 3)^2 + 16 = 0$

▶ **20.** $(x + 2)^2 + 28 = 0$

In Exercises 21 to 32, solve each quadratic equation by completing the square.

21. $x^2 + 6x + 1 = 0$

22. $x^2 + 8x - 10 = 0$

23. $x^2 - 2x - 15 = 0$

24. $x^2 + 2x - 8 = 0$

25. $x^2 + 4x + 5 = 0$

▶ **26.** $x^2 - 6x + 10 = 0$

27. $x^2 + 3x - 1 = 0$

28. $x^2 + 7x - 2 = 0$

29. $2x^2 + 4x - 1 = 0$

▶ **30.** $2x^2 + 10x - 3 = 0$

31. $3x^2 - 8x = -1$

32. $4x^2 - 4x = -15$

In Exercises 33 to 46, solve each quadratic equation by using the quadratic formula.

33. $x^2 - 2x = 15$

34. $x^2 - 5x = 24$

35. $x^2 = -x + 1$

36. $x^2 = -x - 1$

37. $2x^2 + 4x = -1$

▶ **38.** $2x^2 + 4x = 1$

39. $3x^2 - 5x + 3 = 0$

40. $3x^2 - 5x + 4 = 0$

41. $\dfrac{1}{2}x^2 + \dfrac{3}{4}x - 1 = 0$

42. $\dfrac{2}{3}x^2 - 5x + \dfrac{1}{2} = 0$

43. $24x^2 = 22x + 35$

44. $72x^2 + 13x = 15$

45. $0.5x^2 + 0.6x = 0.8$

46. $1.2x^2 + 0.4x - 0.5 = 0$

In Exercises 47 to 56, determine the discriminant of the quadratic equation, and then state the number of real solutions of the equation. Do not solve the equation.

47. $2x^2 - 5x - 7 = 0$

▶ **48.** $x^2 + 3x - 11 = 0$

49. $3x^2 - 2x + 10 = 0$

50. $x^2 + 3x + 3 = 0$

51. $x^2 - 20x + 100 = 0$

52. $4x^2 + 12x + 9 = 0$

53. $24x^2 = -10x + 21$

54. $32x^2 - 44x = -15$

55. $12x^2 + 15x = -7$

56. $8x^2 = 5x - 3$

57. GEOMETRY The length of each side of an equilateral triangle is 31 centimeters. Find the altitude of the triangle. Round to the nearest tenth of a centimeter.

▶ **58.** DIMENSIONS OF A BASEBALL DIAMOND How far, to the nearest tenth of a foot, is it from home plate to second base on a baseball diamond? (*Hint:* The bases in a baseball diamond form a square that measures 90 feet on each side.)

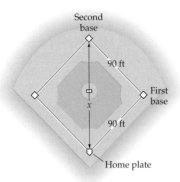

59. DIMENSIONS OF A TELEVISION SCREEN A television screen measures 54 inches diagonally, and its aspect ratio is 4 to 3. Find the width and the height of the screen.

60. PUBLISHING COSTS The cost, in dollars, of publishing x books is $C(x) = 40{,}000 + 20x + 0.0001x^2$. How many books can be published for $250{,}000$?

61. COST OF A WEDDING The average cost of a wedding, in dollars, is modeled by

$$C(t) = 38t^2 + 291t + 15{,}208$$

where $t = 0$ represents the year 1990 and $0 \le t \le 14$. Use the model to determine the year during which the average cost of a wedding first reached $19{,}000$.

62. REVENUE The demand for a certain product is given by $p = 26 - 0.01x$, where x is the number of units sold per month and p is the price, in dollars, at which each item is sold. The monthly revenue is given by $R = xp$. What number of items sold produces a monthly revenue of $16{,}500$?

63. PROFIT A company has determined that the profit, in dollars, it can expect from the manufacture and sale of x tennis racquets is given by

$$P = -0.01x^2 + 168x - 120{,}000$$

How many racquets should the company manufacture and sell to earn a profit of $518{,}000$?

64. QUADRATIC GROWTH A plant's ability to create food through the process of photosynthesis depends on the surface area of its leaves. A biologist has determined that the surface area A of a maple leaf can be closely approximated by the formula $A = 0.72(1.28)h^2$, where h is the height of the leaf in inches.

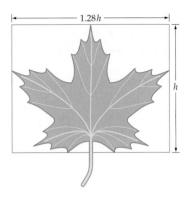

a. Find the surface area of a maple leaf with a height of 7 inches. Round to the nearest tenth of a square inch.

b. Find the height of a maple leaf with an area of 92 square inches. Round to the nearest tenth of an inch.

65. DIMENSIONS OF AN ANIMAL ENCLOSURE A veterinarian wishes to use 132 feet of chain-link fencing to enclose a rectangular region and subdivide the region into two smaller rectangular regions, as shown in the following figure. If the total enclosed area is 576 square feet, find the dimensions of the enclosed region.

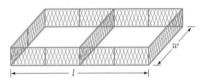

66. CONSTRUCTION OF A BOX A square piece of cardboard is formed into a box by cutting out 3-inch squares from each of the corners and folding up the sides, as shown in the following figure. If the volume of the box needs to be 126.75 cubic inches, what size square piece of cardboard is needed?

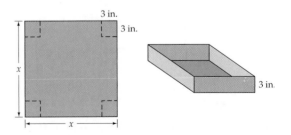

67. POPULATION DENSITY OF A CITY The population density D (in people per square mile) of a city is related to the horizontal distance x, in miles, from the center of the city by $D = -45x^2 + 190x + 200$, $0 < x < 5$. At what distances from the center of the city does the population density equal 250 people per square mile? Round each result to the nearest tenth of a mile.

68. TRAFFIC CONTROL Traffic engineers install "flow lights" at the entrances of freeways to control the number of cars entering the freeway during times of heavy traffic. For a particular freeway entrance, the number of cars N waiting to enter the freeway during the morning hours can be approximated by $N = -5t^2 + 80t - 280$, where t is the time of the day and $6 \leq t \leq 10.5$. According to this model, when will there be 35 cars waiting to enter the freeway?

69. 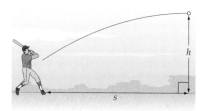 **DAREDEVIL MOTORCYCLE JUMP** In March of 2000, Doug Danger made a successful motorcycle jump over an L-1011 jumbo jet. The horizontal distance of his jump was 160 feet, and his height, in feet, during the jump was approximated by $h = -16t^2 + 25.3t + 20$, $t \geq 0$. He left the takeoff ramp at a height of 20 feet, and he landed on the landing ramp at a height of about 17 feet. How long, to the nearest tenth of a second, was he in the air?

▶ **70. DIMENSIONS OF A CANDY BAR** At the present time a company makes rectangular solid candy bars that measure 5 inches by 2 inches by 0.5 inch. Due to difficult financial times, the company has decided to keep the price of the candy bar fixed and reduce the volume of the bar by 20%. What should be the dimensions, to the nearest tenth of an inch, of the new candy bar if it is decided to keep the height at 0.5 inch and to make the length of the new candy bar 2.5 times longer than its width?

71. HEIGHT OF A ROCKET A model rocket is launched upward with an initial velocity of 220 feet per second. The height, in feet, of the rocket t seconds after the launch is given by $h = -16t^2 + 220t$. How many seconds after the launch will the rocket be 350 feet above the ground? Round to the nearest tenth of a second.

▶ **72. BASEBALL** The height h, in feet, of a baseball above the ground t seconds after it is hit is given by $h = -16t^2 + 52t + 4.5$. Use this equation to determine the number of seconds, to the nearest tenth of a second, from the time the ball is hit until the ball hits the ground.

73. BASEBALL Two equations can be used to track the position of a baseball t seconds after it is hit. For instance, suppose $h = -16t^2 + 50t + 4.5$ gives the height, in feet, of a baseball t seconds after it is hit, and $s = 103.9t$ gives the horizontal distance, in feet, the ball is from home plate t seconds after it is hit. See the figure at the top of the next column. Use these equations to determine whether this particular baseball will clear a 10-foot fence positioned 360 feet from home plate.

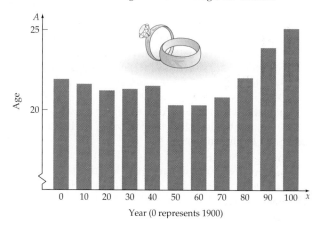

74. BASKETBALL Michael Jordan was known for his "hang time," which is the amount of time a player is in the air when making a jump toward the basket. An equation that approximates the height s, in inches, of one of Jordan's jumps is given by $s = -16t^2 + 26.6t$, where t is time in seconds. Use this equation to determine Michael Jordan's hang time, to the nearest tenth of a second, for this jump.

75. NUMBER OF HANDSHAKES If everyone in a group of n people shakes hands with everyone other than themselves, then the total number of handshakes h is given by

$$h = \frac{1}{2}n(n - 1)$$

The total number of handshakes that are exchanged by a group of people is 36. How many people are in the group?

76. 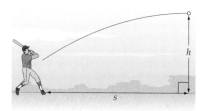 **MEDIAN AGE AT FIRST MARRIAGE** During the first 60 years of the 20th century, couples tended to marry at younger and younger ages. During the last 40 years, that trend was reversed. The median age A, in years, at first marriage for women can be modeled by

$$A = 0.0013x^2 - 0.1048x + 22.5256$$

where $x = 0$ represents the year 1900 and $x = 100$ represents the year 2000. Use the model to predict in what year in the future the median age at first marriage for women will first reach 26 years. (*Source:* U.S. Census Bureau, www.Census.gov.)

Median Age at First Marriage, for Women

77. PERCENT OF DIVORCED CITIZENS The percent P of U.S. citizens who are divorced can be closely approximated by $P = -0.0016t^2 + 0.225t + 6.201$, where t is time in years, with $t = 0$ representing 1980. Use this model to predict in what year the percent of U.S. citizens who are divorced will first reach 11.0%. (*Source:* U.S. Census Bureau, www.Census.gov.)

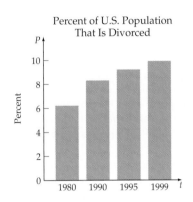

Percent of U.S. Population That Is Divorced

78. AUTOMOTIVE ENGINEERING The number of feet N that a car needs to stop on a certain road surface is given by $N = -0.015v^2 + 3v$, $0 \le v \le 90$, where v is the speed of the car in miles per hour when the driver applies the brakes. What is the maximum speed, to the nearest mile per hour, that a motorist can be traveling and stop the car within 100 feet?

79. ORBITAL DEBRIS The amount of space debris orbiting Earth has been increasing at an alarming rate. In 1995, there were about 14 million pounds of debris orbiting Earth, and by the year 2000, the amount of debris had increased to over 25 million pounds. (*Source:* http://orbitaldebris.jsc.nasa.gov/.)

The equation $A = 0.05t^2 + 2.25t + 14$ closely models the amount of debris orbiting Earth, where A is the amount of debris in millions of pounds and t is the time in years, with $t = 0$ representing the year 1995. Use the equation to

a. estimate the amount of orbital debris we can expect in the year 2006

b. estimate in what year the amount of orbital debris will first reach 50 million pounds

---

CONNECTING CONCEPTS

80. a. Show that the equation $x^2 + bx - 4 = 0$ always has two distinct real number solutions, regardless of the value of b.

b. For what values of k does $x^2 - 6x + k = 0$ have two distinct real number solutions?

81. GOLDEN RECTANGLES A rectangle is called a *golden rectangle* provided its length l and its width w satisfy the equation

$$\frac{l}{w} = \frac{l + w}{l}$$

a. Solve this formula for l. (*Hint:* Multiply both sides of the equation by wl, and then use the quadratic formula to solve for l in terms of w. Because l must be positive, state only the positive solution.)

b. If the width of a golden rectangle measures 101 feet, what is the length of the rectangle? Round to the nearest tenth of a foot.

c. Measure the width and the length of a credit card. Would you say that the credit card closely approximates a golden rectangle?

The following theorem is known as the *sum and product of the roots theorem.*

Let $ax^2 + bx + c = 0$, $a \ne 0$, be a quadratic equation. Then r_1 and r_2 are roots of the equation if and only if

$$r_1 + r_2 = -\frac{b}{a} \quad \text{and} \quad r_1 r_2 = \frac{c}{a}$$

In Exercises 82 to 86, use the sum and product of the roots theorem to determine whether the given numbers are roots of the quadratic equation.

82. $x^2 + 4x - 21 = 0$; $-7, 3$

83. $2x^2 - 7x - 30 = 0$; $-\dfrac{5}{2}, 6$

84. $9x^2 - 12x - 1 = 0$; $\dfrac{2 + \sqrt{5}}{3}, \dfrac{2 - \sqrt{5}}{3}$

85. $x^2 - 2x + 2 = 0$; $1 + i, 1 - i$

86. $x^2 - 4x + 12 = 0$; $2 + 3i, 2 - 3i$

PREPARE FOR SECTION 1.4

87. Factor: $x^3 - 16x$ [P.3]

88. Factor: $x^4 - 36x^2$ [P.3]

89. Evaluate: $8^{2/3}$ [P.2]

90. Evaluate: $16^{3/2}$ [P.2]

91. Find $\left(1 + \sqrt{x - 5}\right)^2, x > 5$ [P.2/P.3]

92. Find $\left(2 - \sqrt{x + 3}\right)^2, x > -3$ [P.2/P.3]

PROJECTS

1. THE SUM AND PRODUCT OF THE ROOTS THEOREM Use the quadratic formula to prove the sum and product of the roots theorem stated just before Exercise 82.

2. VISUAL INSIGHT

President James A. Garfield is credited with the following proof of the Pythagorean Theorem. Write the supporting reasons for each of the steps in this proof.

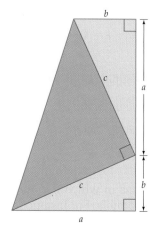

$$\text{Area} = \frac{1}{2}c^2 + 2\left(\frac{1}{2}ab\right) = \frac{1}{2}(\text{height})(\text{sum of bases})$$

$$\frac{1}{2}c^2 + ab = \frac{1}{2}(a + b)(a + b)$$

$$\frac{1}{2}c^2 = \frac{1}{2}a^2 + \frac{1}{2}b^2$$

$$c^2 = a^2 + b^2$$

SECTION 1.4   # OTHER TYPES OF EQUATIONS

- POLYNOMIAL EQUATIONS
- RATIONAL EQUATIONS
- RADICAL EQUATIONS
- EQUATIONS THAT ARE QUADRATIC IN FORM

• POLYNOMIAL EQUATIONS

Some polynomial equations that are neither linear nor quadratic can be solved by the various techniques presented in this section. For instance, the **third-degree equation**, or **cubic equation**, in Example 1 can be solved by factoring the polynomial and using the zero product property.

EXAMPLE 1   Solve a Polynomial Equation by Factoring

Solve: $x^3 - 16x = 0$

Solution

$$x^3 - 16x = 0$$
$$x(x^2 - 16) = 0 \qquad \text{• Factor out the GCF, } x.$$
$$x(x + 4)(x - 4) = 0 \qquad \text{• Factor the difference of squares.}$$

Set each factor equal to zero.

$$x = 0 \quad \text{or} \quad x + 4 = 0 \quad \text{or} \quad x - 4 = 0$$
$$x = 0 \quad \text{or} \quad x = -4 \quad \text{or} \quad x = 4$$

A check will show that $-4, 0,$ and 4 are roots of the original equation.

▶ **TRY EXERCISE 6, PAGE 119**

take note

If you attempt to solve Example 1 by dividing each side by x, you will produce the equation $x^2 - 16 = 0$, which has roots of only -4 and 4. In this case the division of each side of the equation by the variable x does not produce an equivalent equation. To avoid this common mistake, factor out any variable factors that are common to each term instead of dividing each side of the equation by the factor.

• RATIONAL EQUATIONS

A **rational equation** is an equation that involves fractions in which the numerators and/or the denominators of the fractions are polynomials. For instance,

$$\frac{x}{x - 3} = \frac{9}{x - 3} - 5$$

is a rational equation.

Many rational equations can be solved by multiplying each side of the equation by a variable expression to produce a polynomial equation. When we multiply each side of an equation by a variable expression, we restrict the variable so that the expression is not equal to zero. Example 2b illustrates the fact that you may produce incorrect results if you fail to restrict the variable.

EXAMPLE 2   Solve Rational Equations

Solve each equation.

a. $\dfrac{x}{x - 3} = \dfrac{9}{x - 3} - 5$ b. $1 + \dfrac{x}{x - 5} = \dfrac{5}{x - 5}$

Solution

a. First, note that the denominator $x - 3$ would equal zero if x were 3. To produce a simpler equivalent equation, multiply each side by $x - 3$, with the restriction that $x \neq 3$.

$$(x - 3)\left(\frac{x}{x - 3}\right) = (x - 3)\left(\frac{9}{x - 3} - 5\right) \qquad \bullet\, x \neq 3.$$

$$x = (x - 3)\left(\frac{9}{x - 3}\right) - (x - 3)5$$

$$x = 9 - 5x + 15$$

$$6x = 24$$

$$x = 4$$

Substituting 4 for x in the original equation establishes that 4 is the solution.

b. To produce a simpler equivalent equation, multiply each side of the equation by $x - 5$, with the restriction that $x \neq 5$.

$$(x - 5)\left(1 + \frac{x}{x - 5}\right) = (x - 5)\left(\frac{5}{x - 5}\right) \qquad \bullet\, x \neq 5.$$

$$(x - 5)1 + (x - 5)\left(\frac{x}{x - 5}\right) = 5$$

$$x - 5 + x = 5$$

$$2x = 10$$

$$x = 5$$

Although we have obtained 5 as a proposed solution, 5 is *not* a solution of the original equation because it contradicts our restriction $x \neq 5$. Substitution of 5 for x in the original equation results in denominators of 0. In this case the original equation has no solution.

▶ **TRY EXERCISE 14, PAGE 119**

take note

When we multiply both sides of an equation by $x - a$, we assume that $x \neq a$.

MATH MATTERS

Determine the incorrect step in the following "proof" that $2 = 1$.

$a = b$	Given.
$a^2 = ab$	Multiply by a.
$a^2 - b^2 = ab - b^2$	Subtract b^2.
$(a + b)(a - b) = b(a - b)$	Factor.
$\dfrac{(a + b)(a - b)}{(a - b)} = \dfrac{b(a - b)}{(a - b)}$	Divide by $a - b$.
$a + b = b$	Simplify.
$b + b = b$	Substitute b for a.
$2b = b$	Simplify.
$2 = 1$	Divide by b.

EXAMPLE 3   A Medical Application

Young's rule is often used by physicians to determine what portion of the recommended adult dosage of a medication should be administered to a child. In equation form, Young's rule is given by

$$\text{Portion of an adult dosage} = \frac{x}{x + 12}$$

where x represents the age, in years, of the child. Using Young's rule, a physician has determined that Elizabeth should receive only $\frac{1}{7}$ the recommended adult dosage of a medication. Determine Elizabeth's age.

Continued ▶

Solution

Elizabeth is to receive $\dfrac{1}{7}$ of an adult dosage of a particular medication. Thus we need to solve the following rational equation for x.

$$\frac{1}{7} = \frac{x}{x + 12}$$

$$7(x + 12)\frac{1}{7} = 7(x + 12)\left(\frac{x}{x + 12}\right) \qquad \text{• Multiply each side by } 7(x + 12).$$

$$x + 12 = 7x \qquad \text{• Simplify.}$$

$$12 = 6x \qquad \text{• Solve for } x.$$

$$x = 2$$

Elizabeth is 2 years old.

▶ **TRY EXERCISE 60, PAGE 120**

● RADICAL EQUATIONS

Some equations that involve radical expressions can be solved by using the following result.

The Power Principle

If P and Q are algebraic expressions and n is a positive integer, then every solution of $P = Q$ is a solution of $P^n = Q^n$.

EXAMPLE 4   **Solve a Radical Equation**

Use the power principle to solve $\sqrt{x + 4} = 3$.

Solution

$$\sqrt{x + 4} = 3$$

$$\left(\sqrt{x + 4}\right)^2 = 3^2 \qquad \text{• Square each side of the equation. (Apply the power principle with } n = 2.\text{)}$$

$$x + 4 = 9$$

$$x = 5$$

$$\text{Check: } \sqrt{x + 4} = 3$$

$$\sqrt{5 + 4} \overset{?}{=} 3 \qquad \text{• Substitute 5 for } x.$$

$$\sqrt{9} \overset{?}{=} 3$$

$$3 = 3 \qquad \text{• 5 checks.}$$

The only solution is 5.

▶ **TRY EXERCISE 28, PAGE 120**

? QUESTION Does squaring both sides of $x + b = c$ produce the equation $x^2 + b^2 = c^2$?

EXAMPLE 5   **An Application Concerning Reading Levels**

The SMOG (Simplified Measure of Gobbledygook) readability formula estimates the reading grade level required by a person if he or she is to fully understand the written material being assessed. The SMOG formula is

$$\text{SMOG reading grade level} = \sqrt{w} + 3$$

where w is the number of words that have three or more syllables in a sample of 30 sentences.

An author of books for adolescents has decided to write a book that can be fully understood by adolescents at the ninth grade reading level. According to the SMOG reading level formula, what is the maximum number of words with three or more syllables that should appear in a sample of 30 sentences of this book?

Solution

$9 = \sqrt{w} + 3$	• Substitute 9 for the **SMOG** reading grade level.
$6 = \sqrt{w}$	• Subtract 3 from each side of the equation.
$36 = w$	• Square both sides to solve for w.

To produce a ninth grade reading level, the author should strive to use a maximum of 36 words with three or more syllables in any sample of 30 sentences.

▶ **TRY EXERCISE 62, PAGE 120**

Some care must be taken when using the power principle because the equation $P^n = Q^n$ may have more solutions than the original equation $P = Q$. As an example, consider $x = 3$. The only solution is the real number 3. Square each side of the equation to produce $x^2 = 9$, which has both 3 and -3 as solutions. The -3 is called an *extraneous solution* because it is not a solution of the original equation $x = 3$.

Extraneous Solutions

Any solution of $P^n = Q^n$ that is not a solution of $P = Q$ is called an **extraneous solution.** Extraneous solutions *may* be introduced whenever we raise each side of an equation to an *even* power.

? ANSWER No. Squaring both sides of the equation produces $x^2 + 2bx + b^2 = c^2$.

EXAMPLE 6   Solve a Radical Equation

Solve $x = 2 + \sqrt{2 - x}$. Check all proposed solutions.

Solution

$$x = 2 + \sqrt{2 - x}$$

$$x - 2 = \sqrt{2 - x} \qquad \bullet \text{ Isolate the radical.}$$

$$(x - 2)^2 = \left(\sqrt{2 - x}\right)^2 \qquad \bullet \text{ Square each side of the equation.}$$

$$x^2 - 4x + 4 = 2 - x$$

$$x^2 - 3x + 2 = 0 \qquad \bullet \text{ Collect and combine like terms.}$$

$$(x - 2)(x - 1) = 0 \qquad \bullet \text{ Factor.}$$

$$x - 2 = 0 \quad \text{or} \quad x - 1 = 0$$

$$x = 2 \quad \text{or} \qquad x = 1 \qquad \bullet \text{ Proposed solutions}$$

Check for $x = 2$: $x = 2 + \sqrt{2 - x}$

$$2 \overset{?}{=} 2 + \sqrt{2 - (2)} \qquad \bullet \text{ Substitute 2 for } x.$$

$$2 \overset{?}{=} 2 + \sqrt{0}$$

$$2 = 2 \qquad \bullet \text{ 2 is a solution.}$$

Check for $x = 1$: $x = 2 + \sqrt{2 - x}$

$$1 \overset{?}{=} 2 + \sqrt{2 - (1)} \qquad \bullet \text{ Substitute 1 for } x.$$

$$1 \overset{?}{=} 2 + \sqrt{1}$$

$$1 \neq 3 \qquad \bullet \text{ 1 is not a solution.}$$

The preceding check shows that 1 is not a solution. It is an extraneous solution that was created by squaring each side of the equation. The only solution is 2.

▶ **TRY EXERCISE 30, PAGE 120**

In Example 7 it will be necessary to square $\left(1 + \sqrt{2x - 5}\right)$. Recall the special product formula $(x + y)^2 = x^2 + 2xy + y^2$. Using this special product formula to square $\left(1 + \sqrt{2x - 5}\right)$ produces

$$\left(1 + \sqrt{2x - 5}\right)^2 = 1 + 2\sqrt{2x - 5} + (2x - 5)$$

EXAMPLE 7   Solve a Radical Equation

Solve $\sqrt{x + 1} - \sqrt{2x - 5} = 1$. Check all proposed solutions.

Solution

First write an equivalent equation in which one radical is isolated on one side of the equation.

$$\sqrt{x + 1} - \sqrt{2x - 5} = 1$$

$$\sqrt{x + 1} = 1 + \sqrt{2x - 5}$$

The next step is to square each side. Using the result from the discussion preceding this example, we have

$$\left(\sqrt{x+1}\right)^2 = \left(1 + \sqrt{2x-5}\right)^2$$
$$x + 1 = 1 + 2\sqrt{2x-5} + (2x-5)$$
$$-x + 5 = 2\sqrt{2x-5}$$

• **Isolate the remaining radical.**

The right side still contains a radical, so we square each side again.

$$(-x+5)^2 = \left(2\sqrt{2x+5}\right)^2$$
$$x^2 - 10x + 25 = 4(2x-5)$$
$$x^2 - 10x + 25 = 8x - 20$$
$$x^2 - 18x + 45 = 0$$
$$(x-3)(x-15) = 0$$
$$x = 3 \quad \text{or} \quad x = 15$$

• **Proposed solutions**

3 checks as a solution, but 15 does not. Therefore, 3 is the only solution.

▶ **TRY EXERCISE 34, PAGE 120**

● EQUATIONS THAT ARE QUADRATIC IN FORM

The equation $4x^4 - 25x^2 + 36 = 0$ is said to be **quadratic in form,** which means it can be written in the form

$$au^2 + bu + c = 0 \quad a \neq 0$$

where u is an algebraic expression involving x. For example, if we make the substitution $u = x^2$ (which implies $u^2 = x^4$), then our original equation can be written as

$$4u^2 - 25u + 36 = 0$$

This quadratic equation can be solved for u, and then, using the relationship $u = x^2$, we can find the solutions of the original equation.

EXAMPLE 8   Solve an Equation That Is Quadratic in Form

Solve: $4x^4 - 25x^2 + 36 = 0$

Solution

Make the substitutions $u = x^2$ and $u^2 = x^4$ to produce the quadratic equation $4u^2 - 25u + 36 = 0$. Factor the quadratic polynomial on the left side of the equation.

$$(4u-9)(u-4) = 0$$
$$4u - 9 = 0 \quad \text{or} \quad u - 4 = 0$$
$$u = \frac{9}{4} \quad \text{or} \quad u = 4$$

Continued ▶

Substitute x^2 for u to produce

$$x^2 = \frac{9}{4} \qquad \text{or} \qquad x^2 = 4$$

$$x = \pm\sqrt{\frac{9}{4}} \qquad \text{or} \qquad x = \pm\sqrt{4}$$

$$x = \pm\frac{3}{2} \qquad \text{or} \qquad x = \pm 2 \qquad \text{• Check as before.}$$

The solutions are -2, $-\dfrac{3}{2}$, $\dfrac{3}{2}$, and 2.

▶ **TRY EXERCISE 42, PAGE 120**

Following is a table of equations that are quadratic in form. Each equation is accompanied by an appropriate substitution that will enable it to be written in the form $au^2 + bu + c = 0$.

Equations That Are Quadratic in Form

Original Equation	Substitution	$au^2 + bu + c = 0$ Form
$x^4 - 8x^2 + 15 = 0$	$u = x^2$	$u^2 - 8u + 15 = 0$
$x^6 + x^3 - 12 = 0$	$u = x^3$	$u^2 + u - 12 = 0$
$x^{1/2} - 9x^{1/4} + 20 = 0$	$u = x^{1/4}$	$u^2 - 9u + 20 = 0$
$2x^{2/3} + 7x^{1/3} - 4 = 0$	$u = x^{1/3}$	$2u^2 + 7u - 4 = 0$
$15x^{-2} + 7x^{-1} - 2 = 0$	$u = x^{-1}$	$15u^2 + 7u - 2 = 0$

EXAMPLE 9   **Solve an Equation That Is Quadratic in Form**

Solve: $3x^{2/3} - 5x^{1/3} - 2 = 0$

Solution

Substituting u for $x^{1/3}$ gives us

$$3u^2 - 5u - 2 = 0$$

$$(3u + 1)(u - 2) = 0 \qquad \text{• Factor.}$$

$$3u + 1 = 0 \qquad \text{or} \qquad u - 2 = 0$$

$$u = -\frac{1}{3} \qquad \text{or} \qquad u = 2$$

$$x^{1/3} = -\frac{1}{3} \qquad \text{or} \qquad x^{1/3} = 2 \qquad \text{• Replace } u \text{ with } x^{1/3}.$$

$$x = -\frac{1}{27} \qquad \text{or} \qquad x = 8 \qquad \text{• Cube each side.}$$

A check will verify that both $-\dfrac{1}{27}$ and 8 are solutions.

▶ **TRY EXERCISE 50, PAGE 120**

It is possible to solve equations that are quadratic in form without making a formal substitution. For example, to solve $x^4 + 5x^2 - 36 = 0$, factor the equation and apply the zero product property.

$$x^4 + 5x^2 - 36 = 0$$
$$(x^2 + 9)(x^2 - 4) = 0$$
$$x^2 + 9 = 0 \quad \text{or} \quad x^2 - 4 = 0$$
$$x^2 = -9 \quad \text{or} \quad x^2 = 4$$
$$x = \pm 3i \quad \text{or} \quad x = \pm 2$$

TOPICS FOR DISCUSSION

1. If P and Q are algebraic expressions and n is a positive integer, then the equation $P^n = Q^n$ is equivalent to the equation $P = Q$. Do you agree? Explain.

2. Consider the equation $(x^2 - 1)(x - 2) = 3(x - 2)$. Dividing each side of the equation by $x - 2$ yields $x^2 - 1 = 3$. Is this second equation equivalent to the first equation?

3. A tutor claims that cubing each side of $(4x - 1)^{1/3} = -2$ will not introduce any extraneous solutions. Do you agree?

4. What would be an appropriate substitution that would enable you to write $x^{-2} - \dfrac{2}{x} = 15$ as a quadratic equation?

5. A classmate solves the equation $x^2 + y^2 = 25$ for y and produces the equation $y = \sqrt{25 - x^2}$. Do you agree with this result?

EXERCISE SET 1.4

In Exercises 1 to 12, solve each polynomial equation by factoring and using the zero product principle.

1. $x^3 - 25x = 0$

2. $x^3 - x = 0$

3. $x^3 - 2x^2 - x + 2 = 0$

4. $x^3 - 4x^2 - 2x + 8 = 0$

5. $2x^5 - 18x^3 = 0$

▶ **6.** $x^4 - 36x^2 = 0$

7. $x^4 - 3x^3 - 40x^2 = 0$

8. $x^4 + 3x^3 - 8x - 24 = 0$

9. $x^4 - 16x^2 = 0$

10. $x^4 - 16 = 0$

11. $x^3 - 8 = 0$

12. $x^3 + 8 = 0$

In Exercises 13 to 26, solve each rational equation and check your solution(s).

13. $\dfrac{3}{x + 2} = \dfrac{5}{2x - 7}$

▶ **14.** $\dfrac{4}{y + 2} = \dfrac{7}{y - 4}$

15. $\dfrac{30}{10 + x} = \dfrac{20}{10 - x}$

16. $\dfrac{6}{8 + x} = \dfrac{4}{8 - x}$

17. $\dfrac{3x}{x + 4} = 2 - \dfrac{12}{x + 4}$

18. $\dfrac{8}{2m + 1} - \dfrac{1}{m - 2} = \dfrac{5}{2m + 1}$

19. $2 + \dfrac{9}{r - 3} = \dfrac{3r}{r - 3}$

20. $\dfrac{t}{t - 4} + 3 = \dfrac{4}{t - 4}$

21. $\dfrac{5}{x - 3} - \dfrac{3}{x - 2} = \dfrac{4}{x - 3}$

22. $\dfrac{4}{x - 1} + \dfrac{7}{x + 7} = \dfrac{5}{x - 1}$

23. $\dfrac{x}{x - 3} = \dfrac{x + 4}{x + 2}$

24. $\dfrac{x}{x - 5} = \dfrac{x + 7}{x + 1}$

25. $\dfrac{x + 3}{x + 5} = \dfrac{x - 3}{x - 4}$

26. $\dfrac{x - 6}{x + 4} = \dfrac{x - 1}{x + 2}$

In Exercises 27 to 40, use the power principle to solve each radical equation. Check all proposed solutions.

27. $\sqrt{x - 4} - 6 = 0$

▶ **28.** $\sqrt{10 - x} = 4$

29. $x = 3 + \sqrt{3 - x}$

▶ **30.** $x = \sqrt{5 - x} + 5$

31. $\sqrt{3x - 5} - \sqrt{x + 2} = 1$

32. $\sqrt{6 - x} + \sqrt{5x + 6} = 6$

33. $\sqrt{2x + 11} - \sqrt{2x - 5} = 2$

▶ **34.** $\sqrt{x + 7} - 2 = \sqrt{x - 9}$

35. $\sqrt{x + 7} + \sqrt{x - 5} = 6$

36. $x = \sqrt{12x - 35}$

37. $2x = \sqrt{4x + 15}$

38. $\sqrt[3]{7x - 3} = \sqrt[3]{2x + 7}$

39. $\sqrt[3]{2x^2 + 5x - 3} = \sqrt[3]{x^2 + 3}$

40. $\sqrt[4]{x^2 + 20} = \sqrt[4]{9x}$

In Exercises 41 to 56, find all the real solutions of each equation by first rewriting each equation as a quadratic equation.

41. $x^4 - 9x^2 + 14 = 0$

▶ **42.** $x^4 - 10x^2 + 9 = 0$

43. $2x^4 - 11x^2 + 12 = 0$

44. $6x^4 - 7x^2 + 2 = 0$

45. $x^6 + x^3 - 6 = 0$

46. $6x^6 + x^3 - 15 = 0$

47. $x^{1/2} - 3x^{1/4} + 2 = 0$

48. $2x^{1/2} - 5x^{1/4} - 3 = 0$

49. $3x^{2/3} - 11x^{1/3} - 4 = 0$

▶ **50.** $6x^{2/3} - 7x^{1/3} - 20 = 0$

51. $9x^4 = 30x^2 - 25$

52. $4x^4 - 28x^2 = -49$

53. $x^{2/5} - 1 = 0$

54. $2x^{2/5} - x^{1/5} = 6$

55. $9x - 52\sqrt{x} + 64 = 0$

56. $8x - 38\sqrt{x} + 9 = 0$

57. FENCE CONSTRUCTION A worker can build a fence in 8 hours. With the help of an assistant, the fence can be built in 5 hours. How long should it take the assistant, working alone, to build the fence?

58. ROOF REPAIR A roofer and an assistant can repair a roof together in 6 hours. Working alone the assistant can complete the repair in 14 hours. If both the roofer and the assistant work together for 2 hours and then the assistant is left alone to finish the job, how much longer should the assistant need to finish the repairs?

59. AVERAGE GOLF SCORE Renee has played four rounds of golf this season. Her average score is 92. If she can score 86 on each round she plays in the future, how many more rounds will she need to play to bring her average down to 88? (*Hint:* A player's average golf score is equal to the total number of strokes divided by the total number of rounds played.)

▶ **60. MEDICAL DOSAGE FOR A CHILD** A physician has used Young's rule (see Example 3) to determine that Sandy should receive $\dfrac{1}{2}$ of an adult dose of a medication. How old is Sandy?

61. WRITING FOR A PARTICULAR READING LEVEL A writer of books for adolescents has decided to write a book that can be fully understood by adolescents at the sixth grade level. According to the SMOG reading grade level formula, what is the maximum number of words with three or more syllables that should appear in any sample of 30 sentences of the book? (*Hint:* See Example 5.)

▶ **62. WRITING FOR A PARTICULAR READING LEVEL** A writer of books for children has decided to write a book that can be fully understood by children at the fourth grade level. According to the SMOG reading grade level formula, what is the maximum number of words with three or more syllables that should appear in any sample of 30 sentences of the book? (*Hint:* See Example 5.)

63. RADIUS OF A CONE A conical funnel has a height h of 4 inches and a lateral surface area L of 15π square inches. Find the radius r of the cone. (*Hint:* Use the formula $L = \pi r \sqrt{r^2 + h^2}$.)

64. DIAMETER OF A CONE As flour is poured onto a table, it forms a right circular cone whose height is one-third the diameter of the base. What is the diameter of the base when the cone has a volume of 192 cubic inches? Round to the nearest tenth of an inch.

65. PRECIOUS METALS A solid silver sphere has a diameter of 8 millimeters, and a second silver sphere has a diameter of 12 millimeters. The spheres are melted down and recast to form a single cube. What is the length s of each edge of the cube? Round your answer to the nearest tenth of a millimeter.

66. PENDULUM The period T of a pendulum is the time it takes the pendulum to complete one swing from left to right and back. For a pendulum near the surface of the earth,

$$T = 2\pi\sqrt{\frac{L}{32}}$$

where T is measured in seconds and L is the length of the pendulum in feet. Find the length of a pendulum that has a period of 4 seconds. Round to the nearest tenth of a foot.

67. DISTANCE TO THE HORIZON On a ship, the distance d that you can see to the horizon is given by $d = 1.5\sqrt{h}$, where h is the height of your eye measured in feet above sea level and d is measured in miles. How high is the eye level of a navigator who can see 14 miles to the horizon? Round to the nearest foot.

CONNECTING CONCEPTS

68. RADIUS OF A CIRCLE The radius r of a circle inscribed in a triangle with sides of lengths a, b, and c is given by

$$r = \sqrt{\frac{(s-a)(s-b)(s-c)}{s}}$$

where $s = \frac{1}{2}(a + b + c)$.

a. Find the length of the radius of a circle inscribed in a triangle with sides of 5 inches, 6 inches, and 7 inches. Round to the nearest hundredth of an inch.

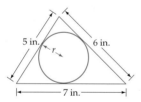

b. The radius of a circle inscribed in an equilateral triangle measures 2 inches. What is the exact length of each side of the equilateral triangle?

69. RADIUS OF A CIRCLE The radius r of a circle that is circumscribed about a triangle with sides of lengths a, b, and c is given by

$$r = \frac{abc}{4\sqrt{s(s-a)(s-b)(s-c)}}$$

where $s = \frac{1}{2}(a + b + c)$.

a. Find the radius of a circle that is circumscribed about a triangle with sides of 7 inches, 10 inches, and 15 inches. Round to the nearest hundredth of an inch.

b. A circle with radius 5 inches is circumscribed about an equilateral triangle (see the following figure). What is the exact length of each side of the equilateral triangle?

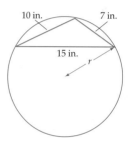

FIGURE FOR EXERCISE 69(b)

In Exercises 70 and 71, the depth s from the opening of a well to the water can be determined by measuring the total time between the instant you drop a stone and the moment you hear it hit the water. The time (in seconds) it takes the stone to hit the water is given by $\sqrt{s}/4$, where s is measured in feet. The time (also in seconds) required for the sound of the impact to travel up to your ears is given by $s/1100$. Thus the total time T (in seconds) between the instant you drop the stone and the moment you hear its impact is

$$T = \frac{\sqrt{s}}{4} + \frac{s}{1100}$$

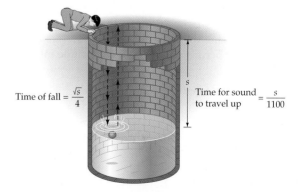

70. TIME OF FALL One of the world's deepest water wells is 7320 feet deep. Find the time between the instant you drop a stone and the time you hear it hit the water if the surface of the water is 7100 feet below the opening of the well. Round your answer to the nearest tenth of a second.

71. Solve $T = \dfrac{\sqrt{s}}{4} + \dfrac{s}{1100}$ for s.

72. DEPTH OF A WELL Use the result of Exercise 71 to determine the depth from the opening of a well to the water level if the time between the instant you drop a stone and the moment you hear its impact is 3 seconds. Round your answer to the nearest foot.

PREPARE FOR SECTION 1.5

73. Find: $\{x \mid x > 2\} \cap \{x \mid x > 5\}$ [P.1]

74. Evaluate $3x^2 - 2x + 5$ for $x = -3$. [P.1]

75. Evaluate $\dfrac{x + 3}{x - 2}$ for $x = 7$. [P.1/P.4]

76. Factor: $10x^2 + 9x - 9$ [P.3]

77. For what value of x is $\dfrac{x - 3}{2x - 7}$ undefined? [P.1/P.4]

78. Solve: $2x^2 - 11x + 15 = 0$ [1.3]

PROJECTS

1. **THE REDUCED CUBIC** The mathematician Francois Vieta knew a method of solving the "reduced cubic" $x^3 + mx + n = 0$ by using the substitution

$$x = \frac{m}{3z} - z.$$

a. Show that this substitution results in the equation
$$z^6 - nz^3 - \frac{m^3}{27} = 0.$$

b. Show that the equation in **a.** is quadratic in form.

c. Solve the equation in **a.** for z.

d. Use your solution from **c.** to find the real solution of the equation $x^3 + 3x = 14$.

2. **FERMAT'S LAST THEOREM** One of the most famous theorems is known as *Fermat's Last Theorem*. Write an essay on Fermat's Last Theorem. Include information about

- the history of Fermat's Last Theorem.

- the relationship between Fermat's Last Theorem and the Pythagorean Theorem.

- Dr. Andrew Wiles's proof of Fermat's Last Theorem.

The following list includes a few of the sources you may wish to consult.

- *Fermat's Enigma*, by Simon Singh. Walker and Company, New York, 1997.

- *The Last Problem*, by Eric Temple Bell. The Mathematical Association of America, 1990.

- "Andrew Wiles: A Math Whiz Battles 350-Year-Old Puzzle," by Gina Kolata, *Math Horizons*, Winter 1993, pp. 8–11. The Mathematical Association of America.

- "Introduction to Fermat's Last Theorem," by David A. Cox, *The American Mathematical Monthly*, vol. 101, no. 1 (January 1994), pp. 3–14.

INEQUALITIES

● PROPERTIES OF INEQUALITIES

In Section P.1 we used inequalities to describe the order of real numbers and to represent subsets of real numbers. In this section we consider inequalities that involve a variable. In particular, we consider how to determine which real numbers make an inequality a true statement.

The **solution set** of an inequality is the set of all real numbers for which the inequality is a true statement. For instance, the solution set of $x + 1 > 4$ is the set of all real numbers greater than 3. Two inequalities are **equivalent inequalities** if they have the same solution set. We can solve many inequalities by producing *simpler* but equivalent inequalities until the solutions are readily apparent. To produce these simpler but equivalent inequalities, we often apply the following properties.

MATH MATTERS

Another property of inequalities, called the *transitive property*, states that for real numbers a, b, and c, if $a > b$ and $b > c$, then $a > c$. We say that the relationship "is greater than" is a transitive relationship.

Not all relationships are transitive relationships. For instance, consider the game of scissors, paper, rock. In this game, scissors wins over paper, paper wins over rock, but scissors does not win over rock!

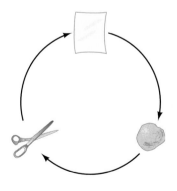

Properties of Inequalities

Let a, b, and c be real numbers.

1. *Addition-Subtraction Property* If the same real number is added to or subtracted from each side of an inequality, the resulting inequality is equivalent to the original inequality.

   $a < b$ and $a + c < b + c$ are equivalent inequalities.

2. *Multiplication-Division Property*

   a. Multiplying or dividing each side of an inequality by the same *positive* real number produces an equivalent inequality.

   If $c > 0$, then $a < b$ and $ac < bc$ are equivalent inequalities.

   b. Multiplying or dividing each side of an inequality by the same *negative* real number produces an equivalent inequality provided the direction of the inequality symbol is *reversed*.

   If $c < 0$, then $a < b$ and $ac > bc$ are equivalent inequalities.

Note the difference between Property 2a and Property 2b. Property 2a states that an equivalent inequality is produced when each side of a given inequality is multiplied (divided) by the same *positive* real number and the inequality symbol is not changed. By contrast, Property 2b states that when each side of a given inequality is multiplied (divided) by a *negative* real number, we must *reverse* the direction of the inequality symbol to produce an equivalent inequality. For instance, multiplying both sides of $-b < 4$ by -1 produces the equivalent inequality $b > -4$. (We multiplied both sides of the first inequality by -1, and we changed the less than symbol to a greater than symbol.)

EXAMPLE 1 **Solve Linear Inequalities**

Solve each of the following inequalities.

a. $2x + 1 < 7$ **b.** $-3x - 2 \le 10$

Solution

a. $2x + 1 < 7$

$\quad\quad 2x < 6$ • Add -1 to each side and keep the inequality symbol as is.

$\quad\quad\quad x < 3$ • Divide each side by 2 and keep the inequality symbol as is.

The inequality $2x + 1 < 7$ is true for all real numbers less than 3. In set-builder notation the solution set is given by $\{x \mid x < 3\}$. In interval notation the solution set is $(-\infty, 3)$. See the following figure.

b. $-3x - 2 \le 10$

$\quad\quad -3x \le 12$ • Add 2 to each side and keep the inequality symbol as is.

$\quad\quad\quad x \ge -4$ • Divide each side by -3 and reverse the direction of the inequality symbol.

The inequality $-3x - 2 \le 10$ is true for all real numbers greater than or equal to -4. In set-builder notation the solution set is given by $\{x \mid x \ge -4\}$. In interval notation the solution set is $[-4, \infty)$. See the following figure.

▶ **Try Exercise 6, page 134**

To review **INTERVAL NOTATION,** *see p. 8.*

> **take note**
>
> Solutions of inequalities are often stated using set-builder notation or interval notation. For instance, the solutions of $2x + 1 < 7$ can be written in set-builder notation as $\{x \mid x < 3\}$ or in interval notation as $(-\infty, 3)$.

● COMPOUND INEQUALITIES

A **compound inequality** is formed by joining two inequalities with the connective word *and* or *or*. The inequalities shown below are compound inequalities.

$$x + 1 > 3 \quad \text{and} \quad 2x - 11 < 7$$
$$x + 3 > 5 \quad \text{or} \quad x - 1 < 9$$

The solution set of a compound inequality with the connective word *or* is the *union* of the solution sets of the two inequalities. The solution set of a compound inequality with the connective word *and* is the *intersection* of the solution sets of the two inequalities.

EXAMPLE 2 **Solve Compound Inequalities**

Solve each compound inequality. Write each solution in set-builder notation.

a. $2x < 10$ or $x + 1 > 9$ **b.** $x + 3 > 4$ and $2x + 1 > 15$

Solution

a. $2x < 10$ or $x + 1 > 9$
     $x < 5$ $x > 8$ • Solve each inequality.
     $\{x \mid x < 5\}$ $\{x \mid x > 8\}$ • Write each solution as
                                                                 a set.

   $\{x \mid x < 5\} \cup \{x \mid x > 8\} = \{x \mid x < 5 \text{ or } x > 8\}$ • Write the union of the
                                                                 solution sets.

b. $x + 3 > 4$ and $2x + 1 > 15$
         $x > 1$ $2x > 14$ • Solve each inequality.
                                     $x > 7$
   $\{x \mid x > 1\}$ $\{x \mid x > 7\}$ • Write each solution as a set.
   $\{x \mid x > 1\} \cap \{x \mid x > 7\} = \{x \mid x > 7\}$ • Write the intersection of the
                                                                 solution sets.

▶ **TRY EXERCISE 10, PAGE 134**

❓ QUESTION What is the solution set of the compound inequality $x > 1$ or
                $x < 3$?

take note

We reserve the notation $a < b < c$ to mean $a < b$ and $b < c$. Thus the solution set of $2 > x > 5$ is the empty set, because there are no numbers less than 2 and greater than 5.

The inequality given by

$$12 < x + 5 < 19$$

is equivalent to the compound inequality $12 < x + 5$ *and* $x + 5 < 19$. You can solve $12 < x + 5 < 19$ by either of the following methods.

Method 1   Find the intersection of the solution sets of the inequalities $12 < x + 5$ and $x + 5 < 19$.

$$12 < x + 5 \quad \text{and} \quad x + 5 < 19$$
$$7 < x \quad \text{and} \quad x < 14$$

The solution set is $\{x \mid x > 7\} \cap \{x \mid x < 14\} = \{x \mid 7 < x < 14\}$.

Method 2   Subtract 5 from each of the three parts of the inequality.

$$12 < \quad x + 5 \quad < 19$$
$$12 - 5 < x + 5 - 5 < 19 - 5$$
$$7 < \quad x \quad < 14$$

take note

The compound inequality $a < b$ and $b < c$ can be written in the compact form $a < b < c$. However, the compound inequality $a < b$ or $b > c$ cannot be expressed in a compact form.

The solution set is $\{x \mid 7 < x < 14\}$.

• ABSOLUTE VALUE INEQUALITIES

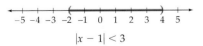

$|x - 1| < 3$

FIGURE 1.6

The solution set of the absolute value inequality $|x - 1| < 3$ is the set of all real numbers whose distance from 1 is *less than* 3. Therefore, the solution set consists of all numbers between -2 and 4. See **Figure 1.6.** In interval notation, the solution set is $(-2, 4)$.

❓ ANSWER   The set of all real numbers. Using interval notation, the solution set is written as $(-\infty, \infty)$.

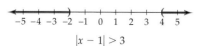

$|x - 1| > 3$

FIGURE 1.7

The solution set of the absolute value inequality $|x - 1| > 3$ is the set of all real numbers whose distance from 1 is *greater than* 3. Therefore, the solution set consists of all real numbers less than -2 *or* greater than 4. See **Figure 1.7.** In interval notation, the solution set is $(-\infty, -2) \cup (4, \infty)$.

The following properties are used to solve absolute value inequalities.

Properties of Absolute Value Inequalities

For any variable expression E and any nonnegative real number k,

$$|E| \le k \quad \text{if and only if} \quad -k \le E \le k$$
$$|E| \ge k \quad \text{if and only if} \quad E \le -k \quad \text{or} \quad E \ge k$$

These properties also hold true when the $<$ symbol is substituted for the $\le$ symbol and when the $>$ symbol is substituted for the $\ge$ symbol.

> **take note**
>
> Some inequalities have a solution set that consists of all real numbers. For example, $|x + 9| \ge 0$ is true for all values of x. Because an absolute value is always nonnegative, the equation is always true.

In Example 3 we make use of the above properties to solve absolute value inequalities.

EXAMPLE 3 Solve Absolute Value Inequalities

Solve each of the following inequalities.

a. $|2 - 3x| < 7$ b. $|4x - 3| \ge 5$

Solution

a. $|2 - 3x| < 7$ if and only if $-7 < 2 - 3x < 7$. Solve this compound inequality.

$$-7 < 2 - 3x < 7$$

$$-9 < -3x < 5$$ • **Subtract 2 from each of the three parts of the inequality.**

$$3 > x > -\frac{5}{3}$$ • **Multiply each part of the inequality by $-\frac{1}{3}$ and reverse the inequality symbols.**

In interval notation, the solution set is given by $\left(-\frac{5}{3}, 3\right)$. See **Figure 1.8.**

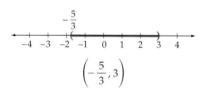

$\left(-\frac{5}{3}, 3\right)$

FIGURE 1.8

b. $|4x - 3| \ge 5$ implies $4x - 3 \le -5$ or $4x - 3 \ge 5$. Solving each of these inequalities produces

$$4x - 3 \le -5 \quad \text{or} \quad 4x - 3 \ge 5$$
$$4x \le -2 \qquad\qquad 4x \ge 8$$
$$x \le -\frac{1}{2} \qquad\qquad x \ge 2$$

The solution set is $\left(-\infty, -\frac{1}{2}\right] \cup [2, \infty)$. See **Figure 1.9.**

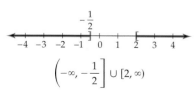

$\left(-\infty, -\frac{1}{2}\right] \cup [2, \infty)$

FIGURE 1.9

▶ TRY EXERCISE 18, PAGE 134

● THE CRITICAL VALUE METHOD

Any value of x that causes a polynomial in x to equal zero is called a **zero of the polynomial.** For example, -4 and 1 are both zeros of the polynomial $x^2 + 3x - 4$, because $(-4)^2 + 3(-4) - 4 = 0$ and $1^2 + 3 \cdot 1 - 4 = 0$.

A Sign Property of Polynomials

Polynomials in x have the property that for all values of x between two consecutive real zeros, all values of the polynomial are positive or all values of the polynomial are negative.

In our work with inequalities that involve polynomials, the real zeros of the polynomial are also referred to as **critical values of the inequality.** On a number line the critical values of an inequality separate the real numbers that make the inequality true from those that make it false. In Example 4, we use critical values and the sign property of polynomials to solve an inequality.

EXAMPLE 4   Solve a Quadratic Inequality

Solve: $x^2 + 3x - 4 < 0$

Solution

Factoring the polynomial $x^2 + 3x - 4$ produces the equivalent inequality

$$(x + 4)(x - 1) < 0$$

FIGURE 1.10

The zeros of the polynomial $x^2 + 3x - 4$ are -4 and 1. They are the critical values of the inequality $x^2 + 3x - 4 < 0$. They separate the real number line into the three intervals shown in **Figure 1.10.**
 To determine the intervals on which $x^2 + 3x - 4 < 0$, pick a number called a **test value** from each of the three intervals and then determine whether $x^2 + 3x - 4 < 0$ for each of these test values. For example, in the interval $(-\infty, -4)$, pick a test value of, say, -5. Then

$$x^2 + 3x - 4 = (-5)^2 + 3(-5) - 4 = 6$$

Because 6 is not less than 0, by the sign property of polynomials, no number in the interval $(-\infty, -4)$ makes $x^2 + 3x - 4 < 0$.
 Now pick a test value from the interval $(-4, 1)$, say, 0. When $x = 0$,

$$x^2 + 3x - 4 = 0^2 + 3(0) - 4 = -4$$

Because -4 is less than 0, by the sign property of polynomials, all numbers in the interval $(-4, 1)$ make $x^2 + 3x - 4 < 0$.
 If we pick a test value of 2 from the interval $(1, \infty)$, then

$$x^2 + 3x - 4 = (2)^2 + 3(2) - 4 = 6$$

Because 6 is not less than 0, by the sign property of polynomials, no number in the interval $(1, \infty)$ makes $x^2 + 3x - 4 < 0$.

Continued ▶

The following table is a summary of our work.

Interval	Test Value x	$x^2 + 3x - 4 \overset{?}{<} 0$
$(-\infty, -4)$	-5	$(-5)^2 + 3(-5) - 4 < 0$ $6 < 0$ False
$(-4, 1)$	0	$(0)^2 + 3(0) - 4 < 0$ $-4 < 0$ True
$(1, \infty)$	2	$(2)^2 + 3(2) - 4 < 0$ $6 < 0$ False

In interval notation, the solution set of $x^2 + 3x - 4 < 0$ is $(-4, 1)$. The solution set is graphed in **Figure 1.11**. Note that in this case the critical values -4 and 1 are not included in the solution set because they do not make $x^2 + 3x - 4$ less than 0.

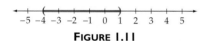

FIGURE 1.11

▶ **TRY EXERCISE 34, PAGE 134**

To avoid the arithmetic in Example 4, we often use a *sign diagram*. For example, note that the factor $(x + 4)$ is negative for all $x < -4$ and positive for all $x > -4$. The factor $(x - 1)$ is negative for all $x < 1$ and positive for all $x > 1$. These results are shown in **Figure 1.12.**

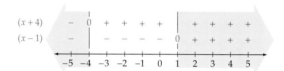

A sign diagram for $(x + 4)$ and $(x - 1)$.

FIGURE 1.12

To determine on which intervals the product $(x + 4)(x - 1)$ is negative, we examine the sign diagram to see where the factors have opposite signs. This occurs only on the interval $(-4, 1)$, where $(x + 4)$ is positive and $(x - 1)$ is negative, so the original equality is true only on the interval $(-4, 1)$.

Following is a summary of the steps used to solve polynomial inequalities by the critical value method.

Solving a Polynomial Inequality by the Critical Value Method

1. Write the inequality so that one side of the inequality is a nonzero polynomial and the other side is 0.

2. Find the real zeros of the polynomial.[3] They are the critical values of the original inequality.

3. Use test values to determine which of the consecutive intervals formed by the critical values are to be included in the solution set.

---

[3] In Chapter 3, additional ways to find the zeros of a polynomial are developed. For the present, however, we will find the zeros by factoring or by using the quadratic formula.

• RATIONAL INEQUALITIES

A rational expression is the quotient of two polynomials. **Rational inequalities** involve rational expressions, and they can be solved by an extension of the critical value method.

Critical Values of a Rational Expression

The **critical values of a rational expression** are the numbers that cause the numerator of the rational expression to equal zero or the denominator of the rational expression to equal zero.

Rational expressions also have the property that they remain either positive for all values of the variable between consecutive critical values or negative for all values of the variable between consecutive critical values.

Following is a summary of the steps used to solve rational inequalities by the critical value method.

Solving a Rational Inequality by the Critical Value Method

1. Write the inequality so that one side of the inequality is a rational expression and the other side is 0.

2. Find the real zeros of the numerator of the rational expression and the real zeros of its denominator. They are the critical values of the inequality.

3. Use test values to determine which of the consecutive intervals formed by the critical values are to be included in the solution set.

EXAMPLE 5   Solve a Rational Inequality

Solve: $\dfrac{3x + 4}{x + 1} \leq 2$

Solution

Write the inequality so that 0 appears on the right side of the inequality.

$$\frac{3x + 4}{x + 1} \leq 2$$

$$\frac{3x + 4}{x + 1} - 2 \leq 0$$

Continued ▶

Write the left side as a rational expression.

$$\frac{3x + 4}{x + 1} - \frac{2(x + 1)}{x + 1} \le 0 \qquad \bullet \text{ The LCD is } x + 1.$$

$$\frac{3x + 4 - 2x - 2}{x + 1} \le 0 \qquad \bullet \text{ Simplify.}$$

$$\frac{x + 2}{x + 1} \le 0$$

The critical values of this inequality are -2 and -1 because the numerator $x + 2$ is equal to zero when $x = -2$, and the denominator $x + 1$ is equal to zero when $x = -1$. The critical values -2 and -1 separate the real number line into the three intervals $(-\infty, -2)$, $(-2, -1)$, and $(-1, \infty)$.

All values of x on the interval $(-2, -1)$ make $\dfrac{x + 2}{x + 1}$ negative, as

desired. On the other intervals, the quotient $\dfrac{x + 2}{x + 1}$ is positive. See the sign

diagram in **Figure 1.13.**

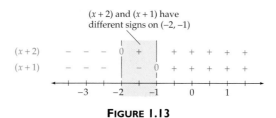

FIGURE 1.13

The solution set is $[-2, -1)$. The graph of the solution set is shown in **Figure 1.14.** Note that -2 is included in the solution set because $\dfrac{x + 2}{x + 1} = 0$ when $x = -2$. However, -1 is not included in the solution set because the denominator $(x + 1)$ is zero when $x = -1$.

FIGURE 1.14

▶ **TRY EXERCISE 46, PAGE 134**

● **APPLICATIONS**

Many applied problems can be solved by using inequalities.

EXAMPLE 6 **Solve an Application Concerning Leases**

A real estate company needs a new copy machine. The company has decided to lease either the model ABC machine for $75 a month plus 5 cents per copy or the model XYZ machine for $210 a month and 2 cents per copy. Under what conditions is it less expensive to lease the XYZ machine?

Solution

Let x represent the number of copies the company produces per month. The dollar costs per month are $75 + 0.05x$ for model ABC and $210 + 0.02x$ for model XYZ. It will be less expensive to lease model XYZ provided

$$210 + 0.02x < 75 + 0.05x$$
$$210 - 0.03x < 75 \qquad \text{• Subtract } 0.05x \text{ from each side.}$$
$$-0.03x < -135 \qquad \text{• Subtract 210 from each side.}$$
$$x > 4500 \qquad \text{• Divide each side by } -0.03. \text{ Reverse the inequality symbol.}$$

The company will find it less expensive to lease model XYZ if it produces over 4500 copies per month.

▶ **TRY EXERCISE 52, PAGE 135**

EXAMPLE 7 Solve an Application Concerning Test Scores

Tyra has test scores of 70 and 81 in her biology class. To receive a C grade, she must obtain an average greater than or equal to 72 but less than 82. What range of test scores on the one remaining test will enable Tyra to get a C for the course?

Solution

The average of three test scores is the sum of the scores divided by 3. Let x represent Tyra's next test score. The requirements for a C grade produce the following inequality:

$$72 \le \frac{70 + 81 + x}{3} < 82$$
$$216 \le 70 + 81 + x < 246 \qquad \text{• Multiply each part of the inequality by 3.}$$
$$216 \le \quad 151 + x \quad < 246 \qquad \text{• Simplify.}$$
$$65 \le \qquad x \qquad < 95 \qquad \text{• Solve for } x \text{ by subtracting 151 from each part of the inequality.}$$

To get a C in the course, Tyra's remaining test score must be in the interval $[65, 95)$.

▶ **TRY EXERCISE 58, PAGE 135**

In many business applications a company is interested in the cost C of manufacturing x items, the revenue R generated by selling all the items, and the profit P made by selling the items.

In the next example the cost of manufacturing x tennis racquets is given by $C = 32x + 120,000$ dollars. The 120,000 represents the fixed cost because it remains constant regardless of how many racquets are manufactured. The $32x$ represents the variable cost because this term varies depending on how many racquets are manufactured. Each additional racquet costs the company an additional $32.

The revenue received from the sale of x tennis racquets is given by $R = x(200 - 0.01x)$ dollars. The quantity $(200 - 0.01x)$ is the price the company charges for each tennis racquet. The price varies depending on the number of racquets that are manufactured. For instance, if the number of racquets x that the company manufactures is small, the company will be able to demand almost $200 for each racquet. As the number of racquets that the company manufactures increases (approaches 20,000), the company will only be able to sell *all* the racquets if it decreases the price of each racquet.

The following profit formula shows the relationship between profit P, revenue R, and cost C:

$$P = R - C$$

EXAMPLE 8   Solve a Business Application

A company determines that the cost C, in dollars, of producing x tennis racquets is $C = 32x + 120,000$. The revenue R, in dollars, from selling all of the tennis racquets is $R = x(200 - 0.01x)$.

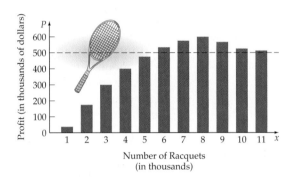

Number of Racquets
(in thousands)

How many racquets should the company manufacture and sell if the company wishes to earn a profit of at least $500,000?

Solution

The profit is given by

$$P = R - C$$
$$= x(200 - 0.01x) - (32x + 120,000)$$
$$= 200x - 0.01x^2 - 32x - 120,000$$
$$= -0.01x^2 + 168x - 120,000$$

The profit will be at least $500,000 provided

$$-0.01x^2 + 168x - 120,000 \geq 500,000$$
$$-0.01x^2 + 168x - 620,000 \geq 0$$

Using the quadratic formula, we find that the approximate critical values of this last inequality are 5474.3 and 11,325.7. Test values show that the inequality is positive only on the interval (5474.3, 11,325.7). The company should manufacture at least 5475 tennis racquets but not more than 11,325 tennis racquets to produce the desired profit.

▶ **TRY EXERCISE 54, PAGE 135**

> EXAMPLE 9   Solve an Application Involving Batting Averages

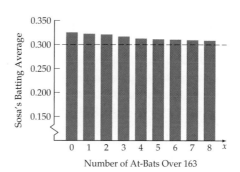

During a recent season, Sammy Sosa had 53 hits out of 163 at-bats. At that time his batting average was approximately 0.325. If Sosa goes into a batting slump in which he gets no hits, how many more at-bats will it take for his batting average to fall below 0.300?

(Chart: vertical axis "Sosa's Batting Average" from 0.150 to 0.350; horizontal axis "Number of At-Bats Over 163" labeled 0 to 8 with variable x; dashed line at 0.300.)

Solution

A baseball player's batting average is determined by dividing the player's number of hits by the number of times the player has been at bat. Let x be the number of additional at-bats that Sosa takes over 163. During this period, his batting average will be $\dfrac{53}{163 + x}$, and we wish to solve

$$\frac{53}{163 + x} < 0.300$$

This rational inequality can be solved by using the critical value method, but there is an easier method. In this application we know that $163 + x$ is positive. Thus, if we multiply each side of the preceding inequality by $163 + x$, we will obtain the linear inequality $53 < 48.9 + 0.300x$, with the condition that x is a positive integer. Solving this inequality produces

$$53 < 48.9 + 0.300x$$
$$4.1 < 0.300x$$
$$x > 13.\overline{6}$$

Because x must be a positive integer, Sosa's average will fall below 0.300 if he goes hitless for 14 or more at-bats.

▶ TRY EXERCISE 66, PAGE 136

 TOPICS FOR DISCUSSION

1. If $x < y$, then $y > x$. Do you agree?

2. Can the solution set of the compound inequality

$$x < -3 \text{ or } x > 5$$

be expressed as $-3 > x > 5$? Explain.

3. If $-a < b$, then it must be true that $a > -b$. Do you agree? Explain.

4. Do the inequalities $x < 4$ and $x^2 < 4^2$ both have the same solution set? Explain.

5. True or false: If $k < 0$, then $|k| = -k$.

EXERCISE SET 1.5

In Exercises 1 to 8, use the properties of inequalities to solve each inequality. Write the solution set using set-builder notation, and graph the solution set.

1. $2x + 3 < 11$

2. $3x - 5 > 16$

3. $x + 4 > 3x + 16$

4. $5x + 6 < 2x + 1$

5. $-3(x + 2) \le 5x + 7$

▶ 6. $-4(x - 5) \ge 2x + 15$

7. $-4(3x - 5) > 2(x - 4)$

8. $3(x + 7) \le 5(2x - 8)$

In Exercises 9 to 16, solve each compound inequality. Write the solution set using set-builder notation, and graph the solution set.

9. $4x + 1 > -2$ and $4x + 1 \le 17$

▶ 10. $2x + 5 > -16$ and $2x + 5 < 9$

11. $10 \ge 3x - 1 \ge 0$

12. $0 \le 2x + 6 \le 54$

13. $x + 2 < -1$ or $x + 3 \ge 2$

14. $x + 1 > 4$ or $x + 2 \le 3$

15. $-4x + 5 > 9$ or $4x + 1 < 5$

16. $2x - 7 \le 15$ or $3x - 1 \le 5$

In Exercises 17 to 28, use interval notation to express the solution set of each inequality.

17. $|2x - 1| > 4$

▶ 18. $|2x - 9| < 7$

19. $|x + 3| \ge 5$

20. $|x - 10| \ge 2$

21. $|3x - 10| \le 14$

22. $|2x - 5| \ge 1$

23. $|4 - 5x| \ge 24$

24. $|3 - 2x| \le 5$

25. $|x - 5| \ge 0$

26. $|x - 7| \ge 0$

27. $|x - 4| \le 0$

28. $|2x + 7| \le 0$

In Exercises 29 to 36, use the critical value method to solve each polynomial inequality. Use interval notation to write each solution set.

29. $x^2 + 7x > 0$

30. $x^2 - 5x \le 0$

31. $x^2 - 16 \le 0$

32. $x^2 - 49 > 0$

33. $x^2 + 7x + 10 < 0$

▶ 34. $x^2 + 5x + 6 < 0$

35. $x^2 - 3x \ge 28$

36. $x^2 < -x + 30$

In Exercises 37 to 50, use the critical value method to solve each rational inequality. Write each solution set in interval notation.

37. $\dfrac{x + 4}{x - 1} < 0$

38. $\dfrac{x - 2}{x + 3} > 0$

39. $\dfrac{x - 5}{x + 8} \ge 3$

40. $\dfrac{x - 4}{x + 6} \le 1$

41. $\dfrac{x}{2x + 7} \ge 4$

42. $\dfrac{x}{3x - 5} \le -5$

43. $\dfrac{(x + 1)(x - 4)}{x - 2} < 0$

44. $\dfrac{x(x - 4)}{x + 5} > 0$

45. $\dfrac{x + 2}{x - 5} \le 2$

▶ 46. $\dfrac{3x + 1}{x - 2} \ge 4$

47. $\dfrac{6x^2 - 11x - 10}{x} > 0$

48. $\dfrac{3x^2 - 2x - 8}{x - 1} \ge 0$

49. $\dfrac{x^2 - 6x + 9}{x - 5} \le 0$

50. $\dfrac{x^2 + 10x + 25}{x + 1} \ge 0$

51. **PERSONAL FINANCE** A bank offers two checking account plans. The monthly fee and charge per check for each plan are shown below. Under what conditions is it less expensive to use the LowCharge plan?

Account Plan	Monthly Fee	Charge per Check
LowCharge	$5.00	$.01
FeeSaver	$1.00	$.08

▶ **52.** **PERSONAL FINANCE** You can rent a car for the day from Company A for $29.00 plus $0.12 a mile. Company B charges $22.00 plus $0.21 a mile. Find the number of miles m (to the nearest mile) per day for which it is cheaper to rent from Company A.

53. **SHIPPING REQUIREMENTS** United Parcel Service (UPS) will only ship packages for which the length is less than or equal to 108 inches and the length plus the girth is less than or equal to 130 inches. The length of a package is defined as the length of the longest side. The girth is defined as twice the width plus twice the height of the package. If a box has a length of 34 inches and a width of 22 inches, determine the possible range of heights h for this package if you wish to ship it by UPS. (*Source:* http://www.iship.com.)

▶ **54.** **MOVIE TICKET PRICES** The average U.S. movie ticket price P, in dollars, can be modeled by

$$P = 0.218t + 4.02, \quad t \geq 0$$

where $t = 0$ represents the year 1994. According to this model, in what year will the average price of a movie ticket first exceed $6.50? (*Source:* National Association of Theatre Owners, http://www.natoonline.org/satistics-tickets.htm.)

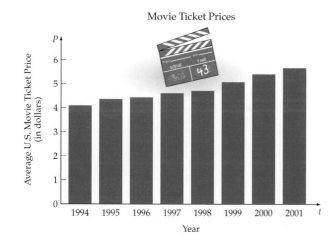

Movie Ticket Prices

55. **PERSONAL FINANCE** A sales clerk has a choice between two payment plans. Plan A pays $100.00 a week plus $8.00 a sale. Plan B pays $250.00 a week plus $3.50 a sale. How many sales per week must be made for plan A to yield the greater paycheck?

56. **PERSONAL FINANCE** A video store offers two rental plans. The yearly membership fee and the daily charge per video for each plan are shown below. How many one-night rentals can be made per year if the No-fee plan is to be the less expensive of the plans?

THE VIDEO STORE		
Rental Plan	**Yearly Fee**	**Daily Charge per Video**
Low-rate	$15.00	$1.49
No-fee	None	$1.99

57. **AVERAGE TEMPERATURES** The average daily minimum-to-maximum temperature range for the city of Palm Springs during the month of September is 68 to 104 degrees Fahrenheit. What is the corresponding temperature range measured on the Celsius temperature scale? (*Hint:* Let F be the average daily temperature. Then $68 \leq F \leq 104$. Now substitute $\frac{9}{5}C + 32$ for F and solve the resulting inequality for C.)

▶ **58.** **AVERAGE TEMPERATURES** The average daily minimum-to-maximum temperature range for the city of Palm Springs during the month of January is 41 to 68 degrees Fahrenheit. What is the corresponding temperature range measured on the Celsius temperature scale? (*Hint:* See Exercise 57.)

59. **CONSECUTIVE EVEN INTEGERS** The sum of three consecutive even integers is between 36 and 54. Find all possible sets of integers that satisfy these conditions.

60. **CONSECUTIVE ODD INTEGERS** The sum of three consecutive odd integers is between 63 and 81. Find all possible sets of integers that satisfy these conditions.

61. **FORENSIC SCIENCE** Forensic specialists can estimate the height of a deceased person from the lengths of the person's bones. These lengths are substituted into mathematical inequalities. For instance, an inequality that relates the height h, in centimeters, of an adult female and the length f, in centimeters, of her femur is

$$|h - (2.47f + 54.10)| \leq 3.72$$

Use this inequality to estimate the possible range of heights, rounded to the nearest 0.1 centimeter, for an adult female whose femur measures 32.24 centimeters.

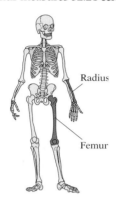

Radius

Femur

62. **FORENSIC SCIENCE** An inequality that is used to calculate the height h of an adult male from the length r of his radius is

$$|h - (3.32r + 85.43)| \leq 4.57$$

where h and r are both in centimeters. Use this inequality to estimate the possible range of heights for an adult male whose radius measures 26.36 centimeters.

63. **REVENUE** The monthly revenue R for a product is given by $R = 420x - 2x^2$, where x is the price in dollars of each unit produced. Find the interval, in terms of x, for which the monthly revenue is greater than zero.

64. **REVENUE** A shoe manufacturer finds that the monthly revenue R from a particular style of aerobics shoe is given by $R = 312x - 3x^2$, where x is the price in dollars of each pair of shoes sold. Find the interval, in terms of x, for which the monthly revenue is greater than or equal to $5925.

65. **PUBLISHING** A publisher has determined that if x books are published, the average cost per book is given by

$$\overline{C} = \frac{14.25x + 350,000}{x}$$

How many books should be published if the company wants to bring the average cost per book below $50?

▶ 66. **MANUFACTURING** A company manufactures running shoes. The company has determined that if it manufactures x pairs of shoes, the average cost, in dollars, per pair is

$$\overline{C} = \frac{0.00014x^2 + 12x + 400,000}{x}$$

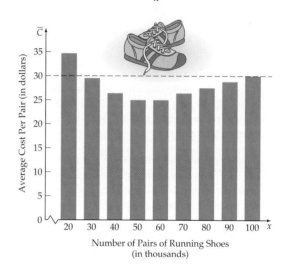

Number of Pairs of Running Shoes
(in thousands)

How many pairs of running shoes should the company manufacture if it wishes to bring the average cost below $30 per pair?

Number of Books Published
(in thousands)

CONNECTING CONCEPTS

67. TOLERANCE A machinist is producing a circular cylinder on a lathe. The circumference of the cylinder must be 28 inches, with a tolerance of 0.15 inch. What maximum and minimum radii (to the nearest 0.001 inch) must the machinist stay between to produce an acceptable cylinder?

68. TOLERANCE A tall, narrow beaker has an inner radius of 2 centimeters. How high h (to the nearest 0.1 centimeter) should we fill the beaker if we need to measure $\frac{3}{4}$ liter (750 cubic centimeters) of a solution with an error of 15 cubic centimeters or less?

2 cm

In Exercises 69 to 72, use the critical value method to solve each inequality. Use interval notation to write each solution set.

69. $\dfrac{(x-3)^2}{(x-6)^2} > 0$

70. $\dfrac{(x-1)^2}{(x-4)^4} \geq 0$

71. $\dfrac{(x-4)^2}{(x+3)^3} \geq 0$

72. $\dfrac{(2x-7)}{(x-1)^2(x+2)^2} \geq 0$

In Exercises 73 to 78, use interval notation to express the solution set of each inequality.

73. $1 < |x| < 5$

74. $2 < |x| < 3$

75. $3 \leq |x| < 7$

76. $0 < |x| \leq 3$

77. $0 < |x - a| < \delta$ $(\delta > 0)$

78. $0 < |x - 5| < 2$

79. HEIGHT OF A PROJECTILE The equation

$$s = -16t^2 + v_0 t + s_0$$

gives the height s, in feet above ground level, at the time t seconds, of an object thrown directly upward from a height s_0 feet above the ground and with an initial velocity of v_0 feet per second. A ball is thrown directly upward from ground level with an initial velocity of 64 feet per second. Find the time interval during which the ball has a height of more than 48 feet.

80. HEIGHT OF A PROJECTILE A ball is thrown directly upward from a height of 32 feet above the ground with an initial velocity of 80 feet per second. Find the time interval during which the ball will be more than 96 feet above the ground. (*Hint:* See Exercise 79.)

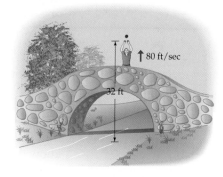

PROJECTS

1. TRIANGLES In any triangle, the sum of the lengths of the two shorter sides must be greater than the length of the longest side. Find all possible values of x if a triangle has sides of lengths

a. $x, x + 5$, and $x + 9$ **b.** $x, x^2 + x$, and $2x^2 + x$

c. $\dfrac{1}{x+2}, \dfrac{1}{x+1}$, and $\dfrac{1}{x}$

2. FAIR COINS A coin is considered a **fair** coin if it has an equal chance of landing heads up or tails up. To decide

whether a coin is a fair coin, a statistician tosses it 1000 times and records the number of tails t. The statistician is prepared to state that the coin is a fair coin if

$$\left|\frac{t - 500}{15.81}\right| \leq 2.33$$

a. Determine what values of t will cause the statistician to state that the coin is a fair coin.

b. Pick a coin and test it according to the criteria above to see whether it is a fair coin.

EXPLORING CONCEPTS
WITH TECHNOLOGY

Use a Graphing Calculator to Solve Equations

Most graphing calculators can be used to solve equations. The following example shows how to solve an equation using the **solve(** feature that is available on a TI-83 graphing calculator.

The calculator display below indicates that the solution of $2x - 17 = 0$ is 8.5.

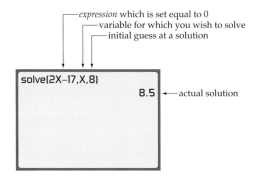

The calculator display above was produced by the following keystrokes. Press $\boxed{\text{2nd}}$ [catalog] S (scroll down to **solve(**) $\boxed{\text{ENTER}}$. Now enter 2 $\boxed{\text{X,T,}\Theta\text{,}n}$ $\boxed{-}$ 17 $\boxed{,}$ $\boxed{\text{X,T,}\Theta\text{,}n}$ $\boxed{,}$ 8 $\boxed{)}$ $\boxed{\text{ENTER}}$. In this example, the 8.5 represents the solution of $2x - 17 = 0$ that is close to our initial guess of 8. Because $2x - 17 = 0$ has only one solution, we are finished. Note that the **solve(** feature can only be used to solve equations of the form

$$Expression = 0$$

Also, you are required to indicate the variable you wish to solve for, and you must enter an initial guess. In the preceding display we entered X as the variable and 8 as our initial guess.

The **solve(** feature can only be used to find *real* solutions. Also, the **solve(** feature finds only one solution each time the solution procedure is applied. If you know that an equation has two real solutions, then you need to apply the solution procedure twice. Each time you must enter an initial guess that is close to the solution you are trying to find. The calculator display below indicates that the solutions of $2x^2 - x - 15 = 0$ are 3 and -2.5. To find these solutions, we first used the **solve(** feature with an initial guess of 2, and we then used the **solve(** feature with an initial guess of -1.

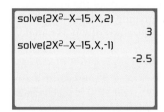

The following chapters will illustrate additional techniques and calculator procedures that can be used to solve equations.

CHAPTER 1 SUMMARY

1.1 Linear and Absolute Value Equations

- A number is said to satisfy an equation if substituting the number for the variable results in an equation that is a true statement. To solve an equation means to find all values of the variable that satisfy the equation. These values that make the equation true are called solutions or roots of the equation. Equivalent equations have the same solution(s).

- A linear equation in the single variable x is an equation that can be written in the form $ax + b = 0$, where a and b are real numbers, with $a \neq 0$.

- The solution set of $|ax + b| = c$, $a \neq 0$, is the union of the solution sets of $ax + b = c$ and $ax + b = -c$.

1.2 Formulas and Applications

- A formula is an equation that expresses known relationships between two or more variables. Application problems are best solved by using the guidelines developed in this section.

1.3 Quadratic Equations

- A quadratic equation in x is an equation that can be written in the form $ax^2 + bx + c = 0$, where $a \neq 0$. If the quadratic polynomial in a quadratic equation is factorable over the set of integers, then the equation can be solved by factoring and using the zero product property. Every quadratic equation can be solved by completing the square or by using the quadratic formula.

- **The Quadratic Formula**

  If $ax^2 + bx + c = 0$, $a \neq 0$, then $x = \dfrac{-b \pm \sqrt{b^2 - 4ac}}{2a}$.

1.4 Other Types of Equations

- **The Power Principle**

  If P and Q are algebraic expressions and n is a positive integer, then every solution of $P = Q$ is a solution of $P^n = Q^n$.

- An equation is said to be quadratic in form if it can be written in the form $au^2 + bu + c = 0$, where $a \neq 0$ and u is an algebraic expression.

1.5 Inequalities

- The set of all solutions of an inequality is the solution set of the inequality. Equivalent inequalities have the same solution set. To solve an inequality, use the properties of inequalities or the critical value method.

- An inequality of the form $|ax + b| > c$, $a \neq 0$, is an absolute value inequality. The inequality symbol $>$ can be replaced by $<$, $\leq$, or $\geq$.

CHAPTER 1 TRUE/FALSE EXERCISES

In Exercises 1 to 10, answer true or false. If the statement is false, state a reason or give an example to show that the statement is false.

1. If $x^2 = 9$, then $x = 3$.

2. The equations

$$x = \sqrt{12 - x} \qquad \text{and} \qquad x^2 = 12 - x$$

   are equivalent equations.

3. Adding the same constant to each side of a given equation produces an equation that is equivalent to the given equation.

4. If $a > b$, then $-a < -b$.

5. If $a \neq 0$, $b \neq 0$, and $a > b$, then $\dfrac{1}{a} > \dfrac{1}{b}$.

6. The discriminant of $ax^2 + bx + c = 0$ is $\sqrt{b^2 - 4ac}$.

7. If $\sqrt{a} + \sqrt{b} = c$, then $a + b = c^2$.

8. The solution set of $|x - a| < b$ with $b > 0$ is given by the interval $(a - b, a + b)$.

9. The only quadratic equation that has roots of 4 and -4 is $x^2 - 16 = 0$.

10. Every quadratic equation $ax^2 + bx + c = 0$ with real coefficients such that $ac < 0$ has two distinct real roots.

CHAPTER 1 REVIEW EXERCISES

In Exercises 1 to 28, solve each equation.

1. $x - 2(5x - 3) = -3(-x + 4)$

2. $3x - 5(2x - 7) = -4(5 - 2x)$

3. $\dfrac{4x}{3} - \dfrac{4x - 1}{6} = \dfrac{1}{2}$

4. $\dfrac{3x}{4} - \dfrac{2x - 1}{8} = \dfrac{3}{2}$

5. $\dfrac{x}{x + 2} + \dfrac{1}{4} = 5$

6. $\dfrac{y - 1}{y + 1} - 1 = \dfrac{2}{y}$

7. $x^2 - 5x + 6 = 0$

8. $6x^2 + x - 12 = 0$

9. $3x^2 - x - 1 = 0$

10. $x^2 - x + 1 = 0$

11. $3x^3 - 5x^2 = 0$

12. $2x^3 - 8x = 0$

13. $6x^4 - 23x^2 + 20 = 0$

14. $3x + 16\sqrt{x} - 12 = 0$

15. $\sqrt{x^2 - 15} = \sqrt{-2x}$

16. $\sqrt{x^2 - 24} = \sqrt{2x}$

17. $\sqrt{3x + 4} + \sqrt{x - 3} = 5$

18. $\sqrt{2x + 2} - \sqrt{x + 2} = \sqrt{x - 6}$

19. $\sqrt{4 - 3x} - \sqrt{5 - x} = \sqrt{5 + x}$

20. $\sqrt{3x + 9} - \sqrt{2x + 4} = \sqrt{x + 1}$

21. $\dfrac{1}{(y + 3)^2} = 1$

22. $\dfrac{1}{(2s - 5)^2} = 4$

23. $|x - 3| = 2$

24. $|x + 5| = 4$

25. $|2x + 1| = 5$

26. $|3x - 7| = 8$

27. $(x + 2)^{1/2} + x(x + 2)^{3/2} = 0$

28. $x^2(3x - 4)^{1/4} + (3x - 4)^{5/4} = 0$

In Exercises 29 to 44, solve each inequality. Express your solution sets by using interval notation.

29. $-3x + 4 \geq -2$

30. $-2x + 7 \leq 5x + 1$

31. $x^2 + 3x - 10 \leq 0$

32. $x^2 - 2x - 3 > 0$

33. $61 \leq \dfrac{9}{5}C + 32 \leq 95$

34. $30 < \dfrac{5}{9}(F - 32) < 65$

35. $x^3 - 7x^2 + 12x \leq 0$

36. $x^3 + 4x^2 - 21x > 0$

37. $\dfrac{x + 3}{x - 4} > 0$

38. $\dfrac{x(x - 5)}{x + 7} \leq 0$

39. $\dfrac{2x}{3 - x} \leq 10$

40. $\dfrac{x}{5 - x} \geq 1$

41. $|3x - 4| < 2$

42. $|2x - 3| \geq 1$

43. $0 < |x - 2| < 1$

44. $0 < |x - a| < b \quad (b > 0)$

In Exercises 45 to 50, solve each equation for the indicated unknown.

45. $V = \pi r^2 h$,   for h

46. $P = \dfrac{A}{1 + rt}$,   for t

47. $A = \dfrac{h}{2}(b_1 + b_2)$,   for b_1

48. $P = 2(l + w)$,   for w

49. $e = mc^2$,   for m

50. $F = G\dfrac{m_1 m_2}{s^2}$,   for m_1

51. UNKNOWN NUMBER One-half of a number minus one-fourth of the number is four more than one-fifth of the number. What is the number?

52. RECTANGULAR REGION The length of a rectangle is 9 feet less than twice the width of the rectangle. The perimeter of the rectangle is 54 feet. Find the width and the length.

53. DISTANCE TO AN ISLAND A motorboat left a harbor and traveled to an island at an average rate of 8 knots. The average speed on the return trip was 6 knots. If the total trip took 7 hours, how far is it from the harbor to the island?

54. PRICE OF SUBSCRIPTION The price of a magazine subscription rose 5% this year. If the subscription now costs $21, how much did the subscription cost last year?

55. INVESTMENT A total of $5500 was deposited into two simple interest accounts. On one account the annual simple interest rate is 4%, and on the second account the annual simple interest rate is 6%. The amount of interest earned for 1 year was $295. How much was invested in each account?

56. INDIVIDUAL PRICE A calculator and a battery together sell for $21. The price of the calculator is $20 more than the price of the battery. Find the price of the calculator and the price of the battery.

57. MAINTENANCE COST Eighteen owners share the maintenance cost of a condominium complex. If six more units are sold, the maintenance cost will be reduced by $12 per month for each of the present owners. What is the total monthly maintenance cost for the condominium complex?

58. RECTANGULAR REGION The perimeter of a rectangle is 40 inches and its area is 96 square inches. Find the length and the width of the rectangle.

59. CONSTRUCTION OF A WALL A mason can build a wall in 9 hours less than an apprentice. Together they can build the wall in 6 hours. How long would it take the apprentice, working alone, to build the wall?

60. COMMERCE An art show brought in $33,196 on the sale of 4526 tickets. The adult tickets sold for $8 and the student tickets sold for $2. How many of each type of ticket were sold?

Art Show

TICKETS

| Adults | $8 |
| Students | $2 |

61. DIAMETER OF A CONE As sand is poured from a chute, it forms a right circular cone whose height is one-fourth the diameter of the base. What is the diameter of the base when the cone has a volume of 144 cubic feet?

62. REVENUE A manufacturer of calculators finds that the monthly revenue R from a particular style of calculator is given by $R = 72x - 2x^2$, where x is the price in dollars of each calculator. Find the interval, in terms of x, for which the monthly revenue is greater than $576.

63. CONSUMER SPENDING The price of a pair of Revo sunglasses and the price of a pair of Bolle sunglasses differ by more than $48. The price of the Revo sunglasses is $218.

  a. Write an absolute value inequality that expresses the relationship between the price B, in dollars, of the Bolle sunglasses and the price, in dollars, of the Revo sunglasses.

  b. Use interval notation to describe the price range, in dollars, that is possible for the Bolle sunglasses.

64. CONSUMER SPENDING Ronda wants to rent one of two apartments. There is less than $150 difference between the monthly rental fees of the apartments. One of the apartments rents for $575 per month. What is the range of possible monthly rental fees for the other apartment?

65. SHIPPING REQUIREMENTS Federal Express (FedEx) will only ship packages for which the length is less than or equal to 119 inches and the length plus the girth is less than or equal to 165 inches. The length of a package is defined as the length of the longest side and the girth is defined as twice the width plus twice the height of the package. If a box has a length of 42 inches and a width of 38 inches, determine the possible range of heights h for this package if you wish to ship it by FedEx. (*Source:* http://www.iship.com)

66. COURSE GRADE An average of 68 to 79 in a biology class receives a C grade. A student has test scores of 82, 72, 64, and 95 on four tests. Find the range of scores on the fifth test that will give the student a C grade for the course.

67. BASKETBALL DIMENSIONS A basketball is to have a circumference of 29.5 inches to 30.0 inches. Find the acceptable range of diameters for the basketball. Round results to the nearest hundredth of an inch.

68. POPULATION DENSITY The population density D, in people per square mile, of a city is related to the horizontal distance x, in miles, from the center of the city by the equation

$$D = -45x^2 + 190x + 200, \qquad 0 < x < 5$$

Describe the region of the city in which the population density exceeds 300 people per square mile. Round critical values to the nearest tenth of a mile.

CHAPTER 1 TEST

1. Solve: $3(2x - 5) + 1 = -2(x - 5)$

2. Solve: $|x - 3| = 8$

3. Solve $6x^2 - 13x - 8 = 0$ by factoring and applying the zero product property.

4. Solve $2x^2 - 8x + 1 = 0$ by completing the square.

5. Use the quadratic formula to solve $3x^2 - 5x - 1 = 0$.

6. Determine the discriminant of $2x^2 + 3x + 1 = 0$ and state the number of real solutions of the equation.

7. Solve $ax - c = c(x - d)$ for x.

8. Solve: $\sqrt{x - 2} - 1 = \sqrt{3 - x}$

9. Solve: $3x^{2/3} + 10x^{1/3} - 8 = 0$

10. Solve: $\dfrac{3}{x + 2} - \dfrac{3}{4} = \dfrac{5}{x + 2}$

11. a. Solve the compound inequality:

$$2x - 5 \leq 11 \quad \text{or} \quad -3x + 2 > 14$$

Write the solution set using set-builder notation.

b. Solve the compound inequality:

$$2x - 1 < 9 \quad \text{and} \quad -3x + 1 \leq 7$$

Write the solution set using interval notation.

12. Solve:

$$\frac{x^2 + x - 12}{x + 1} \geq 0$$

Write the solution set using interval notation.

13. According to the National Collegiate Athletic Association (NCAA), the length x of a football, in inches, must satisfy the following inequality. (*Source:* http://www.infoplease.com.)

$$\left| x - 11\frac{5}{32} \right| \leq \frac{9}{32}$$

Find the acceptable range of lengths for an NCAA football.

14. A boat has a speed of 5 mph in still water. The boat can travel 21 miles with the current in the same time in which it can travel 9 miles against the current. Find the rate of the current.

15. A radiator contains 6 liters of a 20% antifreeze solution. How much should be drained and replaced with pure antifreeze to produce a 50% antifreeze solution?

16. A worker can cover a parking lot with asphalt in 10 hours. With the help of an assistant, the work can be done in 6 hours. How long would it take the assistant, working alone, to cover the parking lot with asphalt?

17. You can rent a car for the day from Company A for $28 plus $0.10 a mile. Company B charges $20 plus $0.18 a mile. At what point, in terms of miles driven per day, is it cheaper to rent from Company A?

18. A football field is built in the shape of a parabolic mound so that water will drain off the field. A model for the parabolic contour of the field is

$$h = -0.0002348x^2 + 0.0375x$$

where h is the height of the field, in feet, at a distance of x feet from one sideline. Describe the portion of the field for which $h > 6$ inches. Round your results to the nearest tenth of a foot.

19. The population density D, in people per square mile, of a city is related to the horizontal distance x, in miles, from the center of the city by the equation

$$D = \frac{4500x}{2x^2 + 25}, \quad 0 < x < 12$$

Describe the region of the city in which the population density exceeds 200 people per square mile.

20. The velocity of a meteorite that approaches the surface of the moon can be approximated by $v = \dfrac{219}{\sqrt{x + 1400}}$, where v is the velocity in miles per second and x is the distance in miles of the meteorite from the surface of the moon. If the velocity of the meteorite is 5 miles per second, find the distance of the meteorite above the surface of the moon. Round to the nearest mile.

CUMULATIVE REVIEW EXERCISES

1. Evaluate: $4 + 3(-5)$

2. Write 0.00017 in scientific notation.

3. Perform the indicated operations and simplify:
$(3x - 5)^2 - (x + 4)(x - 4)$

4. Factor: $8x^2 + 19x - 15$

5. Simplify: $\dfrac{7x - 3}{x - 4} - 5$

6. Simplify: $a^{2/3} \cdot a^{1/4}$

7. Find: $(2 + 5i)(2 - 5i)$

8. Solve: $2(3x - 4) + 5 = 17$

9. Solve $2x^2 - 4x = 3$ by using the quadratic formula.

10. Solve: $|2x - 6| = 4$

11. Solve: $x = 3 + \sqrt{9 - x}$

12. Factor to solve: $x^3 - 36x = 0$

13. Solve: $2x^4 - 11x^2 + 15 = 0$

14. Solve the compound inequality

$$3x - 1 > 2 \quad \text{or} \quad -3x + 5 \geq 8$$

Write the solution set using set-builder notation.

15. Solve $|x - 6| \geq 2$. Write the solution set using interval notation.

16. Solve $\dfrac{x - 2}{2x - 3} \geq 4$. Write the solution set using set-builder notation.

17. A fence built around the border of a rectangular field measures a total of 200 feet. If the length of the field is 16 feet longer than the width, find the dimensions of the field.

18. The revenue, in dollars, earned by selling x inkjet printers is given by $R = 200x - 0.004x^2$. The cost, in dollars, of manufacturing x inkjet printers is $C = 65x + 320,000$. How many printers should be manufactured and sold to earn a profit of at least $600,000?

19. An average score of 80 or above, but less than 90, in a history class receives a B grade. Rebecca has scores of 86, 72, and 94 on three tests. Find the range of scores she could receive on the fourth test that would give her a B grade for the course. Assume that the highest test score she can receive is 100.

20. A highway patrol department estimates that the cost of ticketing p percent of the speeders who travel on a freeway is given by

$$C = \frac{600p}{100 - p}, \qquad 0 < p < 100$$

where C is in thousands of dollars. If the highway patrol department plans to fund its program to ticket speeding drivers with $100,000 to $180,000, what is the range of the percent of speeders the department can expect to ticket? Round your percents to the nearest 0.1%.

CHAPTER 2

FUNCTIONS AND GRAPHS

Functions as Models

The Golden Gate Bridge spans the Golden Gate Strait, which is the entrance to the San Francisco Bay from the Pacific Ocean. Designed by Joseph Strass, the Golden Gate Bridge is a *suspension* bridge. A *quadratic function,* one of the topics of this chapter, can be used to model a cable of this bridge. See **Exercise 74 on page 210.**

Strass had many skeptics who did not believe the bridge could be built. Nonetheless, the bridge opened on May 27, 1937, a little over 4 years after construction began. When it was completed, Strass composed the following poem.

The Mighty Task Is Done

At last the mighty task is done;
Resplendent in the western sun;
The Bridge looms mountain high.

On its broad decks in rightful pride,
The world in swift parade shall ride
Throughout all time to be.

Launched midst a thousand hopes and fears,
Damned by a thousand hostile sneers,
Yet ne'er its course was stayed.
But ask of those who met the foe,
Who stood alone when faith was low,
Ask them the price they paid.
High overhead its lights shall gleam,
Far, far below life's restless stream,
Unceasingly shall flow....

Difference Tables

When devising a plan to solve a problem, it may be helpful to organize information in a table. One particular type of table that can be used to discern some patterns is called a *difference table*. For instance, suppose that we want to determine the number of square tiles in the tenth figure of a pattern whose first four figures are

We begin by creating a difference table by listing the number of tiles in each figure and the differences between the numbers of tiles on the next line.

Tiles	2		5		8		11
Differences		3		3		3	

From the difference table, note that each succeeding figure has three more tiles. Therefore, we can find the number of tiles in the tenth figure by extending the difference table.

Tiles	2		5		8		11		14		17		20		23		26		29
Differences		3		3		3		3		3		3		3		3		3	

There are 29 tiles in the tenth figure.

Sometimes the first differences are not constant as they were in the preceding example. In this case we find the second differences. For instance, consider the pattern at the right.

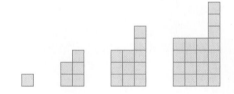

The difference table is shown below.

Tiles		1		5		11		19
First differences			4		6		8	
Second differences				2		2		

In this case the first differences are not constant, but the second differences are constant. With this information we can determine the number of tiles in succeeding figures.

Tiles		1		5		11		19		29		41		55		71		89		109
First differences			4		6		8		10		12		14		16		18		20	
Second differences				2		2		2		2		2		2		2		2		

In this case there are 109 tiles in the tenth figure.

If second differences are not constant, try third differences.[1] If third differences are not constant, try fourth differences, and so on.

---

[1] Not all lists of numbers will end with a difference row of constants. For instance, consider 1, 1, 2, 3, 5, 8, 13, 21, 34,....

A TWO-DIMENSIONAL COORDINATE SYSTEM AND GRAPHS

- CARTESIAN COORDINATE SYSTEMS
- THE DISTANCE AND MIDPOINT FORMULAS
- GRAPH OF AN EQUATION
- INTERCEPTS
- CIRCLES, THEIR EQUATIONS, AND THEIR GRAPHS

---

take note

Abscissa comes from the same root word as scissors. An open pair of scissors looks like an x.

MATH MATTERS

The concepts of *analytic geometry* developed over an extended period of time, culminating in 1637 with the publication of two works: *Discourse on the Method for Rightly Directing One's Reason and Searching for Truth in the Sciences* by René Descartes (1596–1650) and *Introduction to Plane and Solid Loci* by Pierre de Fermat. Each of these works was an attempt to integrate the study of geometry with the study of algebra. Of the two mathematicians, Descartes is usually given most of the credit for developing analytic geometry. In fact, Descartes became so famous in La Haye, the city in which he was born, that it was renamed La Haye-Descartes.

CARTESIAN COORDINATE SYSTEMS

Each point on a coordinate axis is associated with a number called its **coordinate**. Each point on a flat, two-dimensional surface, called a **coordinate plane** or *xy*-plane, is associated with an **ordered pair** of numbers called **coordinates** of the point. Ordered pairs are denoted by (a, b), where the real number a is the **x-coordinate** or **abscissa** and the real number b is the **y-coordinate** or **ordinate**.

The coordinates of a point are determined by the point's position relative to a horizontal coordinate axis called the **x-axis** and a vertical coordinate axis called the **y-axis**. The axes intersect at the point $(0, 0)$, called the **origin**. In **Figure 2.1,** the axes are labeled such that positive numbers appear to the right of the origin on the *x*-axis and above the origin on the *y*-axis. The four regions formed by the axes are called **quadrants** and are numbered counterclockwise. This two-dimensional coordinate system is referred to as a **Cartesian coordinate system** in honor of René Descartes.

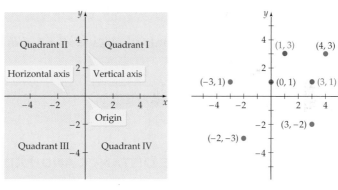

FIGURE 2.1   **FIGURE 2.2**

To **plot a point** $P(a, b)$ means to draw a dot at its location in the coordinate plane. In **Figure 2.2** we have plotted the points $(4, 3)$, $(-3, 1)$, $(-2, -3)$, $(3, -2)$, $(0, 1)$, $(1, 3)$, and $(3, 1)$. The order in which the coordinates of an ordered pair are listed is important. **Figure 2.2** shows that $(1, 3)$ and $(3, 1)$ do not denote the same point.

Data often are displayed in visual form as a set of points called a *scatter diagram* or *scatter plot*. For instance, the scatter diagram in **Figure 2.3** shows the number of Internet virus incidents from 1993 to 2003. The point whose coordinates are approximately $(7, 21,000)$ means that in the year 2000 there were approximately 21,000 Internet virus incidents. The line segments that connect the points in **Figure 2.3** help illustrate trends.

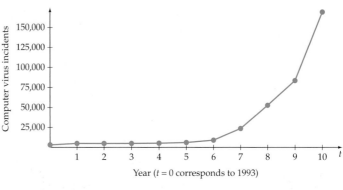

FIGURE 2.3

Source: www.cert.org

? QUESTION If the trend in **Figure 2.3** continues, will the number of virus incidents in 2004 be more or less than 200,000?

In some instances, it is important to know when two ordered pairs are equal.

Equality of Ordered Pairs

The ordered pairs (a, b) and (c, d) are equal if and only if $a = c$ and $b = d$.

For instance, if $(3, y) = (x, -2)$, then $x = 3$ and $y = -2$.

• THE DISTANCE AND MIDPOINT FORMULAS

The Cartesian coordinate system makes it possible to combine the concepts of algebra and geometry into a branch of mathematics called *analytic geometry*.

The distance between two points on a horizontal line is the absolute value of the difference between the x-coordinates of the two points. The distance between two points on a vertical line is the absolute value of the difference between the y-coordinates of the two points. For example, as shown in **Figure 2.4**, the distance d between the points with coordinates $(1, 2)$ and $(1, -3)$ is $d = |2 - (-3)| = 5$.

If two points are not on a horizontal or vertical line, then a *distance formula* for the distance between the two points can be developed as follows.

The distance between the points $P_1(x_1, y_1)$ and $P_2(x_2, y_2)$ in **Figure 2.5** is the length of the hypotenuse of a right triangle whose sides are horizontal and verti-

FIGURE 2.4

? ANSWER More. The increase between 2002 and 2003 was more than 70,000. If this trend continues, the increase between 2003 and 2004 will be at least 70,000 more than 150,000. That is, the number of virus incidents in 2004 will be at least 220,000.

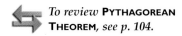

To review **PYTHAGOREAN THEOREM**, *see p. 104.*

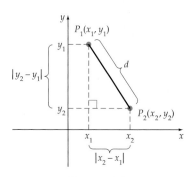

FIGURE 2.5

cal line segments that measure $|x_2 - x_1|$ and $|y_2 - y_1|$, respectively. Applying the Pythagorean Theorem to this triangle produces

$$d^2 = |x_2 - x_1|^2 + |y_2 - y_1|^2$$
$$d = \sqrt{|x_2 - x_1|^2 + |y_2 - y_1|^2}$$

• **The square root theorem. Because *d* is nonnegative, the negative root is not listed.**

$$= \sqrt{(x_2 - x_1)^2 + (y_2 - y_1)^2}$$

• **Because $|x_2 - x_1|^2 = (x_2 - x_1)^2$ and $|y_2 - y_1|^2 = (y_2 - y_1)^2$**

Thus we have established the following theorem.

The Distance Formula

The distance d between the points $P_1(x_1, y_1)$ and $P_2(x_2, y_2)$ is

$$d = \sqrt{(x_2 - x_1)^2 + (y_2 - y_1)^2}$$

The distance d between the points whose coordinates are $P_1(x_1, y_1)$ and $P_2(x_2, y_2)$ is denoted by $d(P_1, P_2)$. To find the distance $d(P_1, P_2)$ between the points $P_1(-3, 4)$ and $P_2(7, 2)$, we apply the distance formula with $x_1 = -3$, $y_1 = 4$, $x_2 = 7$, and $y_2 = 2$.

$$d(P_1, P_2) = \sqrt{(x_2 - x_1)^2 + (y_2 - y_1)^2}$$
$$= \sqrt{[7 - (-3)]^2 + (2 - 4)^2}$$
$$= \sqrt{104} = 2\sqrt{26} \approx 10.2$$

The **midpoint** M of a line segment is the point on the line segment that is equidistant from the endpoints $P_1(x_1, y_1)$ and $P_2(x_2, y_2)$ of the segment. See **Figure 2.6**.

The Midpoint Formula

The midpoint M of the line segment from $P_1(x_1, y_1)$ to $P_2(x_2, y_2)$ is given by

$$\left(\frac{x_1 + x_2}{2}, \frac{y_1 + y_2}{2} \right)$$

FIGURE 2.6

The midpoint formula states that the x-coordinate of the midpoint of a line segment is the *average* of the x-coordinates of the endpoints of the line segment and that the y-coordinate of the midpoint of a line segment is the *average* of the y-coordinates of the endpoints of the line segment.

The midpoint M of the line segment connecting $P_1(-2, 6)$ and $P_2(3, 4)$ is

$$M = \left(\frac{x_1 + x_2}{2}, \frac{y_1 + y_2}{2} \right) = \left(\frac{(-2) + 3}{2}, \frac{6 + 4}{2} \right) = \left(\frac{1}{2}, 5 \right)$$

EXAMPLE 1 **Find the Midpoint and Length of a Line Segment**

Find the midpoint and the length of the line segment connecting the points whose coordinates are $P_1(-4, 3)$ and $P_2(4, -2)$.

Solution

$$\text{Midpoint} = \left(\frac{x_1 + x_2}{2}, \frac{y_1 + y_2}{2} \right)$$

$$= \left(\frac{-4 + 4}{2}, \frac{3 + (-2)}{2} \right)$$

$$= \left(0, \frac{1}{2} \right)$$

$$d(P_1, P_2) = \sqrt{(x_2 - x_1)^2 + (y_2 - y_1)^2}$$

$$= \sqrt{(4 - (-4))^2 + (-2 - 3)^2} = \sqrt{(8)^2 + (-5)^2}$$

$$= \sqrt{64 + 25} = \sqrt{89}$$

▶ **TRY EXERCISE 6, PAGE 158**

● GRAPH OF AN EQUATION

The equations below are equations in two variables.

$$y = 3x^3 - 4x + 2 \qquad x^2 + y^2 = 25 \qquad y = \frac{x}{x + 1}$$

The solution of an equation in two variables is an ordered pair (x, y) whose coordinates satisfy the equation. For instance, the ordered pairs $(3, 4)$, $(4, -3)$, and $(0, 5)$ are some of the solutions of $x^2 + y^2 = 25$. Generally, there are an infinite number of solutions of an equation in two variables. These solutions can be displayed in a *graph*.

Graph of an Equation
The **graph of an equation** in the two variables x and y is the set of all points whose coordinates satisfy the equation.

Consider $y = 2x - 1$. Substituting various values of x into the equation and solving for y produces some of the ordered pairs of the equation. It is convenient to record the results in a table similar to the one shown on the following page. The graph of the ordered pairs is shown in **Figure 2.7**.

x	y = 2x − 1	y	(x, y)
−2	2(−2) − 1	−5	(−2, −5)
−1	2(−1) − 1	−3	(−1, −3)
0	2(0) − 1	−1	(0, −1)
1	2(1) − 1	1	(1, 1)
2	2(2) − 1	3	(2, 3)

Choosing some noninteger values of x produces more ordered pairs to graph, such as $\left(-\dfrac{3}{2}, -4\right)$ and $\left(\dfrac{5}{2}, 4\right)$, as shown in **Figure 2.8**. Using still other values of x would result in more and more ordered pairs being graphed. The result would be so many dots that the graph would appear as the straight line shown in **Figure 2.9**, which is the graph of $y = 2x - 1$.

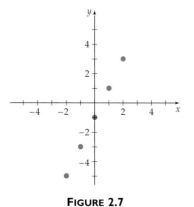

FIGURE 2.7

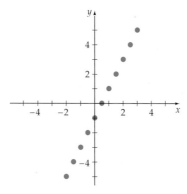

FIGURE 2.8

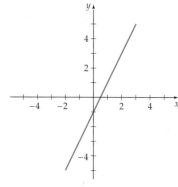

FIGURE 2.9

EXAMPLE 2   **Draw a Graph by Plotting Points**

Graph: $-x^2 + y = 1$

Solution

Solve the equation for y.

$$y = x^2 + 1$$

Select values of x and use the equation to calculate y. Choose enough values of x so that an accurate graph can be drawn. Plot the points and draw a curve through them. See **Figure 2.10**.

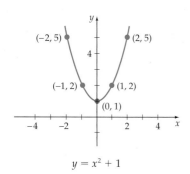

$y = x^2 + 1$

FIGURE 2.10

x	y = x² + 1	y	(x, y)
−2	(−2)² + 1	5	(−2, 5)
−1	(−1)² + 1	2	(−1, 2)
0	(0)² + 1	1	(0, 1)
1	(1)² + 1	2	(1, 2)
2	(2)² + 1	5	(2, 5)

▶ **TRY EXERCISE 26, PAGE 159**

MATH MATTERS

Maria Agnesi (1718–1799) wrote *Foundations of Analysis for the Use of Italian Youth*, one of the most successful textbooks of the 18th century. The French Academy authorized a translation into French in 1749, noting that "there is no other book, in any language, which would enable a reader to penetrate as deeply, or as rapidly, into the fundamental concepts of analysis." A curve that she discusses in her text is given by the equation $y = \dfrac{a^3}{x^2 + a^2}$.

Unfortunately, due to a translation error from Italian to English, the curve became known as the "witch of Agnesi."

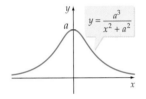

INTEGRATING TECHNOLOGY

Some graphing calculators, such as the *TI-83*, have a TABLE feature that allows you to create a table similar to the one shown in Example 2. Enter the equation to be graphed, the first value for x, and the increment (the difference between successive values of x). For instance, entering $y_1 = x^2 + 1$, an initial value for x of -2, and an increment of 1 yields a display similar to the one in **Figure 2.11**. Changing the initial value to -6 and the increment to 2 gives the table in **Figure 2.12**.

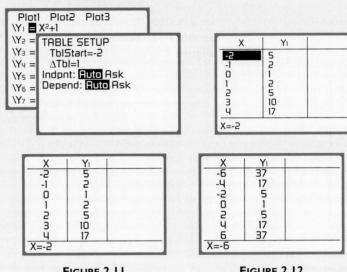

FIGURE 2.11   **FIGURE 2.12**

With some calculators, you may scroll through the table by using the up- or down-arrow keys. In this way, you can determine many more ordered pairs of the graph.

EXAMPLE 3   Graph by Plotting Points

Graph: $y = |x - 2|$

Solution

This equation is already solved for y, so start by choosing an x value and using the equation to determine the corresponding y value. For example, if $x = -3$, then $y = |(-3) - 2| = |-5| = 5$. Continuing in this manner produces the following table:

When x is	−3	−2	−1	0	1	2	3	4	5
y is	5	4	3	2	1	0	1	2	3

Now plot the points listed in the table. Connecting the points forms a V shape, as shown in **Figure 2.13**.

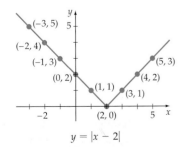

$y = |x - 2|$

FIGURE 2.13

► **TRY EXERCISE 30, PAGE 159**

EXAMPLE 4   **Graph by Plotting Points**

Graph: $y^2 = x$

Solution

Solve the equation for y.

$$y^2 = x$$
$$y = \pm\sqrt{x}$$

Choose several x values, and use the equation to determine the corresponding y values.

When x is	0	1	4	9	16
y is	0	± 1	± 2	± 3	± 4

Plot the points as shown in **Figure 2.14.** The graph is a *parabola*.

▶ **TRY EXERCISE 32, PAGE 159**

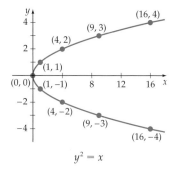

$$y^2 = x$$

FIGURE 2.14

 INTEGRATING TECHNOLOGY

A graphing calculator or computer graphing software can be used to draw the graphs in Examples 3 and 4. These graphing utilities graph a curve in much the same way as you would, by selecting values of x and calculating the corresponding values of y. A curve is then drawn through the points.

If you use a graphing utility to graph $y = |x - 2|$, you will need to use the *absolute value* function that is built into the utility. The equation you enter will look similar to Y₁=abs(X–2).

To graph the equation in Example 4, you will enter two equations. The equations you enter will be similar to

$$Y_1 = \sqrt{(X)}$$
$$Y_2 = -\sqrt{(X)}$$

The graph of the first equation will be the top half of the parabola; the graph of the second equation will be the bottom half.

● **INTERCEPTS**

Any point that has an x- or a y-coordinate of zero is called an **intercept** of the graph of an equation because it is at these points that the graph intersects the x- or the y-axis.

> ### Definition of x-Intercepts and y-Intercepts
>
> If $(x_1, 0)$ satisfies an equation, then the point $(x_1, 0)$ is called an **x-intercept** of the graph of the equation.
>
> If $(0, y_1)$ satisfies an equation, then the point $(0, y_1)$ is called a **y-intercept** of the graph of the equation.

To find the x-intercepts of the graph of an equation, let $y = 0$, and solve the equation for x. To find the y-intercepts of the graph of an equation, let $x = 0$, and solve the equation for y.

EXAMPLE 5   **Find x- and y-Intercepts**

Find the x- and y-intercepts of the graph of $y = x^2 - 2x - 3$.

Algebraic Solution

To find the y-intercept, let $x = 0$ and solve for y.

$$y = 0^2 - 2(0) - 3 = -3$$

To find the x-intercepts, let $y = 0$ and solve for x.

$$0 = x^2 - 2x - 3$$
$$0 = (x - 3)(x + 1)$$
$$(x - 3) = 0 \quad \text{or} \quad (x + 1) = 0$$
$$x = 3 \quad \text{or} \qquad x = -1$$

Because $y = -3$ when $x = 0$, $(0, -3)$ is a y-intercept. Because $x = 3$ or -1 when $y = 0$, $(3, 0)$ and $(-1, 0)$ are x-intercepts. **Figure 2.15** confirms that these three points are intercepts.

Visualize the Solution

The graph of $y = x^2 - 2x - 3$ is shown below. Observe that the graph intersects the x-axis at $(-1, 0)$ and $(3, 0)$, the x-intercepts. The graph also intersects the y-axis at $(0, -3)$, the y-intercept.

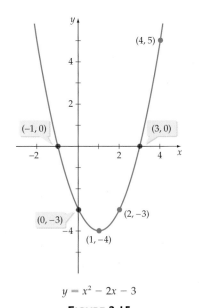

$$y = x^2 - 2x - 3$$

FIGURE 2.15

▶ **TRY EXERCISE 40, PAGE 159**

In Example 5 it was possible to find the x-intercepts by solving a quadratic equation. In some instances, however, solving an equation to find the intercepts may be very difficult. In these cases, a graphing calculator can be used to estimate the x-intercepts.

The x-intercepts of the graph of $y = x^3 + x + 4$ can be estimated using the INTERCEPT feature of a TI-83 calculator. The keystrokes and some sample screens for this procedure are shown below.

Press [Y=]. Now enter X^3+X+4. Press [ZOOM] and select the standard viewing window.

Press [2nd] CALC to access the CALCULATE menu. The y-coordinate of an x-intercept is zero. Therefore, select 2:zero. Press [ENTER].

The "Left Bound?" shown on the bottom of the screen means to move the cursor until it is to the left of an x-intercept. Press [ENTER].

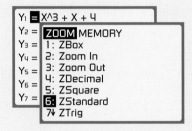

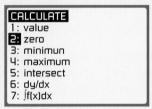

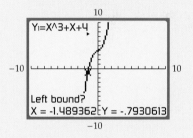

The "Right Bound?" shown on the bottom of the screen means to move the cursor until it is to the right of the desired x-intercept. Press [ENTER].

"Guess?" is shown on the bottom of the screen. Move the cursor until it is approximately on the x-intercept. Press [ENTER].

The "Zero" shown on the bottom of the screen means that the value of y is 0 when $x = -1.378797$. The x-intercept is about $(-1.378797, 0)$.

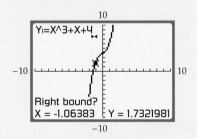

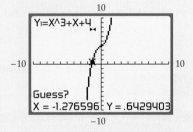

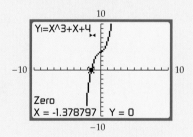

● CIRCLES, THEIR EQUATIONS, AND THEIR GRAPHS

Frequently you will sketch graphs by plotting points. However, some graphs can be sketched merely by recognizing the form of the equation. A *circle* is an example of a curve whose graph you can sketch after you have inspected its equation.

Definition of a Circle

A **circle** is the set of points in a plane that are a fixed distance from a specified point. The distance is the **radius** of the circle, and the specified point is the **center** of the circle.

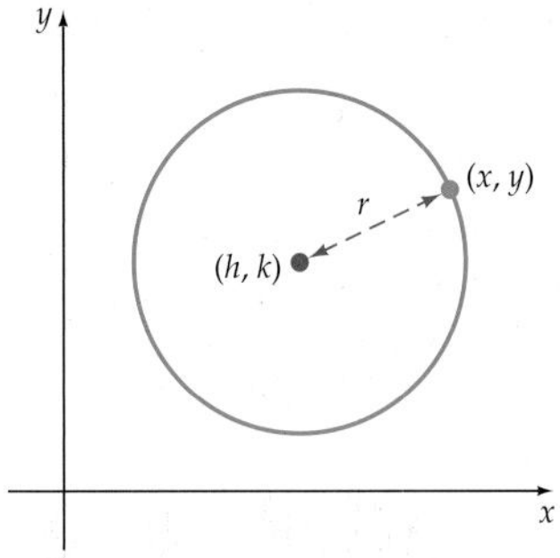

FIGURE 2.16

The standard form of the equation of a circle is derived by using this definition. To derive the standard form, we use the distance formula. **Figure 2.16** is a circle with center (h, k) and radius r. The point (x, y) is on the circle if and only if it is a distance of r units from the center (h, k). Thus (x, y) is on the circle if and only if

$$\sqrt{(x - h)^2 + (y - k)^2} = r$$
$$(x - h)^2 + (y - k)^2 = r^2 \qquad \bullet \text{ Square each side.}$$

Standard Form of the Equation of a Circle

The **standard form of the equation of a circle** with center at (h, k) and radius r is

$$(x - h)^2 + (y - k)^2 = r^2$$

For example, the equation $(x - 3)^2 + (y + 1)^2 = 4$ is the equation of a circle. The standard form of the equation is

$$(x - 3)^2 + (y - (-1))^2 = 2^2$$

from which it can be determined that $h = 3$, $k = -1$, and $r = 2$. Thus the graph is a circle centered at $(3, -1)$ with a radius of 2.

If a circle is centered at the origin $(0, 0)$ (that is, if $h = 0$ and $k = 0$), then the standard form of the equation of the circle simplifies to

$$x^2 + y^2 = r^2$$

For example, the graph of $x^2 + y^2 = 9$ is a circle with center at the origin and radius of 3.

? QUESTION What are the radius and the coordinates of the center of the circle with equation $x^2 + (y - 2)^2 = 10$?

EXAMPLE 6 Find the Standard Form of the Equation of a Circle

Find the standard form of the equation of the circle that has center $C(-4, -2)$ and contains the point $P(-1, 2)$.

Solution

See the graph of the circle in **Figure 2.17.** Because the point P is on the circle, the radius r of the circle must equal the distance from C to P. Thus

$$r = \sqrt{(-1 - (-4))^2 + (2 - (-2))^2}$$
$$= \sqrt{9 + 16} = \sqrt{25} = 5$$

Using the standard form with $h = -4$, $k = -2$, and $r = 5$, we obtain

$$(x + 4)^2 + (y + 2)^2 = 5^2$$

▶ **TRY EXERCISE 64, PAGE 159**

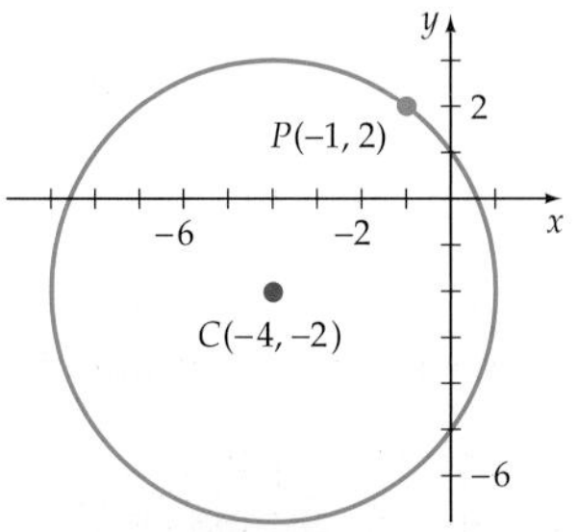

$(x + 4)^2 + (y + 2)^2 = 5^2$

FIGURE 2.17

? ANSWER The radius is $\sqrt{10}$ and the coordinates of the center are $(0, 2)$.

If we rewrite $(x + 4)^2 + (y + 2)^2 = 5^2$ by squaring and combining like terms, we produce

$$x^2 + 8x + 16 + y^2 + 4y + 4 = 25$$
$$x^2 + y^2 + 8x + 4y - 5 = 0$$

This form of the equation is known as the **general form of the equation of a circle.** By completing the square, it is always possible to write the general form $x^2 + y^2 + Ax + By + C = 0$ in the standard form

$$(x - h)^2 + (y - k)^2 = s$$

for some number s. If $s > 0$, the graph is a circle with radius $r = \sqrt{s}$. If $s = 0$, the graph is the point (h, k), and if $s < 0$, the equation has no real solutions and there is no graph.

EXAMPLE 7   **Find the Center and Radius of a Circle by Completing the Square**

Find the center and the radius of the circle that is given by

$$x^2 + y^2 - 6x + 4y - 3 = 0$$

Solution

To review **COMPLETING THE SQUARE**, *see p. 99.*

First rearrange and group the terms as shown.

$$(x^2 - 6x) + (y^2 + 4y) = 3$$

Now complete the squares of $(x^2 - 6x)$ and $(y^2 + 4y)$.

$$(x^2 - 6x + 9) + (y^2 + 4y + 4) = 3 + 9 + 4 \qquad \text{• Add 9 and 4 to each side of the equation.}$$

$$(x - 3)^2 + (y + 2)^2 = 16$$
$$(x - 3)^2 + (y - (-2))^2 = 4^2$$

This equation is the standard form of the equation of a circle and indicates that the graph of the original equation is a circle centered at $(3, -2)$ with radius 4. See **Figure 2.18.**

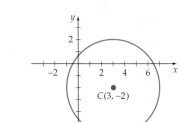

$$x^2 + y^2 - 6x + 4y - 3 = 0$$

FIGURE 2.18

▶ **TRY EXERCISE 66, PAGE 159**

TOPICS FOR DISCUSSION

1. The distance formula states that the distance d between the points $P_1(x_1, y_1)$ and $P_2(x_2, y_2)$ is $d = \sqrt{(x_2 - x_1)^2 + (y_2 - y_1)^2}$. Can the distance formula also be written as follows? Explain.

$$d = \sqrt{(x_1 - x_2)^2 + (y_1 - y_2)^2}$$

2. Does the equation $(x - 3)^2 + (y + 4)^2 = -6$ have a graph that is a circle? Explain.

3. Explain why the graph of $|x| + |y| = 1$ does not contain any points that have

   a. a y-coordinate that is greater than 1 or less than -1

   b. an x-coordinate that is greater than 1 or less than -1

4. Discuss the graph of $xy = 0$.

5. Explain how to determine the x- and y-intercepts of a graph defined by an equation (without using the graph).

EXERCISE SET 2.1

In Exercises 1 and 2, plot the points whose coordinates are given on a Cartesian coordinate system.

1. $(2, 4), (0, -3), (-2, 1), (-5, -3)$

2. $(-3, -5), (-4, 3), (0, 2), (-2, 0)$

3. ⬤ **HEALTH** A study at the Ohio State University measured the changes in heart rates of students doing stepping exercises. Students stepped onto a platform that was approximately 11 inches high at a rate of 14 steps per minute. The heart rate, in beats per minute, before and after the exercise is given in the table below.

Before	After	Before	After
63	84	96	141
72	99	69	93
87	111	81	96
90	129	75	90
90	108	84	90

a. Draw a scatter diagram for these data.

b. For these students, what is the average increase in heart rate?

4. ⬤ **AVERAGE INCOME** The following graph, based on data from the Bureau of Economic Analysis, shows per capita personal income in the United States.

a. From the data, does it appear that per capita personal income is increasing, decreasing, or remaining the same?

b. If per capita personal income continues to increase by the same percent as the percent increase between 2001 and 2002, what will be the per capita income in 2004?

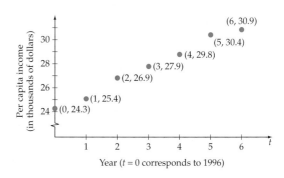

In Exercises 5 to 16, find the distance between the points whose coordinates are given.

5. $(6, 4), (-8, 11)$ ▶ **6.** $(-5, 8), (-10, 14)$

7. $(-4, -20), (-10, 15)$ **8.** $(40, 32), (36, 20)$

9. $(5, -8), (0, 0)$ **10.** $(0, 0), (5, 13)$

11. $(\sqrt{3}, \sqrt{8}), (\sqrt{12}, \sqrt{27})$ **12.** $(\sqrt{125}, \sqrt{20}), (6, 2\sqrt{5})$

13. $(a, b), (-a, -b)$ **14.** $(a - b, b), (a, a + b)$

15. $(x, 4x), (-2x, 3x)$ given that $x < 0$

16. $(x, 4x), (-2x, 3x)$ given that $x > 0$

17. Find all points on the x-axis that are 10 units from $(4, 6)$. (*Hint:* First write the distance formula with $(4, 6)$ as one of the points and $(x, 0)$ as the other point.)

18. Find all points on the y-axis that are 12 units from $(5, -3)$.

In Exercises 19 to 24, find the midpoint of the line segment with the following endpoints.

19. $(1, -1), (5, 5)$ **20.** $(-5, -2), (6, 10)$

21. $(6, -3), (6, 11)$ **22.** $(4, 7), (-10, 7)$

23. $(1.75, 2.25), (-3.5, 5.57)$ **24.** $(-8.2, 10.1), (-2.4, -5.7)$

In Exercises 25 to 38, graph each equation by plotting points that satisfy the equation.

25. $x - y = 4$ ▶ **26.** $2x + y = -1$

27. $y = 0.25x^2$ **28.** $3x^2 + 2y = -4$

29. $y = -2|x - 3|$ ▶ **30.** $y = |x + 3| - 2$

31. $y = x^2 - 3$ ▶ **32.** $y = x^2 + 1$

33. $y = \dfrac{1}{2}(x - 1)^2$ **34.** $y = 2(x + 2)^2$

35. $y = x^2 + 2x - 8$ **36.** $y = x^2 - 2x - 8$

37. $y = -x^2 + 2$ **38.** $y = -x^2 - 1$

In Exercises 39 to 48, find the x- and y-intercepts of the graph of each equation. Use the intercepts and additional points as needed to draw the graph of the equation.

39. $2x + 5y = 12$ ▶ **40.** $3x - 4y = 15$

41. $x = -y^2 + 5$ **42.** $x = y^2 - 6$

43. $x = |y| - 4$ **44.** $x = y^3 - 2$

45. $x^2 + y^2 = 4$ **46.** $x^2 = y^2$

47. $|x| + |y| = 4$ **48.** $|x - 4y| = 8$

In Exercises 49 to 56, determine the center and radius of the circle with the given equation.

49. $x^2 + y^2 = 36$ **50.** $x^2 + y^2 = 49$

51. $(x - 1)^2 + (y - 3)^2 = 49$ **52.** $(x - 2)^2 + (y - 4)^2 = 25$

53. $(x + 2)^2 + (y + 5)^2 = 25$

54. $(x + 3)^2 + (y + 5)^2 = 121$

55. $(x - 8)^2 + y^2 = \dfrac{1}{4}$ **56.** $x^2 + (y - 12)^2 = 1$

In Exercises 57 to 64, find an equation of a circle that satisfies the given conditions. Write your answer in standard form.

57. Center $(4, 1)$, radius $r = 2$

58. Center $(5, -3)$, radius $r = 4$

59. Center $\left(\dfrac{1}{2}, \dfrac{1}{4}\right)$, radius $r = \sqrt{5}$

60. Center $\left(0, \dfrac{2}{3}\right)$, radius $r = \sqrt{11}$

61. Center $(0, 0)$, passing through $(-3, 4)$

62. Center $(0, 0)$, passing through $(5, 12)$

63. Center $(1, 3)$, passing through $(4, -1)$

▶ **64.** Center $(-2, 5)$, passing through $(1, 7)$

In Exercises 65 to 72, find the center and the radius of the graph of the circle. The equations of the circles are written in the general form.

65. $x^2 + y^2 - 6x + 5 = 0$

▶ **66.** $x^2 + y^2 - 6x - 4y + 12 = 0$

67. $x^2 + y^2 - 14x + 8y + 56 = 0$

68. $x^2 + y^2 - 10x + 2y + 25 = 0$

69. $4x^2 + 4y^2 + 4x - 63 = 0$

70. $9x^2 + 9y^2 - 6y - 17 = 0$

71. $x^2 + y^2 - x + 3y - \dfrac{15}{4} = 0$

72. $x^2 + y^2 + 3x - 5y + \dfrac{25}{4} = 0$

73. Find an equation of a circle that has a diameter with endpoints $(2, 3)$ and $(-4, 11)$. Write your answer in standard form.

74. Find an equation of a circle that has a diameter with endpoints $(7, -2)$ and $(-3, 5)$. Write your answer in standard form.

75. Find an equation of a circle that has its center at $(7, 11)$ and is tangent to the x-axis. Write your answer in standard form.

76. Find an equation of a circle that has its center at $(-2, 3)$ and is tangent to the y-axis. Write your answer in standard form.

CONNECTING CONCEPTS

In Exercises 77 to 86, graph the set of all points whose
x- and y-coordinates satisfy the given conditions.

77. $x = 1, y \geq 1$

78. $y = -3, x \geq -2$

79. $y \leq 3$

80. $x \geq 2$

81. $xy \geq 0$

82. $|y| \geq 1, \dfrac{x}{y} \leq 0$

83. $|x| = 2, |y| = 3$

84. $|x| = 4, |y| = 1$

85. $|x| \leq 2, y \geq 2$

86. $x \geq 1, |y| \leq 3$

In Exercises 87 to 90, find the other endpoint of the line
segment that has the given endpoint and midpoint.

87. Endpoint $(5, 1)$, midpoint $(9, 3)$

88. Endpoint $(4, -6)$, midpoint $(-2, 11)$

89. Endpoint $(-3, -8)$, midpoint $(2, -7)$

90. Endpoint $(5, -4)$, midpoint $(0, 0)$

91. Find a formula for the set of all points (x, y) for which the distance from (x, y) to $(3, 4)$ is 5.

92. Find a formula for the set of all points (x, y) for which the distance from (x, y) to $(-5, 12)$ is 13.

93. Find a formula for the set of all points (x, y) for which the sum of the distances from (x, y) to $(4, 0)$ and from (x, y) to $(-4, 0)$ is 10.

94. Find a formula for the set of all points for which the absolute value of the difference of the distances from (x, y) to $(0, 4)$ and from (x, y) to $(0, -4)$ is 6.

95. Find an equation of a circle that is tangent to both axes, has its center in the second quadrant, and has a radius of 3.

96. Find an equation of a circle that is tangent to both axes, has its center in the third quadrant, and has a diameter of $\sqrt{5}$.

PREPARE FOR SECTION 2.2

---

97. Evaluate $x^2 + 3x - 4$ when $x = -3$. [P.1]

98. From the set of ordered pairs $A = \{(-3, 2), (-2, 4), (-1, 1), (0, 4), (2, 5)\}$, create two new sets, D and R, where D is the set of the first coordinates of the ordered pairs of A and R is the set of the second coordinates of the ordered pairs of A. [P.1]

99. Find the length of the line segment connecting $P_1(-4, 1)$ and $P_2(3, -2)$. [2.1]

100. For what values of x is $\sqrt{2x - 6}$ a real number? [P.5/1.5]

101. For what values of x is $\dfrac{x + 3}{x^2 - x - 6}$ not a real number? [P.4]

102. If $a = 3x + 4$ and $a = 6x - 5$, find the value of a. [1.1]

PROJECTS

---

1. VERIFY A GEOMETRIC THEOREM Use the midpoint formula and the distance formula to prove that the midpoint M of the hypotenuse of a right triangle is equidistant from each of the vertices of the triangle. (*Hint:* Label the vertices of the triangle as shown in the figure at the right.)

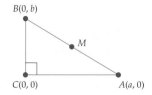

2. SOLVE A QUADRATIC EQUATION GEOMETRICALLY In the 17th century, Descartes (and others) solved equations by using both algebra and geometry. This project outlines the method Descartes used to solve certain quadratic equations.

a. Consider the equation $x^2 = 2ax + b^2$. Construct a right triangle ABC with $d(A, C) = a$ and $d(C, B) = b$. Now draw a circle with center at A and radius a. Let P be the point at which the circle intersects the hypotenuse of the right triangle and Q the point at which an extension of the hypotenuse intersects the circle. Your drawing should be similar to the one at the right.

b. Show that a solution of the equation $x^2 = 2ax + b^2$ is $d(Q, B)$.

c. Show that $d(P, B)$ is a solution of the equation $x^2 = -2ax + b^2$.

d. Construct a line parallel to AC and passing through B. Let S and T be the points at which the line intersects the circle. Show that $d(S, B)$ and $d(T, B)$ are solutions of the equation $x^2 = 2ax - b^2$.

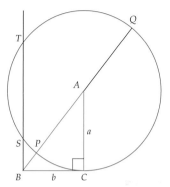

INTRODUCTION TO FUNCTIONS

● RELATIONS

In many situations in science, business, and mathematics, a correspondence exists between two sets. The correspondence is often defined by a *table*, an *equation*, or a *graph*, each of which can be viewed from a mathematical perspective as a set of ordered pairs. In mathematics, any set of ordered pairs is called a **relation.**

Table 2.1 defines a correspondence between a set of percent scores and a set of letter grades. For each score from 0 to 100, there corresponds only one letter grade. The score 94% corresponds to the letter grade of A. Using ordered-pair notation, we record this correspondence as (94, A).

The *equation* $d = 16t^2$ indicates that the distance d that a rock falls (neglecting air resistance) corresponds to the time t that it has been falling. For each nonnegative value t, the equation assigns only one value for the distance d. According to this equation, in 3 seconds a rock will fall 144 feet, which we record as (3, 144). Some of the other ordered pairs determined by $d = 16t^2$ are (0, 0), (1, 16), (2, 64), and (2.5, 100).

$$\text{Equation:} \qquad d = 16t^2$$
$$\text{If } t = 3, \text{ then} \quad d = 16(3)^2 = 144$$

The *graph* in **Figure 2.19** defines a correspondence between the length of a pendulum and the time it takes the pendulum to complete one oscillation. For each nonnegative pendulum length, the graph yields only one time. According to the graph, a pendulum length of 2 feet yields an oscillation time of 1.6 seconds, and a length of 4 feet yields an oscillation time of 2.2 seconds, where the time is measured to the nearest tenth of a second. These results can be recorded as the ordered pairs (2, 1.6) and (4, 2.2).

TABLE 2.1

Score	Grade
[90, 100]	A
[80, 90)	B
[70, 80)	C
[60, 70)	D
[0, 60)	F

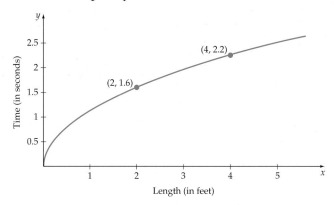

Graph: A pendulum's oscillation time

FIGURE 2.19

● FUNCTIONS

The preceding table, equation, and graph each determines a special type of relation called a *function.*

MATH MATTERS

It is generally agreed among historians that Leonhard Euler (1707–1783) was the first person to use the word *function.* His definition of function occurs in his book *Introduction to Analysis of the Infinite,* published in 1748. Euler contributed to many areas of mathematics and was one of the most prolific expositors of mathematics.

Definition of a Function

A **function** is a set of ordered pairs in which no two ordered pairs have the same first coordinate and different second coordinates.

Although every function is a relation, not every relation is a function. For instance, consider (94, A) from the grading correspondence. The first coordinate, 94, is paired with a second coordinate of A. It would not make sense to have 94 paired with A, (94, A), and 94 paired with B, (94, B). The same first coordinate would be paired with two different second coordinates. This would mean that two students with the same score received different grades, one student an A and the other a B!

Functions may have ordered pairs with the same second coordinate. For instance, (94, A) and (95, A) are both ordered pairs that belong to the function defined by **Table 2.1.** A function may have different first coordinates and the same second coordinate.

The equation $d = 16t^2$ represents a function because for each value of t there is only one value of d. Not every equation, however, represents a function. For instance, $y^2 = 25 - x^2$ does not represent a function. The ordered pairs $(-3, 4)$ and $(-3, -4)$ are both solutions of the equation. But these ordered pairs do not satisfy the definition of a function; there are two ordered pairs with the same first coordinate but *different* second coordinates.

❓ **QUESTION** Does the set $\{(0, 0), (1, 0), (2, 0), (3, 0), (4, 0)\}$ define a function?

The **domain** of a function is the set of all the first coordinates of the ordered pairs. The **range** of a function is the set of all the second coordinates. In the func-

❓ **ANSWER** Yes. There are no two ordered pairs with the same first coordinate that have different second coordinates.

tion determined by the grading correspondence in **Table 2.1,** the domain is the interval [0, 100]. The range is {A, B, C, D, F}. In a function, each domain element is paired with one and only one range element.

If a function is defined by an equation, the variable that represents elements of the domain is the **independent variable.** The variable that represents elements of the range is the **dependent variable.** In the free-fall experiment, we used the equation $d = 16t^2$. The elements of the domain represented the time the rock fell, and the elements of the range represented the distance the rock fell. Thus, in $d = 16t^2$, the independent variable is t and the dependent variable is d.

The specific letters used for the independent and the dependent variable are not important. For example, $y = 16x^2$ represents the same function as $d = 16t^2$. Traditionally, x is used for the independent variable and y for the dependent variable. Anytime we use the phrase "y is a function of x" or a similar phrase with different letters, the variable that follows "function of" is the independent variable.

● FUNCTIONAL NOTATION

Functions can be named by using a letter or a combination of letters, such as f, g, A, log, or tan. If x is an element of the domain of f, then $f(x)$, which is read "f of x" or "the value of f at x," is the element in the range of f that corresponds to the domain element x. The notation "f" and the notation "$f(x)$" mean different things. "f" is the name of the function, whereas "$f(x)$" is the value of the function at x. Finding the value of $f(x)$ is referred to as *evaluating f* at x. To evaluate $f(x)$ at $x = a$, substitute a for x, and simplify.

EXAMPLE 1   **Evaluate Functions**

Let $f(x) = x^2 - 1$, and evaluate.

  a. $f(-5)$ **b.** $f(3b)$ **c.** $3f(b)$ **d.** $f(a + 3)$ **e.** $f(a) + f(3)$

Solution

  a. $f(-5) = (-5)^2 - 1 = 25 - 1 = 24$ • Substitute **−5** for x, and simplify.

  b. $f(3b) = (3b)^2 - 1 = 9b^2 - 1$ • Substitute **3b** for x, and simplify.

  c. $3f(b) = 3(b^2 - 1) = 3b^2 - 3$ • Substitute **b** for x, and simplify.

  d. $f(a + 3) = (a + 3)^2 - 1$ • Substitute **a + 3** for x.
  $\qquad\quad\ = a^2 + 6a + 8$ • Simplify.

  e. $f(a) + f(3) = (a^2 - 1) + (3^2 - 1)$ • Substitute **a** for x; substitute **3** for x.

  $\qquad\qquad\quad = a^2 + 7$ • Simplify.

▶ **TRY EXERCISE 2, PAGE 174**

---

take note

In Example 1, observe that

$$f(3b) \neq 3f(b)$$

and that

$$f(a + 3) \neq f(a) + f(3)$$

Piecewise-defined functions are functions represented by more than one expression. The function shown below is an example of a piecewise-defined function.

$$f(x) = \begin{cases} 2x, & x < -2 \\ x^2, & -2 \leq x < 1 \\ 4 - x, & x \geq 1 \end{cases}$$

• This function is made up of different *pieces*, $2x$, x^2, and $4 - x$, depending on the value of x.

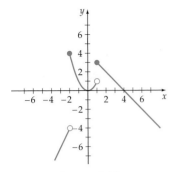

FIGURE 2.20

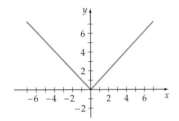

FIGURE 2.21

The expression that is used to evaluate this function depends on the value of x. For instance, to find $f(-3)$, we note that $-3 < -2$ and therefore use the expression $2x$ to evaluate the function.

$$f(-3) = 2(-3) = -6 \qquad \bullet \textbf{When } x < -2, \textbf{ use the expression } 2x.$$

Here are some additional instances of evaluating this function:

$$f(-1) = (-1)^2 = 1 \qquad \bullet \textbf{When } x \textbf{ satisfies } -2 \le x < 1,$$
$$\textbf{use the expression } x^2.$$

$$f(4) = 4 - 4 = 0 \qquad \bullet \textbf{When } x \ge 1, \textbf{ use the expression } 4 - x.$$

The graph of this function is shown in **Figure 2.20.** Note the use of the open and closed circles at the endpoints of the intervals. These circles are used to show the evaluation of the function at the endpoints of each interval. For instance, because -2 is in the interval $-2 \le x < 1$, the value of the function at -2 is 4 $[f(-2) = (-2)^2 = 4]$. Therefore a closed dot is placed at $(-2, 4)$. Similarly, when $x = 1$, because 1 is in the interval $x \ge 1$, the value of the function at 1 is 3 $(f(1) = 4 - 1 = 3)$.

? QUESTION Evaluate the function f defined at the bottom of page 163 when $x = 0.5$.

The absolute value function is another example of a piecewise-defined function. Below is the definition of this function, which is sometimes abbreviated $\text{abs}(x)$. Its graph **(Figure 2.21)** is shown at the left.

$$\text{abs}(x) = \begin{cases} -x, & x < 0 \\ x, & x \ge 0 \end{cases}$$

EXAMPLE 2   **Evaluate a Piecewise-Defined Function**

The number of monthly spam email attacks is shown in **Figure 2.22.**

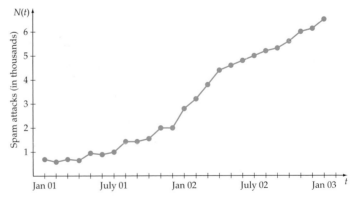

FIGURE 2.22

Source: www.brightmail.com

? ANSWER 0.5 is in the interval $-2 \le x < 1$. Therefore, $f(0.5) = 0.5^2 = 0.25$.

The data in the graph can be approximated by

$$N(t) = \begin{cases} 24.68t^2 - 170.47t + 957.73, & 0 \le t < 17 \\ 196.9t + 1164.6, & 17 \le t \le 26 \end{cases}$$

where $N(t)$ is the number of spam attacks in thousands for month t, where $t = 0$ corresponds to January 2001. Use this function to estimate, to the nearest hundred thousand, the number of monthly spam attacks for the following months.

a.   October 2001   b.   December 2002

Solution

a.   The month October 2001 corresponds to $t = 9$. Because $t = 9$ is in the interval $0 \le t < 17$, evaluate $24.68t^2 - 170.47t + 957.73$ at $t = 9$.

$$24.68t^2 - 170.47t + 957.73$$
$$24.68(9)^2 - 170.47(9) + 957.73 = 1422.58$$

There were approximately 1,423,000 spam attacks in October 2001.

b.   The month December 2002 corresponds to $t = 23$. Because $t = 23$ is in the interval $17 \le t \le 26$, evaluate $196.9t + 1164.6$ at 23.

$$196.9t + 1164.6$$
$$196.9(23) + 1164.6 = 5693.3$$

There were approximately 5,693,000 spam attacks in December 2002.

▶ **TRY EXERCISE 10, PAGE 175**

● IDENTIFYING FUNCTIONS

Recall that although every function is a relation, not every relation is a function. In the next example we examine four relations to determine which are functions.

EXAMPLE 3   **Identify Functions**

Which relations define y as a function of x?

a. $\{(2, 3), (4, 1), (4, 5)\}$ b. $3x + y = 1$ c. $-4x^2 + y^2 = 9$

d.   The correspondence between the x values and the y values in **Figure 2.23.**

Solution

a.   There are two ordered pairs, $(4, 1)$ and $(4, 5)$, with the same first coordinate and different second coordinates. This set does not define y as a function of x.

b.   Solving $3x + y = 1$ for y yields $y = -3x + 1$. Because $-3x + 1$ is a unique real number for each x, this equation defines y as a function of x.

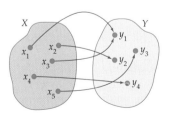

FIGURE 2.23

Continued ▶

c. Solving $-4x^2 + y^2 = 9$ for y yields $y = \pm\sqrt{4x^2 + 9}$. The right side $\pm\sqrt{4x^2 + 9}$ produces two values of y for each value of x. For example, when $x = 0$, $y = 3$ or $y = -3$. Thus $-4x^2 + y^2 = 9$ does not define y as a function of x.

d. Each x is paired with one and only one y. The correspondence in **Figure 2.23** defines y as a function of x.

▶ **TRY EXERCISE 14, PAGE 175**

---

take note

You may indicate the domain of a function using set-builder notation or interval notation. For instance, the domain of $f(x) = \sqrt{x - 3}$ may be given in each of the following ways:

Set-builder notation: $\{x \mid x \geq 3\}$

Interval notation: $[3, \infty)$

Sometimes the domain of a function is stated explicitly. For example, each of f, g, and h below is given by an equation, followed by a statement that indicates the domain of the function.

$$f(x) = x^2, x > 0 \qquad g(t) = \frac{1}{t^2 + 4}, 0 \leq t \leq 5 \qquad h(x) = x^2, x = 1, 2, 3$$

Although f and h have the same equation, they are different functions because they have different domains. If the domain of a function is not explicitly stated, then its domain is determined by the following convention.

Domain of a Function

Unless otherwise stated, the domain of a function is the set of all real numbers for which the function makes sense and yields real numbers.

EXAMPLE 4 Determine the Domain of a Function

Determine the domain of each function.

a. $G(t) = \dfrac{1}{t - 4}$ b. $f(x) = \sqrt{x + 1}$

c. $A(s) = s^2$, where $A(s)$ is the area of a square whose sides are s units.

Solution

a. The number 4 is not an element of the domain because G is undefined when the denominator $t - 4$ equals 0. The domain of G is all real numbers except 4. In interval notation the domain is $(-\infty, 4) \cup (4, \infty)$.

b. The radical $\sqrt{x + 1}$ is a real number only when $x + 1 \geq 0$ or when $x \geq -1$. Thus, in set-builder notation, the domain of f is $\{x \mid x \geq -1\}$.

c. Because s represents the length of the side of a square, s must be positive. In interval notation the domain of A is $(0, \infty)$.

▶ **TRY EXERCISE 28, PAGE 175**

● GRAPHS OF FUNCTIONS

If a is an element of the domain of a function, then $(a, f(a))$ is an ordered pair that belongs to the function.

Graph of a Function

The **graph of a function** is the graph of all the ordered pairs that belong to the function.

EXAMPLE 5   **Graph a Function by Plotting Points**

Graph each function. State the domain and the range of each function.

a. $f(x) = |x - 1|$ **b.** $n(x) = \begin{cases} 2, & \text{if } x \le 1 \\ x, & \text{if } x > 1 \end{cases}$

Solution

a. The domain of f is the set of all real numbers. Write the function as $y = |x - 1|$. Evaluate the function for several domain values. We have used $x = -3, -2, -1, 0, 1, 2, 3,$ and 4.

x	-3	-2	-1	0	1	2	3	4
$y = |x - 1|$	4	3	2	1	0	1	2	3

Plot the points determined by the ordered pairs. Connect the points to form the graph in **Figure 2.24.**
   Because $|x - 1| \ge 0$, we can conclude that the graph of f extends from a height of 0 upward, so the range is $\{y \mid y \ge 0\}$.

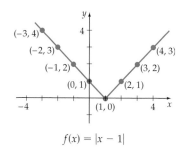

$f(x) = |x - 1|$

FIGURE 2.24

b. The domain is the union of the inequalities $x \le 1$ and $x > 1$. Thus the domain of n is the set of all real numbers. For $x \le 1$, graph $n(x) = 2$. This results in the horizontal ray in **Figure 2.25.** The solid circle indicates that the point $(1, 2)$ *is* part of the graph. For $x > 1$, graph $n(x) = x$. This produces the second ray in **Figure 2.25.** The open circle indicates that the point $(1, 1)$ *is not* part of the graph.
   Examination of the graph shows that it includes only points whose y values are greater than 1. Thus the range of n is $\{y \mid y > 1\}$.

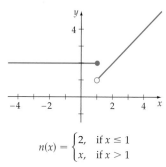

$n(x) = \begin{cases} 2, & \text{if } x \le 1 \\ x, & \text{if } x > 1 \end{cases}$

FIGURE 2.25

▶ **TRY EXERCISE 40, PAGE 175**

INTEGRATING TECHNOLOGY

A graphing utility also can be used to draw the graph of a function. For instance, to graph $f(x) = x^2 - 1$, you will enter an equation similar to Y1=x²–1. The graph is shown in **Figure 2.26.**

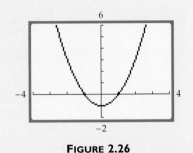

FIGURE 2.26

The definition that a function is a set of ordered pairs in which no two ordered pairs that have the same first coordinate have different second coordinates implies that any vertical line intersects the graph of a function at no more than one point. This is known as the *vertical line test.*

The Vertical Line Test for Functions

A graph is the graph of a function if and only if no vertical line intersects the graph at more than one point.

EXAMPLE 6   **Apply the Vertical Line Test**

Which of the following graphs are graphs of functions?

a. 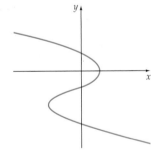   **b.**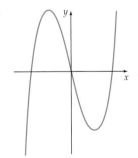

Solution

a. This graph *is not* the graph of a function because some vertical lines intersect the graph in more than one point.

b. This graph *is* the graph of a function because every vertical line intersects the graph in at most one point.

▶ **TRY EXERCISE 50, PAGE 176**

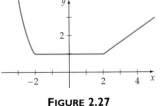

FIGURE 2.27

Consider the graph in **Figure 2.27.** As a point on the graph moves from left to right, this graph falls for values of $x \leq -2$, it remains the same height from

$x = -2$ to $x = 2$, and it rises for $x \geq 2$. The function represented by the graph is said to be *decreasing* on the interval $(-\infty, -2]$, *constant* on the interval $[-2, 2]$, and *increasing* on the interval $[2, \infty)$.

Definition of Increasing, Decreasing, and Constant Functions

If a and b are elements of an interval I that is a subset of the domain of a function f, then

- f is **increasing** on I if $f(a) < f(b)$ whenever $a < b$.
- f is **decreasing** on I if $f(a) > f(b)$ whenever $a < b$.
- f is **constant** on I if $f(a) = f(b)$ for all a and b.

Recall that a function is a relation in which no two ordered pairs that have the same first coordinate have different second coordinates. This means that given any x, there is only one y that can be paired with that x. A **one-to-one function** satisfies the additional condition that given any y, there is only one x that can be paired with that given y. In a manner similar to applying the vertical line test, we can apply a *horizontal line test* to identify one-to-one functions.

Horizontal Line Test for a One-To-One Function

If every horizontal line intersects the graph of a function at most once, then the graph is the graph of a one-to-one function.

For example, some horizontal lines intersect the graph in **Figure 2.28** at more than one point. It is *not* the graph of a one-to-one function. Every horizontal line intersects the graph in **Figure 2.29** at most once. This is the graph of a one-to-one function.

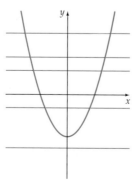

FIGURE 2.28

Some horizontal lines intersect this graph at more than one point. It is *not* the graph of a one-to-one function.

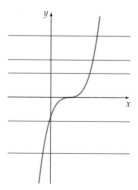

FIGURE 2.29

Every horizontal line intersects this graph at most once. It is the graph of a one-to-one function.

● THE GREATEST INTEGER FUNCTION (FLOOR FUNCTION)

To this point, the graphs of the functions have not had any breaks or gaps. These functions whose graphs can be drawn without lifting the pencil off the paper are called *continuous functions*. The graphs of some functions do have breaks or *discontinuities*. One such function is the **greatest integer function** or **floor function**. This function is denoted by various symbols such as $[\![x]\!]$, $\lfloor x \rfloor$, and int(x).

The value of the greatest integer function at x is the greatest integer that is less than or equal to x. For instance,

$$\lfloor -1.1 \rfloor = -2 \qquad [\![-3]\!] = -3 \qquad \text{int}\left(\frac{5}{2}\right) = 2 \qquad \lfloor 5 \rfloor = 5 \qquad [\![\pi]\!] = 3$$

INTEGRATING TECHNOLOGY

Many graphing calculators use the notation int(x) for the greatest integer function. Here is a screen from a TI-83 Plus.

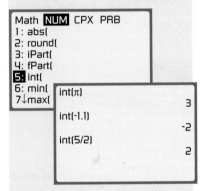

? QUESTION Evaluate. **a.** $\text{int}\left(-\frac{3}{2}\right)$ **b.** $\lfloor 2 \rfloor$

To graph the floor function, first observe that the value of the floor function is constant between any two consecutive integers. For instance, between the integers 1 and 2, we have

$$\text{int}(1.1) = 1 \qquad \text{int}(1.35) = 1 \qquad \text{int}(1.872) = 1 \qquad \text{int}(1.999) = 1$$

Between -3 and -2, we have

$$\text{int}(-2.98) = -3 \qquad \text{int}(-2.4) = -3 \qquad \text{int}(-2.35) = -3 \qquad \text{int}(-2.01) = -3$$

Using this property of the floor function, we can create a table of values and then graph the floor function.

x	$y = \text{int}(x)$
$-5 \le x < 4$	-5
$-4 \le x < -3$	-4
$-3 \le x < -2$	-3
$-2 \le x < -1$	-2
$-1 \le x < 0$	-1
$0 \le x < 1$	0
$1 \le x < 2$	1
$2 \le x < 3$	2
$3 \le x < 4$	3
$4 \le x < 5$	4

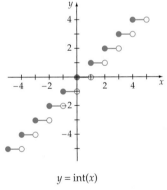

$y = \text{int}(x)$

FIGURE 2.30

The graph of the floor function has discontinuities (breaks) whenever x is an integer. The domain of the floor function is the set of real numbers; the range is the set of integers. Because the graph appears to be a series of steps, sometimes the floor function is referred to as a **step function**.

? ANSWER **a.** Because -2 is the greatest integer that is less than $-\frac{3}{2}$, $\text{int}\left(-\frac{3}{2}\right) = -2$.

  b. Because 2 is the greatest integer less than or *equal* to 2, $\lfloor 2 \rfloor = 2$.

INTEGRATING TECHNOLOGY

A graphing calculator also can be used to graph the floor function. The graph in **Figure 2.31** was drawn in "connected" mode. This graph does not show the discontinuities that occur whenever x is an integer.

The graph in **Figure 2.32** was constructed by graphing the floor function in "dot" mode. In this case the discontinuities at the integers are apparent.

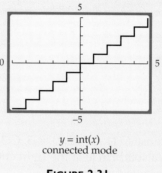

$y = \text{int}(x)$
connected mode

FIGURE 2.31

$y = \text{int}(x)$
dot mode

FIGURE 2.32

EXAMPLE 7   **Use the Greatest Integer Function to Model Expenses**

The cost of parking in a garage is $3 for the first hour or any part of the hour and $2 for each additional hour or any part of the hour thereafter. If x is the time in hours that you park your car, then the cost is given by

$$C(x) = 3 - 2\,\text{int}(1 - x), \quad x > 0$$

a.   Evaluate $C(2)$ and $C(2.5)$.   **b.**   Graph $y = C(x)$ for $0 < x \le 5$.

Solution

a. $C(2) = 3 - 2\,\text{int}(1 - 2)$ $C(2.5) = 3 - 2\,\text{int}(1 - 2.5)$

$\quad\quad = 3 - 2\,\text{int}(-1)$ $\quad\quad = 3 - 2\,\text{int}(-1.5)$

$\quad\quad = 3 - 2(-1)$ $\quad\quad = 3 - 2(-2)$

$\quad\quad = \$5$ $\quad\quad = \$7$

b.   To graph $C(x)$ for $0 < x \le 5$, consider the value of $\text{int}(1 - x)$ for each of the intervals $0 < x \le 1$, $1 < x \le 2$, $2 < x \le 3$, $3 < x \le 4$, and $4 < x \le 5$. For instance, when $0 < x \le 1$, $0 \le 1 - x < 1$. Thus $\text{int}(1 - x) = 0$ when $0 < x \le 1$. Now consider $1 < x \le 2$. When $1 < x \le 2$, $1 \le 1 - x < 2$. Thus $\text{int}(1 - x) = 1$ when $1 < x \le 2$. Applying the same reasoning to

Continued ▶

each of the other intervals gives the following table of values and the corresponding graph of C.

x	C(x) = 3 − 2 int(1 − x)
0 < x ≤ 1	3
1 < x ≤ 2	5
2 < x ≤ 3	7
3 < x ≤ 4	9
4 < x ≤ 5	11

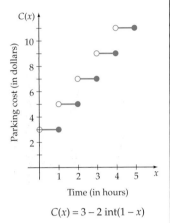

$C(x) = 3 - 2 \text{ int}(1 - x)$

FIGURE 2.33

Because $C(1) = 3$, $C(2) = 5$, $C(3) = 7$, $C(4) = 9$, and $C(5) = 11$, we can use a solid circle at the right endpoint of each "step" and an open circle at each left endpoint.

▶  **TRY EXERCISE 48, PAGE 176**

INTEGRATING TECHNOLOGY

Example 7 illustrates that a graphing calculator may not produce a graph that is a good representation of a function. You may be required to *make adjustments* in the MODE, SET UP, or WINDOW of the graphing calculator so that it will produce a better representation of the function. Some graphs may also require some *fine tuning*, such as open or solid circles at particular points, to accurately represent the function.

● **APPLICATIONS**

EXAMPLE 8   **Solve an Application**

A car was purchased for $16,500. Assuming the car depreciates at a constant rate of $2200 per year (*straight-line depreciation*) for the first 7 years, write the value v of the car as a function of time, and calculate the value of the car 3 years after purchase.

Solution

Let t represent the number of years that have passed since the car was purchased. Then $2200t$ is the amount that the car has depreciated after t years. The value of the car at time t is given by

$$v(t) = 16{,}500 - 2200t, \quad 0 \le t \le 7$$

When $t = 3$, the value of the car is

$$v(3) = 16{,}500 - 2200(3) = 16{,}500 - 6600 = \$9900$$

▶ **TRY EXERCISE 66, PAGE 177**

Often in applied mathematics, formulas are used to determine the functional relationship that exists between two variables.

EXAMPLE 9   Solve an Application

A lighthouse is 2 miles south of a port. A ship leaves port and sails east at a rate of 7 mph. Express the distance d between the ship and the lighthouse as a function of time, given that the ship has been sailing for t hours.

Solution

Draw a diagram and label it as shown in **Figure 2.34.** Note that because distance = (rate)(time) and the rate is 7, in t hours the ship has sailed a distance of $7t$.

$$[d(t)]^2 = (7t)^2 + 2^2 \qquad \bullet \textbf{ The Pythagorean Theorem}$$
$$[d(t)]^2 = 49t^2 + 4$$
$$d(t) = \sqrt{49t^2 + 4} \qquad \bullet \textbf{ The } \pm \textbf{ sign is not used because}$$
$$\qquad\qquad\qquad\qquad\qquad d \textbf{ must be nonnegative.}$$

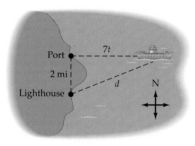

FIGURE 2.34

▶ **TRY EXERCISE 72, PAGE 178**

EXAMPLE 10   Solve an Application

An open box is to be made from a square piece of cardboard that measures 40 inches on each side. To construct the box, squares that measure x inches on each side are cut from each corner of the cardboard as shown in **Figure 2.35.**

a.   Express the volume V of the box as a function of x.

b.   Determine the domain of V.

Solution

a.   The length l of the box is $40 - 2x$. The width w is also $40 - 2x$. The height of the box is x. The volume V of a box is the product of its length, its width, and its height. Thus

$$V = (40 - 2x)^2 x$$

b.   The squares that are cut from each corner require x to be larger than 0 inches but less than 20 inches. Thus the domain is $\{x \mid 0 < x < 20\}$.

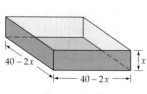

FIGURE 2.35

▶ **TRY EXERCISE 68, PAGE 177**

TOPICS FOR DISCUSSION

1. Discuss the definition of *function*. Give some examples of relationships that are functions and some that are not functions.

2. What is the difference between the domain and range of a function?

3. How many y-intercepts can a function have? How many x-intercepts can a function have?

4. Discuss how the vertical line test is used to determine whether or not a graph is the graph of a function. Explain why the vertical line test works.

5. What is the domain of $f(x) = \dfrac{\sqrt{1-x}}{x^2 - 9}$? Explain.

6. Is 2 in the range of $g(x) = \dfrac{6x - 5}{3x + 1}$? Explain the process you used to make your decision.

7. Suppose that f is a function and that $f(a) = f(b)$. Does this imply that $a = b$? Explain your answer.

EXERCISE SET 2.2

In Exercises 1 to 8, evaluate each function.

1. Given $f(x) = 3x - 1$, find

   a. $f(2)$ **b.** $f(-1)$ **c.** $f(0)$

   d. $f\left(\dfrac{2}{3}\right)$ **e.** $f(k)$ **f.** $f(k + 2)$

▶ **2.** Given $g(x) = 2x^2 + 3$, find

   a. $g(3)$ **b.** $g(-1)$ **c.** $g(0)$

   d. $g\left(\dfrac{1}{2}\right)$ **e.** $g(c)$ **f.** $g(c + 5)$

3. Given $A(w) = \sqrt{w^2 + 5}$, find

   a. $A(0)$ **b.** $A(2)$ **c.** $A(-2)$

   d. $A(4)$ **e.** $A(r + 1)$ **f.** $A(-c)$

4. Given $J(t) = 3t^2 - t$, find

   a. $J(-4)$ **b.** $J(0)$ **c.** $J\left(\dfrac{1}{3}\right)$

   d. $J(-c)$ **e.** $J(x + 1)$ **f.** $J(x + h)$

5. Given $f(x) = \dfrac{1}{|x|}$, find

   a. $f(2)$ **b.** $f(-2)$ **c.** $f\left(\dfrac{-3}{5}\right)$

   d. $f(2) + f(-2)$ **e.** $f(c^2 + 4)$ **f.** $f(2 + h)$

6. Given $T(x) = 5$, find

   a. $T(-3)$ **b.** $T(0)$ **c.** $T\left(\dfrac{2}{7}\right)$

   d. $T(3) + T(1)$ **e.** $T(x + h)$ **f.** $T(3k + 5)$

7. Given $s(x) = \dfrac{x}{|x|}$, find

   a. $s(4)$ **b.** $s(5)$ **c.** $s(-2)$

   d. $s(-3)$ **e.** $s(t), t > 0$ **f.** $s(t), t < 0$

8. Given $r(x) = \dfrac{x}{x + 4}$, find

   a. $r(0)$ **b.** $r(-1)$ **c.** $r(-3)$

d. $r\left(\dfrac{1}{2}\right)$ **e.** $r(0.1)$ **f.** $r(10,000)$

In Exercises 9 and 10, evaluate each piecewise-defined function for the indicated values.

9. $P(x) = \begin{cases} 3x + 1, & \text{if } x < 2 \\ -x^2 + 11, & \text{if } x \ge 2 \end{cases}$

  a. $P(-4)$ **b.** $P(\sqrt{5})$

  c. $P(c), \quad c < 2$ **d.** $P(k + 1), \quad k \ge 1$

▶ **10.** $Q(t) = \begin{cases} 4, & \text{if } 0 \le t \le 5 \\ -t + 9, & \text{if } 5 < t \le 8 \\ \sqrt{t - 7}, & \text{if } 8 < t \le 11 \end{cases}$

  a. $Q(0)$ **b.** $Q(e), \quad 6 < e < 7$

  c. $Q(n), \quad 1 < n < 2$ **d.** $Q(m^2 + 7), \quad 1 < m \le 2$

In Exercises 11 to 20, identify the equations that define y as a function of x.

11. $2x + 3y = 7$ **12.** $5x + y = 8$

13. $-x + y^2 = 2$ ▶ **14.** $x^2 - 2y = 2$

15. $y = 4 \pm \sqrt{x}$ **16.** $x^2 + y^2 = 9$

17. $y = \sqrt[3]{x}$ **18.** $y = |x| + 5$

19. $y^2 = x^2$ **20.** $y^3 = x^3$

In Exercises 21 to 26, identify the sets of ordered pairs (x, y) that define y as a function of x.

21. $\{(2, 3), (5, 1), (-4, 3), (7, 11)\}$

22. $\{(5, 10), (3, -2), (4, 7), (5, 8)\}$

23. $\{(4, 4), (6, 1), (5, -3)\}$

24. $\{(2, 2), (3, 3), (7, 7)\}$

25. $\{(1, 0), (2, 0), (3, 0)\}$

26. $\left\{\left(-\dfrac{1}{3}, \dfrac{1}{4}\right), \left(-\dfrac{1}{4}, \dfrac{1}{3}\right), \left(\dfrac{1}{4}, \dfrac{2}{3}\right)\right\}$

In Exercises 27 to 38, determine the domain of the function represented by the given equation.

27. $f(x) = 3x - 4$ ▶ **28.** $f(x) = -2x + 1$

29. $f(x) = x^2 + 2$ **30.** $f(x) = 3x^2 + 1$

31. $f(x) = \dfrac{4}{x + 2}$ **32.** $f(x) = \dfrac{6}{x - 5}$

33. $f(x) = \sqrt{7 + x}$ **34.** $f(x) = \sqrt{4 - x}$

35. $f(x) = \sqrt{4 - x^2}$ **36.** $f(x) = \sqrt{12 - x^2}$

37. $f(x) = \dfrac{1}{\sqrt{x + 4}}$ **38.** $f(x) = \dfrac{1}{\sqrt{5 - x}}$

In Exercises 39 to 46, graph each function. Insert solid circles or hollow circles where necessary to indicate the true nature of the function.

39. $f(x) = \begin{cases} |x|, & \text{if } x \le 1 \\ 2, & \text{if } x > 1 \end{cases}$

▶ **40.** $g(x) = \begin{cases} -4, & \text{if } x \le 0 \\ x^2 - 4, & \text{if } 0 < x \le 1 \\ -x, & \text{if } x > 1 \end{cases}$

41. $J(x) = \begin{cases} 4, & \text{if } x \le -1 \\ x^2, & \text{if } -1 < x < 1 \\ -x + 5, & \text{if } x \ge 1 \end{cases}$

42. $K(x) = \begin{cases} 1, & \text{if } x \le -2 \\ x^2 - 4, & \text{if } -2 < x < 2 \\ \dfrac{1}{2}x, & \text{if } x \ge 2 \end{cases}$

43. $L(x) = \left[\!\left[\dfrac{1}{3}x\right]\!\right] \quad \text{for } -6 \le x \le 6$

44. $M(x) = [\![x]\!] + 2 \quad \text{for } 0 \le x \le 4$

45. $N(x) = \text{int}(-x) \quad \text{for } -3 \le x \le 3$

46. $P(x) = \text{int}(x) + x \quad \text{for } 0 \le x \le 4$

47.   **FIRST-CLASS MAIL** In 2003, the cost to mail a first-class letter is given by

$$C(w) = 0.37 - 0.34 \text{ int}(1 - w), \quad w > 0$$

where C is in dollars and w is the weight of the letter in ounces.

a. What is the cost to mail a letter that weighs 2.8 ounces?

b. Graph C for $0 < w \le 5$.

▶ **48.** **INCOME TAX** The amount of federal income tax $T(x)$ a person owed in 2003 is given by

$$T(x) = \begin{cases} 0.10x, & 0 \le x < 6000 \\ 0.15(x - 6000) + 600, & 6000 \le x < 27{,}950 \\ 0.27(x - 27{,}950) + 3892.50, & 27{,}950 \le x < 67{,}700 \\ 0.30(x - 67{,}700) + 14{,}625, & 67{,}700 \le x < 141{,}250 \\ 0.35(x - 141{,}250) + 36{,}690, & 141{,}250 \le x < 307{,}050 \\ 0.386(x - 307{,}050) + 94{,}720, & x \ge 307{,}050 \end{cases}$$

where x is the adjusted gross income of the taxpayer.

a. What is the domain of this function?

b. Find the income tax owed by a taxpayer whose adjusted gross income was $31,250.

c. Find the income tax owed by a taxpayer whose adjusted gross income was $72,000.

49. Use the vertical line test to determine which of the following graphs are graphs of functions.

a. 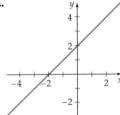 **b.**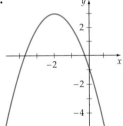

c. 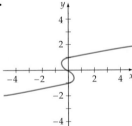 **d.**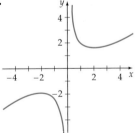

▶ **50.** Use the vertical line test to determine which of the following graphs are graphs of functions.

a. 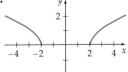 **b.**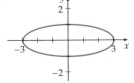

c.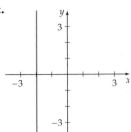

d.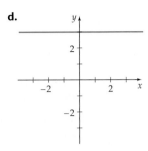

In Exercises 51 to 60, use the indicated graph to identify the intervals over which the function is increasing, constant, or decreasing.

51. 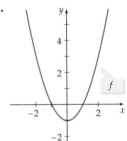 **52.**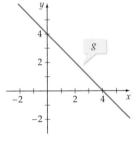

53. 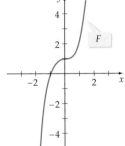 **54.**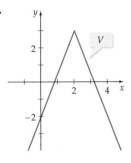

55. 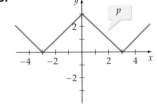 **56.**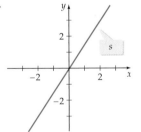

57.

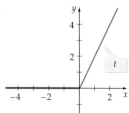

58.

59.

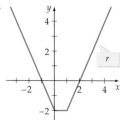

60.

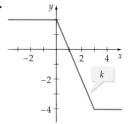

61. Use the horizontal line test to determine which of the following functions are one-to-one.

f as shown in Exercise 51
g as shown in Exercise 52
F as shown in Exercise 53
V as shown in Exercise 54
p as shown in Exercise 55

62. Use the horizontal line test to determine which of the following functions are one-to-one.

s as shown in Exercise 56
t as shown in Exercise 57
m as shown in Exercise 58
r as shown in Exercise 59
k as shown in Exercise 60

63. A rectangle has a length of l feet and a perimeter of 50 feet.

a. Write the width w of the rectangle as a function of its length.

b. Write the area A of the rectangle as a function of its length.

64. The sum of two numbers is 20. Let x represent one of the numbers.

a. Write the second number y as a function of x.

b. Write the product P of the two numbers as a function of x.

65. DEPRECIATION A bus was purchased for $80,000. Assuming the bus depreciates at a rate of $6500 per year (*straight-line depreciation*) for the first 10 years, write the value v of the bus as a function of the time t (measured in years) for $0 \le t \le 10$.

▶ **66. DEPRECIATION** A boat was purchased for $44,000. Assuming the boat depreciates at a rate of $4200 per year (*straight-line depreciation*) for the first 8 years, write the value v of the boat as a function of the time t (measured in years) for $0 \le t \le 8$.

67. COST, REVENUE, AND PROFIT A manufacturer produces a product at a cost of $22.80 per unit. The manufacturer has a fixed cost of $400.00 per day. Each unit retails for $37.00. Let x represent the number of units produced in a 5-day period.

a. Write the total cost C as a function of x.

b. Write the revenue R as a function of x.

c. Write the profit P as a function of x. (*Hint:* The profit function is given by $P(x) = R(x) - C(x)$.)

▶ **68. VOLUME OF A BOX** An open box is to be made from a square piece of cardboard having dimensions 30 inches by 30 inches by cutting out squares of area x^2 from each corner, as shown in the figure.

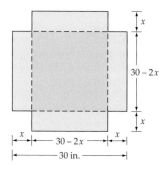

a. Express the volume V of the box as a function of x.

b. State the domain of V.

69. HEIGHT OF AN INSCRIBED CYLINDER A cone has an altitude of 15 centimeters and a radius of 3 centimeters. A right circular cylinder of radius r and height h is inscribed in the cone as shown in the figure. Use similar triangles to write h as a function of r.

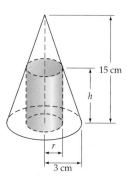

70. VOLUME OF WATER Water is flowing into a conical drinking cup that has an altitude of 4 inches and a radius of 2 inches, as shown in the figure.

a. Write the radius r of the water as a function of its depth h.

b. Write the volume V of the water as a function of its depth h.

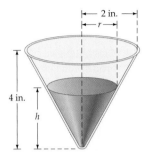

71. DISTANCE FROM A BALLOON For the first minute of flight, a hot air balloon rises vertically at a rate of 3 meters per second. If t is the time in seconds that the balloon has been airborne, write the distance d between the balloon and a point on the ground 50 meters from the point of lift-off as a function of t.

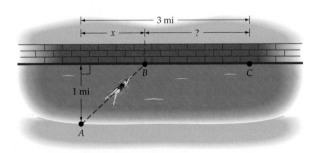

▶ **72. TIME FOR A SWIMMER** An athlete swims from point A to point B at the rate of 2 mph and runs from point B to point C at a rate of 8 mph. Use the dimensions in the figure to write the time t required to reach point C as a function of x.

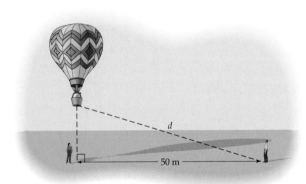

73. DISTANCE BETWEEN SHIPS At 12:00 noon Ship A is 45 miles due south of ship B and is sailing north at a rate of 8 mph. Ship B is sailing east at a rate of 6 mph. Write the distance d between the ships as a function of the time t, where $t = 0$ represents 12:00 noon.

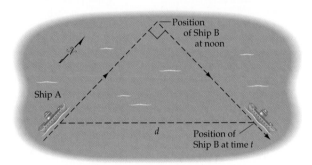

74. AREA A rectangle is bounded by the x- and y-axes and the graph of $y = -\dfrac{1}{2}x + 4$.

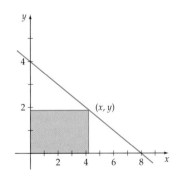

a. Find the area of the rectangle as a function of x.

b. Complete the table below.

x	Area
1	
2	
4	
6	
7	

c. What is the domain of this function?

75. AREA A piece of wire 20 centimeters long is cut at a point x centimeters from the left end. The left-hand piece is formed into the shape of a circle and the right-hand piece is formed into a square.

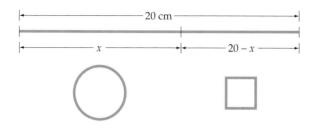

a. Find the area enclosed by the two figures as a function of x.

b. Complete the table below. Round the area to the nearest hundredth.

x	Total Area Enclosed
0	
4	
8	
12	
16	
20	

c. What is the domain of this function?

76. AREA A triangle is bounded by the x- and y-axes and must pass through $P(2, 2)$, as shown below.

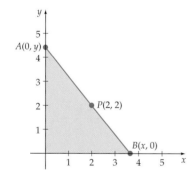

a. Find the area of the triangle as a function of x. (*Suggestion:* The slope of the line between points A and P equals the slope of the line between P and B.)

b. What is the domain of the function you found in part **a.**?

77. LENGTH Two guy wires are attached to utility poles that are 40 feet apart, as shown in the following diagram.

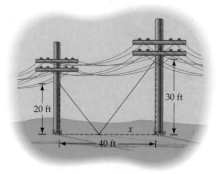

a. Find the total length of the two guy wires as a function of x.

b. Complete the table below. Round the length to the nearest hundredth.

x	Total Length of Wires
0	
10	
20	
30	
40	

c. What is the domain of this function?

78. SALES VS. PRICE A business finds that the number of feet f of pipe it can sell per week is a function of the price p in cents per foot as given by

$$f(p) = \frac{320{,}000}{p + 25}, \quad 40 \le p \le 90$$

Complete the following table by evaluating f (to the nearest hundred feet) for the indicated values of p.

p	40	50	60	75	90
f(p)					

79. MODEL YIELD The yield Y of apples per tree is related to the amount x of a particular type of fertilizer applied (in pounds per year) by the function

$$Y(x) = 400[1 - 5(x - 1)^{-2}], \quad 5 \le x \le 20$$

Complete the following table by evaluating Y (to the nearest apple) for the indicated applications.

x	5	10	12.5	15	20
Y(x)					

80. **MODEL COST** A manufacturer finds that the cost C in dollars of producing x items of a product is given by

$$C(x) = \left(225 + 1.4\sqrt{x}\right)^2, \quad 100 \le x \le 1000$$

Complete the following table by evaluating C (to the nearest dollar) for the indicated numbers of items.

x	100	200	500	750	1000
$C(x)$					

81. If $f(x) = x^2 - x - 5$ and $f(c) = 1$, find c.

82. If $g(x) = -2x^2 + 4x - 1$ and $g(c) = -4$, find c.

83. Determine whether 1 is in the range of $f(x) = \dfrac{x-1}{x+1}$.

84. Determine whether 0 is in the range of $g(x) = \dfrac{1}{x-3}$.

In Exercises 85 to 90, use a graphing utility.

85. Graph $f(x) = \dfrac{[\![x]\!]}{|x|}$ for $-4.7 \le x \le 4.7$ and $x \ne 0$.

86. Graph $f(x) = \dfrac{[\![2x]\!]}{|x|}$ for $-4 \le x \le 4$ and $x \ne 0$.

87. Graph: $f(x) = x^2 - 2|x| - 3$

88. Graph: $f(x) = x^2 - |2x - 3|$

89. Graph: $f(x) = |x^2 - 1| - |x - 2|$

90. Graph: $f(x) = |x^2 - 2x| - 3$

CONNECTING CONCEPTS

The notation $f(x)\big|_a^b$ is used to denote the difference $f(b) - f(a)$. That is,

$$f(x)\big|_a^b = f(b) - f(a)$$

In Exercises 91 to 94, evaluate $f(x)\big|_a^b$ for the given function f and the indicated values of a and b.

91. $f(x) = x^2 - x;\, f(x)\big|_2^3$

92. $f(x) = -3x + 2;\, f(x)\big|_4^7$

93. $f(x) = 2x^3 - 3x^2 - x;\, f(x)\big|_0^2$

94. $f(x) = \sqrt{8 - x};\, f(x)\big|_0^8$

In Exercises 95 to 98, each function has two or more independent variables.

95. Given $f(x, y) = 3x + 5y - 2$, find

  a. $f(1, 7)$ **b.** $f(0, 3)$ **c.** $f(-2, 4)$

  d. $f(4, 4)$ **e.** $f(k, 2k)$ **f.** $f(k + 2, k - 3)$

96. Given $g(x, y) = 2x^2 - |y| + 3$, find

  a. $g(3, -4)$ **b.** $g(-1, 2)$

  c. $g(0, -5)$ **d.** $g\left(\dfrac{1}{2}, -\dfrac{1}{4}\right)$

  e. $g(c, 3c), c > 0$ **f.** $g(c + 5, c - 2), c < 0$

97. **AREA OF A TRIANGLE** The area of a triangle with sides a, b, and c is given by the function

$$A(a, b, c) = \sqrt{s(s - a)(s - b)(s - c)}$$

where s is the semiperimeter

$$s = \frac{a + b + c}{2}$$

Find $A(5, 8, 11)$.

98. **COST OF A PAINTER** The cost in dollars to hire a house painter is given by the function

$$C(h, g) = 15h + 14g$$

where h is the number of hours it takes to paint the house and g is the number of gallons of paint required to paint the house. Find $C(18, 11)$.

A *fixed point* of a function is a number a such that $f(a) = a$. In Exercises 99 and 100, find all fixed points for the given function.

99. $f(x) = x^2 + 3x - 3$

100. $g(x) = \dfrac{x}{x + 5}$

In Exercises 101 and 102, sketch the graph of the piecewise-defined function.

101. $s(x) = \begin{cases} 1 & \text{if } x \text{ is an integer} \\ 2 & \text{if } x \text{ is not an integer} \end{cases}$

102. $v(x) = \begin{cases} 2x - 2 & \text{if } x \neq 3 \\ 1 & \text{if } x = 3 \end{cases}$

PREPARE FOR SECTION 2.3

103. Find the distance on a real number line between the points whose coordinates are -2 and 5. [P.1]

104. Find the product of a nonzero number and its negative reciprocal. [P.4]

105. Given the points $P_1(-3, 4)$ and $P_2(2, -4)$, evaluate $\dfrac{y_2 - y_1}{x_2 - x_1}$. [P.1]

106. Solve $y - 3 = -2(x - 3)$ for y. [1.1]

107. Solve $3x - 5y = 15$ for y. [1.1]

108. Given $y = 3x - 2(5 - x)$, find the value of x for which $y = 0$. [1.1]

PROJECTS

1. **DAY OF THE WEEK** A formula known as Zeller's Congruence makes use of the greatest integer function $[\![x]\!]$ to determine the day of the week on which a given day fell or will fall. To use Zeller's Congruence, we first compute the integer z given by

$$z = \left[\!\left[\frac{13m - 1}{5}\right]\!\right] + \left[\!\left[\frac{y}{4}\right]\!\right] + \left[\!\left[\frac{c}{4}\right]\!\right] + d + y - 2c$$

The variables c, y, d, and m are defined as follows:

$c = $ the century

$y = $ the year of the century

$d = $ the day of the month

$m = $ the month, using 1 for March, 2 for April, . . . , 10 for December. January and February are assigned the values 11 and 12 of the previous year.

For example, for the date September 12, 2001, we use $c = 20$, $y = 1$, $d = 12$, and $m = 7$. The remainder of z divided by 7 gives the day of the week. A remainder of 0 represents a Sunday, a remainder of 1 a Monday, . . . , and a remainder of 6 a Saturday.

a. Verify that December 7, 1941 was a Sunday.

b. Verify that January 1, 2010 will fall on a Friday.

c. Determine on what day of the week Independence Day (July 4, 1776) fell.

d. Determine on what day of the week you were born.

| SECTION 2.3 | **LINEAR FUNCTIONS** |

- SLOPES OF LINES
- FIND THE EQUATION OF A LINE
- APPLICATIONS
- PARALLEL AND PERPENDICULAR LINES

The following function has many applications.

Definition of a Linear Function

A function of the form

$$f(x) = mx + b, \quad m \neq 0$$

where m and b are real numbers, is a **linear function** of x.

• SLOPES OF LINES

The graph of $f(x) = mx + b$, or $y = mx + b$, is a nonvertical straight line.

The graphs shown in **Figure 2.36** are the graphs of $f(x) = mx + b$ for various values of m. The graphs intersect at the point $(-2, -1)$, but they differ in *steepness*. The steepness of a line is called the *slope* of the line and is denoted by the symbol m. The slope of a line is the ratio of the change in the y values of any two points on the line to the change in the x values of the same two points. For example, the graph of the line L_1 in **Figure 2.36** passes through the points $(-2, -1)$ and $(3, 5)$. The change in the y values is determined by subtracting the two y-coordinates.

$$\text{Change in } y = 5 - (-1) = 6$$

The change in the x values is determined by subtracting the two x-coordinates.

$$\text{Change in } x = 3 - (-2) = 5$$

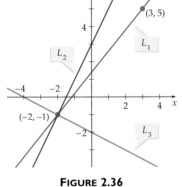

FIGURE 2.36

The slope m of L_1 is the ratio of the change in the y values of the two points to the change in the x values of the two points. That is,

$$m = \frac{\text{change in } y}{\text{change in } x} = \frac{6}{5}$$

Because the slope of a nonvertical line can be calculated by using any two arbitrary points on the line, we have the following formula.

Slope of a Nonvertical Line

The **slope** m of the line passing through the points $P_1(x_1, y_1)$ and $P_2(x_2, y_2)$ with $x_1 \neq x_2$ is given by

$$m = \frac{y_2 - y_1}{x_2 - x_1}$$

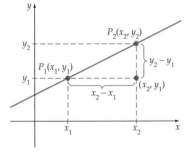

FIGURE 2.37

Because the numerator $y_2 - y_1$ is the vertical **rise** and the denominator $x_2 - x_1$ is the horizontal **run** from P_1 to P_2, slope is often referred to as the *rise over the run* or the *change in y divided by the change in x*. See **Figure 2.37.** Lines that have a positive slope slant upward from left to right. Lines that have a negative slope slant downward from left to right.

▞ **EXAMPLE I**   **Find the Slope of a Line**

Find the slope of the line passing through the points whose coordinates are given.

a. $(1, 2)$ and $(3, 6)$ **b.** $(-3, 4)$ and $(1, -2)$

Solution

a. The slope of the line passing through $(1, 2)$ and $(3, 6)$ is

$$m = \frac{y_2 - y_1}{x_2 - x_1} = \frac{6 - 2}{3 - 1} = \frac{4}{2} = 2$$

Because $m > 0$, the line slants upward from left to right. See **Figure 2.38.**

b. The slope of the line passing through $(-3, 4)$ and $(1, -2)$ is

$$m = \frac{y_2 - y_1}{x_2 - x_1} = \frac{-2 - 4}{1 - (-3)} = \frac{-6}{4} = -\frac{3}{2}$$

Because $m < 0$, the line slants downward from left to right. See **Figure 2.39.**

▶ **TRY EXERCISE 2, PAGE 191**

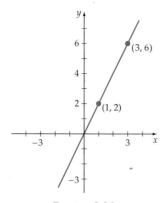

FIGURE 2.38

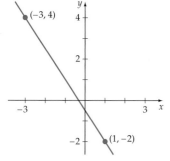

FIGURE 2.39

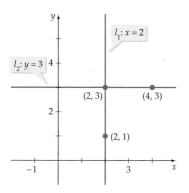

FIGURE 2.40

The definition of slope does not apply to vertical lines. Consider, for example, the points $(2, 1)$ and $(2, 3)$ on the vertical line l_1 in **Figure 2.40.** Applying the definition of slope to this line produces

$$m = \frac{3 - 1}{2 - 2}$$

which is undefined because it requires division by zero. Because division by zero is undefined, we say that the slope of any vertical line is undefined.

Every point on the vertical line through $(a, 0)$ has an x-coordinate of a. The equation of the vertical line through $(a, 0)$ is $x = a$. See **Figure 2.41.**

❓ **QUESTION** Is the graph of a vertical line the graph of a function?

All horizontal lines have 0 slope. For example, the line l_2 through $(2, 3)$ and $(4, 3)$ in **Figure 2.40** is a horizontal line. Its slope is given by

$$m = \frac{3 - 3}{4 - 2} = \frac{0}{2} = 0$$

When computing the slope of a line, it does not matter which point we label P_1 and which P_2 because

$$\frac{y_2 - y_1}{x_2 - x_1} = \frac{y_1 - y_2}{x_1 - x_2}$$

❓ **ANSWER**   No. For example, the vertical line passing through $x = 2$ contains the ordered pairs $(2, 3)$ and $(2, -5)$. Thus there are two ordered pairs with the same first coordinate but different second coordinates.

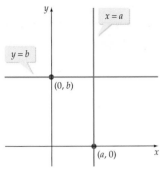

FIGURE 2.41

In functional notation, the points P_1 and P_2 can be represented by

$$(x_1, f(x_1)) \quad \text{and} \quad (x_2, f(x_2))$$

In this notation, the slope formula

$$m = \frac{y_2 - y_1}{x_2 - x_1} \quad \text{is expressed as} \quad m = \frac{f(x_2) - f(x_1)}{x_2 - x_1} \tag{1}$$

If $m = 0$, then $f(x) = mx + b$ can be written as $f(x) = b$, or $y = b$. The graph of $y = b$ is the horizontal line through $(0, b)$. See **Figure 2.41.** Because every point on the graph of $y = b$ has a y-coordinate of b, the function $f(x) = b$ is called a **constant function.**

Horizontal Lines and Vertical Lines

The graph of $x = a$ is a vertical line through $(a, 0)$.

The graph of $y = b$ is a horizontal line through $(0, b)$.

The equation $f(x) = mx + b$ is called the **slope-intercept form** of the equation of a line because of the following theorem.

Slope-Intercept Form

The graph of $f(x) = mx + b$ is a line with slope m and y-intercept $(0, b)$.

Proof   The slope of the graph of $f(x) = mx + b$ is given by Equation (1).

$$\frac{f(x_2) - f(x_1)}{x_2 - x_1} = \frac{(mx_2 + b) - (mx_1 + b)}{x_2 - x_1} = \frac{m(x_2 - x_1)}{x_2 - x_1} = m, \quad x_1 \neq x_2$$

The y-intercept of the graph of $f(x) = mx + b$ is found by letting $x = 0$.

$$f(0) = m(0) + b = b$$

Thus $(0, b)$ is the y-intercept, and m is the slope, of the graph of $f(x) = mx + b$. ◆

If a function is written in the form $f(x) = mx + b$, then its graph can be drawn by first plotting the y-intercept $(0, b)$ and then using the slope m to determine another point on the line.

EXAMPLE 2   Graph a Linear Function

Graph: $f(x) = 2x - 1$

Solution

The equation $y = 2x - 1$ is in slope-intercept form, with $b = -1$ and $m = 2$. Thus the y-intercept is $(0, -1)$, and the slope is 2. Write the slope as

$$m = \frac{2}{1} = \frac{\text{change in } y}{\text{change in } x}$$

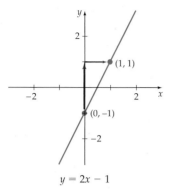

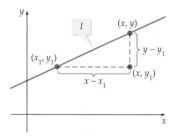

$y = 2x - 1$

FIGURE 2.42

The slope of line l is $m = \dfrac{y - y_1}{x - x_1}$.

FIGURE 2.43

To graph the equation, first plot the y-intercept, and then use the slope to plot a second point. This second point is 2 units up (change in y) and 1 unit to the right (change in x) of the y-intercept. See **Figure 2.42**.

▶ TRY EXERCISE 16, PAGE 192

● FIND THE EQUATION OF A LINE

We can find an equation of a line provided we know its slope and at least one point on the line. **Figure 2.43** suggests that if (x_1, y_1) is a point on a line l of slope m, and (x, y) is *any other* point on the line, then

$$\frac{y - y_1}{x - x_1} = m, \quad x \neq x_1$$

Multiplying each side by $x - x_1$ produces $y - y_1 = m(x - x_1)$. This equation is called the **point-slope form** of the equation of line l.

Point-Slope Form

The graph of

$$y - y_1 = m(x - x_1)$$

is a line that has slope m and passes through (x_1, y_1).

EXAMPLE 3   **Use the Point-Slope Form**

Find an equation of the line with slope -3 that passes through $(-1, 4)$.

Solution

Use the point-slope form with $m = -3$, $x_1 = -1$, and $y_1 = 4$.

$$
\begin{aligned}
y - y_1 &= m(x - x_1) \\
y - 4 &= -3[x - (-1)] \qquad &\text{• Substitute.} \\
y - 4 &= -3x - 3 \qquad &\text{• Solve for y.} \\
y &= -3x + 1 \qquad &\text{• Slope-intercept form}
\end{aligned}
$$

▶ TRY EXERCISE 28, PAGE 192

take note

To determine an equation of a nonvertical line that passes through two points, first determine the slope of the line and then use the coordinates of either one of the points in the point-slope form.

An equation of the form $Ax + By = C$, where A, B, and C are real numbers and both A and B are not zero, is called the **general form of the equation of a line**. For example, the equation $y = -3x + 1$ in Example 3 can be written in general form as $3x + y = 1$.

One way to graph a linear equation that is written in general form is to first solve the equation for y in terms of x. For instance, to graph $3x - 2y = 4$, solve for y.

$$3x - 2y = 4$$
$$-2y = -3x + 4$$
$$y = \frac{3}{2}x - 2$$

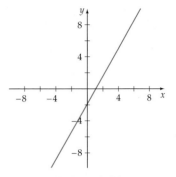

FIGURE 2.44

The y-intercept is $(0, -2)$ and the slope is $\frac{3}{2}$. The graph is shown in **Figure 2.44**.

By solving a first-degree equation, the specific relationship between an element of the domain and an element of the range of a linear function can be determined.

EXAMPLE 4   Find the Value in the Domain of f for which $f(x) = b$

Find the value x in the domain of $f(x) = 3x - 4$ for which $f(x) = 5$.

Algebraic Solution

$$f(x) = 3x - 4$$
$$5 = 3x - 4 \qquad \text{• Replace } f(x) \text{ by 5 and solve for } x.$$
$$9 = 3x$$
$$3 = x$$

When $x = 3$, $f(x) = 5$. This means that 3 in the domain of f is paired with 5 in the range of f. Another way of stating this is that the ordered pair $(3, 5)$ is an element of f.

Visualize the Solution

By graphing $y = 5$ and $f(x) = 3x - 4$, we can see that $f(x) = 5$ when $x = 3$.

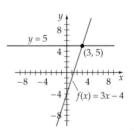

▶ **TRY EXERCISE 42, PAGE 192**

Although we are mainly concerned with linear functions in this section, the following theorem applies to all functions. It illustrates a powerful relationship between the real solutions of $f(x) = 0$ and the x-intercepts of the graph of $y = f(x)$.

Real Solutions and x-Intercepts Theorem

For every function f, the real number c is a solution of $f(x) = 0$ if and only if $(c, 0)$ is an x-intercept of the graph of $y = f(x)$.

❓ **QUESTION**   Is $(-2, 0)$ an x-intercept of $f(x) = x^3 - x + 6$?

The real solutions and x-intercepts theorem tells us that we can find real solutions of $f(x) = 0$ by graphing. The following example illustrates the theorem for a linear function of x.

❓ **ANSWER**   Yes. $f(-2) = (-2)^3 - (-2) + 6 = 0$

EXAMPLE 5   **Verify the Real Solutions and x-Intercepts Theorem**

Let $f(x) = -2x + 6$. Find the real solution of $f(x) = 0$ and then graph $f(x)$.
Compare the solution of $f(x) = 0$ with the x-intercept of the graph of f.

Algebraic Solution

To find the real solution of $f(x) = 0$, replace $f(x)$ by $-2x + 6$ and solve
for x.

$$f(x) = 0$$
$$-2x + 6 = 0$$
$$-2x = -6$$
$$x = 3$$

The x-coordinate of the x-intercept is 3. The real solution of $f(x) = 0$ is 3.

Visualize the Solution

Graph $f(x) = -2x + 6$ (see
Figure 2.45).

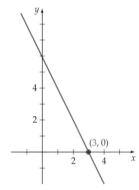

FIGURE 2.45
The x-intercept is $(3, 0)$.

▶ **TRY EXERCISE 46, PAGE 192**

EXAMPLE 6   **Solve $f_1(x) = f_2(x)$**

Let $f_1(x) = 2x - 1$ and $f_2(x) = -x + 11$. Find the values x for which $f_1(x) = f_2(x)$.

Algebraic Solution

$$f_1(x) = f_2(x)$$
$$2x - 1 = -x + 11$$
$$3x = 12$$
$$x = 4$$

When $x = 4$, $f_1(x) = f_2(x)$.

Visualize the Solution

The graphs of $y = f_1(x)$ and
$y = f_2(x)$ are shown on the
same coordinate axes (see
Figure 2.46). Note that the
point of intersection is $(4, 7)$.

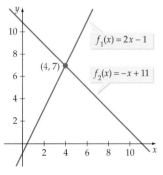

FIGURE 2.46

▶ **TRY EXERCISE 50, PAGE 192**

● **APPLICATIONS**

EXAMPLE 7   **Find a Linear Model of Data**

The bar graph in **Figure 2.47** is based on data from the Nevada Department of Motor Vehicles. The graph illustrates the distance (in feet) a car travels between the time (in seconds) a driver recognizes an emergency and the time the brakes are applied.

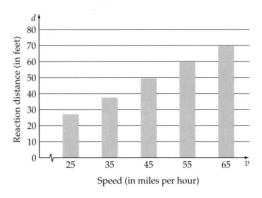

FIGURE 2.47

a. Find a linear function that models the reaction distance in terms of the speed of the car by using the ordered pairs $(25, 27)$ and $(55, 60)$.

b. What reaction distance does the model predict for a car traveling at 50 miles per hour?

Solution

a. First, calculate the slope of the line. Then use the point-slope formula to find the equation of the line.

$$m = \frac{d_2 - d_1}{v_2 - v_1} = \frac{60 - 27}{55 - 25} = \frac{33}{30} = 1.1 \qquad \text{• Find the slope.}$$

$$d - d_1 = m(v - v_1) \qquad \text{• Use the point-slope formula.}$$
$$d - 27 = 1.1(v - 25) \qquad \text{• } d_1 = 27, v_1 = 25, m = 1.1.$$
$$d = 1.1v - 0.5$$

In functional notation, the linear model is $d(v) = 1.1v - 0.5$.

b. To find the reaction distance for a car traveling at 50 miles per hour, evaluate $d(v)$ when $v = 50$.

$$d(v) = 1.1v - 0.5$$
$$d(50) = 1.1(50) - 0.5$$
$$= 54.5$$

The reaction distance is 54.5 feet.

▶ **TRY EXERCISE 56, PAGE 193**

If a manufacturer produces x units of a product that sells for p dollars per unit, then the **cost function** C, the **revenue function** R, and the **profit function** P are defined as follows:

$$C(x) = \text{cost of producing and selling } x \text{ units}$$

$$R(x) = xp = \text{revenue from the sale of } x \text{ units at } p \text{ dollars each}$$

$$P(x) = \text{profit from selling } x \text{ units}$$

Because profit equals the revenue less the cost, we have

$$P(x) = R(x) - C(x)$$

The value of x for which $R(x) = C(x)$ is called the **break-even point.** At the break-even point, $P(x) = 0$.

EXAMPLE 8   **Find the Profit Function and the Break-even Point**

A manufacturer finds that the costs incurred in the manufacture and sale of a particular type of calculator are $180,000 plus $27 per calculator.

a. Determine the profit function P, given that x calculators are manufactured and sold at $59 each.

b. Determine the break-even point.

Solution

a. The cost function is $C(x) = 27x + 180,000$. The revenue function is $R(x) = 59x$. Thus the profit function is

$$P(x) = R(x) - C(x)$$
$$= 59x - (27x + 180,000)$$
$$= 32x - 180,000, \quad x \geq 0 \text{ and } x \text{ is an integer}$$

b. At the break-even point, $R(x) = C(x)$.

$$59x = 27x + 180,000$$
$$32x = 180,000$$
$$x = 5625$$

The manufacturer will break even when 5625 calculators are sold.

▶ **TRY EXERCISE 66, PAGE 195**

---

take note

The graphs of C, R, and P are shown below. Observe that the graphs of C and R intersect at the break-even point, where $x = 5625$ and $P(5625) = 0$.

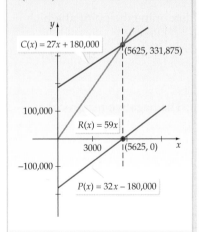

● PARALLEL AND PERPENDICULAR LINES

Two nonintersecting lines in a plane are **parallel.** All vertical lines are parallel to each other. All horizontal lines are parallel to each other.

Two lines are **perpendicular** if and only if they intersect and form adjacent angles, each of which measures 90°. In a plane, vertical and horizontal lines are perpendicular to one another.

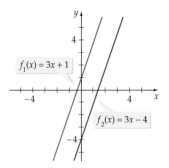

FIGURE 2.48

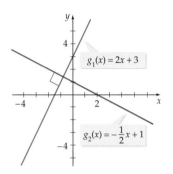

FIGURE 2.49

Parallel and Perpendicular Lines

Let l_1 be the graph of $f_1(x) = m_1x + b$ and l_2 be the graph of $f_2(x) = m_2x + b$. Then

- l_1 and l_2 are parallel if and only if $m_1 = m_2$.

- l_1 and l_2 are perpendicular if and only if $m_1 = -\dfrac{1}{m_2}$.

The graphs of $f_1(x) = 3x + 1$ and $f_2(x) = 3x - 4$ are shown in **Figure 2.48.** Because $m_1 = m_2 = 3$, the lines are parallel.

If $m_1 = -\dfrac{1}{m_2}$, then m_1 and m_2 are negative reciprocals of each other. The graphs of $g_1(x) = 2x + 3$ and $g_2(x) = -\dfrac{1}{2}x + 1$ are shown in **Figure 2.49.** Because 2 and $-\dfrac{1}{2}$ are negative reciprocals of each other, the lines are perpendicular. The symbol $\lnot$ indicates an angle of 90°. In **Figure 2.49** it is used to indicate that the lines are perpendicular.

EXAMPLE 9   Determine a Point of Impact

A rock attached to a string is whirled horizontally in a circular counter-clockwise path about the origin. When the string breaks, the rock travels on a linear path perpendicular to the radius $\overline{OP}$ and hits a wall located at

$$y = x + 12 \qquad (2)$$

If the string breaks when the rock is at $P(4, 3)$, determine the point at which the rock hits the wall. See **Figure 2.50.**

Solution

The slope of the radius from $(0, 0)$ to $(4, 3)$ is $\dfrac{3}{4}$. The negative reciprocal of $\dfrac{3}{4}$ is $-\dfrac{4}{3}$. Therefore, the linear path of the rock is given by

$$y - 3 = -\frac{4}{3}(x - 4)$$

$$y = -\frac{4}{3}x + \frac{25}{3} \qquad (3)$$

To find the point at which the rock hits the wall, set the right side of Equation (2) equal to the right side of Equation (3) and solve for x. This is the procedure explained in Example 6.

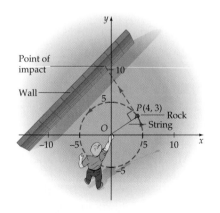

FIGURE 2.50

$$-\frac{4}{3}x + \frac{25}{3} = x + 12$$

$$-4x + 25 = 3x + 36 \qquad \text{• Multiply all terms by 3.}$$

$$-7x = 11$$

$$x = -\frac{11}{7}$$

For every point on the wall, x and y are related by $y = x + 12$. Therefore, substituting $-\frac{11}{7}$ for x in $y = x + 12$ yields $y = -\frac{11}{7} + 12 = \frac{73}{7}$, and the rock hits the wall at $\left(-\frac{11}{7}, \frac{73}{7}\right)$.

▶ **TRY EXERCISE 78, PAGE 196**

TOPICS FOR DISCUSSION

1. Can the graph of a linear function contain points in only one quadrant? only two quadrants? only four quadrants?

2. Is a "break-even point" a point or a number? Explain.

3. Some perpendicular lines do not have the property that their slopes are negative reciprocals of each other. Characterize these lines.

4. Does the real solutions and x-intercepts theorem apply only to linear functions?

5. Explain why the function $f(x) = x$ is referred to as the identity function.

EXERCISE SET 2.3

In Exercises 1 to 10, find the slope of the line that passes through the given points.

1. $(3, 4)$ and $(1, 7)$

▶ **2.** $(-2, 4)$ and $(5, 1)$

3. $(4, 0)$ and $(0, 2)$

4. $(-3, 4)$ and $(2, 4)$

5. $(0, 0)$ and $(0, 4)$

6. $(0, 0)$ and $(3, 0)$

7. $(-3, 4)$ and $(-4, -2)$

8. $(-5, -1)$ and $(-3, 4)$

9. $\left(-4, \frac{1}{2}\right)$ and $\left(\frac{7}{3}, \frac{7}{2}\right)$

10. $\left(\frac{1}{2}, 4\right)$ and $\left(\frac{7}{4}, 2\right)$

In Exercises 11 to 14, find the slope of the line that passes through the given points.

11. $(3, f(3))$ and $(3 + h, f(3 + h))$

12. $(-2, f(-2 + h))$ and $(-2 + h, f(-2 + h))$

13. $(0, f(0))$ and $(h, f(h))$

14. $(a, f(a))$ and $(a + h, f(a + h))$

In Exercises 15 to 26, graph y as a function of x by finding the slope and y-intercept of each line.

15. $y = 2x - 4$

▶ 16. $y = -x + 1$

17. $y = -\dfrac{1}{3}x + 4$

18. $y = \dfrac{2}{3}x - 2$

19. $y = 3$

20. $y = x$

21. $y = 2x$

22. $y = -3x$

23. $2x + y = 5$

24. $x - y = 4$

25. $4x + 3y - 12 = 0$

26. $2x + 3y + 6 = 0$

In Exercises 27 to 38, find the equation of the indicated line. Write the equation in the form y = mx + b.

27. y-intercept $(0, 3)$, slope 1

▶ 28. y-intercept $(0, 5)$, slope -2

29. y-intercept $\left(0, \dfrac{1}{2}\right)$, slope $\dfrac{3}{4}$

30. y-intercept $\left(0, \dfrac{3}{4}\right)$, slope $-\dfrac{2}{3}$

31. y-intercept $(0, 4)$, slope 0

32. y-intercept $(0, -1)$, slope $\dfrac{1}{2}$

33. Through $(-3, 2)$, slope -4

34. Through $(-5, -1)$, slope -3

35. Through $(3, 1)$ and $(-1, 4)$

36. Through $(5, -6)$ and $(2, -8)$

37. Through $(7, 11)$ and $(2, -1)$

38. Through $(-5, 6)$ and $(-3, -4)$

39. Find the value of x in the domain of $f(x) = 2x + 3$ for which $f(x) = -1$.

40. Find the value of x in the domain of $f(x) = 4 - 3x$ for which $f(x) = 7$.

41. Find the value of x in the domain of $f(x) = 1 - 4x$ for which $f(x) = 3$.

▶ 42. Find the value of x in the domain of $f(x) = \dfrac{2x}{3} + 2$ for which $f(x) = 4$.

43. Find the value of x in the domain of $f(x) = 3 - \dfrac{x}{2}$ for which $f(x) = 5$.

44. Find the value of x in the domain of $f(x) = 4x - 3$ for which $f(x) = -2$.

In Exercises 45 to 48, find the solution f(x) = 0. Verify that the solution of f(x) = 0 is the same as the x-coordinate of the x-intercept of the graph of y = f(x).

45. $f(x) = 3x - 12$

▶ 46. $f(x) = -2x - 4$

47. $f(x) = \dfrac{1}{4}x + 5$

48. $f(x) = -\dfrac{1}{3}x + 2$

In Exercises 49 to 52, solve $f_1(x) = f_2(x)$ by an algebraic method and by graphing.

49. $f_1(x) = 4x + 5$ $f_2(x) = x + 6$

▶ 50. $f_1(x) = -2x - 11$ $f_2(x) = 3x + 7$

51. $f_1(x) = 2x - 4$ $f_2(x) = -x + 12$

52. $f_1(x) = \dfrac{1}{2}x + 5$ $f_2(x) = \dfrac{2}{3}x - 7$

53. **OCEANOGRAPHY** The graph below shows the relationship between the speed of sound in water and the temperature of the water. Find the slope of this line, and write a sentence that explains the meaning of the slope in the context of this problem.

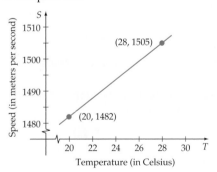

54. **COMPUTER SCIENCE** The graph on the following page shows the relationship between the time, in seconds, it takes to download a file and the size of the file in megabytes. Find the slope of the line between the two points shown on the graph. Write a sentence that states the meaning of the slope in the context of this problem.

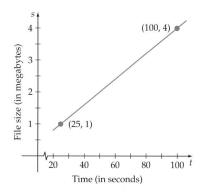

Time (in seconds)

55. AUTOMOTIVE TECHNOLOGY The table below shows the EPA fuel economy values for selected two-seater cars for the 2003 model year. (*Source:* www.fueleconomy.gov.)

EPA Fuel Economy Values for Selected Two-Seater Cars

Car	City mpg	Highway mpg
Audi, TT Roadster	20	29
BMW, Z8	13	21
Ferrari, 360 Spider	11	16
Lamborghini, L-174	9	13
Lotus, Esprit V8	15	22
Maserati, Spider GT	11	17

a. Using the data for the Lamborghini and the Audi, find a linear function that predicts highway miles per gallon in terms of city miles per gallon. Round the slope to the nearest hundredth.

b. Using your model, predict the highway miles per gallon for a Porsche Boxer, whose city fuel efficiency is 18 miles per gallon. Round to the nearest whole number.

▶ **56.** CONSUMER CREDIT The amount of revolving consumer credit (such as credit cards and department store cards) for the years 1997 to 2003 is given in the table below. (*Source:* www.nber.org, Board of Governor's of the Federal Reserve System.)

Year	Consumer Credit (in billions of $)
1997	531.0
1998	562.5
1999	598.0
2000	667.4
2001	701.3
2002	712.0
2003	725.0

a. Using the data for 1997 and 2003, find a linear model that predicts the amount of revolving consumer credit (in billions) for year t. Round the slope to the nearest tenth.

b. Using this model, in what year would consumer credit first exceed $850 billion?

57. LABOR MARKET According to the Bureau of Labor Statistics (BLS), there were 38,000 desktop publishing jobs in the United States in the year 2000. The BLS projects that there will be 63,000 desktop publishing jobs in 2010.

a. Using the BLS data, find the number of desktop publishing jobs as a linear function of the year.

b. Using your model, in what year will the number of desktop publishing jobs first exceed 60,000?

58. POTTERY A piece of pottery is removed from a kiln and allowed to cool in a controlled environment. The temperature (in degrees Fahrenheit) of the pottery after it is removed from the kiln for various times (in minutes) is shown in the table below.

Time, min	Temperature, °F
15	2200
20	2150
30	2050
60	1750

a. Find a linear model for the temperature of the pottery after t minutes.

b. Explain the meaning of the slope of this line in the context of the problem.

c. Assuming temperature continues to decrease at the same rate, what will be the temperature of the pottery in 3 hours?

59. LUMBER INDUSTRY For a log, the number of board-feet (bf) that can be obtained from the log depends on the diameter, in inches, of the log and its length. The table below shows the number of board-feet of lumber that can be obtained from a log that is 32 feet long.

Diameter, inches	bf
16	180
18	240
20	300
22	360

a. Find a linear model for the number of board-feet as a function of tree diameter.

b. Write a sentence explaining the meaning of the slope of this line in the context of the problem.

c. Using this model, how many board-feet of lumber can be obtained from a log 32 feet long with a diameter of 19 inches?

60. ECOLOGY The rate at which water evaporates from a certain reservoir depends on air temperature. The table below shows the number of acre-feet (af) of water per day that evaporate from the reservoir for various temperatures in degrees Fahrenheit.

Temperature, °F	af
40	800
60	1640
70	2060
85	2690

a. Find a linear model for the number of acre-feet of water that evaporate as a function of temperature.

b. Write a sentence that explains the meaning of the slope of this line in the context of this problem.

c. Assuming that water continues to evaporate at the same rate, how many acre-feet of water will evaporate per day when the temperature is 75°F?

61. CYCLING SPEEDS Michelle and Amanda start from the same place on a cycling course. Michelle is riding at 15 miles per hour and Amanda is cycling at 12 miles per hour. The graphs below show the total distance traveled by each cyclist and the total distance between Michelle and Amanda after t hours. Which graphs represent which distances?

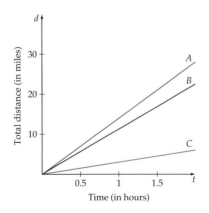

62. TEMPERATURE The graph below shows the temperature changes, in degrees Fahrenheit, over a 12-hour period at a weather station.

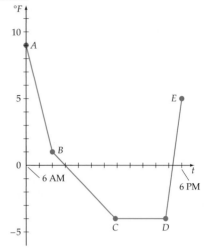

a. How many degrees per hour did the temperature change between A and B?

b. Between which two points did the temperature change most rapidly?

c. Between which two points was the temperature constant?

63. **HEALTH** Framingham Heart Study is an ongoing research project that is attempting to identify risk factors associated with heart disease. Selected blood pressure data from that study is shown in the table at the right.

Selected Framingham Blood Pressure Statistics

Diastolic	Systolic
100	135
88	154
80	110
70	110
80	114
108	180
85	135
75	115

a. Find the equation of a linear model of this data, given that the graph of the line passes through $P_1(70, 110)$ and $P_2(108, 180)$.

b. What systolic blood pressure does the model you found in part **a.** predict for a diastolic pressure of 90?

64. **HEALTH** The table on the following page shows average remaining lifetime, by age, of all people in the United States in 1997. (*Source:* National Institutes of Health)

Average Remaining Lifetime by Age in the United States

Current Age	Remaining Years
0	76.5
15	62.3
35	43.4
65	17.7
75	11.2

a. Find the equation of a linear model of this data, given that the graph of the line passes through $(0, 76.5)$ and $(75, 11.2)$.

b. Based on your model, what is the average remaining lifetime of a person whose current age is 25?

In Exercises 65 to 68, determine the profit function for the given revenue function and cost function. Also determine the break-even point.

65. $R(x) = 92.50x; C(x) = 52x + 1782$

▶ **66.** $R(x) = 124x; C(x) = 78.5x + 5005$

67. $R(x) = 259x; C(x) = 180x + 10{,}270$

68. $R(x) = 14{,}220x; C(x) = 8010x + 1{,}602{,}180$

69. MARGINAL COST In business, *marginal cost* is a phrase used to represent the rate of change or slope of a cost function that relates the cost C to the number of units x produced. If a cost function is given by $C(x) = 8x + 275$, find

a. $C(0)$ **b.** $C(1)$ **c.** $C(10)$ **d.** marginal cost

70. MARGINAL REVENUE In business, *marginal revenue* is a phrase used to represent the rate of change or slope of a revenue function that relates the revenue R to the number of units x sold. If a revenue function is given by the function $R(x) = 210x$, find

a. $R(0)$ **b.** $R(1)$ **c.** $R(10)$ **d.** marginal revenue

71. BREAK-EVEN POINT FOR A RENTAL TRUCK A rental company purchases a truck for $19,500. The truck requires an average of $6.75 per day in maintenance.

a. Find the linear function that expresses the total cost C of owning the truck after t days.

b. The truck rents for $55.00 a day. Find the linear function that expresses the revenue R when the truck has been rented for t days.

c. The profit after t days, $P(t)$, is given by the function $P(t) = R(t) - C(t)$. Find the linear function $P(t)$.

d. Use the function $P(t)$ that you obtained in **c.** to determine how many days it will take the company to break even on the purchase of the truck. Assume that the truck is always in use.

72. BREAK-EVEN POINT FOR A PUBLISHER A magazine company had a profit of $98,000 per year when it had 32,000 subscribers. When it obtained 35,000 subscribers, it had a profit of $117,500. Assume that the profit P is a linear function of the number of subscribers s.

a. Find the function P.

b. What will the profit be if the company obtains 50,000 subscribers?

c. What is the number of subscribers needed to break even?

In Exercises 73 to 76, find the equation of the indicated line. Write the equation in the form $y = mx + b$.

73. Through $(1, 3)$ and parallel to $3x + 4y = -24$

74. Through $(2, -1)$ and parallel to $x + y = 10$

75. Through $(1, 2)$ and perpendicular to $x + y = 4$

76. Through $(-3, 4)$ and perpendicular to $2x - y = 7$

77. POINT OF IMPACT A rock attached to a string is whirled horizontally, in a counterclockwise circular path with radius 5 feet, about the origin. When the string breaks, the rock travels on a linear path perpendicular to the radius $\overline{OP}$ and hits a wall located at $y = 10$ feet.

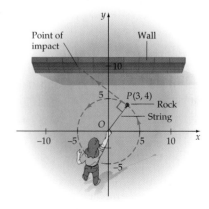

If the string breaks when the rock is at $P(3$ feet, 4 feet$)$, find the x-coordinate of the point at which the rock hits the wall.

▶ **78.** POINT OF IMPACT A rock attached to a string is whirled horizontally, in a counterclockwise circular path with radius 4 feet, about the origin. When the string breaks, the rock travels on a linear path perpendicular to the radius $\overline{OP}$ and hits a wall located at $y = 14$ feet. If the string breaks when the rock is at $P(\sqrt{15}$ feet, 1 foot$)$, find the x-coordinate of the point at which the rock hits the wall.

79. SLOPE OF A SECANT LINE The graph of $y = x^2 + 1$ is shown below with $P(2, 5)$ and $Q(2 + h, [2 + h]^2 + 1)$ points on the graph.

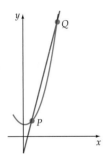

a. If $h = 1$, determine the coordinates of Q and the slope of the line PQ.

b. If $h = 0.1$, determine the coordinates of Q and the slope of the line PQ.

c. If $h = 0.01$, determine the coordinates of Q and the slope of the line PQ.

d. As h approaches 0, what value does the slope of the line PQ seem to be approaching?

e. Verify that the slope of the line passing through $(2, 5)$ and $(2 + h, [2 + h]^2 + 1)$ is $4 + h$.

80. SLOPE OF A SECANT LINE The graph of $y = 3x^2$ is shown below with $P(-1, 3)$ and $Q(-1 + h, 3[-1 + h]^2)$ points on the graph.

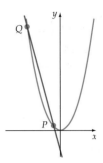

a. If $h = 1$, determine the coordinates of Q and the slope of the line PQ.

b. If $h = 0.1$, determine the coordinates of Q and the slope of the line PQ.

c. If $h = 0.01$, determine the coordinates of Q and the slope of the line PQ.

d. As h approaches 0, what value does the slope of the line PQ seem to be approaching?

e. Verify that the slope of the line passing through $(-1, 3)$ and $(-1 + h, 3[-1 + h]^2)$ is $-6 + 3h$.

81. Verify that the slope of the line passing through (x, x^2) and $(x + h, [x + h]^2)$ is $2x + h$.

82. Verify that the slope of the line passing through $(x, 4x^2)$ and $(x + h, 4[x + h]^2)$ is $8x + 4h$.

CONNECTING CONCEPTS

83. THE TWO-POINT FORM Use the point-slope form to derive the following equation, which is called the two-point form.

$$y - y_1 = \left(\frac{y_2 - y_1}{x_2 - x_1}\right)(x - x_1)$$

84. THE INTERCEPT FORM Use the two-point form from Exercise 83 to show that the line with intercepts $(a, 0)$ and $(0, b)$, $a \neq 0$ and $b \neq 0$, has the equation

$$\frac{x}{a} + \frac{y}{b} = 1$$

In Exercises 85 and 86, use the two-point form to find an equation of the line that passes through the indicated points. Write your answers in slope-intercept form.

85. $(5, 1)$, $(4, 3)$ **86.** $(2, 7)$, $(-1, 6)$

In Exercises 87 to 90, use the equation from Exercise 84 (called the intercept form) to write an equation of the line with the indicated intercepts.

87. x-intercept $(3, 0)$, y-intercept $(0, 5)$

88. x-intercept $(-2, 0)$, y-intercept $(0, 7)$

89. x-intercept $(a, 0)$, y-intercept $(0, 3a)$, point on the line $(5, 2)$, $a \neq 0$

90. x-intercept $(-b, 0)$, y-intercept $(0, 2b)$, point on the line $(-3, 10)$, $b \neq 0$

91. Verify that the slope of the line passing through $(1, 3)$ and $(1 + h, 3[1 + h]^3)$ is $9 + 9h + 3h^2$.

92. Find the two points on the circle given by $x^2 + y^2 = 25$ such that the slope of the radius from $(0, 0)$ to each point is 0.5.

93. Find a point $P(x, y)$ on the graph of the equation $y = x^2$ such that the slope of the line through the point $(3, 9)$ and P is $\dfrac{15}{2}$.

94. Determine whether there is a point $P(x, y)$ on the graph of the equation $y = \sqrt{x + 1}$ such that the slope of the line through the point $(3, 2)$ and P is $\dfrac{3}{8}$.

PREPARE FOR SECTION 2.4

95. Factor: $3x^2 + 10x - 8$ [P.3]

96. Complete the square of $x^2 - 8x$. Write the resulting trinomial as the square of a binomial. [1.3]

97. Find $f(-3)$ for $f(x) = 2x^2 - 5x - 7$. [2.2]

In Exercises 98 and 99, solve for x.

98. $2x^2 - x = 1$ [1.3]

99. $x^2 + 3x - 2 = 0$ [1.3]

100. Suppose that $h = -16t^2 + 64t + 5$. Find two values of t for which $h = 53$. [1.3]

PROJECTS

1. VISUAL INSIGHT

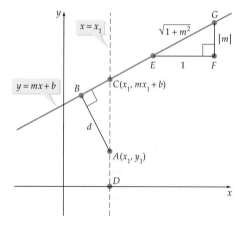

The distance d between the point $A(x_1, y_1)$ and the

line given by $y = mx + b$ is $d = \dfrac{|mx_1 + b - y_1|}{\sqrt{1 + m^2}}$

Write a paragraph that explains how to make use of the figure above to verify the formula for the distance d.

2. VERIFY GEOMETRIC THEOREMS

a. Prove that in any triangle, the line segment that joins the midpoints of two sides of the triangle is parallel to the third side. (*Hint:* Assign coordinates to the vertices of the triangle as shown in the figure at the left below.)

b. Prove that in any square, the diagonals are perpendicular bisectors of each other. (*Hint:* Assign coordinates to the vertices of the square as shown in the figure at the right below.)

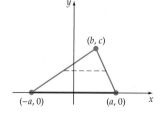

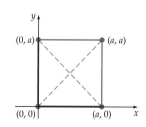

| SECTION 2.4 | # QUADRATIC FUNCTIONS |

- **VERTEX OF A PARABOLA**
- **MAXIMUM AND MINIMUM OF A QUADRATIC FUNCTION**
- **APPLICATIONS**

Some applications can be modeled by a *quadratic function*.

Definition of a Quadratic Function

A **quadratic function** of x is a function that can be represented by an equation of the form

$$f(x) = ax^2 + bx + c$$

where a, b, and c are real numbers and $a \neq 0$.

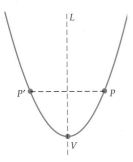

FIGURE 2.51

The graph of $f(x) = ax^2 + bx + c$ is a *parabola*. The graph opens up if $a > 0$, and it opens down if $a < 0$. The **vertex of a parabola** is the lowest point on a parabola that opens up or the highest point on a parabola that opens down. Point V is the vertex of the parabola in **Figure 2.51.**

The graph of $f(x) = ax^2 + bx + c$ is *symmetric* with respect to a vertical line through its vertex.

Definition of Symmetry with Respect to a Line

A graph is **symmetric with respect to a line** L if for each point P on the graph there is a point P' on the graph such that the line L is the perpendicular bisector of the line segment PP'.

take note

The axis of symmetry is a line. When asked to determine the axis of symmetry, the answer is an equation, not just a number.

In **Figure 2.51,** the parabola is symmetric with respect to the line L. The line L is called the **axis of symmetry**. The points P and P' are reflections or images of each other with respect to the axis of symmetry.

If $b = 0$ and $c = 0$, then $f(x) = ax^2 + bx + c$ simplifies to $f(x) = ax^2$. The graph of $f(x) = ax^2$ ($a \neq 0$) is a parabola with its vertex at the origin, and the y-axis is its axis of symmetry. The graph of $f(x) = ax^2$ can be constructed by plotting a few points and drawing a smooth curve that passes through these points, with the origin as the vertex and the y-axis as its axis of symmetry. The graphs of $f(x) = x^2$, $g(x) = 2x^2$, and $h(x) = -\dfrac{1}{2}x^2$ are shown in **Figure 2.52.**

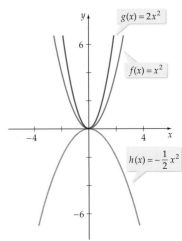

FIGURE 2.52

take note

The equation $z = x^2 - y^2$ defines z as a quadratic function of x and y. You might think that the graph of every quadratic function is a parabola. However, the graph of $z = x^2 - y^2$ is the saddle shown in the figure below. You will study quadratic functions involving two or more independent variables in calculus.

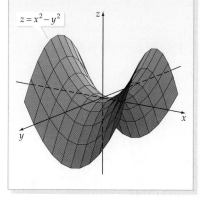

$z = x^2 - y^2$

Standard Form of Quadratic Functions

Every quadratic function f given by $f(x) = ax^2 + bx + c$ can be written in the **standard form of a quadratic function:**

$$f(x) = a(x - h)^2 + k, \quad a \neq 0$$

The graph of f is a parabola with vertex (h, k). The parabola opens up if $a > 0$, and it opens down if $a < 0$. The vertical line $x = h$ is the axis of symmetry of the parabola.

The standard form is useful because it readily gives information about the vertex of the parabola and its axis of symmetry. For example, note that the graph of $f(x) = 2(x - 4)^2 - 3$ is a parabola. The coordinates of the vertex are $(4, -3)$, and the line $x = 4$ is its axis of symmetry. Because a is the positive number 2, the parabola opens upward.

EXAMPLE I   Find the Standard Form of a Quadratic Function

Use the technique of completing the square to find the standard form of $g(x) = 2x^2 - 12x + 19$. Sketch the graph.

Solution

$$
\begin{aligned}
g(x) &= 2x^2 - 12x + 19 \\
&= 2(x^2 - 6x) + 19 && \text{• Factor 2 from the variable terms.} \\
&= 2(x^2 - 6x + 9 - 9) + 19 && \text{• Complete the square.} \\
&= 2(x^2 - 6x + 9) - 2(9) + 19 && \text{• Regroup.} \\
&= 2(x - 3)^2 - 18 + 19 && \text{• Factor and simplify.} \\
&= 2(x - 3)^2 + 1 && \text{• Standard form}
\end{aligned}
$$

The vertex is $(3, 1)$. The axis of symmetry is $x = 3$. Because $a > 0$, the parabola opens up. See **Figure 2.53**.

▶ **TRY EXERCISE 10, PAGE 206**

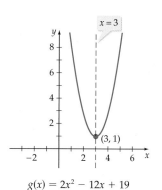

$x = 3$

$(3, 1)$

$g(x) = 2x^2 - 12x + 19$

FIGURE 2.53

● VERTEX OF A PARABOLA

We can write $f(x) = ax^2 + bx + c$ in standard form by completing the square of $ax^2 + bx + c$. This will allow us to derive a general expression for the x- and y-coordinates of the graph of $f(x) = ax^2 + bx + c$.

$$
\begin{aligned}
f(x) &= ax^2 + bx + c \\
&= a\left(x^2 + \frac{b}{a}x\right) + c && \text{• Factor } a \text{ from } ax^2 + bx. \\
&= a\left(x^2 + \frac{b}{a}x + \frac{b^2}{4a^2}\right) + c - \frac{b^2}{4a} && \text{• Complete the square by adding and} \\
& && \text{subtracting } \left(\frac{1}{2} \cdot \frac{b}{a}\right)^2 = \frac{b^2}{4a^2}. \\
&= a\left(x + \frac{b}{2a}\right)^2 + \frac{4ac - b^2}{4a} && \text{• Factor and simplify.}
\end{aligned}
$$

Thus $f(x) = ax^2 + bx + c$ in standard form is $f(x) = a\left(x + \dfrac{b}{2a}\right)^2 + \dfrac{4ac - b^2}{4a}$.

Comparing this last expression with $f(x) = a(x - h)^2 + k$, we see that the coordinates of the vertex are $\left(-\dfrac{b}{2a}, \dfrac{4ac - b^2}{4a}\right)$.

Note that by evaluating $f(x) = a\left(x + \dfrac{b}{2a}\right)^2 + \dfrac{4ac - b^2}{4a}$ at $x = -\dfrac{b}{2a}$, we have

$$f(x) = a\left(x + \frac{b}{2a}\right)^2 + \frac{4ac - b^2}{4a}$$

$$f\left(-\frac{b}{2a}\right) = a\left(-\frac{b}{2a} + \frac{b}{2a}\right)^2 + \frac{4ac - b^2}{4a} = a(0) + \frac{4ac - b^2}{4a}$$

$$= \frac{4ac - b^2}{4a}$$

That is, the y-coordinate of the vertex is $f\left(-\dfrac{b}{2a}\right)$. This is summarized by the following formula.

Vertex Formula

The coordinates of the vertex of $f(x) = ax^2 + bx + c$ are $\left(-\dfrac{b}{2a}, f\left(-\dfrac{b}{2a}\right)\right)$.

The vertex formula can be used to write the standard form of the equation of a parabola. We have

$$h = -\frac{b}{2a} \qquad \text{and} \qquad k = f\left(-\frac{b}{2a}\right)$$

EXAMPLE 2 **Find the Vertex and Standard Form of a Quadratic Function**

Use the vertex formula to find the vertex and standard form of $f(x) = 2x^2 - 8x + 3$. See **Figure 2.54**.

Solution

$$f(x) = 2x^2 - 8x + 3 \qquad \bullet \; a = 2, b = -8, c = 3$$

$$h = -\frac{b}{2a} = -\frac{-8}{2(2)} = 2 \qquad \bullet \; \textbf{\textit{x}-coordinate of the vertex}$$

$$k = f\left(-\frac{b}{2a}\right) = 2(2)^2 - 8(2) + 3 = -5 \qquad \bullet \; \textbf{\textit{y}-coordinate of the vertex}$$

The vertex is $(2, -5)$. Substituting into the standard form equation $f(x) = a(x - h)^2 + k$ yields the standard form $f(x) = 2(x - 2)^2 - 5$.

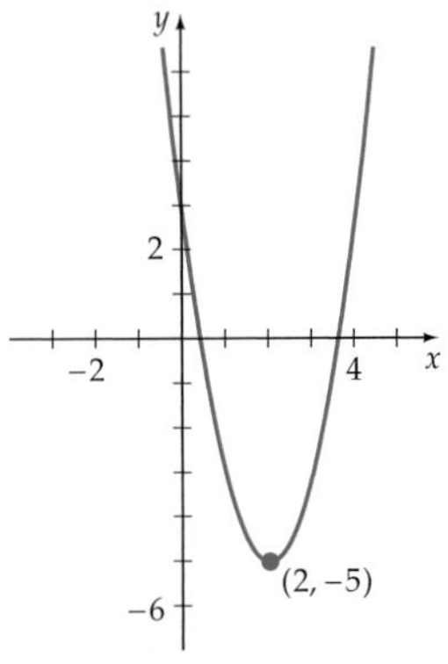

▶ **TRY EXERCISE 20, PAGE 207**

$f(x) = 2x^2 - 8x + 3$

FIGURE 2.54

● MAXIMUM AND MINIMUM OF A QUADRATIC FUNCTION

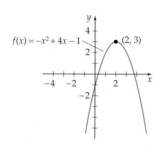

$f(x) = -x^2 + 4x - 1$

Note from Example 2 that the graph of the parabola opens up, and the vertex is the *lowest* point on the graph of the parabola. Therefore, the y-coordinate of the vertex is the *minimum* value of that function. This information can be used to determine the range of $f(x) = 2x^2 - 8x + 3$. The range is $\{y \mid y \geq -5\}$. Similarly, if the graph of a parabola opened down, the vertex would be the *highest* point on the graph, and the y-coordinate of the vertex would be the *maximum* value of the function. For instance, the maximum value of $f(x) = -x^2 + 4x - 1$, graphed at the left, is 3, the y-coordinate of the vertex. The range of the function is $\{y \mid y \leq 3\}$. For the function in Example 2 and the function whose graph is shown at the left, the domain is the set of real numbers.

EXAMPLE 3   Find the Range of $f(x) = ax^2 + bx + c$

Find the range of $f(x) = -2x^2 - 6x - 1$. Determine the values of x for which $f(x) = 3$.

Algebraic Solution

To find the range of f, determine the y-coordinate of the vertex of the graph of f.

$$f(x) = -2x^2 - 6x - 1$$

• $a = -2$, $b = -6$, $c = -1$

$$h = -\frac{b}{2a} = -\frac{-6}{2(-2)} = -\frac{3}{2}$$

• Find the x-coordinate of the vertex.

$$k = f\left(-\frac{3}{2}\right) = -2\left(-\frac{3}{2}\right)^2 - 6\left(-\frac{3}{2}\right) - 1 = \frac{7}{2}$$

• Find the y-coordinate of the vertex.

The vertex is $\left(-\frac{3}{2}, \frac{7}{2}\right)$. Because the parabola opens down, $\frac{7}{2}$ is the maximum value of f. Therefore, the range of f is $\left\{y \mid y \leq \frac{7}{2}\right\}$.

To determine the values of x for which $f(x) = 3$, replace $f(x)$ by $-2x^2 - 6x - 1$ and solve for x.

$$f(x) = 3$$
$$-2x^2 - 6x - 1 = 3$$

• Replace $f(x)$ by $-2x^2 - 6x - 1$.

$$-2x^2 - 6x - 4 = 0$$

• Solve for x.

$$-2(x + 1)(x + 2) = 0$$

• Factor.

$$x + 1 = 0 \quad \text{or} \quad x + 2 = 0$$

• Use the Principle of Zero Products to solve for x.

$$x = -1 \qquad x = -2$$

The values of x for which $f(x) = 3$ are -1 and -2.

Visualize the Solution

The graph of f is shown below. The vertex of the graph is $\left(-\frac{3}{2}, \frac{7}{2}\right)$. Note that the line $y = 3$ intersects the graph of f when $x = -2$ and when $x = -1$.

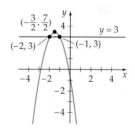

▶ **TRY EXERCISE 32, PAGE 207**

The following theorem can be used to determine the maximum value or the minimum value of a quadratic function.

Maximum or Minimum Value of a Quadratic Function

If $a > 0$, then the vertex (h, k) is the lowest point on the graph of $f(x) = a(x - h)^2 + k$, and the y-coordinate k of the vertex is the **minimum value** of the function f. See **Figure 2.55a**.

If $a < 0$, then the vertex (h, k) is the highest point on the graph of $f(x) = a(x - h)^2 + k$, and the y-coordinate k is the **maximum value** of the function f. See **Figure 2.55b**.

In either case, the maximum or minimum is achieved when $x = h$.

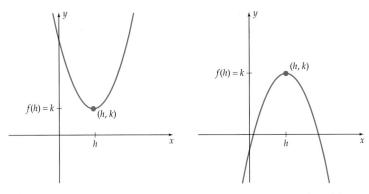

a. k is the minimum value of f.   **b.** k is the maximum value of f.

FIGURE 2.55

Find the Maximum or Minimum of a Quadratic Function

EXAMPLE 4

Find the maximum or minimum value of each quadratic function. State whether the value is a maximum or a minimum.

a. $F(x) = -2x^2 + 8x - 1$ **b.** $G(x) = x^2 - 3x + 1$

Solution

The maximum or minimum value of a quadratic function is the y-coordinate of the vertex of the graph of the function.

a. $h = -\dfrac{b}{2a} = -\dfrac{8}{2(-2)} = 2$ • **x-coordinate of the vertex**

$k = F\left(-\dfrac{b}{2a}\right) = -2(2)^2 + 8(2) - 1 = 7$ • **y-coordinate of the vertex**

Because $a < 0$, the function has a maximum value but no minimum value. The maximum value is 7. See **Figure 2.56**.

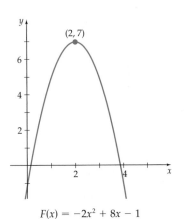

$F(x) = -2x^2 + 8x - 1$

FIGURE 2.56

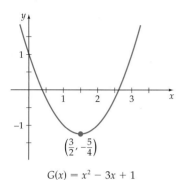

$G(x) = x^2 - 3x + 1$

FIGURE 2.57

b. $h = -\dfrac{b}{2a} = -\dfrac{-3}{2(1)} = \dfrac{3}{2}$ • *x*-coordinate of the vertex

$k = G\left(-\dfrac{b}{2a}\right) = \left(\dfrac{3}{2}\right)^2 - 3\left(\dfrac{3}{2}\right) + 1$

$\qquad\qquad\qquad = -\dfrac{5}{4}$ • *y*-coordinate of the vertex

Because $a > 0$, the function has a minimum value but no maximum value. The minimum value is $-\dfrac{5}{4}$. See **Figure 2.57**.

▶ **TRY EXERCISE 36, PAGE 207**

● **APPLICATIONS**

EXAMPLE 5 Find the Maximum of a Quadratic Function

A long sheet of tin 20 inches wide is to be made into a trough by bending up two sides until they are perpendicular to the bottom. How many inches should be turned up so that the trough will achieve its maximum carrying capacity?

Solution

The trough is shown in **Figure 2.58.** If x is the number of inches to be turned up on each side, then the width of the base is $20 - 2x$ inches. The maximum carrying capacity of the trough will occur when the cross-sectional area is a maximum. The cross-sectional area $A(x)$ is given by

$$A(x) = x(20 - 2x) \qquad \text{• Area = (length)(width)}$$
$$= -2x^2 + 20x$$

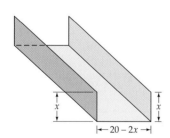

FIGURE 2.58

To find the point at which A obtains its maximum value, find the *x*-coordinate of the vertex of the graph of A. Using the vertex formula with $a = -2$ and $b = 20$, we have

$$x = -\dfrac{b}{2a} = -\dfrac{20}{2(-2)} = 5$$

Therefore, the maximum carrying capacity will be achieved when 5 inches are turned up. See **Figure 2.59.**

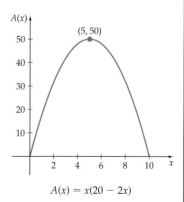

$A(x) = x(20 - 2x)$

FIGURE 2.59

▶ **TRY EXERCISE 46, PAGE 207**

EXAMPLE 6   Solve a Business Application

The owners of a travel agency have determined that they can sell all 160 tickets for a tour if they charge \$8 (their cost) for each ticket. For each \$0.25 increase in the price of a ticket, they estimate they will sell 1 ticket less. A business manager determines that their cost function is $C(x) = 8x$ and that the customer's price per ticket is

$$p(x) = 8 + 0.25(160 - x) = 48 - 0.25x$$

where x represents the number of tickets sold. Determine the maximum profit and the cost per ticket that yields the maximum profit.

Solution

The profit from selling x tickets is $P(x) = R(x) - C(x)$, where P, R, and C are the profit function, the revenue function, and the cost function as defined in Section 2.3. Thus

$$\begin{aligned} P(x) &= R(x) - C(x) \\ &= x[p(x)] - C(x) \\ &= x(48 - 0.25x) - 8x \\ &= 40x - 0.25x^2 \end{aligned}$$

The graph of the profit function is a parabola that opens down. Thus the maximum profit occurs when

$$x = -\frac{b}{2a} = -\frac{40}{2(-0.25)} = 80$$

The maximum profit is determined by evaluating $P(x)$ with $x = 80$.

$$P(80) = 40(80) - 0.25(80)^2 = 1600$$

The maximum profit is \$1600.
To find the price per ticket that yields the maximum profit, we evaluate $p(x)$ with $x = 80$.

$$p(80) = 48 - 0.25(80) = 28$$

Thus the travel agency can expect a maximum profit of \$1600 when 80 people take the tour at a ticket price of \$28 per person. The graph of the profit function is shown in **Figure 2.60.**

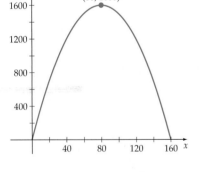

$P(x) = 40x - 0.25x^2$

FIGURE 2.60

> **? QUESTION** In **Figure 2.60,** why have we shown only the portion of the graph that lies in quadrant I?

▶ **TRY EXERCISE 68, PAGE 209**

? ANSWER Since x represents the number of tickets sold, x must be greater than or equal to zero but less than or equal to 160. $P(x)$ is nonnegative for $0 \le x \le 160$.

EXAMPLE 7 Solve a Projectile Application

In **Figure 2.61,** a ball is thrown vertically upward with an initial velocity of 48 feet per second. If the ball started its flight at a height of 8 feet, then its height s at time t can be determined by $s(t) = -16t^2 + 48t + 8$, where $s(t)$ is measured in feet above ground level and t is the number of seconds of flight.

a. Determine the time it takes the ball to attain its maximum height.

b. Determine the maximum height the ball attains.

c. Determine the time it takes the ball to hit the ground.

Solution

a. The graph of $s(t) = -16t^2 + 48t + 8$ is a parabola that opens downward. See **Figure 2.62.** Therefore, s will attain its maximum value at the vertex of its graph. Using the vertex formula with $a = -16$ and $b = 48$, we get

$$t = -\frac{b}{2a} = -\frac{48}{2(-16)} = \frac{3}{2}$$

Therefore, the ball attains its maximum height $1\frac{1}{2}$ seconds into its flight.

b. When $t = \frac{3}{2}$, the height of the ball is

$$s\left(\frac{3}{2}\right) = -16\left(\frac{3}{2}\right)^2 + 48\left(\frac{3}{2}\right) + 8 = 44 \text{ feet}$$

c. The ball will hit the ground when its height $s(t) = 0$. Therefore, solve $-16t^2 + 48t + 8 = 0$ for t.

$$-16t^2 + 48t + 8 = 0$$
$$-2t^2 + 6t + 1 = 0 \qquad \bullet \text{ Divide each side by 8.}$$
$$t = \frac{-(6) \pm \sqrt{6^2 - 4(-2)(1)}}{2(-2)} \qquad \bullet \text{ Use the quadratic formula.}$$
$$= \frac{-6 \pm \sqrt{44}}{-4} = \frac{-3 \pm \sqrt{11}}{-2}$$

Using a calculator to approximate the positive root, we find that the ball will hit the ground in $t \approx 3.16$ seconds. This is also the value of the t-coordinate of the t-intercept in **Figure 2.62.**

▶ **TRY EXERCISE 70, PAGE 209**

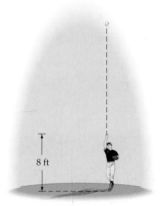

8 ft

FIGURE 2.61

 To review **QUADRATIC FORMULA,** *see p. 101.*

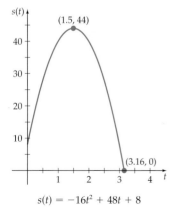

$s(t) = -16t^2 + 48t + 8$

FIGURE 2.62

TOPICS FOR DISCUSSION

1. Does the graph of every quadratic function of the form

$$f(x) = ax^2 + bx + c$$

have a y-intercept? If so, what are the coordinates of the y-intercept?

2. The graph of $f(x) = -x^2 + 6x + 11$ has a vertex of $(3, 20)$. Is this vertex point the highest point or the lowest point on the graph of f?

3. A tutor states that the graph of $f(x) = ax^2 + bx + c$ $(a \neq 0)$ is a parabola and that its axis of symmetry is $y = -\dfrac{b}{2a}$. Do you agree?

4. Every quadratic function of the form $f(x) = ax^2 + bx + c$ has a domain of all real numbers. Do you agree?

5. A classmate states that the graph of every quadratic function of the form

$$f(x) = ax^2 + bx + c$$

must contain points from at least two quadrants. Do you agree?

EXERCISE SET 2.4

In Exercises 1 to 8, match each graph in _a._ through _h._ with the proper quadratic function.

1. $f(x) = x^2 - 3$

2. $f(x) = x^2 + 2$

3. $f(x) = (x - 4)^2$

4. $f(x) = (x + 3)^2$

5. $f(x) = -2x^2 + 2$

6. $f(x) = -\dfrac{1}{2}x^2 + 3$

7. $f(x) = (x + 1)^2 + 3$

8. $f(x) = -2(x - 2)^2 + 2$

e.
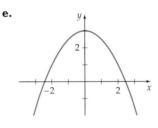

f.
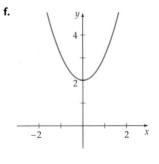

a.
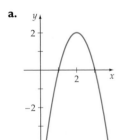

b.
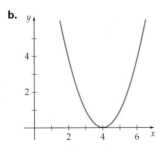

g.
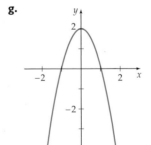

h.
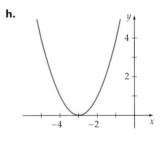

c. **d.**
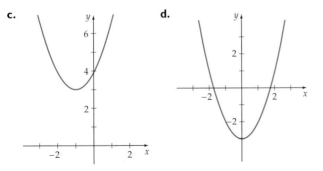

In Exercises 9 to 18, use the method of completing the square to find the standard form of the quadratic function, and then sketch its graph. Label its vertex and axis of symmetry.

9. $f(x) = x^2 + 4x + 1$

▶ **10.** $f(x) = x^2 + 6x - 1$

11. $f(x) = x^2 - 8x + 5$

12. $f(x) = x^2 - 10x + 3$

13. $f(x) = x^2 + 3x + 1$

14. $f(x) = x^2 + 7x + 2$

15. $f(x) = -x^2 + 4x + 2$ **16.** $f(x) = -x^2 - 2x + 5$

17. $f(x) = -3x^2 + 3x + 7$ **18.** $f(x) = -2x^2 - 4x + 5$

In Exercises 19 to 28, use the vertex formula to determine the vertex of the graph of the function and write the function in standard form.

19. $f(x) = x^2 - 10x$ ▶ **20.** $f(x) = x^2 - 6x$

21. $f(x) = x^2 - 10$ **22.** $f(x) = x^2 - 4$

23. $f(x) = -x^2 + 6x + 1$ **24.** $f(x) = -x^2 + 4x + 1$

25. $f(x) = 2x^2 - 3x + 7$ **26.** $f(x) = 3x^2 - 10x + 2$

27. $f(x) = -4x^2 + x + 1$ **28.** $f(x) = -5x^2 - 6x + 3$

29. Find the range of $f(x) = x^2 - 2x - 1$. Determine the values of x in the domain of f for which $f(x) = 2$.

30. Find the range of $f(x) = -x^2 - 6x - 2$. Determine the values of x in the domain of f for which $f(x) = 3$.

31. Find the range of $f(x) = -2x^2 + 5x - 1$. Determine the values of x in the domain of f for which $f(x) = 2$.

▶ **32.** Find the range of $f(x) = 2x^2 + 6x - 5$. Determine the values of x in the domain of f for which $f(x) = 15$.

33. Is 3 in the range of $f(x) = x^2 + 3x + 6$? Explain your answer.

34. Is -2 in the range of $f(x) = -2x^2 - x + 1$? Explain your answer.

In Exercises 35 to 44, find the maximum or minimum value of the function. State whether this value is a maximum or a minimum.

35. $f(x) = x^2 + 8x$ ▶ **36.** $f(x) = -x^2 - 6x$

37. $f(x) = -x^2 + 6x + 2$ **38.** $f(x) = -x^2 + 10x - 3$

39. $f(x) = 2x^2 + 3x + 1$ **40.** $f(x) = 3x^2 + x - 1$

41. $f(x) = 5x^2 - 11$ **42.** $f(x) = 3x^2 - 41$

43. $f(x) = -\dfrac{1}{2}x^2 + 6x + 17$

44. $f(x) = -\dfrac{3}{4}x^2 - \dfrac{2}{5}x + 7$

45. **HEIGHT OF AN ARCH** The height of an arch is given by the equation

$$h(x) = -\frac{3}{64}x^2 + 27, \quad -24 \le x \le 24$$

where $|x|$ is the horizontal distance in feet from the center of the arch.

a. What is the maximum height of the arch?

b. What is the height of the arch 10 feet to the right of center?

c. How far from the center is the arch 8 feet tall?

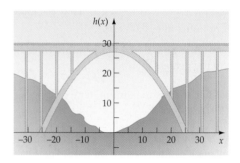

46. The sum of the length l and the width w of a rectangular area is 240 meters.

a. Write w as a function of l.

b. Write the area A as a function of l.

c. Find the dimensions that produce the greatest area.

47. **RECTANGULAR ENCLOSURE** A veterinarian uses 600 feet of chain-link fencing to enclose a rectangular region and also to subdivide the region into two smaller rectangular regions by placing a fence parallel to one of the sides, as shown in the figure.

a. Write the width w as a function of the length l.

b. Write the total area A as a function of l.

c. Find the dimensions that produce the greatest enclosed area.

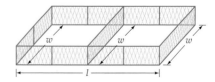

48. RECTANGULAR ENCLOSURE A farmer uses 1200 feet of fence to enclose a rectangular region and also to subdivide the region into three smaller rectangular regions by placing the fences parallel to one of the sides. Find the dimensions that produce the greatest enclosed area.

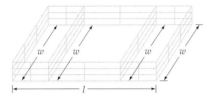

49. TEMPERATURE FLUCTUATIONS The temperature $T(t)$, in degrees Fahrenheit, during the day can be modeled by the equation $T(t) = -0.7t^2 + 9.4t + 59.3$, where t is the number of hours after 6:00 A.M.

   a. At what time is the temperature a maximum? Round to the nearest minute.

   b. What is the maximum temperature? Round to the nearest degree.

50. LARVAE SURVIVAL Soon after insect larvae are hatched, they must begin to search for food. The survival rate of the larvae depends on many factors, but the temperature of the environment is one of the most important. For a certain species of insect, a model of the number of larvae, $N(T)$, that survive this searching period is given by

$$N(T) = -0.6T^2 + 32.1T - 350$$

where T is the temperature in degrees Celsius.

   a. At what temperature will the maximum number of larvae survive? Round to the nearest degree.

   b. What is the maximum number of surviving larvae? Round to the nearest whole number.

   c. Find the x-intercepts, to the nearest whole number, for the graph of this function.

   d. Write a sentence that describes the meaning of the x-intercepts in the context of this problem.

51. REAL ESTATE The number of California homes that have sold for over $1,000,000 between 1989 and 2002 can be modeled by

$$N(t) = 1.43t^2 - 11.44t + 47.68$$

where $N(t)$ is the number (in hundreds) of homes that were sold in year t, with $t = 0$ corresponding to 1989. According to this model, in what year were the least number of million-dollar homes sold? How many million-dollar homes, to the nearest hundred, were sold that year?

52. GEOLOGY In June 2001, Mt. Etna in Sicily, Italy erupted, sending volcanic bombs (a mass of molten lava ejected from the volcano) into the air. A model of the height h, in meters, of a volcanic bomb above the crater of the volcano t seconds after the eruption is given by $h(t) = -9.8t^2 + 100t$. Find the maximum height of a volcanic bomb above the crater for this eruption. Round to the nearest meter.

53. SPORTS For a serve to be legal in tennis, the ball must be at least 3 feet high when it is 39 feet from the server, and it must land in a spot that is less than 60 feet from the server. Does the path of a ball given by $h(x) = -0.002x^2 - 0.03x + 8$, where $h(x)$ is the height of the ball (in feet) x feet from the server, satisfy the conditions of a legal serve?

54. SPORTS A pitcher releases a baseball 6 feet above the ground at a speed of 132 feet per second (90 miles per hour) toward home plate, which is 60.5 feet away. The height $h(x)$, in feet, of the ball x feet from home plate can be approximated by $h(x) = -0.0009x^2 + 6$. To be considered a strike, the ball must cross home plate and be at least 2.5 feet high and less than 5.4 feet high. Assuming the ball crosses home plate, is this particular pitch a strike? Explain.

55. AUTOMOTIVE ENGINEERING The fuel efficiency for a certain midsize car is given by

$$E(v) = -0.018v^2 + 1.476v + 3.4$$

where $E(v)$ is the fuel efficiency in miles per gallon for a car traveling v miles per hour.

   a. What speed will yield the maximum fuel efficiency? Round to the nearest mile per hour.

   b. What is the maximum fuel efficiency for this car? Round to the nearest mile per gallon.

56. SPORTS Some football fields are built in a parabolic mound shape so that water will drain off the field. A model for the shape of a certain field is given by

$$h(x) = -0.0002348x^2 + 0.0375x$$

where $h(x)$ is the height, in feet, of the field at a distance of x feet from one sideline. Find the maximum height of the field. Round to the nearest tenth of a foot.

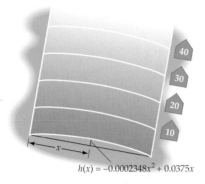

$$h(x) = -0.0002348x^2 + 0.0375x$$

In Exercises 57 to 60, determine the y- and x-intercepts (if any) of the quadratic function.

57. $f(x) = x^2 + 6x$

58. $f(x) = -x^2 + 4x$

59. $f(x) = -3x^2 + 5x - 6$

60. $f(x) = 2x^2 + 3x + 4$

In Exercises 61 and 62, determine the number of units x that produce a maximum revenue for the given revenue function. Also determine the maximum revenue.

61. $R(x) = 296x - 0.2x^2$

62. $R(x) = 810x - 0.6x^2$

In Exercises 63 and 64, determine the number of units x that produce a maximum profit for the given profit function. Also determine the maximum profit.

63. $P(x) = -0.01x^2 + 1.7x - 48$

64. $P(x) = -\dfrac{x^2}{14,000} + 1.68x - 4000$

In Exercises 65 and 66, determine the profit function for the given revenue function and cost function. Also determine the break-even point(s).

65. $R(x) = x(102.50 - 0.1x); C(x) = 52.50x + 1840$

66. $R(x) = x(210 - 0.25x); C(x) = 78x + 6399$

67. TOUR COST A charter bus company has determined that the cost of providing x people a tour is

$$C(x) = 180 + 2.50x$$

A full tour consists of 60 people. The ticket price per person is $15 plus $0.25 for each unsold ticket. Determine

a. the revenue function   **b.** the profit function

c. the company's maximum profit

d. the number of ticket sales that yields the maximum profit

▶ **68. DELIVERY COST** An air freight company has determined that the cost, in dollars, of delivering x parcels per flight is

$$C(x) = 2025 + 7x$$

The price per parcel, in dollars, the company charges to send x parcels is

$$p(x) = 22 - 0.01x$$

Determine

a. the revenue function   **b.** the profit function

c. the company's maximum profit

d. the price per parcel that yields the maximum profit

e. the minimum number of parcels the air freight company must ship to break even

69. PROJECTILE If the initial velocity of a projectile is 128 feet per second, then its height h in feet is a function of time t in seconds given by the equation $h(t) = -16t^2 + 128t$.

a. Find the time t when the projectile achieves its maximum height.

b. Find the maximum height of the projectile.

c. Find the time t when the projectile hits the ground.

▶ **70. PROJECTILE** The height in feet of a projectile with an initial velocity of 64 feet per second and an initial height of 80 feet is a function of time t in seconds given by

$$h(t) = -16t^2 + 64t + 80$$

a. Find the maximum height of the projectile.

b. Find the time t when the projectile achieves its maximum height.

c. Find the time t when the projectile has a height of 0 feet.

71. FIRE MANAGEMENT The height of a stream of water from the nozzle of a fire hose can be modeled by

$$y(x) = -0.014x^2 + 1.19x + 5$$

where $y(x)$ is the height, in feet, of the stream x feet from the firefighter. What is the maximum height that the stream of water from this nozzle can reach? Round to the nearest foot.

72. OLYMPIC SPORTS In 1988, Louise Ritter of the United States set the women's Olympic record for the high jump. A mathematical model that approximates her jump is given by

$$h(t) = -204.8t^2 + 256t$$

where $h(t)$ is her height in inches t seconds after beginning her jump. Find the maximum height of her jump.

73. NORMAN WINDOW A Norman window has the shape of a rectangle surmounted by a semicircle. The exterior perimeter of the window shown in the figure is 48 feet. Find the height h and the radius r that will allow the maximum amount of light to enter the window. (*Hint:* Write the area of the window as a quadratic function of the radius r.)

74. GOLDEN GATE BRIDGE The suspension cables of the main span of the Golden Gate Bridge are in the shape of a parabola. If a coordinate system is drawn as shown, find the quadratic function that models a suspension cable for the main span of the bridge.

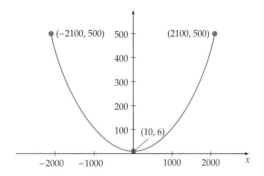

CONNECTING CONCEPTS

75. Let $f(x) = x^2 - (a + b)x + ab$, where a and b are real numbers.

   a. Show that the x-intercepts are $(a, 0)$ and $(b, 0)$.

   b. Show that the minimum value of the function occurs at the x-value of the midpoint of the line segment defined by the x-intercepts.

76. Let $f(x) = ax^2 + bx + c$, where a, b, and c are real numbers.

   a. What conditions must be imposed on the coefficients so that f has a maximum?

   b. What conditions must be imposed on the coefficients so that f has a minimum?

   c. What conditions must be imposed on the coefficients so that the graph of f intersects the x-axis?

77. Find the quadratic function of x whose graph has a minimum at $(2, 1)$ and passes through $(0, 4)$.

78. Find the quadratic function of x whose graph has a maximum at $(-3, 2)$ and passes through $(0, -5)$.

79. AREA OF A RECTANGLE A wire 32 inches long is bent so that it has the shape of a rectangle. The length of the rectangle is x and the width is w.

   a. Write w as a function of x.

   b. Write the area A of the rectangle as a function of x.

80. **MAXIMIZE AREA** Use the function A from **b.** in Exercise 79 to prove that the area A is greatest if the rectangle is a square.

81. Show that the function $f(x) = x^2 + bx - 1$ has a real zero for any value b.

82. Show that the function $g(x) = -x^2 + bx + 1$ has a real zero for any value b.

83. What effect does increasing the constant c have on the graph of $f(x) = ax^2 + bx + c$?

84. If $a > 0$, what effect does decreasing the coefficient a have on the graph of $f(x) = ax^2 + bx + c$?

85. Find two numbers whose sum is 8 and whose product is a maximum.

86. Find two numbers whose difference is 12 and whose product is a minimum.

87. Verify that the slope of the line passing through (x, x^3) and $(x + h, [x + h]^3)$ is $3x^2 + 3xh + h^2$.

88. Verify that the slope of the line passing through $(x, 4x^3 + x)$ and $(x + h, 4[x + h]^3 + [x + h])$ is given by $12x^2 + 12xh + 4h^2 + 1$.

PREPARE FOR SECTION 2.5

89. For the graph of the parabola whose equation is $f(x) = x^2 + 4x - 6$, what is the equation of the axis of symmetry? [2.4]

90. For $f(x) = \dfrac{3x^4}{x^2 + 1}$, show that $f(-3) = f(3)$. [2.2]

91. For $f(x) = 2x^3 - 5x$, show that $f(-2) = -f(2)$. [2.2]

92. Let $f(x) = x^2$ and $g(x) = x + 3$. Find $f(a) - g(a)$ for $a = -2, -1, 0, 1, 2$. [2.2]

93. What is the midpoint of the line segment between $P(-a, b)$ and $Q(a, b)$? [2.1]

94. What is the midpoint of the line segment between $P(-a, -b)$ and $Q(a, b)$? [2.1]

PROJECTS

1. **THE CUBIC FORMULA** Write an essay on the development of the cubic formula. An excellent source of information is the chapter "Cardano and the Solution of the Cubic" in *Journey Through Genius*, by William Dunham (New York: Wiley, 1990).

2. **SIMPSON'S RULE** In calculus a procedure known as *Simpson's Rule* is often used to approximate the area under a curve. The figure at the right shows the graph of a parabola that passes through $P_0(-h, y_0)$, $P_1(0, y_1)$, and $P_2(h, y_2)$. The equation of the parabola is of the form $y = Ax^2 + Bx + C$. Using calculus procedures, we can show that the area bounded by the parabola, the x-axis, and the vertical lines $x = -h$ and $x = h$ is

$$\frac{h}{3}(2Ah^2 + 6C)$$

Use algebra to show that $y_0 + 4y_1 + y_2 = 2Ah^2 + 6C$, from which we can deduce that the area of the bounded region can also be written as

$$\frac{h}{3}(y_0 + 4y_1 + y_2)$$

(*Hint:* Evaluate $Ax^2 + Bx + C$ at $x = -h$, $x = 0$, and $x = h$ to determine values of y_0, y_1, and y_2, respectively. Then compute $y_0 + 4y_1 + y_2$.)

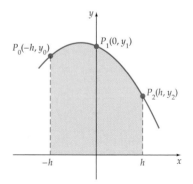

PROPERTIES OF GRAPHS

• SYMMETRY

The graph in **Figure 2.63** is symmetric with respect to the line *l*. Note that the graph has the property that if the paper is folded along the dotted line *l*, the point *A'* will coincide with the point *A*, the point *B'* will coincide with the point *B*, and the point *C'* will coincide with the point *C*. One part of the graph is a *mirror image* of the rest of the graph across the line *l*.

A graph is **symmetric with respect to the y-axis** if, whenever the point given by (x, y) is on the graph, then $(-x, y)$ is also on the graph. The graph in **Figure 2.64** is symmetric with respect to the *y*-axis. A graph is **symmetric with respect to the x-axis** if, whenever the point given by (x, y) is on the graph, then $(x, -y)$ is also on the graph. The graph in **Figure 2.65** is symmetric with respect to the *x*-axis.

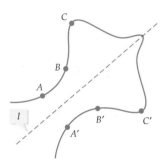

FIGURE 2.63

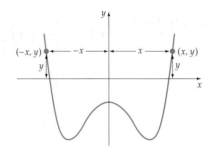

FIGURE 2.64
Symmetry with respect to the *y*-axis

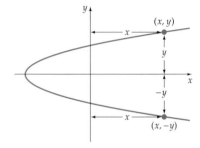

FIGURE 2.65
Symmetry with respect to the *x*-axis

Tests for Symmetry with Respect to a Coordinate Axis

The graph of an equation is symmetric with respect to

- the *y*-axis if the replacement of x with $-x$ leaves the equation unaltered.
- the *x*-axis if the replacement of y with $-y$ leaves the equation unaltered.

❓ **QUESTION** Which of the graphs below, I, II, or III, is **a.** symmetric with respect to the *y*-axis? **b.** symmetric with respect to the *x*-axis?

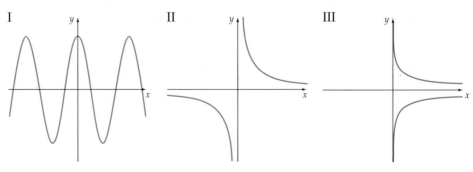

❓ **ANSWER** **a.** I is symmetric with respect to the *y*-axis.
   b. III is symmetric with respect to the *x*-axis.

EXAMPLE 1   **Determine Symmetries of a Graph**

Determine whether the graph of the given equation has symmetry with respect to either the x- or the y-axis.

a. $y = x^2 + 2$ b. $x = |y| - 2$

Solution

a.   The equation $y = x^2 + 2$ *is unaltered* by the replacement of x with $-x$. That is, the simplification of $y = (-x)^2 + 2$ yields the original equation $y = x^2 + 2$. Thus the graph of $y = x^2 + 2$ is symmetric with respect to the y-axis. However, the equation $y = x^2 + 2$ *is altered* by the replacement of y with $-y$. That is, the simplification of $-y = x^2 + 2$, which is $y = -x^2 - 2$, *does not* yield the original equation $y = x^2 + 2$. The graph of $y = x^2 + 2$ is not symmetric with respect to the x-axis. See **Figure 2.66.**

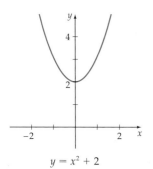

$y = x^2 + 2$

FIGURE 2.66

b.   The equation $x = |y| - 2$ *is altered* by the replacement of x with $-x$. That is, the simplification of $-x = |y| - 2$, which is $x = -|y| + 2$, *does not* yield the original equation $x = |y| - 2$. This implies that the graph of $x = |y| - 2$ is not symmetric with respect to the y-axis. However, the equation $x = |y| - 2$ *is unaltered* by the replacement of y with $-y$. That is, the simplification of $x = |-y| - 2$ yields the original equation $x = |y| - 2$. The graph of $x = |y| - 2$ is symmetric with respect to the x-axis. See **Figure 2.67.**

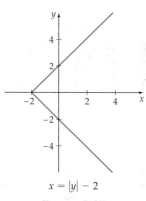

$x = |y| - 2$

FIGURE 2.67

▶ TRY EXERCISE 14, PAGE 222

Symmetry with Respect to a Point

A graph is **symmetric with respect to a point** Q if for each point P on the graph there is a point P' on the graph such that Q is the midpoint of the line segment PP'.

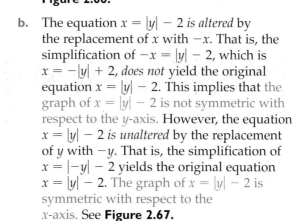

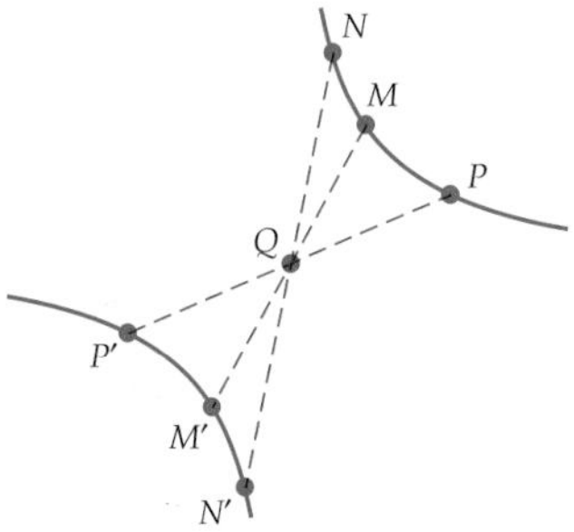

FIGURE 2.68

The graph in **Figure 2.68** is symmetric with respect to the point Q. For any point P on the graph, there exists a point P' on the graph such that Q is the midpoint of $P'P$.

When we discuss symmetry with respect to a point, we frequently use the origin. A graph is symmetric with respect to the origin if, whenever the point given by (x, y) is on the graph, then $(-x, -y)$ is also on the graph. The graph in **Figure 2.69** is symmetric with respect to the origin.

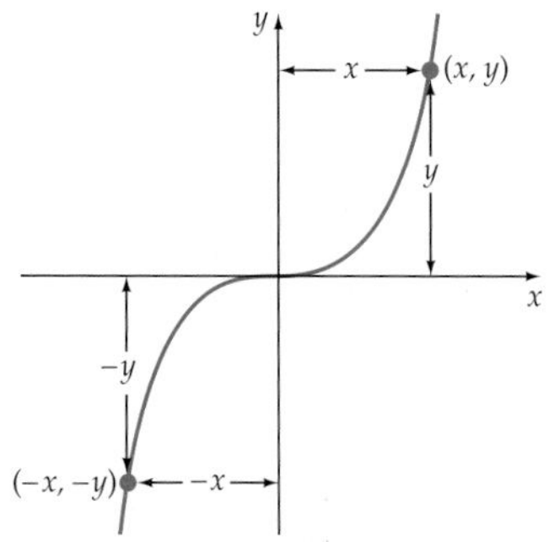

Symmetry with respect to the origin

FIGURE 2.69

Test for Symmetry with Respect to the Origin

The graph of an equation is symmetric with respect to the origin if the replacement of x with $-x$ and of y with $-y$ leaves the equation unaltered.

EXAMPLE 2   **Determine Symmetry with Respect to the Origin**

Determine whether the graph of each equation has symmetry with respect to the origin.

a. $xy = 4$ b. $y = x^3 + 1$

Solution

a.   The equation $xy = 4$ is unaltered by the replacement of x with $-x$ and of y with $-y$. That is, the simplification of $(-x)(-y) = 4$ yields the original equation $xy = 4$. Thus the graph of $xy = 4$ is symmetric with respect to the origin. See **Figure 2.70.**

b.   The equation $y = x^3 + 1$ *is altered* by the replacement of x with $-x$ and of y with $-y$. That is, the simplification of $-y = (-x)^3 + 1$, which is $y = x^3 - 1$, *does not* yield the original equation $y = x^3 + 1$. Thus the graph of $y = x^3 + 1$ is not symmetric with respect to the origin. See **Figure 2.71.**

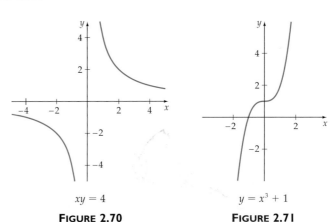

$xy = 4$

FIGURE 2.70

$y = x^3 + 1$

FIGURE 2.71

▶ **TRY EXERCISE 24, PAGE 222**

Some graphs have more than one symmetry. For example, the graph of $|x| + |y| = 2$ has symmetry with respect to the x-axis, the y-axis, and the origin. **Figure 2.72** is the graph of $|x| + |y| = 2$.

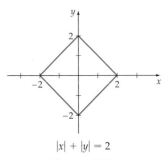

$|x| + |y| = 2$

FIGURE 2.72

● **EVEN AND ODD FUNCTIONS**

Some functions are classified as either *even* or *odd*.

Definition of Even and Odd Functions

The function f is an **even function** if

$$f(-x) = f(x) \quad \text{for all } x \text{ in the domain of } f$$

The function f is an **odd function** if

$$f(-x) = -f(x) \quad \text{for all } x \text{ in the domain of } f$$

EXAMPLE 3 **Identify Even or Odd Functions**

Determine whether each function is even, odd, or neither.

a. $f(x) = x^3$ **b.** $F(x) = |x|$ **c.** $h(x) = x^4 + 2x$

Solution

Replace x with $-x$ and simplify.

a. $f(-x) = (-x)^3 = -x^3 = -(x^3) = -f(x)$
Because $f(-x) = -f(x)$, this function is an odd function.

b. $F(-x) = |-x| = |x| = F(x)$
Because $F(-x) = F(x)$, this function is an even function.

c. $h(-x) = (-x)^4 + 2(-x) = x^4 - 2x$
This function is neither an even nor an odd function because

$$h(-x) = x^4 - 2x,$$

which is not equal to either $h(x)$ or $-h(x)$.

▶ **TRY EXERCISE 44, PAGE 222**

The following properties are a result of the tests for symmetry:

● The graph of an even function is symmetric with respect to the y-axis.

● The graph of an odd function is symmetric with respect to the origin.

The graph of f in **Figure 2.73** is symmetric with respect to the y-axis. It is the graph of an even function. The graph of g in **Figure 2.74** is symmetric with respect to the origin. It is the graph of an odd function. The graph of h in **Figure 2.75** is not symmetric with respect to the y-axis and is not symmetric with respect to the origin. It is neither an even nor an odd function.

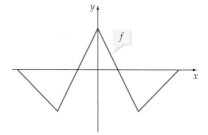

FIGURE 2.73

The graph of an even function is symmetric with respect to the y-axis.

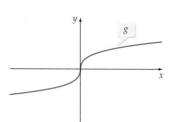

FIGURE 2.74

The graph of an odd function is symmetric with respect to the origin.

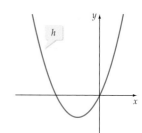

FIGURE 2.75

If the graph of a function is not symmetric to the y-axis or to the origin, then the function is neither even nor odd.

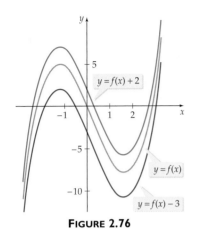

FIGURE 2.76

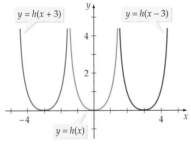

FIGURE 2.77

● TRANSLATIONS OF GRAPHS

The shape of a graph may be exactly the same as the shape of another graph; only their positions in the xy-plane may differ. For example, the graph of $y = f(x) + 2$ is the graph of $y = f(x)$ with each point moved up vertically 2 units. The graph of $y = f(x) - 3$ is the graph of $y = f(x)$ with each point moved down vertically 3 units. See **Figure 2.76.**

The graphs of $y = f(x) + 2$ and $y = f(x) - 3$ in **Figure 2.76** are called *vertical translations* of the graph of $y = f(x)$.

Vertical Translations

If f is a function and c is a positive constant, then the graph of

● $y = f(x) + c$ is the graph of $y = f(x)$ shifted up *vertically c* units.

● $y = f(x) - c$ is the graph of $y = f(x)$ shifted down *vertically c* units.

In **Figure 2.77,** the graph of $y = h(x + 3)$ is the graph of $y = h(x)$ with each point shifted to the left horizontally 3 units. Similarly, the graph of $y = h(x - 3)$ is the graph of $y = h(x)$ with each point shifted to the right horizontally 3 units.

The graphs of $y = h(x + 3)$ and $y = h(x - 3)$ in **Figure 2.77** are called *horizontal translations* of the graph of $y = h(x)$.

Horizontal Translations

If f is a function and c is a positive constant, then the graph of

● $y = f(x + c)$ is the graph of $y = f(x)$ shifted left *horizontally c* units.

● $y = f(x - c)$ is the graph of $y = f(x)$ shifted right *horizontally c* units.

INTEGRATING TECHNOLOGY

A graphing calculator can be used to draw the graphs of a *family* of functions. For instance, $f(x) = x^2 + c$ constitutes a family of functions with **parameter** c. The only feature of the graph that changes is the value of c.

A graphing calculator can be used to produce the graphs of a family of curves for specific values of the parameter. The LIST feature of the calculator can be used. For instance, to graph $f(x) = x^2 + c$ for $c = -2, 0$, and 1, we will create a list and use that list to produce the family of curves. The keystrokes for a TI-83 calculator are given below.

Now use the ⌊Y=⌋ key to enter

Y= X x² + 2nd L1 ZOOM 6

Sample screens for the keystrokes and graphs are shown here. You can use similar keystrokes for Exercises 75–82 of this section.

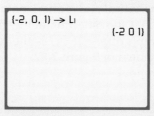

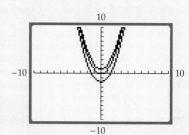

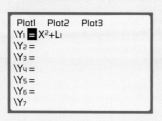

EXAMPLE 4 **Graph by Using Translations**

Use vertical and horizontal translations of the graph of $f(x) = x^3$, shown in **Figure 2.78,** to graph

a. $g(x) = x^3 - 2$ **b.** $h(x) = (x + 1)^3$

Solution

a. The graph of $g(x) = x^3 - 2$ is the graph of $f(x) = x^3$ shifted down vertically 2 units. See **Figure 2.79.**

b. The graph of $h(x) = (x + 1)^3$ is the graph of $f(x) = x^3$ shifted to the left horizontally 1 unit. See **Figure 2.80.**

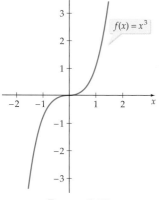

FIGURE 2.78

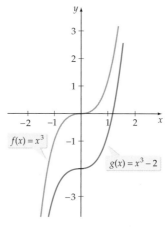

FIGURE 2.79

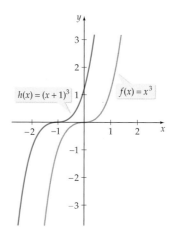

FIGURE 2.80

▶ TRY EXERCISE 58, PAGE 223

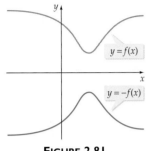

FIGURE 2.81

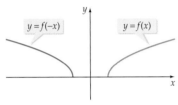

FIGURE 2.82

• REFLECTIONS OF GRAPHS

The graph of $y = -f(x)$ cannot be obtained from the graph of $y = f(x)$ by a combination of vertical and/or horizontal shifts. **Figure 2.81** illustrates that the graph of $y = -f(x)$ is the reflection of the graph of $y = f(x)$ across the x-axis.

The graph of $y = f(-x)$ is the reflection of the graph of $y = f(x)$ across the y-axis as, shown in **Figure 2.82.**

Reflections

The graph of

- $y = -f(x)$ is the graph of $y = f(x)$ reflected across the x-axis.

- $y = f(-x)$ is the graph of $y = f(x)$ reflected across the y-axis.

EXAMPLE 5   **Graph by Using Reflections**

Use reflections of the graph of $f(x) = \sqrt{x - 1} + 1$, shown in **Figure 2.83,** to graph

a. $g(x) = -\left(\sqrt{x - 1} + 1\right)$ **b.** $h(x) = \sqrt{-x - 1} + 1$

Solution

a. Because $g(x) = -f(x)$, the graph of g is the graph of f reflected across the x-axis. See **Figure 2.84.**

b. Because $h(x) = f(-x)$, the graph of h is the graph of f reflected across the y-axis. See **Figure 2.85.**

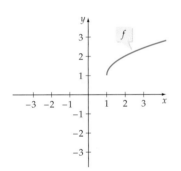

FIGURE 2.83

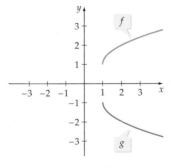

FIGURE 2.84

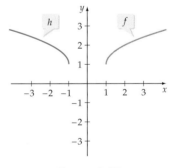

FIGURE 2.85

▶ **TRY EXERCISE 68, PAGE 224**

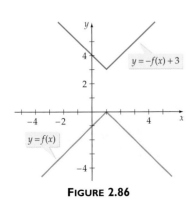

FIGURE 2.86

Some graphs of functions can be constructed by using a combination of translations and reflections. For instance, the graph of $y = -f(x) + 3$ in **Figure 2.86** was obtained by reflecting the graph of $y = f(x)$ in **Figure 2.86** across the x-axis and then shifting that graph up vertically 3 units.

• COMPRESSING AND STRETCHING OF GRAPHS

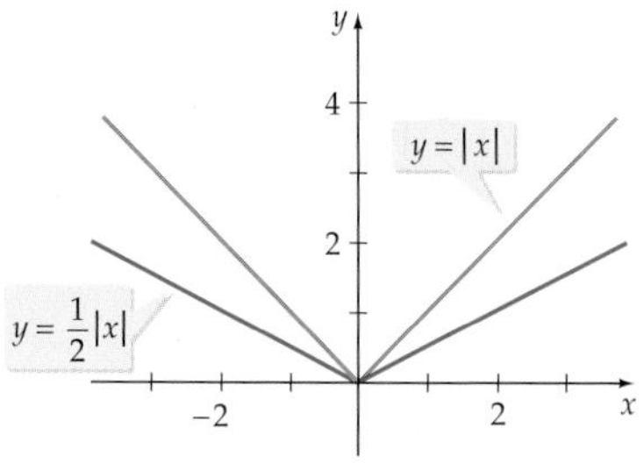

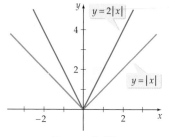

FIGURE 2.87

The graph of the equation $y = c \cdot f(x)$ for $c \neq 1$ vertically compresses or stretches the graph of $y = f(x)$. To determine the points on the graph of $y = c \cdot f(x)$, multiply each y-coordinate of the points on the graph of $y = f(x)$ by c. For example, **Figure 2.87** shows that the graph of $y = \dfrac{1}{2}|x|$ can be obtained by plotting points that have a y-coordinate that is one-half of the y-coordinate of those found on the graph of $y = |x|$.

If $0 < c < 1$, then the graph of $y = c \cdot f(x)$ is obtained by *compressing* the graph of $y = f(x)$. **Figure 2.87** illustrates the vertical compressing of the graph of $y = |x|$ toward the x-axis to form the graph of $y = \dfrac{1}{2}|x|$.

If $c > 1$, then the graph of $y = c \cdot f(x)$ is obtained by *stretching* the graph of $y = f(x)$. For example, if $f(x) = |x|$, then we obtain the graph of

$$y = 2f(x) = 2|x|$$

by stretching the graph of f away from the x-axis. See **Figure 2.88**.

FIGURE 2.88

Vertical Stretching and Compressing of Graphs

If f is a function and c is a positive constant, then

- if $c > 1$, the graph of $y = c \cdot f(x)$ is the graph of $y = f(x)$ *stretched* vertically by a factor of c away from the x-axis.

- if $0 < c < 1$, the graph of $y = c \cdot f(x)$ is the graph of $y = f(x)$ *compressed* vertically by a factor of c toward the x-axis.

EXAMPLE 6   Graph by Using Vertical Compressing and Shifting

Graph: $H(x) = \dfrac{1}{4}|x| - 3$

Solution

The graph of $y = |x|$ has a V shape that has its lowest point at $(0, 0)$ and passes through $(4, 4)$ and $(-4, 4)$. The graph of $y = \dfrac{1}{4}|x|$ is a compressing of the graph of $y = |x|$. The y-coordinates of the ordered pairs $(0, 0)$, $(4, 1)$, and $(-4, 1)$ are obtained by multiplying the y-coordinates of the ordered pairs $(0, 0)$, $(4, 4)$, and $(-4, 4)$ by $\dfrac{1}{4}$. To find the points on the graph of H, we still need to subtract 3 from each y-coordinate. Thus the graph of H is a V shape that has its lowest point at $(0, -3)$ and passes through $(4, -2)$ and $(-4, -2)$. See **Figure 2.89**.

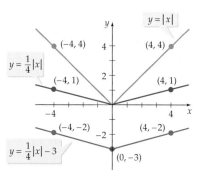

FIGURE 2.89

▶ **TRY EXERCISE 70, PAGE 224**

Some functions can be graphed by using a horizontal compressing or stretching of a given graph. The procedure makes use of the following concept.

Horizontal Compressing and Stretching of Graphs

If f is a function and c is a positive constant, then

- if $c > 1$, the graph of $y = f(c \cdot x)$ is the graph of $y = f(x)$ *compressed* horizontally by a factor of $\dfrac{1}{c}$ toward the y-axis.

- if $0 < c < 1$, the graph of $y = f(c \cdot x)$ is the graph of $y = f(x)$ *stretched* horizontally by a factor of $\dfrac{1}{c}$ away from the y-axis.

If the point (x, y) is on the graph of $y = f(x)$, then the graph of $y = f(cx)$ will contain the point $\left(\dfrac{1}{c}x, y \right)$.

EXAMPLE 7   **Graph by Using Horizontal Compressing and Stretching**

Use the graph of $y = f(x)$ shown in **Figure 2.90** to graph

a. $y = f(2x)$ **b.** $y = f\left(\dfrac{1}{3}x \right)$

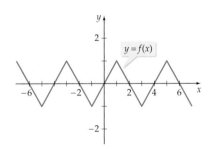

FIGURE 2.90

Solution

a. Because $2 > 1$, the graph of $y = f(2x)$ is a horizontal compression of the graph of $y = f(x)$ by a factor of $\dfrac{1}{2}$. For example, the point $(2, 0)$ on the graph of $y = f(x)$ becomes the point $(1, 0)$ on the graph of $y = f(2x)$. See **Figure 2.91.**

b. Since $0 < \dfrac{1}{3} < 1$, the graph of $y = f\left(\dfrac{1}{3}x \right)$ is a horizontal stretching of the graph of $y = f(x)$ by a factor of 3. For example, the point $(1, 1)$ on the graph of $y = f(x)$ becomes the point (3, 1) on the graph of $y = f\left(\dfrac{1}{3}x\right)$.

See **Figure 2.92.**

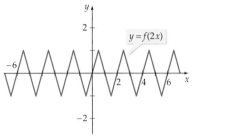

FIGURE 2.91

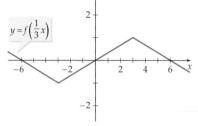

FIGURE 2.92

▶ **TRY EXERCISE 72, PAGE 224**

TOPICS FOR DISCUSSION

1. Discuss the meaning of symmetry of a graph with respect to a line. How do you determine whether a graph has symmetry with respect to the *x*-axis? with respect to the *y*-axis?

2. Discuss the meaning of symmetry of a graph with respect to a point. How do you determine whether a graph has symmetry with respect to the origin?

3. What does it mean to reflect a graph across the *x*-axis or across the *y*-axis?

4. Explain how the graphs of $y_1 = 2x^3 - x^2$ and $y_2 = 2(-x)^3 - (-x)^2$ are related.

5. Given the graph of $y_3 = f(x)$, explain how to obtain the graph of $y_4 = f(x - 3) + 1$.

6. The graph of the *step function* $y_5 = [\![x]\!]$ has steps that are 1 unit wide. Determine how wide the steps are in the graph of $y_6 = \left[\!\!\left[\dfrac{1}{3}x\right]\!\!\right]$.

EXERCISE SET 2.5

In Exercises 1 to 6, plot the image of the given point with respect to
a. **the y-axis. Label this point A.**
b. **the x-axis. Label this point B.**
c. **the origin. Label this point C.**

1. $P(5, -3)$ **2.** $Q(-4, 1)$ **3.** $R(-2, 3)$

4. $S(-5, 3)$ **5.** $T(-4, -5)$ **6.** $U(5, 1)$

In Exercises 7 and 8, sketch a graph that is symmetric to the given graph with respect to the x-axis.

7.
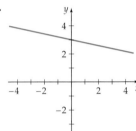

8.
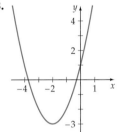

In Exercises 9 and 10, sketch a graph that is symmetric to the given graph with respect to the y-axis.

9.
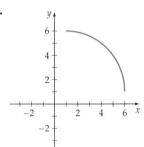

10.
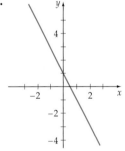

In Exercises 11 and 12, sketch a graph that is symmetric to the given graph with respect to the origin.

11.
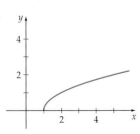

12.
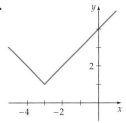

In Exercises 13 to 21, determine whether the graph of each equation is symmetric with respect to the a. **x-axis,** b. **y-axis.**

13. $y = 2x^2 - 5$ ▶ **14.** $x = 3y^2 - 7$ **15.** $y = x^3 + 2$

16. $y = x^5 - 3x$ **17.** $x^2 + y^2 = 9$ **18.** $x^2 - y^2 = 10$

19. $x^2 = y^4$ **20.** $xy = 8$ **21.** $|x| - |y| = 6$

In Exercises 22 to 30, determine whether the graph of each equation is symmetric with respect to the origin.

22. $y = x + 1$ **23.** $y = 3x - 2$ ▶ **24.** $y = x^3 - x$

25. $y = -x^3$ **26.** $y = \dfrac{9}{x}$ **27.** $x^2 + y^2 = 10$

28. $x^2 - y^2 = 4$ **29.** $y = \dfrac{x}{|x|}$ **30.** $|y| = |x|$

In Exercises 31 to 42, graph the given equations. Label each intercept. Use the concept of symmetry to confirm that the graph is correct.

31. $y = x^2 - 1$ **32.** $x = y^2 - 1$

33. $y = x^3 - x$ **34.** $y = -x^3$

35. $xy = 4$ **36.** $xy = -8$

37. $y = 2|x - 4|$ **38.** $y = |x - 2| - 1$

39. $y = (x - 2)^2 - 4$ **40.** $y = (x - 1)^2 - 4$

41. $y = x - |x|$ **42.** $|y| = |x|$

In Exercises 43 to 56, identify whether the given function is an even function, an odd function, or neither.

43. $g(x) = x^2 - 7$ ▶ **44.** $h(x) = x^2 + 1$

45. $F(x) = x^5 + x^3$ **46.** $G(x) = 2x^5 - 10$

47. $H(x) = 3|x|$ **48.** $T(x) = |x| + 2$

49. $f(x) = 1$ **50.** $k(x) = 2 + x + x^2$

51. $r(x) = \sqrt{x^2 + 4}$ **52.** $u(x) = \sqrt{3 - x^2}$

53. $s(x) = 16x^2$ **54.** $v(x) = 16x^2 + x$

55. $w(x) = 4 + \sqrt[3]{x}$ **56.** $z(x) = \dfrac{x^3}{x^2 + 1}$

57. Use the graph of f to sketch the graph of

    a. $y = f(x) + 3$ **b.** $y = f(x - 3)$

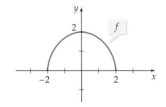

▶ **58.** Use the graph of g to sketch the graph of

    a. $y = g(x) - 2$ **b.** $y = g(x - 3)$

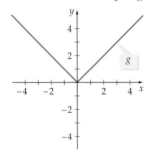

59. Use the graph of f to sketch the graph of

    a. $y = f(x + 2)$ **b.** $y = f(x) + 2$

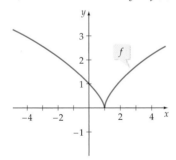

60. Use the graph of g to sketch the graph of

    a. $y = g(x - 1)$ **b.** $y = g(x) - 1$

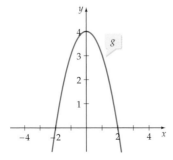

61. Let f be a function such that $f(-2) = 5$, $f(0) = -2$, and $f(1) = 0$. Give the coordinates of three points on the graph of

    a. $y = f(x + 3)$ **b.** $y = f(x) + 1$

62. Let g be a function such that $g(-3) = -1$, $g(1) = -3$, and $g(4) = 2$. Give the coordinates of three points on the graph of

    a. $y = g(x - 2)$ **b.** $y = g(x) - 2$

63. Use the graph of f to sketch the graph of

    a. $y = f(-x)$ **b.** $y = -f(x)$

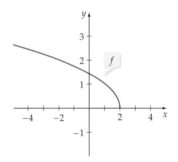

64. Use the graph of g to sketch the graph of

    a. $y = -g(x)$ **b.** $y = g(-x)$

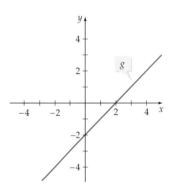

65. Let f be a function such that $f(-1) = 3$ and $f(2) = -4$. Give the coordinates of two points on the graph of

    a. $y = f(-x)$ **b.** $y = -f(x)$

66. Let g be a function such that $g(4) = -5$ and $g(-3) = 2$. Give the coordinates of two points on the graph of

    a. $y = -g(x)$ **b.** $y = g(-x)$

67. Use the graph of F to sketch the graph of

a. $y = -F(x)$ **b.** $y = F(-x)$

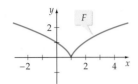

▶ **68.** Use the graph of E to sketch the graph of

a. $y = -E(x)$ **b.** $y = E(-x)$

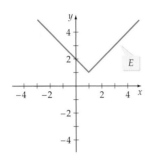

69. Use the graph of $m(x) = x^2 - 2x - 3$ to sketch the graph of $y = -\dfrac{1}{2}m(x) + 3$.

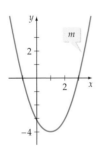

▶ **70.** Use the graph of $n(x) = -x^2 - 2x + 8$ to sketch the graph of $y = \dfrac{1}{2}n(x) + 1$.

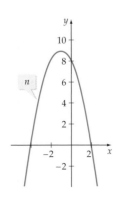

71. Use the graph of $y = f(x)$ to sketch the graph of

a. $y = f(2x)$ **b.** $y = f\left(\dfrac{1}{3}x\right)$

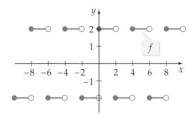

▶ **72.** Use the graph of $y = g(x)$ to sketch the graph of

a. $y = g(2x)$ **b.** $y = g\left(\dfrac{1}{2}x\right)$

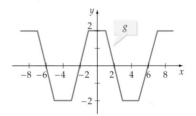

73. Use the graph of $y = h(x)$ to sketch the graph of

a. $y = h(2x)$ **b.** $y = h\left(\dfrac{1}{2}x\right)$

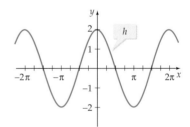

74. Use the graph of $y = j(x)$ to sketch the graph of

a. $y = j(2x)$ **b.** $y = j\left(\dfrac{1}{3}x\right)$

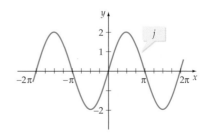

In Exercises 75 to 82, use a graphing utility.

75. On the same coordinate axes, graph
$$G(x) = \sqrt[3]{x} + c$$
for $c = 0, -1$, and 3.

76. On the same coordinate axes, graph
$$H(x) = \sqrt[3]{x + c}$$
for $c = 0, -1$, and 3.

77. On the same coordinate axes, graph
$$J(x) = |2(x + c) - 3| - |x + c|$$
for $c = 0, -1$, and 2.

78. On the same coordinate axes, graph
$$K(x) = |x - 1| - |x| + c$$
for $c = 0, -1$, and 2.

79. On the same coordinate axes, graph
$$L(x) = cx^2$$
for $c = 1, \dfrac{1}{2}$, and 2.

80. On the same coordinate axes, graph
$$M(x) = c\sqrt{x^2 - 4}$$
for $c = 1, \dfrac{1}{3}$, and 3.

81. On the same coordinate axes, graph
$$S(x) = c(|x - 1| - |x|)$$
for $c = 1, \dfrac{1}{4}$, and 4.

82. On the same coordinate axes, graph
$$T(x) = c\left(\dfrac{x}{|x|}\right)$$
for $c = 1, \dfrac{2}{3}$, and $\dfrac{3}{2}$.

83. Graph $V(x) = [\![cx]\!], 0 \le x \le 6$, for each value of c.

   a. $c = 1$ **b.** $c = \dfrac{1}{2}$ **c.** $c = 2$

84. Graph $W(x) = [\![cx]\!] - cx, 0 \le x \le 6$, for each value of c.

   a. $c = 1$ **b.** $c = \dfrac{1}{3}$ **c.** $c = 3$

CONNECTING CONCEPTS

85. Use the graph of $f(x) = 2/(x^2 + 1)$ to determine equations for the graphs shown in **a.** and **b.**

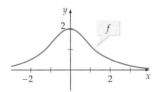

a.
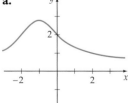

b.
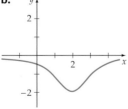

86. Use the graph of $f(x) = x\sqrt{2 + x}$ to determine equations for the graphs shown in **a.** and **b.**

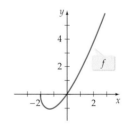

a.
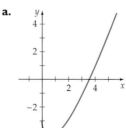

b.
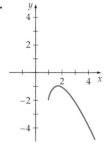

PREPARE FOR SECTION 2.6
---

87. Subtract: $(2x^2 + 3x - 4) - (x^2 + 3x - 5)$ [P.3]

88. Multiply: $(3x^2 - x + 2)(2x - 3)$ [P.3]

In Exercises 89 and 90, find each of the following for
$f(x) = 2x^2 - 5x + 2.$

89. $f(3a)$ [2.2]

90. $f(2 + h)$ [2.2]

In Exercises 91 and 92, find the domain of each function.

91. $F(x) = \dfrac{x}{x - 1}$ [2.2]

92. $r(x) = \sqrt{2x - 8}$ [2.2]

PROJECTS
---

1. **DIRICHLET FUNCTION** We owe our present-day definition of a function to the German mathematician Peter Gustav Dirichlet (1805–1859). He created the following unusual function, which is now known as the *Dirichlet function.*

$$f(x) = \begin{cases} 0, & \text{if } x \text{ is a rational number} \\ 1, & \text{if } x \text{ is an irrational number} \end{cases}$$

Answer the following questions about the Dirichlet function.

a. What is its domain?   **b.** What is its range?

c. What are its x-intercepts?

d. What is its y-intercept?

e. Is it an even or an odd function?

f. Explain why a graphing calculator cannot be used to produce an accurate graph of the function.

g. Write a sentence or two that describes its graph.

2. **ISOLATED POINT** Consider the function given by

$$y = \sqrt{(x - 1)^2(x - 2)} + 1$$

Verify that the point $(1, 1)$ is a solution of the equation. Now use a graphing utility to graph the function. Does your graph include the isolated point at $(1, 1)$, as shown at the right? If the graphing utility you used failed to include the point $(1, 1)$, explain at least one reason for the omission of this isolated point.

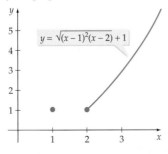

$$y = \sqrt{(x-1)^2(x-2)} + 1$$

3. **A LINE WITH A HOLE** The function

$$f(x) = \frac{(x - 2)(x + 1)}{(x - 2)}$$

graphs as a line with a y-intercept of 1, a slope of 1, and a hole at $(2, 3)$. Use a graphing utility to graph f. Explain why a graphing utility might not show the hole at $(2, 3)$.

4. **FINDING A COMPLETE GRAPH** Use a graphing utility to graph the function $f(x) = 3x^{5/3} - 6x^{4/3} + 2$ for $-2 \le x \le 10$. Compare your graph with the graph below. Does your graph include the part to the left of the y-axis? If not, how might you enter the function so that the graphing utility you used would include this part?

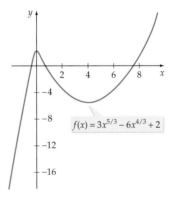

$$f(x) = 3x^{5/3} - 6x^{4/3} + 2$$

THE ALGEBRA OF FUNCTIONS

• OPERATIONS ON FUNCTIONS

Functions can be defined in terms of other functions. For example, the function defined by $h(x) = x^2 + 8x$ is the sum of

$$f(x) = x^2 \quad \text{and} \quad g(x) = 8x$$

Thus, if we are given any two functions f and g, we can define the four new functions $f + g$, $f - g$, fg, and $\dfrac{f}{g}$ as follows.

Operations on Functions

For all values of x for which both $f(x)$ and $g(x)$ are defined, we define the following functions.

Sum	$(f + g)(x) = f(x) + g(x)$
Difference	$(f - g)(x) = f(x) - g(x)$
Product	$(fg)(x) = f(x) \cdot g(x)$
Quotient	$\left(\dfrac{f}{g}\right)(x) = \dfrac{f(x)}{g(x)}, \quad g(x) \neq 0$

Domain of $f + g$, $f - g$, f/g

For the given functions f and g, the domains of $f + g$, $f - g$, and $f \cdot g$ consist of all real numbers formed by the intersection of the domains of f and g. The domain of $\dfrac{f}{g}$ is the set of all real numbers formed by the intersection of the domains of f and g, except for those real numbers x such that $g(x) = 0$.

EXAMPLE I Determine the Domain of a Function

If $f(x) = \sqrt{x - 1}$ and $g(x) = x^2 - 4$, find the domain of $f + g$, of $f - g$, of fg, and of $\dfrac{f}{g}$.

Solution

Note that f has the domain $\{x \mid x \geq 1\}$ and g has the domain of all real numbers. Therefore, the domain of $f + g$, $f - g$, and fg is $\{x \mid x \geq 1\}$. Because $g(x) = 0$ when $x = -2$ or $x = 2$, neither -2 nor 2 is in the domain of $\dfrac{f}{g}$. The domain of $\dfrac{f}{g}$ is $\{x \mid x \geq 1 \text{ and } x \neq 2\}$.

▶ **TRY EXERCISE 10, PAGE 235**

EXAMPLE 2   **Evaluate Functions**

Let $f(x) = x^2 - 9$ and $g(x) = 2x + 6$. Find

a. $(f + g)(5)$ b. $(fg)(-1)$ c. $\left(\dfrac{f}{g}\right)(4)$

Solution

a. $(f + g)(x) = f(x) + g(x) = (x^2 - 9) + (2x + 6) = x^2 + 2x - 3$
    Therefore, $(f + g)(5) = (5)^2 + 2(5) - 3 = 25 + 10 - 3 = 32$.

b. $(fg)(x) = f(x) \cdot g(x) = (x^2 - 9)(2x + 6) = 2x^3 + 6x^2 - 18x - 54$
    Therefore, $(fg)(-1) = 2(-1)^3 + 6(-1)^2 - 18(-1) - 54$
    $$= -2 + 6 + 18 - 54 = -32.$$

c. $\left(\dfrac{f}{g}\right)(x) = \dfrac{f(x)}{g(x)} = \dfrac{x^2 - 9}{2x + 6} = \dfrac{(x + 3)(x - 3)}{2(x + 3)} = \dfrac{x - 3}{2}, \quad x \neq -3$

    Therefore, $\left(\dfrac{f}{g}\right)(4) = \dfrac{4 - 3}{2} = \dfrac{1}{2}$.

▶ **TRY EXERCISE 14, PAGE 235**

● THE DIFFERENCE QUOTIENT

The expression

$$\dfrac{f(x + h) - f(x)}{h}, \quad h \neq 0$$

is called the **difference quotient** of f. It enables us to study the manner in which a function changes in value as the independent variable changes.

EXAMPLE 3   **Determine a Difference Quotient**

Determine the difference quotient of $f(x) = x^2 + 7$.

Solution

$\dfrac{f(x + h) - f(x)}{h} = \dfrac{[(x + h)^2 + 7] - [x^2 + 7]}{h}$ • Apply the difference quotient.

$= \dfrac{[x^2 + 2xh + h^2 + 7] - [x^2 + 7]}{h}$

$= \dfrac{x^2 + 2xh + h^2 + 7 - x^2 - 7}{h}$

$= \dfrac{2xh + h^2}{h} = \dfrac{h(2x + h)}{h} = 2x + h$

▶ **TRY EXERCISE 30, PAGE 235**

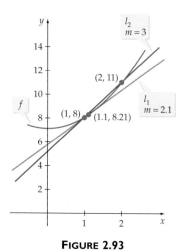

FIGURE 2.93

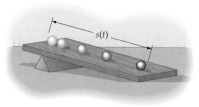

FIGURE 2.94

The difference quotient $2x + h$ of $f(x) = x^2 + 7$ from Example 3 is the slope of the secant line through the points

$$(x, f(x)) \qquad \text{and} \qquad (x + h, f(x + h))$$

For instance, let $x = 1$ and $h = 1$. Then the difference quotient is

$$2x + h = 2(1) + 1 = 3$$

This is the slope of the secant line l_2 through $(1, 8)$ and $(2, 11)$, as shown in **Figure 2.93.** If we let $x = 1$ and $h = 0.1$, then the difference quotient is

$$2x + h = 2(1) + 0.1 = 2.1$$

This is the slope of the secant line l_1 through $(1, 8)$ and $(1.1, 8.21)$.

The difference quotient

$$\frac{f(x + h) - f(x)}{h}$$

can be used to compute *average velocities*. In such cases it is traditional to replace f with s (for distance), the variable x with the variable a (for the time at the start of an observed interval of time), and the variable h with Δt (read as "delta t"), where Δt is the difference between the time at the end of an interval and the time at the start of the interval. For example, if an experiment is observed over the time interval from $t = 3$ seconds to $t = 5$ seconds, then the time interval is denoted as $[3, 5]$ with $a = 3$ and $\Delta t = 5 - 3 = 2$. Thus if the distance traveled by a ball that rolls down a ramp is given by $s(t)$, where t is the time in seconds after the ball is released (see **Figure 2.94**), then the **average velocity** of the ball over the interval $t = a$ to $t = a + \Delta t$ is the difference quotient

$$\frac{s(a + \Delta t) - s(a)}{\Delta t}$$

EXAMPLE 4   Evaluate Average Velocities

The distance traveled by a ball rolling down a ramp is given by $s(t) = 4t^2$, where t is the time in seconds after the ball is released, and $s(t)$ is measured in feet. Evaluate the average velocity of the ball for each time interval.

a. $[3, 5]$ **b.** $[3, 4]$ **c.** $[3, 3.5]$ **d.** $[3, 3.01]$

Solution

a. In this case, $a = 3$ and $\Delta t = 2$. Thus the average velocity over this interval is

$$\frac{s(a + \Delta t) - s(a)}{\Delta t} = \frac{s(3 + 2) - s(3)}{2} = \frac{s(5) - s(3)}{2} = \frac{100 - 36}{2}$$

$$= 32 \text{ feet per second}$$

b. Let $a = 3$ and $\Delta t = 4 - 3 = 1$.

$$\frac{s(a + \Delta t) - s(a)}{\Delta t} = \frac{s(3 + 1) - s(3)}{1} = \frac{s(4) - s(3)}{1} = \frac{64 - 36}{1}$$

$$= 28 \text{ feet per second}$$

Continued ▶

c. Let $a = 3$ and $\Delta t = 3.5 - 3 = 0.5$.

$$\frac{s(a + \Delta t) - s(a)}{\Delta t} = \frac{s(3 + 0.5) - s(3)}{0.5} = \frac{49 - 36}{0.5} = 26 \text{ feet per second}$$

d. Let $a = 3$ and $\Delta t = 3.01 - 3 = 0.01$.

$$\frac{s(a + \Delta t) - s(a)}{\Delta t} = \frac{s(3 + 0.01) - s(3)}{0.01} = \frac{36.2404 - 36}{0.01}$$

$$= 24.04 \text{ feet per second}$$

▶ **TRY EXERCISE 72, PAGE 237**

● COMPOSITION OF FUNCTIONS

Composition of functions is another way in which functions can be combined. This method of combining functions uses the output of one function as the input for a second function.

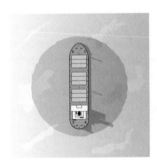

Suppose that the spread of oil from a leak in a tanker can be approximated by a circle with the tanker at its center. The radius r (in feet) of the spill t hours after the leak begins is given by $r(t) = 150\sqrt{t}$. The area of the spill is the area of a circle and is given by the formula $A(r) = \pi r^2$. To find the area of the spill 4 hours after the leak begins, we first find the radius of the spill and then use that number to find the area of the spill.

$$
\begin{array}{ll}
r(t) = 150\sqrt{t} & A(r) = \pi r^2 \\
r(4) = 150\sqrt{4} \quad \bullet\, t = 4 \text{ hours} & A(300) = \pi(300^2) \quad \bullet\, r = 300 \text{ feet} \\
\quad = 150(2) & \quad = 90{,}000\pi \\
\quad = 300 & \quad \approx 283{,}000
\end{array}
$$

The area of the spill after 4 hours is approximately 283,000 square feet.

There is an alternative way to solve this problem. Because the area of the spill depends on the radius and the radius depends on the time, there is a relationship between area and time. We can determine this relationship by evaluating the formula for the area of a circle using $r(t) = 150\sqrt{t}$. This will give the area of the spill as a function of time.

$$
\begin{array}{ll}
A(r) = \pi r^2 & \\
A[r(t)] = \pi [r(t)]^2 & \bullet \text{ Replace } r \text{ by } r(t). \\
\quad = \pi \left[150\sqrt{t}\,\right]^2 & \bullet\, r(t) = 150\sqrt{t} \\
A(t) = 22{,}500\pi t & \bullet \text{ Simplify.}
\end{array}
$$

The area of the spill as a function of time is $A(t) = 22{,}500\pi t$. To find the area of the oil spill after 4 hours, evaluate this function at $t = 4$.

$$
\begin{array}{ll}
A(t) = 22{,}500\pi t & \\
A(4) = 22{,}500\pi(4) & \bullet\, t = 4 \text{ hours} \\
\quad = 90{,}000\pi & \\
\quad \approx 283{,}000 &
\end{array}
$$

This is the same result we calculated earlier.

The function $A(t) = 22{,}500\pi t$ is referred to as the *composition* of A with r. The notation $A \circ r$ is used to denote this composition of functions. That is,

$$(A \circ r)(t) = 22{,}500\pi t$$

Definition of the Composition of Two Functions

Let f and g be two functions such that $g(x)$ is in the domain of f for all x in the domain of g. Then the composition of the two functions, denoted by $f \circ g$, is the function whose value at x is given by $(f \circ g)(x) = f[g(x)]$.

The function defined by $(f \circ g)(x)$ is also called the *composite* of f and g. We read $(f \circ g)(x)$ as "f circle g of x" and $f[g(x)]$ as "f of g of x."

Consider the functions $f(x) = 2x - 1$ and $g(x) = x^2 - 3$. The expression $(f \circ g)(-1)$ (or, equivalently, $f[g(-1)]$) means to evaluate the function f at $g(-1)$.

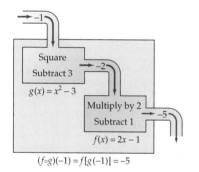

$(f \circ g)(-1) = f[g(-1)] = -5$

FIGURE 2.95

$$g(x) = x^2 - 3$$
$$g(-1) = (-1)^2 - 3 \qquad \text{• Evaluate } g \text{ at } -1.$$
$$= -2$$

$$f(x) = 2x - 1$$
$$f(-2) = 2(-2) - 1 = -5 \qquad \text{• Evaluate } f \text{ at } g(-1) = -2.$$

A graphical depiction of the composition $(f \circ g)(-1)$ would look something like **Figure 2.95.**

The requirement in the definition of the composition of two functions that $g(x)$ be in the domain of f for all x in the domain of g is important. For instance, let

$$f(x) = \frac{1}{x - 1} \qquad \text{and} \qquad g(x) = 3x - 5$$

When $x = 2$,

$$g(2) = 3(2) - 5 = 1$$
$$f[g(2)] = f(1) = \frac{1}{1 - 1} = \frac{1}{0} \qquad \text{• Undefined}$$

In this case, $g(2)$ is not in the domain of f. Thus the composition $(f \circ g)(x)$ is not defined at 2.

We can find a general expression for $f[g(x)]$ by evaluating f at $g(x)$. For instance, using $f(x) = 2x - 1$ and $g(x) = x^2 - 3$ as in **Figure 2.95,** we have

$$f(x) = 2x - 1$$
$$f[g(x)] = 2[g(x)] - 1 \qquad \text{• Replace } x \text{ by } g(x).$$
$$= 2[x^2 - 3] - 1 \qquad \text{• Replace } g(x) \text{ by } x^2 - 3.$$
$$= 2x^2 - 7 \qquad \text{• Simplify.}$$

In general, the composition of functions is not a commutative operation. That is, $(f \circ g)(x) \neq (g \circ f)(x)$. To verify this, we will compute the composition

$(g \circ f)(x) = g[f(x)]$, again using the functions $f(x) = 2x - 1$ and $g(x) = x^2 - 3$.

$$g(x) = x^2 - 3$$
$$g[f(x)] = [f(x)]^2 - 3 \qquad \text{• Replace } x \text{ by } f(x).$$
$$= [2x - 1]^2 - 3 \qquad \text{• Replace } f(x) \text{ by } 2x - 1.$$
$$= 4x^2 - 4x - 2 \qquad \text{• Simplify.}$$

Thus $f[g(x)] = 2x^2 - 7$, which is not equal to $g[f(x)] = 4x^2 - 4x - 2$. Therefore, $(f \circ g)(x) \neq (g \circ f)(x)$ and composition is not a commutative operation.

? QUESTION   Let $f(x) = x - 1$ and $g(x) = x + 1$. Then $f[g(x)] = g[f(x)]$. (You should verify this statement.) Does this contradict the statement we made that composition is not a commutative operation?

EXAMPLE 5   **Form Composite Functions**

If $f(x) = x^2 - 3x$ and $g(x) = 2x + 1$, find

a. $(g \circ f)$ **b.** $(f \circ g)$

Solution

a.
$$(g \circ f) = g[f(x)] = 2(f(x)) + 1 \qquad \text{• Substitute } f(x) \text{ for } x \text{ in } g.$$
$$= 2(x^2 - 3x) + 1 \qquad \text{• } f(x) = x^2 - 3x$$
$$= 2x^2 - 6x + 1$$

b.
$$(f \circ g) = f[g(x)] = (g(x))^2 - 3(g(x)) \qquad \text{• Substitute } g(x) \text{ for } x \text{ in } f.$$
$$= (2x + 1)^2 - 3(2x + 1) \qquad \text{• } g(x) = 2x + 1$$
$$= 4x^2 - 2x - 2$$

▶ **TRY EXERCISE 38, PAGE 236**

Note that in this example $(f \circ g) \neq (g \circ f)$. In general, the composition of functions is not a commutative operation.

Caution   Some care must be used when forming the composition of functions. For instance, if $f(x) = x + 1$ and $g(x) = \sqrt{x - 4}$, then

$$(g \circ f)(2) = g[f(2)] = g(3) = \sqrt{3 - 4} = \sqrt{-1}$$

which is not a real number. We can avoid this problem by imposing suitable restrictions on the domain of f so that the range of f is part of the domain of g. If the

---

? ANSWER   No. When we say that composition is not a commutative operation, we mean that generally, given any two functions, $(f \circ g)(x) \neq (g \circ f)(x)$. However, there may be particular instances in which $(f \circ g)(x) = (g \circ f)(x)$. It turns out that these particular instances are quite important, as we shall see later.

domain of f is restricted to $[3, \infty)$, then the range of f is $[4, \infty)$. But this is precisely the domain of g. Note that $2 \notin [3, \infty)$, and thus we avoid the problem of $(g \circ f)(2)$ not being a real number.

To evaluate $(f \circ g)(c)$ for some constant c, you can use either of the following methods.

Method 1   First evaluate $g(c)$. Then substitute this result for x in $f(x)$.

Method 2   First determine $f[g(x)]$ and then substitute c for x.

EXAMPLE 6   **Evaluate a Composite Function**

Evaluate $(f \circ g)(3)$, where $f(x) = 2x - 7$ and $g(x) = x^2 + 4$.

Solution

Method 1 $(f \circ g)(3) = f[g(3)]$

$\qquad\qquad\qquad = f[(3)^2 + 4]$ • Evaluate $g(3)$.

$\qquad\qquad\qquad = f(13)$

$\qquad\qquad\qquad = 2(13) - 7 = 19$ • Substitute 13 for x in f.

Method 2 $(f \circ g)(x) = 2[g(x)] - 7$ • Form $f[g(x)]$.

$\qquad\qquad\qquad = 2[x^2 + 4] - 7$

$\qquad\qquad\qquad = 2x^2 + 1$

$\qquad (f \circ g)(3) = 2(3)^2 + 1 = 19$ • Substitute 3 for x.

▶ **TRY EXERCISE 50, PAGE 236**

take note

In Example 6, both Method 1 and Method 2 produce the same result. Although Method 2 is longer, it is the better method if you must evaluate $(f \circ g)(x)$ for several values of x.

Figures 2.96 and **2.97** graphically illustrate the difference between Method 1 and Method 2.

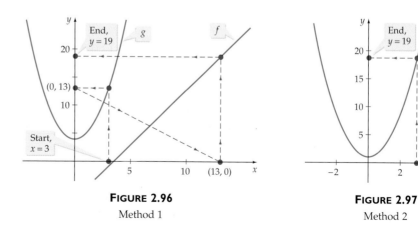

FIGURE 2.96
Method 1

FIGURE 2.97
Method 2

EXAMPLE 7	Use a Composite Function to Solve an Application

A graphic artist has drawn a 3-inch by 2-inch rectangle on a computer screen. The artist has been scaling the size of the rectangle for t seconds in such a way that the upper right corner of the original rectangle is moving to the right at the rate of 0.5 inch per second and downward at the rate of 0.2 inch per second. See **Figure 2.98**.

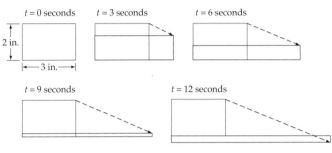

$t = 0$ seconds $t = 3$ seconds $t = 6$ seconds

2 in.

|— 3 in.—|

$t = 9$ seconds $t = 12$ seconds

FIGURE 2.98

a.   Write the length l and the width w of the scaled rectangles as functions of t.

b.   Write the area A of the scaled rectangle as a function of t.

c.   Find the intervals on which A is an increasing function for $0 \le t \le 14$. Also find the intervals on which A is a decreasing function.

d.   Find the value of t (where $0 \le t \le 14$) that maximizes $A(t)$.

Solution

a.   Because *distance* = *rate* · *time*, we see that the change in l is given by $0.5t$. Therefore, the length at any time t is $l = 3 + 0.5t$. For $0 \le t \le 10$, the width is given by $w = 2 - 0.2t$. For $10 < t \le 14$, the width is $w = -2 + 0.2t$. In either case the width can be determined by finding $w = |2 - 0.2t|$. (The absolute value symbol is needed to keep the width positive for $10 < t \le 14$.)

b. $A = lw = (3 + 0.5t)|2 - 0.2t|$

c.   Use a graphing utility to determine that A is increasing on $[0, 2]$ and on $[10, 14]$ and that A is decreasing on $[2, 10]$. See **Figure 2.99**.

d.   The highest point on the graph of A occurs when $t = 14$ seconds. See **Figure 2.99**.

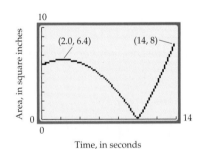

$A = (3 + 0.5t)|2 - 0.2t|$

FIGURE 2.99

▶ **TRY EXERCISE 66, PAGE 236**

You may be inclined to think that if the area of a rectangle is decreasing, then its perimeter is also decreasing, but this is not always the case. For example, the area of the scaled rectangle in Example 7 was shown to decrease on $[2, 10]$ even though its perimeter is always increasing. See Exercise 68 in Exercise Set 2.6.

TOPICS FOR DISCUSSION

1. The domain of $f + g$ consists of all real numbers formed by the *union* of the domain of f and the domain of g. Do you agree?

2. Given $f(x) = 3x - 2$ and $g(x) = \dfrac{1}{3}x + \dfrac{2}{3}$, determine $f \circ g$ and $g \circ f$. Does this show that composition of functions is a commutative operation?

3. A tutor states that the difference quotient of $f(x) = x^2$ and the difference quotient of $g(x) = x^2 + 4$ are the same. Do you agree?

4. A classmate states that the difference quotient of any linear function $f(x) = mx + b$ is always m. Do you agree?

5. When we use a difference quotient to determine an average velocity, we generally replace the variable h with the variable Δt. What does Δt represent?

EXERCISE SET 2.6

In Exercises 1 to 12, use the given functions f and g to find $f + g, f - g, fg$, and $\dfrac{f}{g}$. State the domain of each.

1. $f(x) = x^2 - 2x - 15, \quad g(x) = x + 3$

2. $f(x) = x^2 - 25, \quad g(x) = x - 5$

3. $f(x) = 2x + 8, \quad g(x) = x + 4$

4. $f(x) = 5x - 15, \quad g(x) = x - 3$

5. $f(x) = x^3 - 2x^2 + 7x, \quad g(x) = x$

6. $f(x) = x^2 - 5x - 8, \quad g(x) = -x$

7. $f(x) = 2x^2 + 4x - 7, \quad g(x) = 2x^2 + 3x - 5$

8. $f(x) = 6x^2 + 10, \quad g(x) = 3x^2 + x - 10$

9. $f(x) = \sqrt{x - 3}, \quad g(x) = x$

▶ 10. $f(x) = \sqrt{x - 4}, \quad g(x) = -x$

11. $f(x) = \sqrt{4 - x^2}, \quad g(x) = 2 + x$

12. $f(x) = \sqrt{x^2 - 9}, \quad g(x) = x - 3$

In Exercises 13 to 28, evaluate the indicated function, where $f(x) = x^2 - 3x + 2$ and $g(x) = 2x - 4$.

13. $(f + g)(5)$

▶ 14. $(f + g)(-7)$

15. $(f + g)\left(\dfrac{1}{2}\right)$

16. $(f + g)\left(\dfrac{2}{3}\right)$

17. $(f - g)(-3)$

18. $(f - g)(24)$

19. $(f - g)(-1)$

20. $(f - g)(0)$

21. $(fg)(7)$

22. $(fg)(-3)$

23. $(fg)\left(\dfrac{2}{5}\right)$

24. $(fg)(-100)$

25. $\left(\dfrac{f}{g}\right)(-4)$

26. $\left(\dfrac{f}{g}\right)(11)$

27. $\left(\dfrac{f}{g}\right)\left(\dfrac{1}{2}\right)$

28. $\left(\dfrac{f}{g}\right)\left(\dfrac{1}{4}\right)$

In Exercises 29 to 36, find the difference quotient of the given function.

29. $f(x) = 2x + 4$

▶ 30. $f(x) = 4x - 5$

31. $f(x) = x^2 - 6$

32. $f(x) = x^2 + 11$

33. $f(x) = 2x^2 + 4x - 3$

34. $f(x) = 2x^2 - 5x + 7$

35. $f(x) = -4x^2 + 6$

36. $f(x) = -5x^2 - 4x$

In Exercises 37 to 48, find $g \circ f$ and $f \circ g$ for the given functions f and g.

37. $f(x) = 3x + 5, \quad g(x) = 2x - 7$

▶ **38.** $f(x) = 2x - 7, \quad g(x) = 3x + 2$

39. $f(x) = x^2 + 4x - 1, \quad g(x) = x + 2$

40. $f(x) = x^2 - 11x, \quad g(x) = 2x + 3$

41. $f(x) = x^3 + 2x, \quad g(x) = -5x$

42. $f(x) = -x^3 - 7, \quad g(x) = x + 1$

43. $f(x) = \dfrac{2}{x + 1}, \quad g(x) = 3x - 5$

44. $f(x) = \sqrt{x + 4}, \quad g(x) = \dfrac{1}{x}$

45. $f(x) = \dfrac{1}{x^2}, \quad g(x) = \sqrt{x - 1}$

46. $f(x) = \dfrac{6}{x - 2}, \quad g(x) = \dfrac{3}{5x}$

47. $f(x) = \dfrac{3}{|5 - x|}, \quad g(x) = -\dfrac{2}{x}$

48. $f(x) = |2x + 1|, \quad g(x) = 3x^2 - 1$

In Exercises 49 to 64, evaluate each composite function, where $f(x) = 2x + 3, g(x) = x^2 - 5x$, and $h(x) = 4 - 3x^2$.

49. $(g \circ f)(4)$

▶ **50.** $(f \circ g)(4)$

51. $(f \circ g)(-3)$

52. $(g \circ f)(-1)$

53. $(g \circ h)(0)$

54. $(h \circ g)(0)$

55. $(f \circ f)(8)$

56. $(f \circ f)(-8)$

57. $(h \circ g)\left(\dfrac{2}{5}\right)$

58. $(g \circ h)\left(-\dfrac{1}{3}\right)$

59. $(g \circ f)\left(\sqrt{3}\right)$

60. $(f \circ g)\left(\sqrt{2}\right)$

61. $(g \circ f)(2c)$

62. $(f \circ g)(3k)$

63. $(g \circ h)(k + 1)$

64. $(h \circ g)(k - 1)$

65. WATER TANK A water tank has the shape of a right circular cone, with height 16 feet and radius 8 feet. Water is running into the tank so that the radius r (in feet) of the surface of the water is given by $r = 1.5t$, where t is the time (in minutes) that the water has been running.

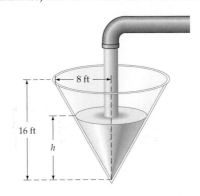

a. The area A of the surface of the water is $A = \pi r^2$. Find $A(t)$ and use it to determine the area of the surface of the water when $t = 2$ minutes.

b. The volume V of the water is given by $V = \dfrac{1}{3}\pi r^2 h$.

Find $V(t)$ and use it to determine the volume of the water when $t = 3$ minutes. (*Hint:* The height of the water in the cone is always twice the radius of the water.)

▶ **66.** **SCALING A RECTANGLE** Work Example 7 of this section with the scaling as follows. The upper right corner of the original rectangle is pulled to the *left* at 0.5 inch per second and downward at 0.2 inch per second.

67. TOWING A BOAT A boat is towed by a rope that runs through a pulley that is 4 feet above the point where the rope is tied to the boat. The length (in feet) of the rope from the boat to the pulley is given by $s = 48 - t$, where t is the time in seconds that the boat has been in tow. The horizontal distance from the pulley to the boat is d.

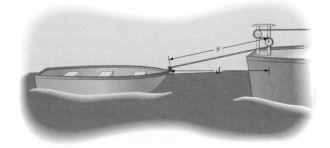

a. Find $d(t)$.   **b.** Evaluate $s(35)$ and $d(35)$.

68. **PERIMETER OF A SCALED RECTANGLE** Show by a graph that the perimeter

$$P = 2(3 + 0.5t) + 2|2 - 0.2t|$$

of the scaled rectangle in Example 7 of this section is an increasing function over $0 \leq t \leq 14$.

69. CONVERSION FUNCTIONS The function $F(x) = \dfrac{x}{12}$ converts x inches to feet. The function $Y(x) = \dfrac{x}{3}$ converts x feet to yards. Explain the meaning of $(Y \circ F)(x)$.

70. CONVERSION FUNCTIONS The function $F(x) = 3x$ converts x yards to feet. The function $I(x) = 12x$ converts x feet to inches. Explain the meaning of $(I \circ F)(x)$.

71. **CONCENTRATION OF A MEDICATION** The concentration $C(t)$ (in milligrams per liter) of a medication in a patient's blood is given by the data in the following table.

Concentration of Medication in Patient's Blood

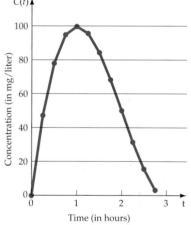

t hours	$C(t)$ mg/l
0	0
0.25	47.3
0.50	78.1
0.75	94.9
1.00	99.8
1.25	95.7
1.50	84.4
1.75	68.4
2.00	50.1
2.25	31.6
2.50	15.6
2.75	4.3

Time (in hours)

The **average rate of change** of the concentration over the time interval from $t = a$ to $t = a + \Delta t$ is

$$\frac{C(a + \Delta t) - C(a)}{\Delta t}$$

Use the data in the table to evaluate the average rate of change for each of the following time intervals.

a. $[0, 1]$ (*Hint:* In this case, $a = 0$ and $\Delta t = 1$.) Compare this result to the slope of the line through $(0, C(0))$ and $(1, C(1))$.

b. $[0, 0.5]$ **c.** $[1, 2]$ **d.** $[1, 1.5]$ **e.** $[1, 1.25]$

f. The data in the table can be modeled by the function $Con(t) = 25t^3 - 150t^2 + 225t$. Use $Con(t)$ to verify that the average rate of change over $[1, 1 + \Delta t]$ is $-75(\Delta t) + 25(\Delta t)^2$. What does the average rate of change over $[1, 1 + \Delta t]$ seem to approach as Δt approaches 0?

▶ **72. BALL ROLLING ON A RAMP** The distance traveled by a ball rolling down a ramp is given by $s(t) = 6t^2$, where t is the time in seconds after the ball is released, and $s(t)$ is measured in feet. The ball travels 6 feet in 1 second and it travels 24 feet in 2 seconds. Use the difference quotient for average velocity given on page 229 to evaluate the average velocity for each of the following time intervals.

a. $[2, 3]$ (*Hint:* In this case, $a = 2$ and $\Delta t = 1$.) Compare this result to the slope of the line through $(2, s(2))$ and $(3, s(3))$.

b. $[2, 2.5]$ **c.** $[2, 2.1]$ **d.** $[2, 2.01]$ **e.** $[2, 2.001]$

f. Verify that the average velocity over $[2, 2 + \Delta t]$ is $24 + 6(\Delta t)$. What does the average velocity seem to approach as Δt approaches 0?

CONNECTING CONCEPTS

In Exercises 73 to 76, show that $(f \circ g)(x) = (g \circ f)(x)$.

73. $f(x) = 2x + 3;$ $g(x) = 5x + 12$

74. $f(x) = 4x - 2;$ $g(x) = 7x - 4$

75. $f(x) = \dfrac{6x}{x - 1};$ $g(x) = \dfrac{5x}{x - 2}$

76. $f(x) = \dfrac{5x}{x + 3};$ $g(x) = -\dfrac{2x}{x - 4}$

In Exercises 77 to 82, show that

$$(g \circ f)(x) = x \quad \text{and} \quad (f \circ g)(x) = x$$

77. $f(x) = 2x + 3, \quad g(x) = \dfrac{x - 3}{2}$

78. $f(x) = 4x - 5, \quad g(x) = \dfrac{x + 5}{4}$

79. $f(x) = \dfrac{4}{x + 1}, \quad g(x) = \dfrac{4 - x}{x}$

80. $f(x) = \dfrac{2}{1 - x}, \quad g(x) = \dfrac{x - 2}{x}$

81. $f(x) = x^3 - 1, \quad g(x) = \sqrt[3]{x + 1}$

82. $f(x) = -x^3 + 2, \quad g(x) = \sqrt[3]{2 - x}$

PROJECTS

1. **A GRAPHING UTILITY PROJECT** For any two different real numbers x and y, the larger of the two numbers is given by

$$\text{Maximum}(x, y) = \frac{x + y}{2} + \frac{|x - y|}{2} \qquad (1)$$

a. Verify Equation (1) for $x = 5$ and $y = 9$.

b. Verify Equation (1) for $x = 201$ and $y = 80$.

For any two different functional values $f(x)$ and $g(x)$, the larger of the two is given by

$$\text{Maximum}(f(x), g(x)) = \frac{f(x) + g(x)}{2} + \frac{|f(x) - g(x)|}{2} \qquad (2)$$

To illustrate how we might make use of Equation (2), consider the functions $y_1 = x^2$ and $y_2 = \sqrt{x}$ on the interval from Xmin $= -1$ to Xmax $= 6$. The graphs of y_1 and y_2 are shown below.

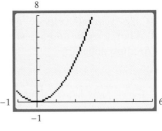

$y_1 = x^2$

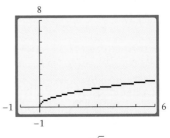

$y_2 = \sqrt{x}$

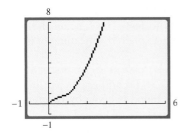

$y_3 = (y_1 + y_2)/2 + (\text{abs } (y_1 - y_2))/2$

Now consider the function

$$y_3 = (y_1 + y_2)/2 + (\text{abs}(y_1 - y_2))/2$$

where "abs" represents the absolute value function. The graph of y_3 is shown above.

c. Write a sentence or two that explains why the graph of y_3 is as shown.

d. What is the domain of y_1? of y_2? of y_3? Write a sentence that explains how to determine the domain of y_3, given the domain of y_1 and the domain of y_2.

e. Determine a formula for the function Minimum($f(x), g(x)$).

2. **THE NEVER-NEGATIVE FUNCTION** The author J. D. Murray describes a function f_+ that is defined in the following manner.[2]

$$f_+ = \begin{cases} f & \text{if } f \geq 0 \\ 0 & \text{if } f < 0 \end{cases}$$

We will refer to this function as a **never-negative** function. Never-negative functions can be graphed by using Equation (2) in Project 1. For example, if we let $g(x) = 0$, then Equation (2) simplifies to

$$\text{Maximum}(f(x), 0) = \frac{f(x)}{2} + \frac{|f(x)|}{2} \qquad (3)$$

[2]*Mathematical Biology* (New York: Springer-Verlag, 1989), p. 101.

The graph of $y = \text{Maximum}(f(x), 0)$ is the graph of $y = f(x)$ provided that $f(x) \geq 0$, and it is the graph of $y = 0$ provided that $f(x) < 0$.

An Application The mosquito population per acre of a large resort is controlled by spraying on a monthly basis. A biologist has determined that the mosquito population can be approximated by the never-negative function M_+ with

$$M(t) = -35{,}400(t - \text{int}(t))^2 + 35{,}400(t - \text{int}(t)) - 4000$$

Here t represents the month, and $t = 0$ corresponds to June 1, 2004.

a. Use a graphing utility to graph M for $0 \leq t \leq 3$.

b. Use a graphing utility to graph M_+ for $0 \leq t \leq 3$.

c. Write a sentence or two that explains how the graph of M_+ differs from the graph of M.

d. What is the maximum mosquito population per acre for $0 \leq t \leq 3$? When does this maximum mosquito population occur?

e. Explain when would be the best time to visit the resort, provided that you wished to minimize your exposure to mosquitos.

EXPLORING CONCEPTS WITH TECHNOLOGY

Graphing Piecewise Functions with a Graphing Calculator

A graphing calculator can be used to graph piecewise functions by including as part of the function the interval on which each piece of the function is defined. The method is based on the fact that a graphing calculator "evaluates" inequalities. For purposes of this Exploration, we will use keystrokes for a TI-83 calculator.

For instance, store 3 in **X** by pressing 3 | STO▶ | | X,T,Θ,*n* | | ENTER |. Now enter the inequality $x > 4$ by pressing | X,T,Θ,*n* | | 2ND | TEST 3 4 | ENTER |. Your screen should look like the one at the left. Note that the value of the inequality is 0. This occurs because the calculator replaced **X** by 3 and then determined whether the inequality $3 > 4$ was true or false. The calculator expresses the fact that the inequality is false by placing a zero on the screen. If we repeat the sequence of steps above, except that we store 5 in **X** instead of 3, the calculator will determine that the inequality is true and place a 1 on the screen.

This property of calculators is used to graph piecewise functions. Graphs of these functions work best when Dot mode rather than Connected mode is used. To switch to Dot mode, select | MODE |, use the arrow keys to highlight | DOT |, and then press | ENTER |.

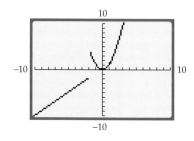

Now we will graph the piecewise function defined by $f(x) = \begin{cases} x, & x \leq -2 \\ x^2, & x > -2 \end{cases}$.

Enter the function[3] as **Y₁=X∗(X≤-2)+X²∗(X>-2)** and graph this in the standard viewing window. Note that you are multiplying each piece of the function by its domain. The graph will appear as shown at the left.

[3]Note that pressing | 2ND | TEST will display the inequality menu.

To understand how the graph is drawn, we will consider two values of x, -8 and 2, and evaluate Y_1 for each of these values.

$Y_1=X*(X\leq-2)+X^2*(X>-2)$

$= -8(-8 \leq -2) + (-8)^2(-8 > -2)$

$= -8(1) + 64(0) = -8$ • When $x = -8$, the value assigned to $-8 \leq -2$ is 1; the value assigned to $-8 > -2$ is 0.

$Y_1=X*(X\leq-2)+X^2*(X>-2)$

$=2(2 \leq -2) + 2^2(2 > -2)$

$=2(0) + 4(1) = 4$ • When $x = 2$, the value assigned to $2 \leq -2$ is 0; the value assigned to $2 > -2$ is 1.

In a similar manner, for any value of x for which $x \leq -2$, the value assigned to $(X\leq-2)$ is 1 and the value assigned to $(X>-2)$ is 0. Thus $Y_1=X*1+X^2*0=X$ on that interval. This means that only the $f(x) = x$ piece of the function is graphed. When $x > -2$, the value assigned to $(X\leq-2)$ is 0 and the value assigned to $(X>-2)$ is 1. Thus $Y_1=X*0+X^2*1=X^2$ on that interval. This means that only the $f(x) = x^2$ piece of the function is graphed on that interval.

1. Graph: $f(x) = \begin{cases} x^2, & x < 2 \\ -x, & x \geq 2 \end{cases}$ 2. Graph: $f(x) = \begin{cases} x^2 - x, & x < 2 \\ -x + 4, & x \geq 2 \end{cases}$

3. Graph: $f(x) = \begin{cases} -x^2 + 1, & x < 0 \\ x^2 - 1, & x \geq 0 \end{cases}$ 4. Graph: $f(x) = \begin{cases} x^3 - 4x, & x < 1 \\ x^2 - x + 2, & x \geq 1 \end{cases}$

CHAPTER 2 SUMMARY

2.1 A Two-Dimensional Coordinate System and Graphs

- *The Distance Formula*   The distance d between the points represented by (x_1, y_1) and (x_2, y_2) is

$$d = \sqrt{(x_2 - x_1)^2 + (y_2 - y_1)^2}$$

- The midpoint of the line segment from $P_1(x_1, y_1)$ to $P_2(x_2, y_2)$ is

$$\left(\frac{x_1 + x_2}{2}, \frac{y_1 + y_2}{2} \right)$$

- The standard form of the equation of a circle with center at (h, k) and radius r is $(x - h)^2 + (y - k)^2 = r^2$.

2.2 Introduction to Functions

- *Definition of a Function*   A function is a set of ordered pairs in which no two ordered pairs that have the same first coordinate have different second coordinates.

- A graph is the graph of a function if and only if no vertical line intersects the graph at more than one point. If every horizontal line intersects the graph of a function at most once, then the graph is the graph of a one-to-one function.

2.3 Linear Functions

- A function is a linear function of x if it can be written in the form $f(x) = mx + b$, where m and b are real numbers and $m \neq 0$.

- The slope m of the line passing through the points $P_1(x_1, y_1)$ and $P_2(x_2, y_2)$ with $x_1 \neq x_2$ is given by

$$m = \frac{y_2 - y_1}{x_2 - x_1}$$

- The graph of the equation $f(x) = mx + b$ has slope m and y-intercept $(0, b)$.

- Two nonvertical lines are parallel if and only if their slopes are equal. Two lines with slopes m_1 and m_2 are perpendicular if and only if $m_1 = -\dfrac{1}{m_2}$.

2.4 Quadratic Functions

- A quadratic function of x is a function that can be represented by an equation of the form $f(x) = ax^2 + bx + c$, where a, b, and c are real numbers and $a \neq 0$.

- The vertex of the graph of $f(x) = ax^2 + bx + c$ is

$$\left(-\frac{b}{2a}, f\left(-\frac{b}{2a}\right)\right)$$

- Every quadratic function $f(x) = ax^2 + bx + c$ can be written in the standard form $f(x) = a(x - h)^2 + k$, $a \neq 0$. The graph of f is a parabola with vertex (h, k). The parabola is symmetric with respect to the vertical line $x = h$, which is called the axis of symmetry of the parabola. The parabola opens up if $a > 0$; it opens down if $a < 0$.

2.5 Properties of Graphs

- The graph of an equation is symmetric with respect to

  the y-axis if the replacement of x with $-x$ leaves the equation unaltered.

  the x-axis if the replacement of y with $-y$ leaves the equation unaltered.

  the origin if the replacement of x with $-x$ and y with $-y$ leaves the equation unaltered.

- If f is a function and c is a positive constant, then

  $y = f(x) + c$ is the graph of $y = f(x)$ shifted up *vertically* c units

  $y = f(x) - c$ is the graph of $y = f(x)$ shifted down *vertically* c units

  $y = f(x + c)$ is the graph of $y = f(x)$ shifted left *horizontally* c units

  $y = f(x - c)$ is the graph of $y = f(x)$ shifted right *horizontally* c units

- The graph of

  $y = -f(x)$ is the graph of $y = f(x)$ reflected across the x-axis.

  $y = f(-x)$ is the graph of $y = f(x)$ reflected across the y-axis.

- If $a > 1$, then the graph of $y = f(ax)$ is a horizontal compressing of $y = f(x)$.

- If $0 < a < 1$, then the graph of $y = f(ax)$ is a horizontal stretching of the graph of $y = f(x)$.

2.6 The Algebra of Functions

- For all values of x for which both $f(x)$ and $g(x)$ are defined, we define the following functions.

  Sum $\quad (f + g)(x) = f(x) + g(x)$

  Difference $\quad (f - g)(x) = f(x) - g(x)$

  Product $\quad (fg)(x) = f(x) \cdot g(x)$

  Quotient $\quad \left(\dfrac{f}{g}\right)(x) = \dfrac{f(x)}{g(x)}, \quad g(x) \neq 0$

- The expression

$$\frac{f(x + h) - f(x)}{h}, \quad h \neq 0$$

  is called the difference quotient of f. The difference quotient is an important function because it can be used to compute the *average rate of change* of f over the time interval $[x, x + h]$.

- For the functions f and g, the composite function, or composition, of f by g is given by $(g \circ f)(x) = g[f(x)]$ for all x in the domain of f such that $f(x)$ is in the domain of g.

CHAPTER 2 TRUE/FALSE EXERCISES

In Exercises 1 to 13, answer true or false. If the statement is false, give an example or a reason to show that the statement is false.

1. Let f be any function. Then $f(a) = f(b)$ implies that $a = b$.

2. If f and g are two functions, then $(f \circ g)(x) = (g \circ f)(x)$.

3. If f is not a one-to-one function, then there are at least two numbers u and v in the domain of f for which $f(u) = f(v)$.

4. Let f be a function such that $f(x) = f(x + 4)$ for all real numbers x. If $f(2) = 3$, then $f(18) = 3$.

5. For all functions f, $[f(x)]^2 = f[f(x)]$.

6. Let f be any function. Then for all a and b in the domain of f such that $f(b) \neq 0$ and $b \neq 0$,

$$\frac{f(a)}{f(b)} = \frac{a}{b}$$

7. The **identity function** $f(x) = x$ is its own inverse.

8. If f is a function, then $f(a + b) = f(a) + f(b)$ for all real numbers a and b in the domain of f.

9. If f is defined by $f(x) = |x|$, then $f(ab) = f(a)f(b)$ for all real numbers a and b.

10. If f is a one-to-one function and a and b are real numbers in the domain of f with $a < b$, then $f(a) \neq f(b)$.

11. The coordinates of a point on the graph of $y = f(x)$ are (a, b). If k is a positive constant, then (a, kb) are the coordinates of a point on the graph of $y = kf(x)$.

12. For every function f, the real number c is a solution of $f(x) = 0$ if and only if $(c, 0)$ is an x-intercept of the graph of $y = f(x)$.

13. The domain of every polynomial function is the set of real numbers.

CHAPTER 2 REVIEW EXERCISES

In Exercises 1 and 2, find the distance between the points whose coordinates are given.

1. $(-3, 2)$ $(7, 11)$
2. $(5, -4)$ $(-3, -8)$

In Exercises 3 and 4, find the midpoint of the line segment with the given endpoints.

3. $(2, 8)$ $(-3, 12)$
4. $(-4, 7)$ $(8, -11)$

In Exercises 5 and 6, determine the center and radius of the circle with the given equation.

5. $(x - 3)^2 + (y + 4)^2 = 81$

6. $x^2 + y^2 + 10x + 4y + 20 = 0$

In Exercises 7 and 8, find the equation in standard form of the circle that satisfies the given conditions.

7. Center $C = (2, -3)$, radius $r = 5$

8. Center $C = (-5, 1)$, passing through $(3, 1)$

9. If $f(x) = 3x^2 + 4x - 5$, find
   a. $f(1)$ **b.** $f(-3)$ **c.** $f(t)$
   d. $f(x + h)$ **e.** $3f(t)$ **f.** $f(3t)$

10. If $g(x) = \sqrt{64 - x^2}$, find
   a. $g(3)$ **b.** $g(-5)$ **c.** $g(8)$
   d. $g(-x)$ **e.** $2g(t)$ **f.** $g(2t)$

11. If $f(x) = x^2 + 4x$ and $g(x) = x - 8$, find
   a. $(f \circ g)(3)$ **b.** $(g \circ f)(-3)$
   c. $(f \circ g)(x)$ **d.** $(g \circ f)(x)$

12. If $f(x) = 2x^2 + 7$ and $g(x) = |x - 1|$, find
   a. $(f \circ g)(-5)$ **b.** $(g \circ f)(-5)$
   c. $(f \circ g)(x)$ **d.** $(g \circ f)(x)$

13. If $f(x) = 4x^2 - 3x - 1$, find the difference quotient
$$\frac{f(x + h) - f(x)}{h}$$

14. If $g(x) = x^3 - x$, find the difference quotient
$$\frac{g(x + h) - g(x)}{h}$$

In Exercises 15 to 20, sketch the graph of f. Find the interval(s) in which f is a. increasing, b. constant, c. decreasing.

15. $f(x) = |x - 3| - 2$ **16.** $f(x) = x^2 - 5$

17. $f(x) = |x + 2| - |x - 2|$ **18.** $f(x) = [\![x + 3]\!]$

19. $f(x) = \dfrac{1}{2}x - 3$ **20.** $f(x) = \sqrt[3]{x}$

In Exercises 21 to 24, determine the domain of the function represented by the given equation.

21. $f(x) = -2x^2 + 3$ **22.** $f(x) = \sqrt{6 - x}$

23. $f(x) = \sqrt{25 - x^2}$ **24.** $f(x) = \dfrac{3}{x^2 - 2x - 15}$

In Exercises 25 and 26, find the slope-intercept form of the equation of the line through the two points.

25. $(-1, 3)$ $(4, -7)$ **26.** $(0, 0)$ $(7, 11)$

27. Find the slope-intercept form of the equation of the line that is parallel to the graph of $3x - 4y = 8$ and passes through $(2, 11)$.

28. Find the slope-intercept form of the equation of the line that is perpendicular to the graph of $2x = -5y + 10$ and passes through $(-3, -7)$.

In Exercises 29 to 34, use the method of completing the square to write each quadratic equation in its standard form.

29. $f(x) = x^2 + 6x + 10$

30. $f(x) = 2x^2 + 4x + 5$

31. $f(x) = -x^2 - 8x + 3$

32. $f(x) = 4x^2 - 6x + 1$

33. $f(x) = -3x^2 + 4x - 5$

34. $f(x) = x^2 - 6x + 9$

In Exercises 35 to 38, find the vertex of the graph of the quadratic function.

35. $f(x) = 3x^2 - 6x + 11$

36. $h(x) = 4x^2 - 10$

37. $k(x) = -6x^2 + 60x + 11$

38. $m(x) = 14 - 8x - x^2$

39. Use the formula

$$d = \frac{|mx_1 + b - y_1|}{\sqrt{1 + m^2}}$$

to find the distance from the point $(1, 3)$ to the line given by $y = 2x - 3$.

40. A freight company has determined that its cost per delivery of delivering x parcels is

$$C(x) = 1050 + 0.5x$$

The price it charges to send a parcel is $13.00 per parcel. Determine

a. the revenue function

b. the profit function

c. the minimum number of parcels the company must ship to break even

In Exercises 41 and 42, sketch a graph that is symmetric to the given graph with respect to the *a.* **x-axis,** *b.* **y-axis,** *c.* **origin.**

41.

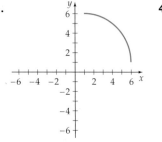

42.

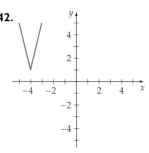

In Exercises 43 to 50, determine whether the graph of each equation is symmetric with respect to the *a.* **x-axis,** *b.* **y-axis,** *c.* **origin.**

43. $y = x^2 - 7$

44. $x = y^2 + 3$

45. $y = x^3 - 4x$

46. $y^2 = x^2 + 4$

47. $\dfrac{x^2}{3^2} + \dfrac{y^2}{4^2} = 1$

48. $xy = 8$

49. $|y| = |x|$

50. $|x + y| = 4$

In Exercises 51 to 56, sketch the graph of g. *a.* **Find the domain and the range of g.** *b.* **State whether g is even, odd, or neither even nor odd.**

51. $g(x) = -x^2 + 4$

52. $g(x) = -2x - 4$

53. $g(x) = |x - 2| + |x + 2|$

54. $g(x) = \sqrt{16 - x^2}$

55. $g(x) = x^3 - x$

56. $g(x) = 2[\![x]\!]$

In Exercises 57 to 62, first write the quadratic function in standard form, and then make use of translations to graph the function.

57. $F(x) = x^2 + 4x - 7$

58. $A(x) = x^2 - 6x - 5$

59. $P(x) = 3x^2 - 4$

60. $G(x) = 2x^2 - 8x + 3$

61. $W(x) = -4x^2 - 6x + 6$

62. $T(x) = -2x^2 - 10x$

63. On the same set of coordinate axes, sketch the graph of $p(x) = \sqrt{x} + c$ for $c = 0, -1$, and 2.

64. On the same set of coordinate axes, sketch the graph of $q(x) = \sqrt{x + c}$ for $c = 0, -1$, and 2.

65. On the same set of coordinate axes, sketch the graph of $r(x) = c\sqrt{9 - x^2}$ for $c = 1, \dfrac{1}{2}$, and -2.

66. On the same set of coordinate axes, sketch the graph of $s(t) = [\![cx]\!]$ for $c = 1, \dfrac{1}{4}$, and 4.

In Exercises 67 and 68, graph each piecewise-defined function.

67. $f(x) = \begin{cases} x, & \text{if } x \le 0 \\ \dfrac{1}{2}x, & \text{if } x > 0 \end{cases}$

68. $g(x) = \begin{cases} -2, & \text{if } x < -3 \\ \dfrac{2}{3}x, & \text{if } -3 \le x \le 3 \\ 2, & \text{if } x > 3 \end{cases}$

In Exercises 69 and 70, use the given functions f and g to find $f + g, f - g, fg,$ and $\dfrac{f}{g}$. State the domain of each.

69. $f(x) = x^2 - 9, \quad g(x) = x + 3$

70. $f(x) = x^3 + 8, \quad g(x) = x^2 - 2x + 4$

71. Find two numbers whose sum is 50 and whose product is a maximum.

72. Find two numbers whose difference is 10 and the sum of whose squares is a minimum.

73. The distance traveled by a ball rolling down a ramp is given by $s(t) = 3t^2$, where t is the time in seconds after the ball is released and $s(t)$ is measured in feet. Evaluate the average velocity of the ball for each of the following time intervals.

a. $[2, 4]$ **b.** $[2, 3]$ **c.** $[2, 2.5]$ **d.** $[2, 2.01]$

e. What appears to be the average velocity of the ball for the time interval $[2, 2 + \Delta t]$ as Δt approaches 0?

74. The distance traveled by a ball that is pushed down a ramp is given by $s(t) = 2t^2 + t$, where t is the time in seconds after the ball is released and $s(t)$ is measured in feet. Evaluate the average velocity of the ball for each of the following time intervals.

a. $[3, 5]$ **b.** $[3, 4]$ **c.** $[3, 3.5]$ **d.** $[3, 3.01]$

e. What appears to be the average velocity of the ball for the time interval $[3, 3 + \Delta t]$ as Δt approaches 0?

CHAPTER 2 TEST

1. Find the midpoint and the length of the line segment with endpoints $(-2, 3)$ and $(4, -1)$.

2. Determine the x- and y-intercepts, and then graph the equation $x = 2y^2 - 4$.

3. Graph the equation $y = |x + 2| + 1$.

4. Find the center and radius of the circle that has the general form $x^2 - 4x + y^2 + 2y - 4 = 0$.

5. Determine the domain of the function
$$f(x) = -\sqrt{x^2 - 16}$$

6. Graph $f(x) = -2|x - 2| + 1$. Identify the intervals over which the function is

a. increasing

b. constant

c. decreasing

7. An air freight company has determined that its cost per flight of delivering x parcels is
$$C(x) = 875 + 0.75x$$

The price it charges to send a parcel is $12.00 per parcel. Determine

a. the revenue function

b. the profit function

c. the minimum number of parcels the company must ship to break even

8. Use the graph of $f(x) = |x|$ to graph $y = -f(x + 2) - 1$.

9. Classify each of the following as an even function, an odd function, or neither an even nor an odd function.

a. $f(x) = x^4 - x^2$ **b.** $f(x) = x^3 - x$

c. $f(x) = x - 1$

10. Find the slope-intercept form of the equation of the line that passes through $(4, -2)$ and is perpendicular to the graph of $3x - 2y = 4$.

11. Find the maximum or minimum value of the function $f(x) = x^2 - 4x - 8$. State whether this value is a maximum or a minimum value.

12. Let $f(x) = x^2 - 1$ and $g(x) = x - 2$. Find $(f + g)$ and (f/g).

13. Find the difference quotient of the function

$$f(x) = x^2 + 1$$

14. Evaluate $(f \circ g)(x)$, where

$$f(x) = x^2 - 2x \quad \text{and} \quad g(x) = 2x + 5$$

15. The distance traveled by a ball rolling down a ramp is given by $s(t) = 5t^2$, where t is the time in seconds after the ball is released and $s(t)$ is measured in feet. Evaluate the average velocity of the ball for each of the following time intervals.

a. $[2, 3]$ **b.** $[2, 2.5]$ **c.** $[2, 2.01]$

CUMULATIVE REVIEW EXERCISES

1. What property of real numbers is demonstrated by $3(a + b) = 3(b + a)$?

2. Which of the numbers $-3, -\dfrac{2}{3}, \dfrac{6}{\pi}, 0, \sqrt{16}$ and $\sqrt{2}$ are not rational numbers?

In Exercises 3 to 8, simplify the expression.

3. $3 + 4(2x - 9)$

4. $(-4xy^2)^3(-2x^2y^4)$

5. $\dfrac{24a^4b^3}{18a^4b^5}$

6. $(2x + 3)(3x - 7)$

7. $\dfrac{x^2 + 6x - 27}{x^2 - 9}$

8. $\dfrac{4}{2x - 1} - \dfrac{2}{x - 1}$

In Exercises 9 to 14, solve for x.

9. $6 - 2(2x - 4) = 14$

10. $x^2 - x - 1 = 0$

11. $(2x - 1)(x + 3) = 4$

12. $3x + 2y = 15$

13. $x^4 - x^2 - 2 = 0$

14. $3x - 1 < 5x + 7$

15. Find the distance between the points $P_1(-2, -4)$ and $P_2(2, -3)$.

16. Given $G(x) = 2x^3 - 4x - 7$, find $G(-2)$.

17. Find the equation of the line between the points $P_1(2, -3)$ and $P_2(-2, -1)$.

18. How many ounces of pure water must be added to 60 ounces of an 8% salt solution to make a 3% salt solution?

19. The path of a tennis ball during a serve is given by $h(x) = -0.002x^2 - 0.03x + 8$. For a serve to be legal in tennis, the ball must be at least 3 feet high when it is 39 feet from the server, and it must land in a spot that is less than 60 feet from the server. Does the path of a ball given by $h(x) = -0.002x^2 - 0.03x + 8$, where $h(x)$ is the height of the ball in feet x feet from the server, satisfy the conditions of a legal serve?

20. A patient with a fever is given a medication to reduce the fever. The equation $T = -0.04t + 104$ models the temperature T, in degrees Fahrenheit, t minutes after taking the medication. What is the rate, in degrees per minute, at which the patient's temperature is decreasing?

POLYNOMIAL AND RATIONAL FUNCTIONS

DVD players have become very popular in the last few years. More than 31 million DVD players have been sold as of January 2002.

Production Cost and Average Cost

In this chapter you will study polynomial and rational functions. These types of functions have many practical applications. For instance, they can be used to model production costs and average costs associated with the manufacture of DVD players.

The cost, in dollars, of producing x DVD players is given by the polynomial function

$$C(x) = 0.001x^2 + 101x + 245{,}000$$

The average cost per DVD player is given by the rational function

$$\overline{C}(x) = \frac{C(x)}{x} = \frac{0.001x^2 + 101x + 245{,}000}{x}$$

The following graph of $\overline{C}$ shows that the minimum average cost per DVD player is obtained by producing 15,652 DVD players.

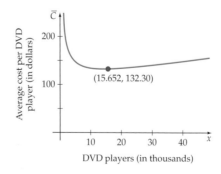

DVD players (in thousands)

Additional average cost applications are given in **Exercises 53 and 54 on page 312.**

Find and Use Clues to Narrow the Search

The game of Clue is a classic whodunit game. At the beginning of the game you are informed that Mr. Boddy has been murdered! It is your job to find and use clues to determine the murderer, the weapon, and the room in which the murder was committed. Was it Miss Scarlet in the billiard room with the revolver? Or did Professor Plum commit the murder in the conservatory with the rope? There are six suspects, six possible murder weapons, and nine rooms in Mr. Boddy's mansion. There are a total of $6 \times 6 \times 9 = 324$ possible solutions to each game.

In this chapter you will often need to find the zeros of a polynomial function. Finding the zeros of a polynomial function can be more complicated than solving a game of Clue. After all, any real or complex number is a possible zero. Quite often the zeros of a polynomial function are found by using several theorems to narrow the search. In most cases no single theorem can be used to find the zeros of a given polynomial function, but by combining the results of several theorems, we are often able to gather enough clues to find the zeros. In **Exercise 52, page 287,** you will apply several theorems from this chapter to find the zeros of a polynomial function.

THE REMAINDER THEOREM AND THE FACTOR THEOREM

If $P(x)$ is a polynomial, then the values of x for which $P(x)$ is equal to 0 are called the **zeros** of $P(x)$. For instance, -1 is a zero of $P(x) = 2x^3 - x + 1$ because

$$P(-1) = 2(-1)^3 - (-1) + 1$$
$$= -2 + 1 + 1$$
$$= 0$$

? **QUESTION** Is 0 a zero of $P(x) = 2x^3 - x + 1$?

Much of the work in this chapter concerns finding the zeros of a polynomial. Sometimes the zeros of a polynomial $P(x)$ are determined by dividing $P(x)$ by another polynomial.

● DIVISION OF POLYNOMIALS

take note

Recall that a fraction bar acts as a grouping symbol. Division of a polynomial by a monomial is an application of the distributive property.

To divide a polynomial by a monomial, divide each term of the polynomial by the monomial. For instance,

$$\frac{16x^3 - 8x^2 + 12x}{4x} = \frac{16x^3}{4x} - \frac{8x^2}{4x} + \frac{12x}{4x}$$

• Divide each term in the numerator by the denominator.

$$= 4x^2 - 2x + 3$$

• Simplify.

To divide a polynomial by a binomial, we use a method similar to that used to divide whole numbers. For instance, consider $(6x^3 - 16x^2 + 23x - 5) \div (3x - 2)$.

$$3x - 2 \overline{)6x^3 - 16x^2 + 23x - 5}$$

$$\begin{array}{r} 2x^2 \\ 3x - 2 \overline{)6x^3 - 16x^2 + 23x - 5} \\ \underline{6x^3 - 4x^2} \\ -12x^2 + 23x \end{array}$$

• Think $\dfrac{6x^3}{3x} = 2x^2$.

• Multiply: $2x^2(3x - 2) = 6x^3 - 4x^2$

• Subtract and bring down the next term, $23x$.

$$\begin{array}{r} 2x^2 - 4x \\ 3x - 2 \overline{)6x^3 - 16x^2 + 23x - 5} \\ \underline{6x^3 - 4x^2} \\ -12x^2 + 23x \\ \underline{-12x^2 + 8x} \\ 15x - 5 \end{array}$$

• Think $\dfrac{-12x^2}{3x} = -4x$.

• Multiply: $-4x(3x - 2) = -12x^2 + 8x$

• Subtract and bring down the next term, -5.

? **ANSWER**   No. $P(0) = 2(0)^3 - 0 + 1 = 1$. Because $P(0) \neq 0$, we know that 0 is not a zero of $P(x)$.

$$
\begin{array}{r}
2x^2 - 4x + 5 \\
3x - 2\overline{)6x^3 - 16x^2 + 23x - 5} \\
\underline{6x^3 - 4x^2} \\
-12x^2 + 23x \\
\underline{-12x^2 + 8x} \\
15x - 5 \\
\underline{15x - 10} \\
5
\end{array}
$$

- **Think** $\dfrac{15x}{3x} = 5$.

- **Multiply:** $5(3x - 2) = 15x - 10$

- **Subtract to produce the remainder, 5.**

Thus $(6x^3 - 16x^2 + 23x - 5) \div (3x - 2) = 2x^2 - 4x + 5$ with a remainder of 5.

Although there is nothing wrong with writing the answer as we did above, it is more common to write the answer as the quotient plus the remainder divided by the divisor. (See the Take Note at the left.) Using this method, we write

$$
\overbrace{\underbrace{\frac{6x^3 - 16x^2 + 23x - 5}{3x - 2}}_{\text{Divisor}}}^{\text{Dividend}} = \overbrace{2x^2 - 4x + 5}^{\text{Quotient}} + \underbrace{\frac{5}{3x - 2}}_{} \quad \begin{matrix} \longleftarrow \text{Remainder} \\ \longleftarrow \text{Divisor} \end{matrix}
$$

In this example, $6x^3 - 16x^2 + 23x - 5$ is called the **dividend**, $3x - 2$ is the **divisor,** $2x^2 - 4x + 5$ is the **quotient,** and 5 is the **remainder.** In every division, the dividend is equal to the product of the quotient and divisor, plus the remainder. That is,

$$
\underbrace{6x^3 - 16x^2 + 23x - 5}_{\text{Dividend}} = \underbrace{(2x^2 - 4x + 5)}_{\text{quotient}} \cdot \underbrace{(3x - 2)}_{\text{divisor}} + \underbrace{5}_{\text{remainder}}
$$

Before dividing polynomials, make sure that each polynomial is written in descending order. In some cases, it is helpful to insert a 0 in the dividend for a missing term (one whose coefficient is 0) so that like terms align in the same column. This is demonstrated in Example 1.

> **? QUESTION** What is the first step you should perform to find the quotient of $(2x + 1 + x^2) \div (x - 1)$?

EXAMPLE 1   **Divide Polynomials**

Divide: $\dfrac{-5x^2 - 8x + x^4 + 3}{x - 3}$

Solution

Write the numerator in descending order. Then divide.

$$
\frac{-5x^2 - 8x + x^4 + 3}{x - 3} = \frac{x^4 - 5x^2 - 8x + 3}{x - 3}
$$

> **? ANSWER** Write the dividend in descending order as $x^2 + 2x + 1$.

$$\begin{array}{r}
x^3 + 3x^2 + 4x + 4 \\
x - 3 \overline{\smash{)}\; x^4 + 0x^3 - 5x^2 - 8x + 3} \\
\underline{x^4 - 3x^3} \\
3x^3 - 5x^2 \\
\underline{3x^3 - 9x^2} \\
4x^2 - 8x \\
\underline{4x^2 - 12x} \\
4x + 3 \\
\underline{4x - 12} \\
15
\end{array}$$

• Inserting $0x^3$ for the missing term helps align like terms in the same column.

Thus $\dfrac{-5x^2 - 8x + x^4 + 3}{x - 3} = x^3 + 3x^2 + 4x + 4 + \dfrac{15}{x - 3}$.

▶ **TRY EXERCISE 2, PAGE 257**

A procedure called **synthetic division** can expedite the division process. To apply the synthetic division procedure, the divisor must be a polynomial of the form $x - c$, where c is a constant. In the synthetic division procedure, the variables that occur in the polynomials are not listed. To understand how synthetic division is performed, examine the following **long division** on the left and the related synthetic division on the right.

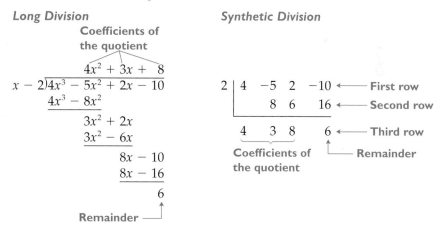

Long Division   *Synthetic Division*

In the long division above, the dividend is $4x^3 - 5x^2 + 2x - 10$, and the divisor is $x - 2$. Because the divisor is of the form $x - c$, with $c = 2$, the division can be performed by the synthetic division procedure. Observe that in the accompanying synthetic division

1. The constant c is listed as the first number in the first row, followed by the coefficients of the dividend.

2. The first number in the third row is the leading coefficient of the dividend.

3. Each number in the second row is determined by computing the product of c and the number in the third row of the preceding column.

4. Each of the numbers in the third row, other than the first number, is determined by adding the numbers directly above it.

The following explanation illustrates the steps used to find the quotient and remainder of $(2x^3 - 8x + 7) \div (x + 3)$ by using synthetic division. The divisor $x + 3$ is written in $x - c$ form as $x - (-3)$, which indicates that $c = -3$. The dividend $2x^3 - 8x + 7$ is missing an x^2 term. If we insert $0x^2$ for the missing term, the dividend becomes $2x^3 + 0x^2 - 8x + 7$.

Coefficients of the dividend

$$
\begin{array}{r|rrrr}
-3 & 2 & 0 & -8 & 7 \\
  &   &   &   &   \\
\hline
  & 2 &   &   &   \\
\end{array}
$$

- **Write the constant c, −3, followed by the coefficients of the dividend. Bring down the first coefficient in the first row, 2, as the first number of the third row.**

$$
\begin{array}{r|rrrr}
-3 & 2 & 0 & -8 & 7 \\
  &   & -6 &   &   \\
\hline
  & 2 & -6 &   &   \\
\end{array}
$$

- **Multiply c times the first number in the third row, 2, to produce the first number of the second row, −6. Add the 0 and the −6 to produce the next number of the third row, −6.**

$$
\begin{array}{r|rrrr}
-3 & 2 & 0 & -8 & 7 \\
  &   & -6 & 18 &   \\
\hline
  & 2 & -6 & 10 &   \\
\end{array}
$$

- **Multiply c times the second number in the third row, −6, to produce the next number of the second row, 18. Add the −8 and the 18 to produce the next number of the third row, 10.**

$$
\begin{array}{r|rrrr}
-3 & 2 & 0 & -8 & 7 \\
  &   & -6 & 18 & -30 \\
\hline
  & 2 & -6 & 10 & -23 \\
\end{array}
$$

- **Multiply c times the third number in the third row, 10, to produce the next number of the second row, −30. Add the 7 and the −30 to produce the last number of the third row, −23.**

Coefficients of the quotient   **Remainder**

The last number in the bottom row, −23, is the remainder. The other numbers in the bottom row are the coefficients of the quotient. The quotient of a synthetic division always has a degree that is *one less* than the degree of the dividend. Thus the quotient in this example is $2x^2 - 6x + 10$. The results of the above synthetic division can be expressed in **fractional form** as

$$
\frac{2x^3 - 8x + 7}{x + 3} = 2x^2 - 6x + 10 + \frac{-23}{x + 3}
$$

or as

$$
2x^3 - 8x + 7 = (x + 3)(2x^2 - 6x + 10) - 23
$$

In Example 2 we illustrate the compact form of synthetic division, obtained by condensing the process explained above.

---

take note

$2x^2 - 6x + 10 + \dfrac{-23}{x + 3}$

can also be written as

$2x^2 - 6x + 10 - \dfrac{23}{x + 3}$

EXAMPLE 2   **Use Synthetic Division to Divide Polynomials**

Use synthetic division to divide $x^4 - 4x^2 + 7x + 15$ by $x + 4$.

Solution

Because the divisor is $x + 4$, we perform synthetic division with $c = -4$.

$$
\begin{array}{r|rrrrr}
-4 & 1 & 0 & -4 & 7 & 15 \\
   &   & -4 & 16 & -48 & 164 \\
\hline
   & 1 & -4 & 12 & -41 & 179 \\
\end{array}
$$

The quotient is $x^3 - 4x^2 + 12x - 41$, and the remainder is 179.

$$\frac{x^4 - 4x^2 + 7x + 15}{x + 4} = x^3 - 4x^2 + 12x - 41 + \frac{179}{x + 4}$$

▶ **TRY EXERCISE 12, PAGE 257**

INTEGRATING TECHNOLOGY

A TI-82/83 synthetic-division program called SYDIV is available on the Internet at math.college.hmco.com. The program prompts you to enter the degree of the dividend, the coefficients of the dividend, and the constant c from the divisor $x - c$. For instance, to perform the synthetic division in Example 2, enter 4 for the degree of the dividend, followed by the coefficients 1, 0, –4, 7, and 15. See **Figure 3.1**. Press ENTER followed by –4 to produce the display in **Figure 3.2**. Press ENTER to produce the display in **Figure 3.3**. Press ENTER again to produce the display in **Figure 3.4**.

```
prgmSYDIV
DEGREE? 4
DIVIDEND COEF
?1
?0
?-4
?7
?15
```

FIGURE 3.1

```
C? -4
```

FIGURE 3.2

```
COEF OF QUOTIENT
                 1
                -4
                12
               -41
```

FIGURE 3.3

```
REMAINDER
                179
QUIT?  PRESS 1
NEW C?  PRESS 2
```

FIGURE 3.4

● **THE REMAINDER THEOREM**

The following theorem shows that synthetic division can be used to determine the value $P(c)$ for a given polynomial P and constant c.

The Remainder Theorem

If a polynomial $P(x)$ is divided by $x - c$, then the remainder equals $P(c)$.

The following example illustrates the Remainder Theorem by showing that the remainder of $(x^2 + 9x - 16) \div (x - 3)$ is the same as $P(x) = x^2 + 9x - 16$ evaluated at $x = 3$.

Let $x = 3$ and $P(x) = x^2 + 9x - 16$.

Then $P(3) = (3)^2 + 9(3) - 16$

$ = 9 + 27 - 16$

$ = 20$

$$\begin{array}{r} x + 12 \\ x - 3\overline{)x^2 + 9x - 16} \\ \underline{x^2 - 3x} \\ 12x - 16 \\ \underline{12x - 36} \\ 20 \end{array}$$

$P(3)$ is equal to the remainder of $P(x)$ divided by $(x - 3)$.

In Example 3 we use synthetic division and the Remainder Theorem to evaluate a polynomial function.

EXAMPLE 3 **Use the Remainder Theorem to Evaluate a Polynomial Function**

Let $P(x) = 2x^3 + 3x^2 + 2x - 2$. Use the Remainder Theorem to find $P(c)$ for $c = -2$ and $c = \dfrac{1}{2}$.

Algebraic Solution

Perform synthetic division with $c = -2$ and $c = \dfrac{1}{2}$ and examine the remainders.

$$
\begin{array}{r|rrrr}
-2 & 2 & 3 & 2 & -2 \\
   &   & -4 & 2 & -8 \\
\hline
   & 2 & -1 & 4 & -10
\end{array}
$$

The remainder is -10. Therefore, $P(-2) = -10$.

$$
\begin{array}{r|rrrr}
\dfrac{1}{2} & 2 & 3 & 2 & -2 \\
   &   & 1 & 2 & 2 \\
\hline
   & 2 & 4 & 4 & 0
\end{array}
$$

The remainder is 0. Therefore, $P\left(\dfrac{1}{2}\right) = 0$.

Visualize the Solution

A graph of P shows that the points $(-2, -10)$ and $\left(\dfrac{1}{2}, 0\right)$ are on the graph.

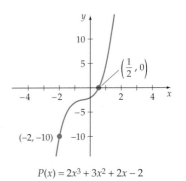

$P(x) = 2x^3 + 3x^2 + 2x - 2$

▶ **TRY EXERCISE 26, PAGE 257**

Using the Remainder Theorem to evaluate a polynomial function is often faster than evaluating the polynomial function by direct substitution. For instance, to evaluate $P(x) = x^5 - 10x^4 + 35x^3 - 50x^2 + 24x$ by substituting 7 for x, we must do the following work.

$$
\begin{aligned}
P(7) &= (7)^5 - 10(7)^4 + 35(7)^3 - 50(7)^2 + 24(7) \\
&= 16{,}807 - 10(2401) + 35(343) - 50(49) + 24(7) \\
&= 16{,}807 - 24{,}010 + 12{,}005 - 2450 + 168 \\
&= 2520
\end{aligned}
$$

> **take note**
>
> Because $P(x)$ has a constant term of 0, we must include 0 as the last number in the first row of the synthetic division at the right.

Using the Remainder Theorem to perform the above evaluation requires only the following work.

$$
\begin{array}{r|rrrrrr}
7 & 1 & -10 & 35 & -50 & 24 & 0 \\
  &   & 7 & -21 & 98 & 336 & 2520 \\
\hline
  & 1 & -3 & 14 & 48 & 360 & 2520 \longleftarrow P(7)
\end{array}
$$

● THE FACTOR THEOREM

Note from Example 3 that $P\left(\dfrac{1}{2}\right) = 0$. Recall that $\dfrac{1}{2}$ is a zero of P because $P(x) = 0$ when $x = \dfrac{1}{2}$.

The following theorem is a direct result of the Remainder Theorem. It points out the important relationship between a zero of a given polynomial function and a factor of the polynomial function.

The Factor Theorem

A polynomial function $P(x)$ has a factor $(x - c)$ if and only if $P(c) = 0$. That is, $(x - c)$ is a factor of $P(x)$ if and only if c is a zero of P.

EXAMPLE 4   Apply the Factor Theorem

Use synthetic division and the Factor Theorem to determine whether $(x + 5)$ or $(x - 2)$ is a factor of $P(x) = x^4 + x^3 - 21x^2 - x + 20$.

Solution

$$\begin{array}{r|rrrrr} -5 & 1 & 1 & -21 & -1 & 20 \\ & & -5 & 20 & 5 & -20 \\ \hline & 1 & -4 & -1 & 4 & 0 \end{array}$$

The remainder of 0 indicates that $(x + 5)$ is a factor of $P(x)$.

$$\begin{array}{r|rrrrr} 2 & 1 & 1 & -21 & -1 & 20 \\ & & 2 & 6 & -30 & -62 \\ \hline & 1 & 3 & -15 & -31 & -42 \end{array}$$

The remainder of -42 indicates that $(x - 2)$ is not a factor of $P(x)$.

▶ **TRY EXERCISE 36, PAGE 257**

❓ **QUESTION** Is -5 a zero of $P(x)$ given in Example 4?

Here is a summary of the important role played by the remainder in the division of a polynomial by $(x - c)$.

The Remainder of a Polynomial Division

In the division of the polynomial function $P(x)$ by $(x - c)$, the remainder is

- equal to $P(c)$.

- 0 if and only if $(x - c)$ is a factor of P.

- 0 if and only if c is a zero of P.

Also, if c is a real number, then the remainder of $P(x) \div (x - c)$ is 0 if and only if $(c, 0)$ is an x-intercept of the graph of P.

❓ **ANSWER** Yes. Because $(x + 5)$ is a factor of $P(x)$, the Factor Theorem states that $P(-5) = 0$, and thus -5 is a zero of $P(x)$.

● REDUCED POLYNOMIALS

In Example 4 we determined that $(x + 5)$ is a factor of the polynomial function $P(x) = x^4 + x^3 - 21x^2 - x + 20$ and that the quotient of $x^4 + x^3 - 21x^2 - x + 20$ divided by $(x + 5)$ is $Q(x) = x^3 - 4x^2 - x + 4$. Thus

$$P(x) = (x + 5)(x^3 - 4x^2 - x + 4)$$

The quotient $Q(x) = x^3 - 4x^2 - x + 4$ is called a **reduced polynomial** or a **depressed polynomial** of $P(x)$ because it is a factor of $P(x)$ and its degree is 1 less than the degree of $P(x)$. Reduced polynomials will play an important role in Sections 3.3 and 3.4.

EXAMPLE 5   **Find a Reduced Polynomial**

Verify that $(x - 3)$ is a factor of $P(x) = 2x^3 - 3x^2 - 4x - 15$, and write $P(x)$ as the product of $(x - 3)$ and the reduced polynomial $Q(x)$.

Solution

$$
\begin{array}{r|rrrr}
3 & 2 & -3 & -4 & -15 \\
  &   & 6 & 9 & 15 \\
\hline
  & 2 & 3 & 5 & 0
\end{array}
$$

Coefficients of the
reduced polynomial $Q(x)$

Thus $(x - 3)$ and the reduced polynomial $2x^2 + 3x + 5$ are both factors of $P(x)$. That is,

$$P(x) = 2x^3 - 3x^2 - 4x - 15 = (x - 3)(2x^2 + 3x + 5)$$

▶ **TRY EXERCISE 56, PAGE 258**

  ### TOPICS FOR DISCUSSION

1. Explain the meaning of the phrase *zero of a polynomial.*

2. If $P(x)$ is a polynomial of degree 3, what is the degree of the quotient of $\dfrac{P(x)}{x - c}$?

3. Discuss how the Remainder Theorem can be used to determine whether a number is a zero of a polynomial.

4. A zero of $P(x) = x^3 - x^2 - 14x + 24$ is -4. Discuss how this information and the Factor Theorem can be used to solve $x^3 - x^2 - 14x + 24 = 0$.

5. Discuss the advantages and disadvantages of using synthetic division rather than substitution to evaluate a polynomial function at $x = c$.

EXERCISE SET 3.1

In Exercises 1 to 10, use long division to divide the first polynomial by the second.

1. $5x^3 + 6x^2 - 17x + 20,\ \ x + 3$

▶ **2.** $6x^3 + 15x^2 - 8x + 2,\ \ x + 4$

3. $x^4 - 5x^2 + 3x - 1,\ \ x - 2$

4. $x^4 - 5x^3 + x - 4,\ \ x - 1$

5. $x^2 + x^3 - 2x - 5,\ \ x - 3$

6. $4x + 3x^2 + x^3 - 5,\ \ x - 2$

7. $x^4 + 3x^3 - 5x + 3x^2 - 1,\ \ x - 4$

8. $2x^4 + x^3 - 5x^2 + 2x - 8,\ \ x + 4$

9. $x^5 + x^4 - 2x^3 + 2x^2 - 3x - 7,\ \ x - 1$

10. $x^5 - 2x^4 - x^3 + 3x^2 - 5x + 8,\ \ x + 4$

In Exercises 11 to 24, use synthetic division to divide the first polynomial by the second.

11. $4x^3 - 5x^2 + 6x - 7,\ \ x - 2$

▶ **12.** $5x^3 + 6x^2 - 8x + 1,\ \ x - 5$

13. $4x^3 - 2x + 3,\ \ x + 1$

14. $6x^3 - 4x^2 + 17,\ \ x + 3$

15. $x^5 - 10x^3 + 5x - 1,\ \ x - 4$

16. $6x^4 - 2x^3 - 3x^2 - x,\ \ x - 5$

17. $x^5 - 1,\ \ x - 1$

18. $x^4 + 1,\ \ x + 1$

19. $8x^3 - 4x^2 + 6x - 3,\ \ x - \dfrac{1}{2}$

20. $12x^3 + 5x^2 + 5x + 6,\ \ x + \dfrac{3}{4}$

21. $x^8 + x^6 + x^4 + x^2 + 4,\ \ x - 2$

22. $-x^7 - x^5 - x^3 - x - 5,\ \ x + 1$

23. $x^6 + x - 10,\ \ x + 3$

24. $2x^5 - 3x^4 - 5x^2 - 10,\ \ x - 4$

In Exercises 25 to 34, use the Remainder Theorem to find $P(c)$.

25. $P(x) = 3x^3 + x^2 + x - 5, c = 2$

▶ **26.** $P(x) = 2x^3 - x^2 + 3x - 1, c = 3$

27. $P(x) = 4x^4 - 6x^2 + 5, c = -2$

28. $P(x) = 6x^3 - x^2 + 4x, c = -3$

29. $P(x) = -2x^3 - 2x^2 - x - 20, c = 10$

30. $P(x) = -x^3 + 3x^2 + 5x + 30, c = 8$

31. $P(x) = -x^4 + 1, c = 3$

32. $P(x) = x^5 - 1, c = 1$

33. $P(x) = x^4 - 10x^3 + 2, c = 3$

34. $P(x) = x^5 + 20x^2 - 1, c = -5$

In Exercises 35 to 44, use synthetic division and the Factor Theorem to determine whether the given binomial is a factor of $P(x)$.

35. $P(x) = x^3 + 2x^2 - 5x - 6, x - 2$

▶ **36.** $P(x) = x^3 + 4x^2 - 27x - 90, x + 6$

37. $P(x) = 2x^3 + x^2 - 3x - 1, x + 1$

38. $P(x) = 3x^3 + 4x^2 - 27x - 36, x - 4$

39. $P(x) = x^4 - 25x^2 + 144, x + 3$

40. $P(x) = x^4 - 25x^2 + 144, x - 3$

41. $P(x) = x^5 + 2x^4 - 22x^3 - 50x^2 - 75x, x - 5$

42. $P(x) = 9x^4 - 6x^3 - 23x^2 - 4x + 4, x + 1$

43. $P(x) = 16x^4 - 8x^3 + 9x^2 + 14x + 4, x - \dfrac{1}{4}$

44. $P(x) = 10x^4 + 9x^3 - 4x^2 + 9x + 6, x + \dfrac{1}{2}$

In Exercises 45 to 54, use synthetic division to show that c is a zero of P(x).

45. $P(x) = 3x^3 - 8x^2 - 10x + 28, c = 2$

46. $P(x) = 4x^3 - 10x^2 - 8x + 6, c = 3$

47. $P(x) = x^4 - 1, c = 1$

48. $P(x) = x^3 + 8, c = -2$

49. $P(x) = 3x^4 + 8x^3 + 10x^2 + 2x - 20, c = -2$

50. $P(x) = x^4 - 2x^2 - 100x - 75, c = 5$

51. $P(x) = 2x^3 - 18x^2 - 50x + 66, c = 11$

52. $P(x) = 2x^4 - 34x^3 + 70x^2 - 153x + 45, c = 15$

53. $P(x) = 3x^2 - 8x + 4, c = \dfrac{2}{3}$

54. $P(x) = 5x^2 + 12x + 4, c = -\dfrac{2}{5}$

In Exercises 55 to 58, verify that the given binomial is a factor of P(x), and write P(x) as the product of the binomial and its reduced polynomial Q(x).

55. $P(x) = x^3 + x^2 + x - 14, x - 2$

▶ **56.** $P(x) = x^4 + 5x^3 + 3x^2 - 5x - 4, x + 1$

57. $P(x) = x^4 - x^3 - 9x^2 - 11x - 4, x - 4$

58. $P(x) = 2x^5 - x^4 - 7x^3 + x^2 + 7x - 10, x - 2$

59. COST OF A WEDDING The average cost of a wedding, in dollars, is modeled by

$$C(t) = 38t^2 + 291t + 15{,}208$$

where $t = 0$ represents the year 1990 and $0 \le t \le 12$. Use the Remainder Theorem to estimate the average cost of a wedding in

a. 1998.

b. 2001.

60. SELECTION OF BRIDESMAIDS A bride-to-be has several girlfriends, but she has decided to have only five bridesmaids, including the maid of honor. The number of different ways n girlfriends can be chosen and assigned a position, such as maid of honor, first matron, second matron, and so on, is given by the polynomial function

$$P(n) = n^5 - 10n^4 + 35n^3 - 50n^2 + 24n, \quad n \ge 5$$

a. Use the Remainder Theorem to determine the number of ways the bride can select her bridesmaids if she chooses from $n = 7$ girlfriends.

b. Evaluate $P(n)$ for $n = 7$ by substituting 7 for n. How does this result compare with the result obtained in part **a.**?

61. SELECTION OF CARDS The number of ways you can select three cards from a stack of n cards, in which the order of selection is important, is given by

$$P(n) = n^3 - 3n^2 + 2n, \quad n \ge 3$$

a. Use the Remainder Theorem to determine the number of ways you can select three cards from a stack of $n = 8$ cards.

b. Evaluate $P(n)$ for $n = 8$ by substituting 8 for n. How does this result compare with the result obtained in part **a.**?

62. ROCKET LAUNCH A model rocket is projected upward from an initial height of 4 feet with an initial velocity of 158 feet per second. The height of the rocket, in feet, is given by $s = -16t^2 + 158t + 4$, where $0 \le t \le 9.9$ seconds. Use the Remainder Theorem to determine the height of the rocket at

a. $t = 5$ seconds.

b. $t = 8$ seconds.

63. HOUSE OF CARDS The number of cards C needed to build a house of cards with r rows (levels) is given by the function $C(r) = 1.5r^2 + 0.5r$.

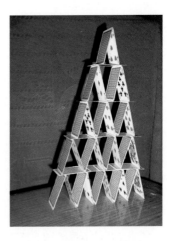

Use the Remainder Theorem to determine the number of cards needed to build a house of cards with

a. $r = 8$ rows.

b. $r = 20$ rows.

64. ELECTION OF CLASS OFFICERS The number of ways a class of n students can elect a president, a vice president, a secretary, and a treasurer is given by the function

$P(n) = n^4 - 6n^3 + 11n^2 - 6n$, where $n \geq 4$. Use the Remainder Theorem to determine the number of ways the class can elect officers if the class consists of

a. $n = 12$ students.

b. $n = 24$ students.

65. **POPULATION DENSITY OF A CITY** The population density D, in people per square mile, of a city is related to the distance x, in miles, from the center of the city by $D = -45x^2 + 190x + 200$, $0 < x < 5$. Use the Remainder Theorem to determine the population density of the city at a distance of

a. $x = 2$ miles.

b. $x = 4$ miles.

66. **VOLUME OF A SOLID** The volume of the solid at the right is given by $V(x) = x^3 + 3x^2$.

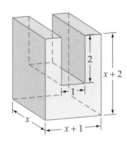

Use the Remainder Theorem to determine the volume of the solid if

a. $x = 7$ inches.

b. $x = 11$ inches.

67. **VOLUME OF A SOLID** The volume of the following solid is given by $V(x) = x^3 + x^2 + 10x - 8$.

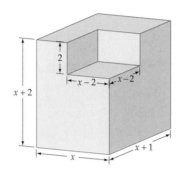

Use the Remainder Theorem to determine the volume of the solid if

a. $x = 6$ inches.

b. $x = 9$ inches.

CONNECTING CONCEPTS

68. Use the Factor Theorem to show that for any positive integer n, $P(x) = x^n - 1$ has $x - 1$ as a factor.

69. Find the remainder of $5x^{48} + 6x^{10} - 5x + 7$ divided by $x - 1$.

70. Find the remainder of $18x^{80} - 6x^{50} + 4x^{20} - 2$ divided by $x + 1$.

71. Determine whether i is a zero of $P(x) = x^3 - 3x^2 + x - 3$.

72. Determine whether $-2i$ is a zero of $P(x) = x^4 - 2x^3 + x^2 - 8x - 12$.

PREPARE FOR SECTION 3.2

73. Find the minimum value of $P(x) = x^2 - 4x + 6$. [2.4]

74. Find the maximum value of $P(x) = -2x^2 - x + 1$. [2.4]

75. Find the interval on which $P(x) = x^2 + 2x + 7$ is increasing. [2.4]

76. Find the interval on which $P(x) = -2x^2 + 4x + 5$ is decreasing. [2.4]

77. Factor: $x^4 - 5x^2 + 4$ [P.3]

78. Find the x-intercepts of the graph of $P(x) = 6x^2 - x - 2$. [2.4]

PROJECTS

I.   **HORNER'S POLYNOMIAL FORM** William Horner (1786–1837) devised a method of writing a polynomial in a form that does not involve any exponents other than 1. For instance, $4x^4 + 2x^3 - 5x^2 + 7x - 11$ can be written in each of the following forms.

$4x^4 + 2x^3 - 5x^2 + 7x - 11$

$= (4x^3 + 2x^2 - 5x + 7)x - 11$ • Factor an x from the first four terms.

$= [(4x^2 + 2x - 5)x + 7]x - 11$ • Factor an x from the first three terms inside the innermost parentheses.

$= \{[(4x + 2)x - 5]x + 7\}x - 11$ • Factor an x from the first two terms inside the innermost parentheses.

Horner's form, $\{[(4x + 2)x - 5]x + 7\}x - 11$, is easier to evaluate than the descending exponent form, $4x^4 + 2x^3 - 5x^2 + 7x - 11$. Horner's form is sometimes used by computer programmers to make their programs run faster.

a. Let $P(x) = 3x^5 - 4x^4 + 5x^3 - 2x^2 + 3x - 8$. Find $P(6)$ by direct substitution.

b. Use Horner's method to write $P(x)$ in a form that does not involve any exponents other than 1. Now use this form to evaluate $P(6)$. Which was easier to perform, the evaluation in part **a.** or in part **b.**?

SECTION 3.2

POLYNOMIAL FUNCTIONS OF HIGHER DEGREE

- **FAR-LEFT AND FAR-RIGHT BEHAVIOR**
- **MAXIMUM AND MINIMUM VALUES**
- **REAL ZEROS OF A POLYNOMIAL FUNCTION**
- **EVEN AND ODD POWERS OF $(x - c)$ THEOREM**
- **A PROCEDURE FOR GRAPHING POLYNOMIAL FUNCTIONS**

Table 3.1 summarizes information developed in Chapter 2 about graphs of polynomial functions of degree 0, 1, or 2.

TABLE 3.1

Polynomial Function $P(x)$	Graph
$P(x) = a$ (degree 0)	Horizontal line through $(0, a)$
$P(x) = ax + b$ (degree 1), $a \neq 0$	Line with y-intercept $(0, b)$ and slope a.
$P(x) = ax^2 + bx + c$ (degree 2), $a \neq 0$	Parabola with vertex $\left(-\dfrac{b}{2a}, P\left(-\dfrac{b}{2a}\right)\right)$

Polynomial functions of degree 3 or higher can be graphed by the technique of plotting points; however, some additional knowledge about polynomial functions will make graphing easier.

All polynomial functions have graphs that are smooth continuous curves. The terms *smooth* and *continuous* are defined rigorously in calculus, but for the

take note

The general form of a polynomial is given by $a_nx^n + a_{n-1}x^{n-1} + \cdots + a_0$. In this text the coefficients $a_n, a_{n-1}, \ldots, a_0$ are all real numbers unless specifically stated otherwise.

present, a smooth curve is a curve that does not have sharp corners such as that shown in **Figure 3.5a.** A continuous curve does not have a break or hole such as those shown in **Figure 3.5b.**

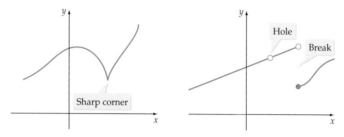

a. Continuous, but not smooth   **b.** Not continuous

FIGURE 3.5

● FAR-LEFT AND FAR-RIGHT BEHAVIOR

take note

The leading term of a polynomial function in x is the nonzero term that contains the largest power of x. The leading coefficient of a polynomial function is the coefficient of the leading term.

The graph of a polynomial function may have several up and down fluctuations; however, the graph of every polynomial function eventually will increase or decrease without bound as $|x|$ becomes large. The **leading term** a_nx^n is said to be the **dominate term** of the polynomial function $P(x) = a_nx^n + a_{n-1}x^{n-1} + \cdots + a_1x + a_0$ because as $|x|$ becomes large, the absolute value of a_nx^n will be much larger than the absolute value of any of the other terms. Because of this condition, you can determine the **far-left and far-right behavior** of the polynomial by examining the **leading coefficient** a_n and the degree n of the polynomial.

Table 3.2 indicates the far-left and far-right behavior of a polynomial function $P(x)$ with leading term a_nx^n.

TABLE 3.2   Far-Right and Far-Left Behavior of the Graph of a Polynomial Function with Leading Term a_nx^n

	n is even	*n* is odd
$a_n > 0$	Up to far left and up to far right	Down to far left and up to far right
$a_n < 0$	Down to far left and down to far right	Up to far left and down to far right

EXAMPLE 1   **Determine the Far-Left and Far-Right Behavior of a Polynomial Function**

Examine the leading term to determine the far-left and far-right behavior of the graph of each polynomial function.

a. $P(x) = x^3 - x$ b. $S(x) = \dfrac{1}{2}x^4 - \dfrac{5}{2}x^2 + 2$

c. $T(x) = -2x^3 + x^2 + 7x - 6$ d. $U(x) = 9 + 8x^2 - x^4$

Continued ▶

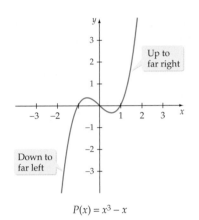

$P(x) = x^3 - x$

FIGURE 3.6

Solution

a. Because $a_n = 1$ is *positive* and $n = 3$ is *odd*, the graph of P goes down to its far left and up to its far right. See **Figure 3.6.**

b. Because $a_n = \dfrac{1}{2}$ is *positive* and $n = 4$ is *even*, the graph of S goes up to its far left and up to its far right. See **Figure 3.7.**

c. Because $a_n = -2$ is *negative* and $n = 3$ is *odd*, the graph of T goes up to its far left and down to its far right. See **Figure 3.8.**

d. The leading term of $U(x)$ is $-x^4$ and the leading coefficient is -1. Because $a_n = -1$ is *negative* and $n = 4$ is *even*, the graph of U goes down to its far left and down to its far right. See **Figure 3.9.**

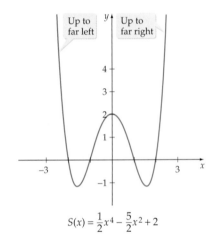

$S(x) = \dfrac{1}{2}x^4 - \dfrac{5}{2}x^2 + 2$

FIGURE 3.7

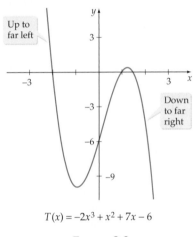

$T(x) = -2x^3 + x^2 + 7x - 6$

FIGURE 3.8

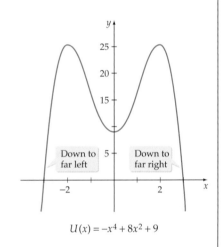

$U(x) = -x^4 + 8x^2 + 9$

FIGURE 3.9

▶ **TRY EXERCISE 2, PAGE 271**

● MAXIMUM AND MINIMUM VALUES

Figure 3.10 illustrates the graph of a polynomial function of degree 3 with two **turning points,** points at which the function changes from an increasing function to a decreasing function, or vice versa. In general, the graph of a polynomial function of degree n has at most $n - 1$ turning points.

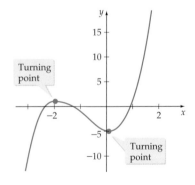

$P(x) = 2x^3 + 5x^2 - x - 5$

FIGURE 3.10

Turning points can be related to the concepts of maximum and minimum values of a function. These concepts were introduced in the discussion of graphs of second-degree equations in two variables earlier in the text. Recall that the minimum value of a function f is the smallest range value of f. It is often called the **absolute minimum**. The maximum value of a function f is the largest range value of f. The maximum value of a function is also called the **absolute maximum**. For the function whose graph is shown in **Figure 3.11**, the y value of point E is the absolute minimum. There are no y values less than y_5.

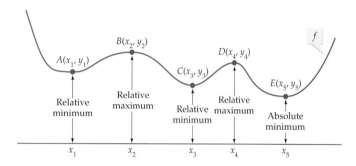

FIGURE 3.11

Now consider y_1, the y value of turning point A in **Figure 3.11**. It is not the smallest y value of every point on the graph of f; however, it is the smallest y value if we *localize* our field of view to a small open interval containing x_1. It is for this reason that we refer to y_1 as a **local minimum**, or **relative minimum**, of f. The y value of point C is also a relative minimum of f.

The function does not have an absolute maximum because it goes up both to its far left and to its far right. The y value of point B is a relative maximum, as is the y value of point D. The formal definitions of **relative maximum** and **relative minimum** are presented below.

Relative Minimum and Relative Maximum

If there is an open interval I containing c on which

• $f(c) \le f(x)$ for all x in I, then $f(c)$ is a **relative minimum** of f.

• $f(c) \ge f(x)$ for all x in I, then $f(c)$ is a **relative maximum** of f.

❓ QUESTION   Is the absolute minimum y_5 shown in **Figure 3.11** also a relative minimum of f?

❓ ANSWER   Yes, the absolute minimum y_5 also satisfies the requirements of a relative minimum.

INTEGRATING
TECHNOLOGY

A graphing utility can estimate the minimum and maximum values of a function. To use a TI-83 calculator to estimate the relative maximum of

$$P(x) = 0.3x^3 - 2.8x^2 + 6.4x + 2$$

use the following steps:

1. Enter the function in the $Y =$ menu. Choose your window settings.

2. Select 4:maximum from the $\boxed{\text{CALC}}$ menu, which is located above the $\boxed{\text{TRACE}}$ key. The graph of Y1 is displayed.

3. PRESS ◄ or ► repeatedly to select an x-value that is to the left of the relative maximum point. Press $\boxed{\text{ENTER}}$. A left bound is displayed in the bottom left corner.

4. Press ► repeatedly to select an x-value that is to the right of the relative maximum point. Press $\boxed{\text{ENTER}}$. A right bound is displayed in the bottom left corner.

5. The word **Guess?** is now displayed in the bottom left corner. Press ◄ repeatedly to move to a point near the maximum point. Press $\boxed{\text{ENTER}}$.

6. The cursor appears on the relative maximum point and the coordinates of the relative maximum point are displayed. In this example, the y value 6.312608 is the approximate relative maximum of the function P. *Note:* If your window settings, bounds, or your guess are different from those shown below, then your final results may differ slightly from the final results shown below in step 6.

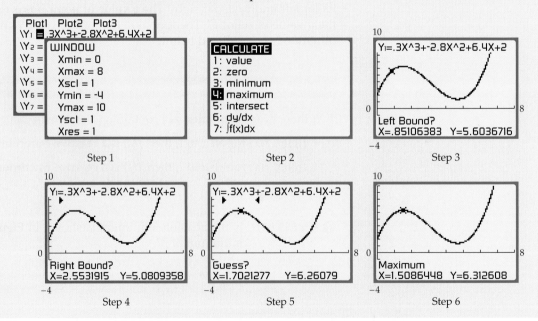

The following example illustrates the role a maximum may play in an application.

EXAMPLE 2   **Solve an Application**

A rectangular piece of cardboard measures 12 inches by 16 inches. An open box is formed by cutting congruent squares that measure x inches by x inches from each of the corners of the cardboard and folding up the sides as shown below.

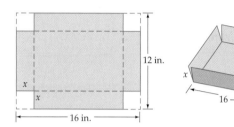

a.   Express the volume V of the box as a function of x.

b.   Determine (to the nearest tenth of an inch) the x value that maximizes the volume.

Solution

a.   The height, width, and length of the open box are x, $12 - 2x$, and $16 - 2x$. The volume is given by

$$V(x) = x(12 - 2x)(16 - 2x)$$
$$= 4x^3 - 56x^2 + 192x$$

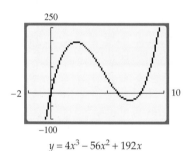

$$y = 4x^3 - 56x^2 + 192x$$

FIGURE 3.12

b.   Use a graphing utility to graph $y = V(x)$. The graph is shown in **Figure 3.12.**

Note that we are interested only in the part of the graph for which $0 < x < 6$. This is so because the length of each side of the box must be positive. In other words,

$$x > 0, \quad 12 - 2x > 0 \quad \text{and} \quad 16 - 2x > 0$$
$$x < 6 \qquad\qquad x < 8$$

The domain of V is the intersection of the solution sets of the three inequalities. Thus the domain is $\{x \,|\, 0 < x < 6\}$.

Now use a graphing utility to find that V attains its maximum of about 194.06736 when $x \approx 2.3$. See **Figure 3.13.**

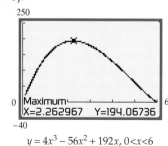

$$y = 4x^3 - 56x^2 + 192x, \, 0<x<6$$

FIGURE 3.13

▶ **TRY EXERCISE 48, PAGE 272**

INTEGRATING TECHNOLOGY

A TI graphing calculator program is available that simulates the construction of a box by cutting out squares from each corner of a rectangular piece of cardboard. This program, CUTOUT, can be found on our website at
math.college.hmco.com

● REAL ZEROS OF A POLYNOMIAL FUNCTION

Sometimes the real zeros of a polynomial function can be determined by using the factoring procedures developed in previous chapters. We illustrate this concept in the next example.

EXAMPLE 3   **Factor to Find the Real Zeros of a Polynomial Function**

Factor to find the three real zeros of $P(x) = x^3 + 3x^2 - 4x$.

Algebraic Solution

$P(x)$ can be factored as shown below.

$$P(x) = x^3 + 3x^2 - 4x$$
$$= x(x^2 + 3x - 4) \qquad \text{• Factor out the common factor } x.$$
$$= x(x - 1)(x + 4) \qquad \text{• Factor the trinomial } x^2 + 3x - 4.$$

The real zeros of $P(x)$ are $x = 0$, $x = 1$, and $x = -4$.

Visualize the Solution

The graph of $P(x)$ has x-intercepts at $(0, 0)$, $(1, 0)$, and $(-4, 0)$.

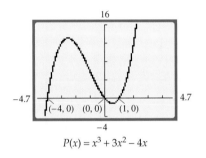

$$P(x) = x^3 + 3x^2 - 4x$$

▶ **TRY EXERCISE 22, PAGE 272**

The graph of every polynomial function P is a smooth continuous curve, and if the value of P changes sign on an interval, then $P(c)$ must equal zero for at least one real number c in the interval. This result is known as the *Zero Location Theorem*.

> **The Zero Location Theorem**
>
> Let $P(x)$ be a polynomial function and let a and b be two distinct real numbers. If $P(a)$ and $P(b)$ have opposite signs, then there is at least one real number c between a and b such that $P(c) = 0$.

For instance, if the value of P is negative at $x = a$ and positive at $x = b$, then there is at least one real number c between a and b such that $P(c) = 0$. See **Figure 3.14**.

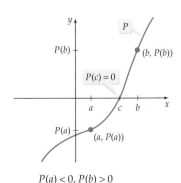

$$P(a) < 0, P(b) > 0$$

FIGURE 3.14

EXAMPLE 4   **Apply the Zero Location Theorem**

Use the Zero Location Theorem to verify that $S(x) = x^3 - x - 2$ has a real zero between 1 and 2.

Algebraic Solution

Use synthetic division to evaluate S for $x = 1$ and $x = 2$. If S changes sign between these two values, then S has a real zero between 1 and 2.

$$
\begin{array}{r|rrrr}
1 & 1 & 0 & -1 & -2 \\
  &   & 1 & 1 & 0 \\
\hline
  & 1 & 1 & 0 & -2
\end{array}
$$

• $S(1)$ is negative.

$$
\begin{array}{r|rrrr}
2 & 1 & 0 & -1 & -2 \\
  &   & 2 & 4 & 6 \\
\hline
  & 1 & 2 & 3 & 4
\end{array}
$$

• $S(2)$ is positive.

The graph of S is continuous because S is a polynomial function. Also, $S(1)$ is negative and $S(2)$ is positive. Thus the Zero Location Theorem indicates that there is a real zero between 1 and 2.

Visualize the Solution

The graph of S crosses the x-axis between $x = 1$ and $x = 2$. Thus S has a real zero between 1 and 2.

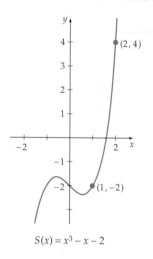

$S(x) = x^3 - x - 2$

▶ **TRY EXERCISE 28, PAGE 272**

The following theorem summarizes important relationships among the real zeros of a polynomial function, the x-intercepts of its graph, and its factors that can be written in the form $(x - c)$, where c is a real number.

Polynomial Functions, Real Zeros, Graphs, and Factors $(x - c)$

If P is a polynomial function and c is a real number, then all the following statements are equivalent in the sense that if any one statement is true, then they are all true, and if any one statement is false, then they are all false.

● $(x - c)$ is a factor of P.

● $x = c$ is a real solution of $P(x) = 0$.

● $x = c$ is a real zero of P.

● $(c, 0)$ is an x-intercept of the graph of $y = P(x)$.

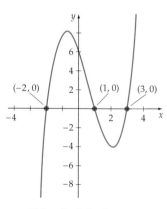

$S(x) = x^3 - 2x^2 - 5x + 6$

FIGURE 3.15

Sometimes it is possible to make use of the preceding theorem and a graph of a polynomial function to find factors of a function. For example, the graph of

$$S(x) = x^3 - 2x^2 - 5x + 6$$

is shown in **Figure 3.15.** The x-intercepts are $(-2, 0)$, $(1, 0)$, and $(3, 0)$. Hence -2, 1, and 3 are zeros of S, and $[x - (-2)]$, $(x - 1)$, and $(x - 3)$ are all factors of S.

● EVEN AND ODD POWERS OF $(x − c)$ THEOREM

Use a graphing utility to graph $P(x) = (x + 3)(x − 4)^2$. Compare your graph with **Figure 3.16.** Examine the graph near the x-intercepts $(−3, 0)$ and $(4, 0)$. Observe that the graph of P

- crosses the x-axis at $(−3, 0)$.

- intersects the x-axis but does not cross the x-axis at $(4, 0)$.

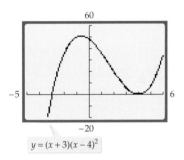

$$y = (x + 3)(x − 4)^2$$

FIGURE 3.16

The following theorem can be used to determine at which x-intercepts the graph of a polynomial function will cross the x-axis and at which x-intercepts the graph will intersect but not cross the x-axis.

Even and Odd Powers of $(x − c)$ Theorem

If c is a real number and the polynomial function $P(x)$ has $(x − c)$ as a factor exactly k times, then the graph of P will

- intersect but not cross the x-axis at $(c, 0)$, provided k is an even positive integer.

- cross the x-axis at $(c, 0)$, provided k is an odd positive integer.

EXAMPLE 5

Apply the Even and Odd Powers of $(x − c)$ Theorem

Determine where the graph of $P(x) = (x + 3)(x − 2)^2(x − 4)^3$ crosses the x-axis and where the graph intersects but does not cross the x-axis.

Solution

The exponents of the factors $(x + 3)$ and $(x − 4)$ are odd integers. Therefore, the graph of P will cross the x-axis at the x-intercepts $(−3, 0)$ and $(4, 0)$.

The exponent of the factor $(x − 2)$ is an even integer. Therefore, the graph of P will intersect but not cross the x-axis at $(2, 0)$.

Use a graphing utility to check these results.

▶ **TRY EXERCISE 34, PAGE 272**

● A Procedure for Graphing Polynomial Functions

You may find that you can sketch the graph of a polynomial function just by plotting several points; however, the following procedure will help you sketch the graph of many polynomial functions in an efficient manner.

A Procedure for Graphing Polynomial Functions

$$P(x) = a_n x^n + a_{n-1} x^{n-1} + \cdots + a_1 x + a_0, \quad a_n \neq 0$$

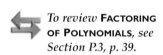

To review **Factoring of Polynomials,** *see Section P.3, p. 39.*

To graph P:

1. *Determine the far-left and the far-right behavior.* Examine the leading coefficient $a_n x^n$ to determine the far-left and the far-right behavior of the graph.

2. *Find the y-intercept.* Determine the y-intercept by evaluating $P(0)$.

3. *Find the x-intercept(s) and determine the behavior of the graph near the x-intercept(s).* If possible, find the x-intercepts by factoring. If $(x - c)$, where c is a real number, is a factor of P, then $(c, 0)$ is an x-intercept of the graph. Use the Even and Odd Powers of $(x - c)$ Theorem to determine where the graph crosses the x-axis and where the graph intersects but does not cross the x-axis.

4. *Find additional points on the graph.* Find a few additional points (in addition to the intercepts).

5. *Check for symmetry.*

    a.   The graph of an even function is symmetric with respect to the y-axis.

    b.   The graph of an odd function is symmetric with respect to the origin.

6. *Sketch the graph.* Use all the information obtained above to sketch the graph of the polynomial function. The graph should be a smooth continuous curve that passes through the points determined in steps 2 to 4. The graph should have a maximum of $n - 1$ turning points.

EXAMPLE 6   **Graph a Polynomial Function**

Sketch the graph of $P(x) = x^3 - 4x^2 + 4x$.

Solution

Step 1   *Determine the far-left and the far-right behavior.* The leading term is $1x^3$. Because the leading coefficient 1 is positive and the degree of the polynomial 3 is odd, the graph of P goes down to its far left and up to its far right.

Step 2   *Find the y-intercept.* $P(0) = 0^3 - 4(0)^2 + 4(0) = 0$. The y-intercept is $(0, 0)$.

Continued ▶

Step 3   ***Find the x-intercept(s) and determine the behavior of the graph near the x-intercept(s).*** Try to factor $x^3 - 4x^2 + 4x$.

$$x^3 - 4x^2 + 4x = x(x^2 - 4x + 4)$$
$$= x(x - 2)(x - 2)$$
$$= x(x - 2)^2$$

Because $(x - 2)$ is a factor of P, the point $(2, 0)$ is an x-intercept of the graph of P. Because x is a factor of P (think of x as $x - 0$), the point $(0, 0)$ is an x-intercept of the graph of P. Applying the Even and Odd Powers of $(x - c)$ Theorem allows us to determine that the graph of P crosses the x-axis at $(0, 0)$ and intersects but does not cross the x-axis at $(2, 0)$.

Step 4   ***Find additional points on the graph.***

x	P(x)
−1	−9
0.5	1.125
1	1
3	3

Step 5   ***Check for symmetry.*** The function P is not an even or an odd function, so the graph of P is *not* symmetric to either the y-axis or the origin.

Step 6   ***Sketch the graph.***

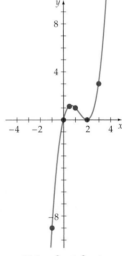

$P(x) = x^3 - 4x^2 + 4x$

 TRY EXERCISE 42, PAGE 272

TOPICS FOR DISCUSSION

1. Give an example of a polynomial function and of a function that is not a polynomial function.

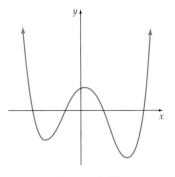

FIGURE 3.17

2. Is it possible for the graph of the polynomial function shown in **Figure 3.17** to be the graph of a polynomial function of odd degree? If so, explain how. If not, explain why not.

3. Explain the difference between a relative minimum and an absolute minimum.

4. Discuss how the Zero Location Theorem can be used to find a real zero of a polynomial function.

5. Let $P(x)$ be a polynomial function with real coefficients. Explain the relationships among the real zeros of the polynomial function, the x-coordinates of the x-intercepts of the graph of the polynomial function, and the solutions of the equation $P(x) = 0$.

EXERCISE SET 3.2

In Exercises 1 to 8, examine the leading term and determine the far-left and far-right behavior of the graph of the polynomial function.

1. $P(x) = 3x^4 - 2x^2 - 7x + 1$

▶ **2.** $P(x) = -2x^3 - 6x^2 + 5x - 1$

3. $P(x) = 5x^5 - 4x^3 - 17x^2 + 2$

4. $P(x) = -6x^4 - 3x^3 + 5x^2 - 2x + 5$

5. $P(x) = 2 - 3x - 4x^2$

6. $P(x) = -16 + x^4$

7. $P(x) = \dfrac{1}{2}(x^3 + 5x^2 - 2)$

8. $P(x) = -\dfrac{1}{4}(x^4 + 3x^2 - 2x + 6)$

9. The following graph is the graph of a third-degree (cubic) polynomial function. What does the far-left and far-right behavior of the graph say about the leading coefficient a?

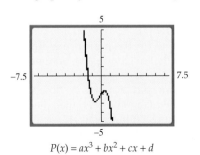

$P(x) = ax^3 + bx^2 + cx + d$

10. The following graph is the graph of a fourth-degree (quartic) polynomial function. What does the far-left and far-right behavior of the graph say about the leading coefficient a?

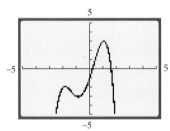

$P(x) = ax^4 + bx^3 + cx^2 + dx + e$

In Exercises 11 to 14, state the vertex of the graph of the function and use your knowledge of the vertex of a parabola to find the maximum or minimum of each function.

11. $P(x) = x^2 + 4x - 1$

12. $P(x) = x^2 + 6x + 1$

13. $P(x) = -x^2 - 8x + 1$

14. $P(x) = -2x^2 + 8x - 1$

In Exercises 15 to 20, use a graphing utility to graph each polynomial. Use the maximum and minimum features of the graphing utility to estimate, to the nearest tenth, the coordinates of the points where $P(x)$ has a relative maximum or a relative minimum. For each point, indicate whether the y value is a relative maximum or a relative minimum. The number in parentheses to the

right of the polynomial is the total number of relative maxima and minima.

15. $P(x) = x^3 + x^2 - 9x - 9$ (2)

16. $P(x) = x^3 + 4x^2 - 4x - 16$ (2)

17. $P(x) = x^3 - 3x^2 - 24x + 3$ (2)

18. $P(x) = -2x^3 - 3x^2 + 12x + 1$ (2)

19. $P(x) = x^4 - 4x^3 - 2x^2 + 12x - 5$ (3)

20. $P(x) = x^4 - 10x^2 + 9$ (3)

In Exercises 21 to 26, find the real zeros of each polynomial function by factoring. The number in parentheses to the right of each polynomial indicates the number of real zeros of the given polynomial function.

21. $P(x) = x^3 - 2x^2 - 15x$ (3)

▶ **22.** $P(x) = x^3 - 6x^2 + 8x$ (3)

23. $P(x) = x^4 - 13x^2 + 36$ (4)

24. $P(x) = 4x^4 - 37x^2 + 9$ (4)

25. $P(x) = x^5 - 5x^3 + 4x$ (5)

26. $P(x) = x^5 - 25x^3 + 144x$ (5)

In Exercises 27 to 32, use the Zero Location Theorem to verify that P has a zero between a and b.

27. $P(x) = 2x^3 + 3x^2 - 23x - 42;\quad a = 3, b = 4$

▶ **28.** $P(x) = 4x^3 - x^2 - 6x + 1;\quad a = 0, b = 1$

29. $P(x) = 3x^3 + 7x^2 + 3x + 7;\quad a = -3, b = -2$

30. $P(x) = 2x^3 - 21x^2 - 2x + 25;\quad a = 1, b = 2$

31. $P(x) = 4x^4 + 7x^3 - 11x^2 + 7x - 15;\quad a = 1, b = 1\frac{1}{2}$

32. $P(x) = 5x^3 - 16x^2 - 20x + 64;\quad a = 3, b = 3\frac{1}{2}$

In Exercises 33 to 40, determine the x-intercepts of the graph of P. For each x-intercept, use the Even and Odd Powers of $(x - c)$ Theorem to determine whether the graph of P crosses the x-axis or intersects but does not cross the x-axis.

33. $P(x) = (x - 1)(x + 1)(x - 3)$

▶ **34.** $P(x) = (x + 2)(x - 6)^2$

35. $P(x) = -(x - 3)^2(x - 7)^5$

36. $P(x) = (x + 2)^3(x - 6)^{10}$

37. $P(x) = (2x - 3)^4(x - 1)^{15}$

38. $P(x) = (5x + 10)^6(x - 2.7)^5$

39. $P(x) = x^3 - 6x^2 + 9x$

40. $P(x) = x^4 + 3x^3 + 4x^2$

In Exercises 41 to 46, sketch the graph of the polynomial function.

41. $P(x) = x^3 - x^2 - 2x$

▶ **42.** $P(x) = x^3 + 2x^2 - 3x$

43. $P(x) = -x^3 - 2x^2 + 5x + 6$ (*Hint:* In factored form $P(x) = (x + 3)(x + 1)(x - 2)$.)

44. $P(x) = -x^3 - 3x^2 + x + 3$ (*Hint:* In factored form $P(x) = (x + 3)(x + 1)(x - 1)$.)

45. $P(x) = x^4 - 4x^3 + 2x^2 + 4x - 3$ (*Hint:* In factored form $P(x) = (x + 1)(x - 1)^2(x - 3)$.)

46. $P(x) = x^4 - 6x^3 + 8x^2$

47. **CONSTRUCTION OF A BOX** A company constructs boxes from rectangular pieces of cardboard that measure 10 inches by 15 inches. An open box is formed by cutting squares that measure x inches by x inches from each corner of the cardboard and folding up the sides, as shown in the following figure.

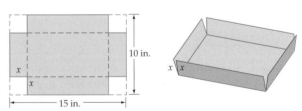

a. Express the volume V of the box as a function of x.

b. Determine (to the nearest hundredth of an inch) the x value that maximizes the volume of the box.

▶ **48.** **MAXIMIZING VOLUME** A closed box is to be constructed from a rectangular sheet of cardboard that measures 18 inches by 42 inches. The box is made by cutting rectangles that measure x inches by $2x$ inches from two of the corners and by cutting two squares that measure x inches by x inches from the top and from the bottom of the rectangle, as shown in the following figure. What value of x (to the nearest thousandth of an inch) will produce a box with maximum volume?

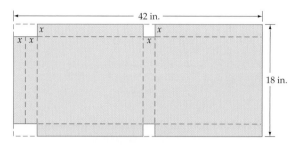

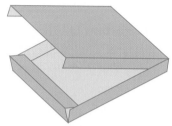

49. MAXIMIZING VOLUME An open box is to be constructed from a rectangular sheet of cardboard that measures 16 inches by 22 inches. To assemble the box, make the four cuts shown in the figure below and then fold on the dashed lines. What value of x (to the nearest thousandth of an inch) will produce a box with maximum volume?

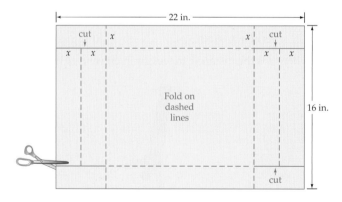

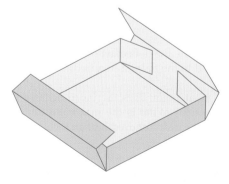

50. PROFIT A software company produces a computer game. The company has determined that its profit P, in dollars, from the manufacture and sale of x games is given by

$$P(x) = -0.000001x^3 + 96x - 98{,}000$$

where $0 < x \le 9000$.

a. What is the maximum profit, to the nearest thousand dollars, the company can expect from the sale of its games?

b. How many games, to the nearest unit, does the company need to produce and sell to obtain the maximum profit?

51. ADVERTISING EXPENSES A company manufactures digital cameras. The company estimates that the profit from camera sales is

$$P(x) = -0.02x^3 + 0.01x^2 + 1.2x - 1.1$$

where P is the profit in millions of dollars and x is the amount, in hundred-thousands of dollars, spent on advertising.

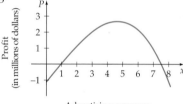

Advertising expenses
(in hundred-thousands of dollars)

Determine the amount, rounded to the nearest thousand dollars, the company needs to spend on advertising if it is to generate the maximum profit.

52. DIVORCE RATE The divorce rate for a given year is defined as the number of divorces per thousand population. The function

$$D(t) = 0.00001807t^4 - 0.001406t^3 + 0.02884t^2 - 0.003466t + 2.1148$$

approximates the U.S. divorce rate for the years 1960 ($t = 0$) to 1999 ($t = 39$). Use $D(t)$ and a graphing utility to estimate

a. the year during which the U.S. divorce rate reached its absolute maximum for the period from 1960 to 1999.

b. the absolute minimum divorce rate, rounded to the nearest 0.1, during the period from 1960 to 1999.

53. MARRIAGE RATE The marriage rate for a given year is defined as the number of marriages per thousand population. The function

$$M(t) = -0.00000115t^4 + 0.000252t^3 - 0.01827t^2 + 0.4438t + 9.1829$$

approximates the U.S. marriage rate for the years 1900 ($t = 0$) to 1999 ($t = 99$).

U.S. Marriage Rate, 1900–1999

Number of marriages per thousand population

Year (00 represents 1900)

Use $M(t)$ and a graphing utility to estimate

a. during what year the U.S. marriage rate reached its maximum for the period from 1900 to 1999.

b. the relative minimum marriage rate, rounded to the nearest 0.1, during the period from 1950 to 1970.

54. 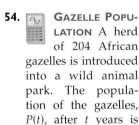 **GAZELLE POPU-LATION** A herd of 204 African gazelles is introduced into a wild animal park. The population of the gazelles, $P(t)$, after t years is given by $P(t) = -0.7t^3 + 18.7t^2 - 69.5t + 204$, where $0 < t \leq 18$.

a. Use a graph of P to determine the absolute minimum gazelle population (rounded to the nearest single gazelle) that is attained during this time period.

b. Use a graph of P to determine the absolute maximum gazelle population (rounded to the nearest single gazelle) that is attained during this time period.

55.  **MEDICATION LEVEL** Pseudoephedrine hydrochloride is an allergy medication. The function

$$L(t) = 0.03t^4 + 0.4t^3 - 7.3t^2 + 23.1t$$

where $0 \leq t \leq 5$, models the level of pseudoephedrine hydrochloride, in milligrams, in the bloodstream of a patient t hours after 30 milligrams of the medication have been taken.

a. Use a graphing utility and the function $L(t)$ to determine the maximum level of pseudoephedrine hydrochloride in the patient's bloodstream. Round your result to the nearest 0.01 milligram.

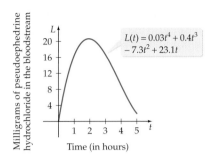

Milligrams of pseudoephedrine hydrochloride in the bloodstream

$L(t) = 0.03t^4 + 0.4t^3 - 7.3t^2 + 23.1t$

Time (in hours)

b. At what time t, to the nearest minute, is this maximum level of pseudoephedrine hydrochloride reached?

56. **SQUIRREL POPULATION** The population P of squirrels in a wilderness area is given by

$$P(t) = 0.6t^4 - 13.7t^3 + 104.5t^2 - 243.8t + 360,$$

where $0 \leq t \leq 12$ years.

a. What is the absolute minimum number of squirrels (rounded to the nearest single squirrel) attained on the interval $0 \leq t \leq 12$?

b. The absolute maximum of P is attained at the endpoint, where $t = 12$. What is this absolute maximum (rounded to the nearest single squirrel)?

57. **BEAM DEFLECTION** The deflection D, in feet, of an 8-foot beam that is center loaded is given by

$$D(x) = (-0.0025)(4x^3 - 3 \cdot 8x^2), \quad 0 < x \leq 4$$

where x is the distance, in feet, from one end of the beam.

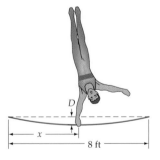

a. Determine the deflection of the beam when $x = 3$ feet. Round to the nearest hundredth of an inch.

b. At what point does the beam achieve its maximum deflection? What is the maximum deflection? Round to the nearest hundredth of an inch.

c. What is the deflection at $x = 5$ feet?

58. ENGINEERING A cylindrical log with a diameter of 22 inches is to be cut so that it will yield a beam that has a rectangular cross section of depth d and width w.

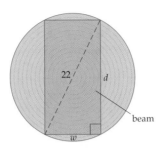

Cross section of a log

An engineer has determined that the stiffness S of the resulting beam is given by $S = 1.15wd^2$, where $0 < w < 22$ inches. Find the width and the depth that will maximize the stiffness of the beam. Round each result to the nearest hundredth of an inch. (*Hint:* Use the Pythagorean Theorem to solve for d^2 in terms of w^2.)

CONNECTING CONCEPTS

59. Use a graph of $P(x) = x^3 - x - 25$ to determine between which two consecutive integers P has a real zero.

60. Use a graph of the polynomial function $P(x) = 4x^4 - 12x^3 + 13x^2 - 12x + 9$ to determine between which two consecutive integers P has a real zero.

61. The point $(2, 0)$ is on the graph of $P(x)$. What point must be on the graph of $P(x - 3)$?

62. The point $(3, 5)$ is on the graph of $P(x)$. What point must be on the graph of $P(x + 1) - 2$?

63. Explain how to use the graph of $y = x^3$ to produce the graph of $P(x) = (x - 2)^3 + 1$.

64. Consider the following conjecture. Let $P(x)$ be a polynomial function. If a and b are real numbers such that $a < b$, $P(a) > 0$, and $P(b) > 0$, then $P(x)$ does not have a real zero between a and b. Is this conjecture true or false? Support your answer.

PREPARE FOR SECTION 3.3

65. Find the zeros of $P(x) = 6x^2 - 25x + 14$. [1.3/2.4]

66. Use synthetic division to divide $2x^3 + 3x^2 + 4x - 7$ by $x + 2$. [3.1]

67. Use synthetic division to divide $3x^4 - 21x^2 - 3x - 5$ by $x - 3$. [3.1]

68. List all natural numbers that are factors of 12. [P.1]

69. List all integers that are factors of 27. [P.1]

70. Given $P(x) = 4x^3 - 3x^2 - 2x + 5$, find $P(-x)$. [2.5]

PROJECTS

1. A student thinks that $P(n) = n^3 - n$ is always a multiple of 6 for all natural numbers n. What do you think? Provide a mathematical argument to show that the student is correct or a counterexample to show that the student is wrong.

ZEROS OF POLYNOMIAL FUNCTIONS

● MULTIPLE ZEROS OF A POLYNOMIAL FUNCTION

Recall that if $P(x)$ is a polynomial function, then the values of x for which $P(x)$ is equal to 0 are called the *zeros* of $P(x)$ or the **roots** of the equation $P(x) = 0$. A zero of a polynomial function may be a **multiple zero**. For example, $P(x) = x^2 + 6x + 9$ can be expressed in factored form as $(x + 3)(x + 3)$. Setting each factor equal to zero yields $x = -3$ in both cases. Thus $P(x) = x^2 + 6x + 9$ has a zero of -3 that occurs twice. The following definition will be most useful when we are discussing multiple zeros.

Definition of Multiple Zeros of a Polynomial Function

If a polynomial function $P(x)$ has $(x - r)$ as a factor exactly k times, then r is a **zero of multiplicity** k of the polynomial function $P(x)$.

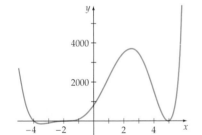

$P(x) = (x - 5)^2(x + 2)^3(x + 4)$

FIGURE 3.18

The graph of the polynomial function

$$P(x) = (x - 5)^2(x + 2)^3(x + 4)$$

is shown in **Figure 3.18.** This polynomial function has

- 5 as a zero of multiplicity 2.

- -2 as a zero of multiplicity 3.

- -4 as a zero of multiplicity 1.

A zero of multiplicity 1 is generally referred to as a **simple zero.**

When searching for the zeros of a polynomial function, it is important that we know how many zeros to expect. This question is answered completely in Section 3.4. For the work in this section, the following result is valuable.

Number of Zeros of a Polynomial Function

A polynomial function P of degree n has at most n zeros, where each zero of multiplicity k is counted k times.

● THE RATIONAL ZERO THEOREM

The rational zeros of polynomial functions with integer coefficients can be found with the aid of the following theorem.

The Rational Zero Theorem

If $P(x) = a_n x^n + a_{n-1} x^{n-1} + \cdots + a_1 x + a_0$ has *integer* coefficients ($a_n \neq 0$) and $\dfrac{p}{q}$ is a rational zero (in lowest terms) of P, then

- p is a factor of the constant term a_0 and
- q is a factor of the leading coefficient a_n.

take note

The Rational Zero Theorem is one of the most important theorems of this chapter. It enables us to narrow the search for rational zeros to a finite list.

The Rational Zero Theorem often is used to make a list of all possible rational zeros of a polynomial function. The list consists of all rational numbers of the form $\dfrac{p}{q}$, where p is an integer factor of the constant term a_0 and q is an integer factor of the leading coefficient a_n.

EXAMPLE 1 **Apply the Rational Zero Theorem**

Use the Rational Zero Theorem to list all possible rational zeros of

$$P(x) = 4x^4 + x^3 - 40x^2 + 38x + 12$$

Solution

List all integers p that are factors of 12 and all integers q that are factors of 4.

$$p: \quad \pm 1, \pm 2, \pm 3, \pm 4, \pm 6, \pm 12$$
$$q: \quad \pm 1, \pm 2, \pm 4$$

Form all possible rational numbers using ± 1, ± 2, ± 3, ± 4, ± 6, and ± 12 for the numerator and ± 1, ± 2, and ± 4 for the denominator. By the Rational Zero Theorem, the possible rational zeros are

$$\pm 1, \pm \frac{1}{2}, \pm \frac{1}{4}, \pm 2, \pm 3, \pm \frac{3}{2}, \pm \frac{3}{4}, \pm 4, \pm 6, \pm 12$$

It is not necessary to list a factor that is already listed in reduced form. For example, $\pm \dfrac{6}{4}$ is not listed because it is equal to $\pm \dfrac{3}{2}$.

▶ **TRY EXERCISE 10, PAGE 286**

take note

The Rational Zero Theorem gives the *possible* rational zeros of a polynomial function. That is, if P has a rational zero, then it must be one indicated by the theorem. However, P may not have any rational zeros. In the case of the polynomial function in Example 1, the only rational zeros are $-\dfrac{1}{4}$ and 2. The remaining rational numbers in the list are not zeros of P.

❓ QUESTION If $P(x) = a_n x^n + a_{n-1} x^{n-1} + \cdots + a_1 x + a_0$ has integer coefficients and a leading coefficient of $a_n = 1$, must all the rational zeros of P be integers?

❓ ANSWER Yes. By the Rational Zero Theorem, the rational zeros of P are of the form $\dfrac{p}{q}$, where p is an integer factor of a_0 and q is an integer factor of a_n. Thus $q = \pm 1$ and $\dfrac{p}{q} = \dfrac{p}{\pm 1} = \pm p$.

● UPPER AND LOWER BOUNDS FOR REAL ZEROS

A real number b is called an **upper bound** of the zeros of the polynomial function P if no zero is greater than b. A real number b is called a **lower bound** of the zeros of P if no zero is less than b. The following theorem is often used to find positive upper bounds and negative lower bounds for the real zeros of a polynomial function.

Upper- and Lower-Bound Theorem

Let $P(x)$ be a polynomial function with real coefficients. Use synthetic division to divide $P(x)$ by $x - b$, where b is a nonzero real number.

Upper bound   **a.** If $b > 0$ and the leading coefficient of P is positive, then b is an upper bound for the real zeros of P provided none of the numbers in the bottom row of the synthetic division are negative.

b. If $b > 0$ and the leading coefficient of P is negative, then b is an upper bound for the real zeros of P provided none of the numbers in the bottom row of the synthetic division are positive.

Lower bound   If $b < 0$ and the numbers in the bottom row of the synthetic division alternate in sign (the number zero can be considered positive or negative as needed to produce an alternating sign pattern), then b is a lower bound for the real zeros of P.

Upper and lower bounds are not unique. For example, if b is an upper bound for the real zeros of P, then any number greater than b is also an upper bound. Likewise, if a is a lower bound for the real zeros of P, then any number less than a is also a lower bound.

EXAMPLE 2   **Find Upper and Lower Bounds**

According to the Upper- and Lower-Bound Theorem, what is the smallest positive integer that is an upper bound and the largest negative integer that is a lower bound of the real zeros of $P(x) = 2x^3 + 7x^2 - 4x - 14$?

Solution

To find the smallest positive-integer upper bound, use synthetic division with $1, 2, \ldots$, as test values.

$$
\begin{array}{r|rrrr}
1 & 2 & 7 & -4 & -14 \\
  &   & 2 & 9 & 5 \\
\hline
  & 2 & 9 & 5 & -9
\end{array}
\qquad
\begin{array}{r|rrrr}
2 & 2 & 7 & -4 & -14 \\
  &   & 4 & 22 & 36 \\
\hline
  & 2 & 11 & 18 & 22
\end{array}
$$
• No negative numbers

Thus 2 is the smallest positive-integer upper bound.

take note

When you check for bounds, you do not need to limit your choices to the possible zeros given by the Rational Zero Theorem. For instance, in Example 2 the integer -4 is a lower bound; however, -4 is not one of the possible zeros of P as given by the Rational Zero Theorem.

Now find the largest negative-integer lower bound.

$$
\begin{array}{r|rrrr}
-1 & 2 & 7 & -4 & -14 \\
   &   & -2 & -5 & 9 \\
\hline
   & 2 & 5 & -9 & -5
\end{array}
\qquad
\begin{array}{r|rrrr}
-2 & 2 & 7 & -4 & -14 \\
   &   & -4 & -6 & 20 \\
\hline
   & 2 & 3 & -10 & 6
\end{array}
$$

$$
\begin{array}{r|rrrr}
-3 & 2 & 7 & -4 & -14 \\
   &   & -6 & -3 & 21 \\
\hline
   & 2 & 1 & -7 & 7
\end{array}
\qquad
\begin{array}{r|rrrr}
-4 & 2 & 7 & -4 & -14 \\
   &   & -8 & 4 & 0 \\
\hline
   & 2 & -1 & 0 & -14
\end{array}
$$

• **Alternating signs**

Thus -4 is the largest negative-integer lower bound.

▶ **TRY EXERCISE 18, PAGE 286**

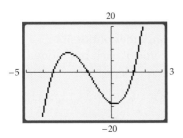

$P(x) = 2x^3 + 7x^2 - 4x - 14$

FIGURE 3.19

INTEGRATING TECHNOLOGY

You can use the Upper- and Lower-Bound Theorem to determine Xmin (the lower bound) and Xmax (the upper bound) for the viewing window of a graphing utility. This will ensure that all the real zeros, which are the x-coordinates of the x-intercepts of the polynomial function, will be shown. Note in **Figure 3.19** that the zeros of $P(x) = 2x^3 + 7x^2 - 4x - 14$ are between -4 (a lower bound) and 2 (an upper bound).

● DESCARTES' RULE OF SIGNS

MATH MATTERS

Descartes' Rule of Signs first appeared in his *La Géométrie* (1673). Although a proof of Descartes' Rule of Signs is beyond the scope of this course, we can see that a polynomial function with no variations in sign cannot have a positive zero. For instance, consider $P(x) = x^3 + x^2 + x + 1$. Each term of P is positive for any positive value of x. Thus P is never zero for $x > 0$.

Descartes' Rule of Signs is another theorem that is often used to obtain information about the zeros of a polynomial function. In Descartes' Rule of Signs, the number of **variations in sign** of the coefficients of $P(x)$ or $P(-x)$ refers to sign changes of the coefficients from positive to negative or from negative to positive that we find when we examine successive terms of the function. The terms are assumed to appear in order of descending powers of x. For example, the polynomial function

$$P(x) = +3x^4 - 5x^3 - 7x^2 + x - 7$$

has three variations in sign. The polynomial function

$$
\begin{aligned}
P(-x) &= +3(-x)^4 - 5(-x)^3 - 7(-x)^2 + (-x) - 7 \\
&= +\ 3x^4\ +\ 5x^3\ -\ 7x^2\ -\ x\ -\ 7
\end{aligned}
$$

has one variation in sign.

Terms that have a coefficient of 0 are not counted as variations in sign and may be ignored. For example,

$$P(x) = -x^5 + 4x^2 + 1$$

has one variation in sign.

Descartes' Rule of Signs

Let $P(x)$ be a polynomial function with real coefficients and with the terms arranged in order of decreasing powers of x.

1. The number of positive real zeros of $P(x)$ is equal to the number of variations in sign of $P(x)$, or to that number decreased by an even integer.

2. The number of negative real zeros of $P(x)$ is equal to the number of variations in sign of $P(-x)$, or to that number decreased by an even integer.

EXAMPLE 3   **Apply Descartes' Rule of Signs**

Use Descartes' Rule of Signs to determine both the number of possible positive and the number of possible negative real zeros of each polynomial function.

a. $P(x) = x^4 - 5x^3 + 5x^2 + 5x - 6$ **b.** $P(x) = 2x^5 + 3x^3 + 5x^2 + 8x + 7$

Solution

a.
$$P(x) = +x^4 - 5x^3 + 5x^2 + 5x - 6$$
$$\underbrace{\qquad}_{1}\ \underbrace{\qquad}_{2}\qquad\underbrace{\qquad}_{3}$$

There are three variations in sign. By Descartes' Rule of Signs, there are either three or one positive real zeros. Now examine the variations in sign of $P(-x)$.

$$P(-x) = x^4 + 5x^3 + 5x^2 - 5x - 6$$
$$\underbrace{\qquad}_{1}$$

There is one variation in sign of $P(-x)$. By Descartes' Rule of Signs, there is one negative real zero.

b. $P(x) = 2x^5 + 3x^3 + 5x^2 + 8x + 7$ has no variation in sign, so there are no positive real zeros.

$$P(-x) = -2x^5 - 3x^3 + 5x^2 - 8x + 7$$
$$\underbrace{\qquad}_{1}\ \underbrace{\qquad}_{2}\ \underbrace{\qquad}_{3}$$

$P(-x)$ has three variations in sign, so there are either three or one negative real zeros.

▶ **TRY EXERCISE 28, PAGE 286**

❓ **QUESTION** If $P(x) = ax^2 + bx + c$ has two variations in sign, must $P(x)$ have two positive real zeros?

❓ **ANSWER** No. According to Descartes' Rule of Signs, $P(x)$ will have either two positive real zeros or no positive real zeros.

In applying Descartes' Rule of Signs, we count each zero of multiplicity k as k zeros. For instance,

$$P(x) = x^2 - 10x + 25$$

has two variations in sign. Thus, by Descartes' Rule of Signs, $P(x)$ must have either two or no positive real zeros. Factoring $P(x)$ produces $(x - 5)^2$, from which it can be observed that 5 is a positive zero of multiplicity 2.

● ZEROS OF A POLYNOMIAL FUNCTION

Guidelines for Finding the Zeros of a Polynomial Function with Integer Coefficients

1. *Gather general information.* Determine the degree n of the polynomial function. The number of distinct zeros of the polynomial function is at most n. Apply Descartes' Rule of Signs to find the possible number of positive zeros and also the possible number of negative zeros.

2. *Check suspects.* Apply the Rational Zero Theorem to list rational numbers that are possible zeros. Use synthetic division to test numbers in your list. If you find an upper or a lower bound, then eliminate from your list any number that is greater than the upper bound or less than the lower bound.

3. *Work with the reduced polynomials.* Each time a zero is found, you obtain a reduced polynomial.

   ● If a reduced polynomial is of degree 2, find its zeros either by factoring or by applying the quadratic formula.

   ● If the degree of a reduced polynomial is 3 or greater, repeat the above steps for this polynomial.

Example 4 illustrates the procedure discussed in the above guidelines.

EXAMPLE 4   Find the Zeros of a Polynomial Function

Find the zeros of $P(x) = 3x^4 + 23x^3 + 56x^2 + 52x + 16$.

Solution

1. *Gather general information.* The degree of P is 4. Thus the number of zeros of P is at most 4. By Descartes' Rule of Signs, there are no positive zeros, and there are either four, two, or no negative zeros.

2. *Check suspects.* By the Rational Zero Theorem, the possible negative rational zeros of P are

$$\frac{p}{q}: \quad -1, -2, -4, -8, -16, -\frac{1}{3}, -\frac{2}{3}, -\frac{4}{3}, -\frac{8}{3}, -\frac{16}{3}$$

Continued ▶

INTEGRATING TECHNOLOGY

If you have a graphing utility, you can produce a graph similar to the one below. By looking at the x-intercepts of the graph, you can reject as possible zeros some of the values suggested by the Rational Zero Theorem. This will reduce the amount of work that is necessary to find the zeros of the polynomial function.

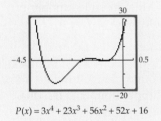

$$P(x) = 3x^4 + 23x^3 + 56x^2 + 52x + 16$$

Use synthetic division to test the possible rational zeros. The following work shows that -4 is a zero of P.

$$
\begin{array}{r|rrrrr}
-4 & 3 & 23 & 56 & 52 & 16 \\
   &   & -12 & -44 & -48 & -16 \\
\hline
   & 3 & 11 & 12 & 4 & 0 \\
\end{array}
$$

Coefficients of the first reduced polynomial

3. *Work with the reduced polynomials.* Because -4 is a zero, $(x + 4)$ and the first reduced polynomial $(3x^3 + 11x^2 + 12x + 4)$ are both factors of P. Thus

$$P(x) = (x + 4)(3x^3 + 11x^2 + 12x + 4)$$

All remaining zeros of P must be zeros of $3x^3 + 11x^2 + 12x + 4$. The Rational Zero Theorem indicates that the only possible negative rational zeros of $3x^3 + 11x^2 + 12x + 4$ are

$$\frac{p}{q}: \quad -1, -2, -4, -\frac{1}{3}, -\frac{2}{3}, -\frac{4}{3}$$

Synthetic division is again used to test possible zeros.

$$
\begin{array}{r|rrrr}
-2 & 3 & 11 & 12 & 4 \\
   &   & -6 & -10 & -4 \\
\hline
   & 3 & 5 & 2 & 0 \\
\end{array}
$$

Coefficients of the second reduced polynomial

Because -2 is a zero, $(x + 2)$ is also a factor of P. Thus

$$P(x) = (x + 4)(x + 2)(3x^2 + 5x + 2)$$

The remaining zeros of P must be zeros of $3x^2 + 5x + 2$.

$$3x^2 + 5x + 2 = 0$$
$$(3x + 2)(x + 1) = 0$$
$$x = -\frac{2}{3} \quad \text{and} \quad x = -1$$

The zeros of $P(x) = 3x^4 + 23x^3 + 56x^2 + 52x + 16$ are $-4, -2, -\frac{2}{3}$, and -1.

▶ **TRY EXERCISE 38, PAGE 286**

● APPLICATIONS OF POLYNOMIAL FUNCTIONS

In the following example we make use of an upper bound to eliminate several of the possible zeros that are given by the Rational Zero Theorem.

> **EXAMPLE 5**   **Solve an Application**

Glasses can be stacked to form a triangular pyramid.

Level 1
Level 2
Level 3
Level 4
Level 5
Level 6

The total number of glasses in one of these pyramids is given by

$$T = \frac{1}{6}(k^3 + 3k^2 + 2k)$$

where k is the number of levels in the pyramid. If 220 glasses are used to form a triangular pyramid, how many levels are in the pyramid?

Solution

We need to solve $220 = \frac{1}{6}(k^3 + 3k^2 + 2k)$ for k. Multiplying each side of the equation by 6 produces $1320 = k^3 + 3k^2 + 2k$, which can be written as $k^3 + 3k^2 + 2k - 1320 = 0$. The number 1320 has many natural number divisors, but we can eliminate many of these by showing that 12 is an upper bound.

$$
\begin{array}{r|rrrr}
12 & 1 & 3 & 2 & -1320 \\
   &   & 12 & 180 & 2184 \\
\hline
   & 1 & 15 & 182 & 864
\end{array}
$$

No number in the bottom row is negative. Thus 12 is an upper bound.

The only natural number divisors of 1320 that are less than 12 are 1, 2, 3, 4, 5, 6, 8, 10, and 11. The following synthetic division shows that 10 is a zero of $k^3 + 3k^2 + 2k - 1320$.

$$
\begin{array}{r|rrrr}
10 & 1 & 3 & 2 & -1320 \\
   &   & 10 & 130 & 1320 \\
\hline
   & 1 & 13 & 132 & 0
\end{array}
$$

The pyramid has 10 levels. There is no need to seek additional solutions, because the number of levels is uniquely determined by the number of glasses.

▶ **TRY EXERCISE 72, PAGE 289**

---

take note

The reduced polynomial $k^2 + 13k + 132$ has zeros of $k = \dfrac{-13 \pm i\sqrt{359}}{2}$. These zeros are not solutions of this application because the number of levels must be a natural number.

The procedures developed in this section will not find all solutions of every polynomial equation. However, a graphing utility can be used to estimate the real solutions of any polynomial equation. In Example 6 we utilize a graphing utility to solve an application.

EXAMPLE 6 **Use a Graphing Utility to Solve an Application**

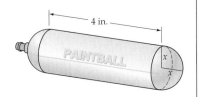

A CO_2 (carbon dioxide) cartridge for a paintball rifle has the shape of a right circular cylinder with a hemisphere at each end. The cylinder is 4 inches long, and the volume of the cartridge is 2π cubic inches (approximately 6.3 cubic inches). In the figure at the right, the common interior radius of the cylinder and the hemispheres is denoted by x. Use a graphing utility to estimate, to the nearest hundredth of an inch, the length of the radius x.

Solution

The volume of the cartridge is equal to the volume of the two hemispheres plus the volume of the cylinder. Recall that the volume of a sphere of radius x is given by $\frac{4}{3}\pi x^3$. Therefore, the volume of a hemisphere is $\frac{1}{2}\left(\frac{4}{3}\pi x^3\right)$.

The volume of a right circular cylinder is $\pi x^2 h$, where x is the radius of the base and h is the height of the cylinder. Thus the volume V of the cartridge is given by

$$V = \frac{1}{2}\left(\frac{4}{3}\pi x^3\right) + \frac{1}{2}\left(\frac{4}{3}\pi x^3\right) + \pi x^2 h$$

$$= \frac{4}{3}\pi x^3 + \pi x^2 h$$

Replacing V with 2π and h with 4 yields

$$2\pi = \frac{4}{3}\pi x^3 + 4\pi x^2$$

$$2 = \frac{4}{3}x^3 + 4x^2 \qquad \text{• Divide by } \pi.$$

$$3 = 2x^3 + 6x^2 \qquad \text{• Multiply by } \frac{3}{2}.$$

Here are two methods that can be used to solve

$$3 = 2x^3 + 6x^2 \qquad\qquad (1)$$

for x with the aid of a graphing utility.

1. **Intersection Method** Use a graphing utility to graph $y = 2x^3 + 6x^2$ and $y = 3$ on the same screen, with $x > 0$. The x-coordinate of the point of intersection of the two graphs is the desired solution. The graphs intersect at $x \approx 0.64$ inch. See the following figures.

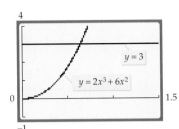
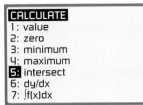
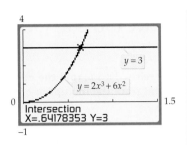

The length of the radius is approximately 0.64 inch.

2. **Intercept Method** Rewrite Equation (1) as $2x^3 + 6x^2 - 3 = 0$. Graph $y = 2x^3 + 6x^2 - 3$ with $x > 0$. Use a graphing utility to find the x-intercept of the graph. This method also shows that $x \approx 0.64$ inch.

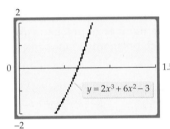
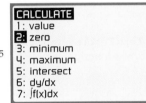
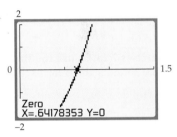

The length of the radius is approximately 0.64 inch.

▶ **TRY EXERCISE 68, PAGE 288**

 TOPICS FOR DISCUSSION

1. What is a multiple zero of a polynomial function? Give an example of a polynomial function that has -2 as a multiple zero.

2. Discuss how the Rational Zero Theorem is used.

3. Let $P(x)$ be a polynomial function with real coefficients. Explain why $(a, 0)$ is an x-intercept of the graph of $P(x)$ if a is a real zero of $P(x)$.

4. Let $P(x)$ be a polynomial function with integer coefficients. Suppose that the Rational Zero Theorem is applied to $P(x)$ and that after testing each possible rational zero, it is determined that $P(x)$ has no rational zeros. Does this mean that all of the zeros of $P(x)$ are irrational numbers?

EXERCISE SET 3.3

In Exercises 1 to 6, find the zeros of the polynomial function and state the multiplicity of each zero.

1. $P(x) = (x - 3)^2(x + 5)$

2. $P(x) = (x + 4)^3(x - 1)^2$

3. $P(x) = x^2(3x + 5)^2$

4. $P(x) = x^3(2x + 1)(3x - 12)^2$

5. $P(x) = (x^2 - 4)(x + 3)^2$

6. $P(x) = (x + 4)^3(x^2 - 9)^2$

In Exercises 7 to 16, use the Rational Zero Theorem to list possible rational zeros for each polynomial function.

7. $P(x) = x^3 + 3x^2 - 6x - 8$

8. $P(x) = x^3 - 19x - 30$

9. $P(x) = 2x^3 + x^2 - 25x + 12$

▶ **10.** $P(x) = 3x^3 + 11x^2 - 6x - 8$

11. $P(x) = 6x^4 + 23x^3 + 19x^2 - 8x - 4$

12. $P(x) = 2x^3 + 9x^2 - 2x - 9$

13. $P(x) = 4x^4 - 12x^3 - 3x^2 + 12x - 7$

14. $P(x) = x^5 - x^4 - 7x^3 + 7x^2 - 12x - 12$

15. $P(x) = x^5 - 32$

16. $P(x) = x^4 - 1$

In Exercises 17 to 26, find the smallest positive integer and the largest negative integer that, by the Upper- and Lower-Bound Theorem, are upper and lower bounds for the real zeros of each polynomial function.

17. $P(x) = x^3 + 3x^2 - 6x - 6$

▶ **18.** $P(x) = x^3 - 19x - 28$

19. $P(x) = 2x^3 + x^2 - 25x + 10$

20. $P(x) = 3x^3 + 11x^2 - 6x - 9$

21. $P(x) = 6x^4 + 23x^3 + 19x^2 - 8x - 4$

22. $P(x) = -2x^3 - 9x^2 + 2x + 9$

23. $P(x) = -4x^4 + 12x^3 + 3x^2 - 12x + 7$

24. $P(x) = x^5 - x^4 - 7x^3 + 7x^2 - 12x - 12$

25. $P(x) = x^5 - 32$

26. $P(x) = x^4 - 1$

In Exercises 27 to 36, use Descartes' Rule of Signs to state the number of possible positive and negative real zeros of each polynomial function.

27. $P(x) = x^3 + 3x^2 - 6x - 8$

▶ **28.** $P(x) = x^3 - 19x - 30$

29. $P(x) = 2x^3 + x^2 - 25x + 12$

30. $P(x) = 3x^3 + 11x^2 - 6x - 8$

31. $P(x) = 6x^4 + 23x^3 + 19x^2 - 8x - 4$

32. $P(x) = 2x^3 + 9x^2 - 2x - 9$

33. $P(x) = 4x^4 - 12x^3 - 3x^2 + 12x - 7$

34. $P(x) = x^5 - x^4 - 7x^3 + 7x^2 - 12x - 12$

35. $P(x) = x^5 - 32$

36. $P(x) = x^4 - 1$

In Exercises 37 to 58, find the zeros of each polynomial function. If a zero is a multiple zero, state its multiplicity.

37. $P(x) = x^3 + 3x^2 - 6x - 8$

▶ **38.** $P(x) = x^3 - 19x - 30$

39. $P(x) = 2x^3 + x^2 - 25x + 12$

40. $P(x) = 3x^3 + 11x^2 - 6x - 8$

41. $P(x) = 6x^4 + 23x^3 + 19x^2 - 8x - 4$

42. $P(x) = 2x^3 + 9x^2 - 2x - 9$

43. $P(x) = 2x^4 - 9x^3 - 2x^2 + 27x - 12$

44. $P(x) = 3x^3 - x^2 - 6x + 2$

45. $P(x) = x^3 - 8x^2 + 8x + 24$

46. $P(x) = x^3 - 7x^2 - 7x + 69$

47. $P(x) = 2x^4 - 19x^3 + 51x^2 - 31x + 5$

48. $P(x) = 4x^4 - 35x^3 + 71x^2 - 4x - 6$

49. $P(x) = 3x^6 - 10x^5 - 29x^4 + 34x^3 + 50x^2 - 24x - 24$

50. $P(x) = 2x^4 + 3x^3 - 4x^2 - 3x + 2$

51. $P(x) = x^3 - 3x - 2$

52. $P(x) = 3x^4 - 4x^3 - 11x^2 + 16x - 4$

53. $P(x) = x^4 - 5x^2 - 2x$

54. $P(x) = x^3 - 2x + 1$

55. $P(x) = x^4 + x^3 - 3x^2 - 5x - 2$

56. $P(x) = 6x^4 - 17x^3 - 11x^2 + 42x$

57. $P(x) = 2x^4 - 17x^3 + 4x^2 + 35x - 24$

58. $P(x) = x^5 + 5x^4 + 10x^3 + 10x^2 + 5x + 1$

59. FIND THE DIMENSIONS A cube measures n inches on each edge. If a slice 2 inches thick is cut from one face of the cube, the resulting solid has a volume of 567 cubic inches. Find n.

60. FIND THE DIMENSIONS A cube measures n units on each edge. If a slice 1 inch thick is cut from one face of the cube, and then a slice 3 inches thick is cut from another face of the cube as shown, the resulting solid has a volume of 1560 cubic inches. Find the dimensions of the original cube.

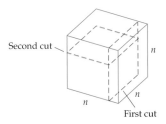

61. DIMENSIONS OF A SOLID For what value of x will the volume of the following solid be 112 cubic inches?

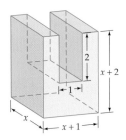

62. DIMENSIONS OF A BOX The length of a rectangular box is 1 inch more than twice the height of the box, and the width is 3 inches more than the height. If the volume of the box is 126 cubic inches, find the dimensions of the box.

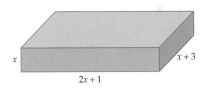

63. PIECES AND CUTS One straight cut through a thick piece of cheese produces two pieces. Two straight cuts can produce a maximum of four pieces. Three straight cuts can produce a maximum of eight pieces.

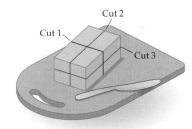

You might be inclined to think that every additional cut doubles the previous number of pieces. However, for four straight cuts, you get a maximum of 15 pieces. The maximum number of pieces P that can be produced by n straight cuts is given by

$$P(n) = \frac{n^3 + 5n + 6}{6}$$

a. Use the above function to determine the maximum number of pieces that can be produced by five straight cuts.

b. What is the fewest number of straight cuts that are needed to produce 64 pieces?

64. 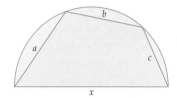 **INSCRIBED QUADRILATERAL** Isaac Newton discovered that if a quadrilateral with sides of lengths a, b, c, and x is inscribed in a semicircle with diameter x, then the lengths of the sides are related by the following equation.

$$x^3 - (a^2 + b^2 + c^2)x - 2abc = 0$$

Given $a = 6$, $b = 5$, and $c = 4$, find x. Round to the nearest hundredth.

65. **CANNONBALL STACKS** Cannonballs can be stacked to form a pyramid with a square base. The total number of cannonballs T in one of these square pyramids is

$$T = \frac{1}{6}(2n^3 + 3n^2 + n)$$

where n is the number of rows (levels). If 140 cannonballs are used to form a square pyramid, how many rows are in the pyramid?

66. 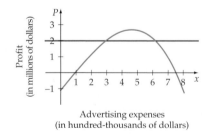 **ADVERTISING EXPENSES** A company manufactures digital cameras. The company estimates that the profit from camera sales is

$$P(x) = -0.02x^3 + 0.01x^2 + 1.2x - 1.1$$

where P is the profit in millions of dollars and x is the amount, in hundred-thousands of dollars, spent on advertising.

Determine the minimum amount, rounded to the nearest thousand dollars, the company needs to spend on advertising if it is to receive a profit of $2,000,000.

67. **COST CUTTING** At the present time, a nutrition bar in the shape of a rectangular solid measures 0.75 inch by 1 inch by 5 inches.

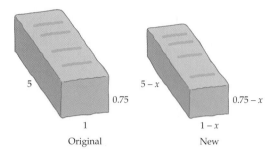

To reduce costs the manufacturer has decided to decrease each of the dimensions of the nutrition bar by x inches. What value of x, rounded to the nearest thousandth of an inch, will produce a new nutrition bar with a volume that is 0.75 cubic inch less than the present bar's volume?

▶ **68.** **PROPANE TANK DIMENSIONS** A propane tank has the shape of a circular cylinder with a hemisphere at each end. The cylinder is 6 feet long and the volume of the tank is 9π cubic feet. Find, to the nearest thousandth of a foot, the length of the radius x.

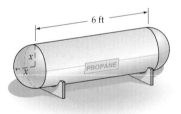

69. **DIVORCE RATE** The divorce rate for a given year is defined as the number of divorces per thousand population. The polynomial function

$$D(t) = 0.00001807t^4 - 0.001406t^3 + 0.02884t^2 - 0.003466t + 2.1148$$

approximates the U.S. divorce rate for the years 1960 ($t = 0$) to 1999 ($t = 39$). Use $D(t)$ and a graphing utility to determine during what years the U.S. divorce rate attained a level of 5.0.

70. **MEDICATION LEVEL** Pseudoephedrine hydrochloride is an allergy medication. The polynomial function

$$L(t) = 0.03t^4 + 0.4t^3 - 7.3t^2 + 23.1t$$

where $0 \le t \le 5$, models the level of pseudoephedrine hydrochloride, in milligrams, in the bloodstream of a patient t hours after 30 milligrams of the medication have been taken.

At what times, to the nearest minute, does the level of pseudoephedrine hydrochloride in the bloodstream reach 12 milligrams?

71. WEIGHT AND HEIGHT OF GIRAFFES A veterinarian at a wild animal park has determined that the average weight w, in pounds, of an adult male giraffe is closely approximated by the function

$$w = 8.3h^3 - 307.5h^2 + 3914h - 15{,}230$$

where h is the giraffe's height in feet, and $15 \leq h \leq 18$. Use the above function to estimate the height of a giraffe that weighs 3150 pounds. Round to the nearest tenth of a foot.

▶ **72.** SELECTION OF CARDS The number of ways one can select three cards from a group of n cards (the order of the selection matters), where $n \geq 3$, is given by $P(n) = n^3 - 3n^2 + 2n$. For a certain card trick a magician has determined that there are exactly 504 ways to choose three cards from a given group. How many cards are in the group?

73. DIGITS OF PI In 1999, Professor Yasumasa Kanada of the University of Tokyo used a supercomputer to compute 206,158,430,000 digits of pi (π). (*Source: Guinness World Records 2001*, Bantam Books, p. 252.) Computer scientists often try to find mathematical models that approximate the time a computer program takes to complete a calculation or mathematical procedure. Procedures for which the completion time can be closely modeled by a polynomial are called *polynomial time procedures*. Here is an example. A student finds that the time, in seconds, required to compute $n \times 10{,}000$ digits of pi on a personal computer using the mathematical program MAPLE is closely approximated by

$$T(n) = 0.23245n^3 + 0.53797n^2 + 7.88932n - 8.53299$$

a. Evaluate $T(n)$ to estimate how long, to the nearest second, the computer takes to compute 50,000 digits of pi.

b. About how many digits of pi can the computer compute in 5 minutes? Round to the nearest thousand digits.

CONNECTING CONCEPTS

74. If p is a prime number, prove that $\sqrt{p}$ is an irrational number. (*Hint:* Start with the equation $x = \sqrt{p}$, and square each side to produce the equivalent equation $x^2 = p$, which can be written as $x^2 - p = 0$. Then apply the Rational Zero Theorem to show that $P(x) = x^2 - p$ has no rational zeros.)

The mathematician **Augustin Louis Cauchy** (1789–1857) proved the following theorem, which can be used to quickly establish a bound B for *all* the zeros (both real and complex) of a given polynomial function.

Cauchy's Bound Theorem

Let $P(x) = a_nx^n + a_{n-1}x^{n-1} + \cdots + a_1 + a_0$ be a polynomial function with complex coefficients. The absolute value of each zero of P is less than

$$B = \left(\frac{\text{maximum of } (|a_{n-1}|, |a_{n-2}|, \cdots, |a_1|, |a_0|)}{|a_n|} + 1 \right)$$

In Exercises 75 to 78, a polynomial function and its zeros are given. For each polynomial function, apply Cauchy's Bound Theorem to determine the bound B for the polynomial and determine whether the absolute value of each of the given zeros is less than B. (*Hint:* $|a + bi| = \sqrt{a^2 + b^2}$)

75. $P(x) = 2x^3 - 5x^2 - 28x + 15$, zeros: $-3, \dfrac{1}{2}, 5$

76. $P(x) = x^3 - 5x^2 + 2x + 8$, zeros: $-1, 2, 4$

77. $P(x) = x^4 - 2x^3 + 9x^2 + 2x - 10$, zeros: $1 + 3i, 1 - 3i, 1, -1$

78. $P(x) = x^4 - 4x^3 + 14x^2 - 4x + 13$, zeros: $2 + 3i, 2 - 3i, i, -i$

PREPARE FOR SECTION 3.4
---

79. What is the conjugate of $3 - 2i$? [P.5]

80. What is the conjugate of $2 + i\sqrt{5}$? [P.5]

81. Find $(x - 1)(x - 3)(x - 4)$. [P.3]

82. Find $[x - (2 + i)][x - (2 - i)]$. [P.3/P.5]

83. Solve: $x^2 + 9 = 0$ [1.3]

84. Solve: $x^2 - x + 5 = 0$ [1.3]

PROJECTS

I. **RELATIONSHIPS BETWEEN ZEROS AND COEFFICIENTS**
Consider the polynomial function

$$P(x) = x^n + C_1 x^{n-1} + C_2 x^{n-2} + \cdots + C_n$$

with zeros $r_1, r_2, r_3, \ldots, r_n$. The following equations illustrate important relationships between the zeros of the polynomial function and the coefficients of the polynomial.

- The sum of the zeros.

$$r_1 + r_2 + r_3 + \cdots + r_{n-1} + r_n = -C_1$$

- The sum of the products of the zeros taken two at a time.

$$r_1 r_2 + r_1 r_3 + \cdots + r_{n-2} r_n + r_{n-1} r_n = C_2$$

- The sum of the products of the zeros taken three at a time.

$$r_1 r_2 r_3 + r_1 r_2 r_4 + \cdots + r_{n-2} r_{n-1} r_n = -C_3$$

$$\vdots$$

- The product of the zeros.

$$r_1 r_2 r_3 r_4 \cdots r_{n-1} r_n = (-1)^n C_n$$

a. Show that each of the previous equations holds true for the polynomial function

$$P(x) = x^3 - 6x^2 + 11x - 6$$

which has zeros of 1, 2, and 3.

b. Create a polynomial function of degree 4 with four real zeros. Illustrate that each of the above equations holds true for your polynomial function. (*Hint:* The polynomial function

$$P(x) = (x - a)(x - b)(x - c)(x - d)$$

has $a, b, c,$ and d as zeros.)

| SECTION 3.4 | THE FUNDAMENTAL THEOREM OF ALGEBRA |

- THE FUNDAMENTAL THEOREM OF ALGEBRA
- THE NUMBER OF ZEROS OF A POLYNOMIAL FUNCTION
- THE CONJUGATE PAIR THEOREM
- FIND A POLYNOMIAL FUNCTION WITH GIVEN ZEROS

• THE FUNDAMENTAL THEOREM OF ALGEBRA

The German mathematician Carl Friedrich Gauss (1777–1855) was the first to prove that every polynomial function has at least one complex zero. This concept is so basic to the study of algebra that it is called the **Fundamental Theorem of Algebra**. The proof of the Fundamental Theorem is beyond the scope of this text; however, it is important to understand the theorem and its consequences. As you consider each of the following theorems, keep in mind that the terms *complex coefficients* and *complex zeros* include real coefficients and real zeros because the set of real numbers is a subset of the set of complex numbers.

MATH MATTERS

Carl Friedrich Gauss (1777–1855) has often been referred to as the Prince of Mathematics. His work covered topics in algebra, calculus, analysis, probability, number theory, non-Euclidean geometry, astronomy, and physics, to name but a few. The following quote by Eric Temple Bell gives credence to the fact that Gauss was one of the greatest mathematicians of all time. "Archimedes, Newton, and Gauss, these three, are in a class by themselves among the great mathematicians, and it is not for ordinary mortals to attempt to range them in order of merit."*

*Men of Mathematics, by E. T. Bell, New York, Simon and Schuster, 1937.

The Fundamental Theorem of Algebra

If $P(x)$ is a polynomial function of degree $n \geq 1$ with complex coefficients, then $P(x)$ has at least one complex zero.

● THE NUMBER OF ZEROS OF A POLYNOMIAL FUNCTION

Let $P(x)$ be a polynomial function of degree $n \geq 1$ with complex coefficients. The Fundamental Theorem implies that $P(x)$ has a complex zero—say, c_1. The Factor Theorem implies that

$$P(x) = (x - c_1)Q(x)$$

where $Q(x)$ is a polynomial of degree one less than the degree of $P(x)$. Recall that the polynomial $Q(x)$ is called a *reduced polynomial*. Assuming that the degree of $Q(x)$ is 1 or more, the Fundamental Theorem implies that it also must have a zero. A continuation of this reasoning process leads to the following theorem.

The Linear Factor Theorem

If $P(x)$ is a polynomial function of degree $n \geq 1$ with leading coefficient $a_n \neq 0$,

$$P(x) = a_n x^n + a_{n-1} x^{n-1} + \cdots + a_1 x^1 + a_0$$

then $P(x)$ has exactly n linear factors

$$P(x) = a_n(x - c_1)(x - c_2) \cdots (x - c_n)$$

where $c_1, c_2, \ldots, c_n$ are complex numbers.

The following theorem follows directly from the Linear Factor Theorem.

The Number of Zeros of a Polynomial Function Theorem

If $P(x)$ is a polynomial function of degree $n \geq 1$, then $P(x)$ has exactly n complex zeros, provided each zero is counted according to its multiplicity.

The Linear Factor Theorem and the Number of Zeros of a Polynomial Function Theorem are referred to as **existence theorems**. They state that an nth degree polynomial will have n linear factors and n complex zeros, but they do not provide any information on how to determine the linear factors or the zeros. In Example 1 we make use of previously developed methods to actually find the linear factors and zeros of some polynomial functions.

EXAMPLE I | Find the Zeros and Linear Factors of a Polynomial Function

Find all the zeros of each of the following polynomial functions, and write each polynomial as a product of linear factors.

a. $P(x) = x^4 - 4x^3 + 8x^2 - 16x + 16$

b. $S(x) = x^4 - 6x^3 + 10x^2 + 2x - 15$

Solution

a.   We know that $P(x)$ will have four zeros and four linear factors. The possible rational zeros are $\pm 1, \pm 2, \pm 4, \pm 8, \pm 16$. Synthetic division can be used to show that 2 is a zero of multiplicity 2.

$$
\begin{array}{r|rrrrr}
2 & 1 & -4 & 8 & -16 & 16 \\
  &   & 2 & -4 & 8 & -16 \\
\hline
  & 1 & -2 & 4 & -8 & 0 \\
\end{array}
$$

$$
\begin{array}{r|rrrr}
2 & 1 & -2 & 4 & -8 \\
  &   & 2 & 0 & 8 \\
\hline
  & 1 & 0 & 4 & 0 \\
\end{array}
$$

To review CONCEPTS INVOLVING COMPLEX NUMBERS, see Section P.5, p. 60.

The final reduced polynomial is $x^2 + 4$. The zeros of $x^2 + 4$ can be found by solving $x^2 + 4 = 0$, as shown below.

$$x^2 + 4 = 0$$
$$x^2 = -4$$
$$x = \pm\sqrt{-4}$$
$$x = \pm 2i$$

Thus the four zeros of $P(x)$ are $2, 2, -2i,$ and $2i$. The linear factored form of $P(x)$ is

$$P(x) = (x - 2)(x - 2)[x - (-2i)][x - 2i]$$

or

$$P(x) = (x - 2)^2(x + 2i)(x - 2i)$$

b.   We know that $S(x)$ will have four zeros and four linear factors. The possible rational zeros are $\pm 1, \pm 3, \pm 5, \pm 15$. Synthetic division can be used to show that 3 and -1 are zeros of $S(x)$.

$$
\begin{array}{r|rrrrr}
3 & 1 & -6 & 10 & 2 & -15 \\
  &   & 3 & -9 & 3 & 15 \\
\hline
  & 1 & -3 & 1 & 5 & 0 \\
\end{array}
$$

$$
\begin{array}{r|rrrr}
-1 & 1 & -3 & 1 & 5 \\
   &   & -1 & 4 & -5 \\
\hline
   & 1 & -4 & 5 & 0 \\
\end{array}
$$

The final reduced polynomial is $x^2 - 4x + 5$. We can find the remaining zeros by using the quadratic formula to solve $x^2 - 4x + 5 = 0$.

$$x = \frac{-(-4) \pm \sqrt{(-4)^2 - 4(1)(5)}}{2(1)}$$

$$= \frac{4 \pm \sqrt{-4}}{2}$$

$$= 2 \pm i$$

Thus the four zeros of $S(x)$ are $3, -1, 2 + i$, and $2 - i$. The linear factored form of $S(x)$ is

$$S(x) = (x - 3)[x - (-1)][x - (2 + i)][x - (2 - i)]$$

or $S(x) = (x - 3)(x + 1)(x - 2 - i)(x - 2 + i)$

▶ **TRY EXERCISE 2, PAGE 297**

● THE CONJUGATE PAIR THEOREM

You may have noticed that the complex zeros of the polynomial function in Example 1 were complex conjugates. The following theorem shows that this is not a coincidence.

For a review of **COMPLEX CONJUGATES**, *see p. 63.*

The Conjugate Pair Theorem

If $a + bi$ $(b \neq 0)$ is a complex zero of a polynomial function *with real coefficients*, then the conjugate $a - bi$ is also a complex zero of the polynomial function.

EXAMPLE 2   Use the Conjugate Pair Theorem to Find Zeros

Find all the zeros of $P(x) = x^4 - 4x^3 + 14x^2 - 36x + 45$ given that $2 + i$ is a zero.

Solution

Because the coefficients are real numbers and $2 + i$ is a zero, the Conjugate Pair Theorem implies that $2 - i$ also must be a zero. Using synthetic division with $2 + i$ and then $2 - i$, we have

$$
\begin{array}{r|rrrrr}
2+i & 1 & -4 & 14 & -36 & 45 \\
 & & 2+i & -5 & 18+9i & -45 \\
\hline
 & 1 & -2+i & 9 & -18+9i & 0 \\
2-i & 1 & -2+i & 9 & -18+9i & \\
 & & 2-i & 0 & 18-9i & \\
\hline
 & 1 & 0 & 9 & 0 & \\
\end{array}
$$

• The coefficients of the reduced polynomial

• The coefficients of the next reduced polynomial

The resulting reduced polynomial is $x^2 + 9$, which has $3i$ and $-3i$ as zeros. Therefore, the four zeros of $x^4 - 4x^3 + 14x^2 - 36x + 45$ are $2 + i, 2 - i, 3i$, and $-3i$.

▶ **TRY EXERCISE 12, PAGE 297**

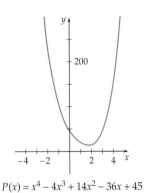

$P(x) = x^4 - 4x^3 + 14x^2 - 36x + 45$

FIGURE 3.20

A graph of $P(x) = x^4 - 4x^3 + 14x^2 - 36x + 45$ is shown in **Figure 3.20.** Because the polynomial in Example 2 is a fourth-degree polynomial and because we have verified that $P(x)$ has four imaginary solutions, it comes as no surprise that the graph does not intersect the x-axis.

When performing synthetic division with complex numbers, it is helpful to write the coefficients of the given polynomial as complex coefficients. For instance, -10 can be written as $-10 + 0i$. This technique is illustrated in the next example.

EXAMPLE 3   Apply the Conjugate Pair Theorem

Find all the zeros of $P(x) = x^5 - 10x^4 + 65x^3 - 184x^2 + 274x - 204$ given that $3 - 5i$ is a zero.

Solution

Because the coefficients are real numbers and $3 - 5i$ is a zero, $3 + 5i$ also must be a zero. Use synthetic division to produce

$3 - 5i$	1	$-10 + 0i$	$65 + 0i$	$-184 + 0i$	$274 + 0i$	-204
		$3 - 5i$	$-46 + 20i$	$157 - 35i$	$-256 + 30i$	204
$3 + 5i$	1	$-7 - 5i$	$19 + 20i$	$-27 - 35i$	$18 + 30i$	0
		$3 + 5i$	$-12 - 20i$	$21 + 35i$	$-18 - 30i$	
	1	-4	7	-6	0	

Descartes' Rule of Signs can be used to show that the reduced polynomial $x^3 - 4x^2 + 7x - 6$ has three or one positive zeros and no negative zeros. Using the Rational Zero Theorem, we have

$$\frac{p}{q} = 1, 2, 3, 6$$

Use synthetic division to determine that 2 is a zero.

2	1	-4	7	-6
		2	-4	6
	1	-2	3	0

Use the quadratic formula to solve $x^2 - 2x + 3 = 0$.

$$x = \frac{-(-2) \pm \sqrt{(-2)^2 - 4(1)(3)}}{2(1)} = \frac{2 \pm \sqrt{-8}}{2} = \frac{2 \pm 2\sqrt{2}i}{2} = 1 \pm \sqrt{2}i$$

The zeros of $P(x) = x^5 - 10x^4 + 65x^3 - 184x^2 + 274x - 204$ are $3 - 5i$, $3 + 5i, 2, 1 + \sqrt{2}i,$ and $1 - \sqrt{2}i.$

▶ **TRY EXERCISE 16, PAGE 297**

INTEGRATING TECHNOLOGY

Many graphing calculators can be used to do computations with complex numbers. The following TI-83 screen display shows that the product of $3 - 5i$ and $-7 - 5i$ is $-46 + 20i$. The i symbol is located above the decimal point key.

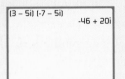

```
(3 - 5i) (-7 - 5i)
              -46 + 20i
```

❓ QUESTION   Is it possible for a third-degree polynomial function with real coefficients to have two real zeros and one complex zero?

❓ ANSWER   No. Because the coefficients of the polynomial are real numbers, the complex zeros of the polynomial function must occur as conjugate pairs.

Recall that the real zeros of a polynomial function P are the x-coordinates of the x-intercepts of the graph of P. This important connection between the real zeros of a polynomial function and the x-intercepts of the graph of the polynomial function is the basis for using a graphing utility to solve equations. Careful analysis of the graph of a polynomial function and your knowledge of the properties of polynomial functions can be used to solve many polynomial equations.

EXAMPLE 4   **Solve a Polynomial Equation**

Solve: $x^4 - 5x^3 + 4x^2 + 3x + 9 = 0$

Solution

Let $P(x) = x^4 - 5x^3 + 4x^2 + 3x + 9$. The x-intercepts of the graph of P are the real solutions of the equation. Use a graphing utility to graph P. See **Figure 3.21.**

From the graph, it appears that $(3, 0)$ is an x-intercept and the only x-intercept. Because the graph of P intersects but does not cross the x-axis at $(3, 0)$, we know that 3 is a multiple zero of P with an even multiplicity.

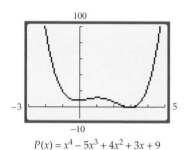

$P(x) = x^4 - 5x^3 + 4x^2 + 3x + 9$

FIGURE 3.21

$$
\begin{array}{r|rrrrr}
3 & 1 & -5 & 4 & 3 & 9 \\
  &   & 3 & -6 & -6 & -9 \\
\hline
  & 1 & -2 & -2 & -3 & 0
\end{array}
$$

• **Coefficients of P**

• The remainder is zero. Thus 3 is a zero.

By the Number of Zeros Theorem, there are three more zeros of P. Use synthetic division to show that 3 is also a zero of the reduced polynomial $x^3 - 2x^2 - 2x - 3$.

$$
\begin{array}{r|rrrr}
3 & 1 & -2 & -2 & -3 \\
  &   & 3 & 3 & 3 \\
\hline
  & 1 & 1 & 1 & 0
\end{array}
$$

• **Coefficients of reduced polynomial**

• The remainder is zero. Thus 3 is a zero of multiplicity 2.

We now have 3 as a double root of the original equation, and from the last line of the preceding synthetic division, the remaining solutions must be solutions of $x^2 + x + 1 = 0$. Use the quadratic formula to solve this equation.

$$
x = \frac{-1 \pm \sqrt{1^2 - 4(1)(1)}}{2(1)} = \frac{-1 \pm \sqrt{-3}}{2} = \frac{-1 \pm i\sqrt{3}}{2}
$$

The solutions of $x^4 - 5x^3 + 4x^2 + 3x + 9 = 0$ are $3, 3, -\dfrac{1}{2} + \dfrac{\sqrt{3}}{2}i$, and $-\dfrac{1}{2} - \dfrac{\sqrt{3}}{2}i$.

▶ **TRY EXERCISE 24, PAGE 297**

● FIND A POLYNOMIAL FUNCTION WITH GIVEN ZEROS

Many of the problems in this section and in Section 3.3 dealt with the process of finding the zeros of a given polynomial function. Example 5 considers the reverse process, finding a polynomial function when the zeros are given.

EXAMPLE 5   **Determine a Polynomial Function Given Its Zeros**

Find each polynomial function.

a. A polynomial function of degree 3 that has 1, 2, and -3 as zeros

b. A polynomial function of degree 4 that has real coefficients and zeros $2i$ and $3 - 7i$

Solution

a. Because 1, 2, and -3 are zeros, $(x - 1)$, $(x - 2)$, and $(x + 3)$ are factors. The product of these factors produces a polynomial function that has the indicated zeros.

$$P(x) = (x - 1)(x - 2)(x + 3) = (x^2 - 3x + 2)(x + 3) = x^3 - 7x + 6$$

b. By the Conjugate Pair Theorem, the polynomial function also must have $-2i$ and $3 + 7i$ as zeros. The product of the factors $x - 2i$, $x - (-2i)$, $x - (3 - 7i)$, and $x - (3 + 7i)$ produces the desired polynomial function.

$$P(x) = (x - 2i)(x + 2i)[x - (3 - 7i)][x - (3 + 7i)]$$
$$= (x^2 + 4)(x^2 - 6x + 58)$$
$$= x^4 - 6x^3 + 62x^2 - 24x + 232$$

▶ **TRY EXERCISE 42, PAGE 297**

A polynomial function that has a given set of zeros is not unique. For example, $P(x) = x^3 - 7x + 6$ has zeros 1, 2, and -3, but so does any nonzero multiple of $P(x)$, such as $S(x) = 2x^3 - 14x + 12$. This concept is illustrated in **Figure 3.22.** The graphs of the two polynomial functions are different, but they have the same x-intercepts.

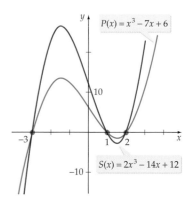

FIGURE 3.22

 TOPICS FOR DISCUSSION

1. What is the Fundamental Theorem of Algebra, and why is this theorem so important?

2. Let $P(x)$ be a polynomial function of degree n with real coefficients. Discuss the number of *possible* real zeros of this polynomial function. Include in your discussion the cases n is even and n is odd.

3. Consider the graph of the polynomial function in **Figure 3.23.** Is it possible that the degree of the polynomial is 3? Explain.

4. If two polynomial functions have exactly the same zeros, do the graphs of the polynomial functions look exactly the same?

5. Does the graph of every polynomial function have at least one x-intercept?

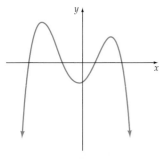

FIGURE 3.23

EXERCISE SET 3.4

In Exercises I to 10, find all the zeros of the polynomial function and write the polynomial as a product of linear factors. (*Hint:* First determine the rational zeros.)

1. $P(x) = x^4 + x^3 - 2x^2 + 4x - 24$

▶ 2. $P(x) = x^3 - 3x^2 + 7x - 5$

3. $P(x) = 2x^4 + x^3 + 39x^2 + 136x - 78$

4. $P(x) = x^3 - 13x^2 + 65x - 125$

5. $P(x) = x^5 - 9x^4 + 34x^3 - 58x^2 + 45x - 13$

6. $P(x) = x^4 - 4x^3 + 53x^2 - 196x + 196$

7. $P(x) = 2x^4 - x^3 - 15x^2 + 23x + 15$

8. $P(x) = 3x^4 - 17x^3 - 39x^2 + 337x + 116$

9. $P(x) = 2x^4 - 14x^3 + 33x^2 - 46x + 40$

10. $P(x) = 3x^4 - 10x^3 + 15x^2 + 20x - 8$

In Exercises II to 22, use the given zero to find the remaining zeros of each polynomial function.

11. $P(x) = 2x^3 - 5x^2 + 6x - 2; \quad 1 + i$

▶ 12. $P(x) = 3x^3 - 29x^2 + 92x + 34; \quad 5 + 3i$

13. $P(x) = x^3 + 3x^2 + x + 3; \quad -i$

14. $P(x) = x^4 - 6x^3 + 71x^2 - 146x + 530; \quad 2 + 7i$

15. $P(x) = x^4 - 4x^3 + 14x^2 - 4x + 13; \quad 2 - 3i$

▶ 16. $P(x) = x^5 - 6x^4 + 22x^3 - 64x^2 + 117x - 90; \quad 3i$

17. $P(x) = x^4 - 4x^3 + 19x^2 - 30x + 50; \quad 1 + 3i$

18. $P(x) = x^5 - x^4 - 4x^3 - 4x^2 - 5x - 3; \quad i$

19. $P(x) = x^5 - 3x^4 + 7x^3 - 13x^2 + 12x - 4; \quad -2i$

20. $P(x) = x^4 - 8x^3 + 18x^2 - 8x + 17; \quad i$

21. $P(x) = x^4 - 17x^3 + 112x^2 - 333x + 377; \quad 5 + 2i$

22. $P(x) = 2x^5 - 8x^4 + 61x^3 - 99x^2 + 12x + 182; \quad 1 - 5i$

In Exercises 23 to 30, use a graph and your knowledge of the zeros of polynomial functions to determine the *exact* values of all the solutions of each equation.

23. $2x^3 - x^2 + x - 6 = 0$

▶ 24. $4x^3 + 3x^2 + 16x + 12 = 0$

25. $24x^3 - 62x^2 - 7x + 30 = 0$

26. $12x^3 - 52x^2 + 27x + 28 = 0$

27. $x^4 - 4x^3 + 5x^2 - 4x + 4 = 0$

28. $x^4 + 4x^3 + 8x^2 + 16x + 16 = 0$

29. $x^4 + 4x^3 - 2x^2 - 12x + 9 = 0$

30. $x^4 + 3x^3 - 6x^2 - 28x - 24 = 0$

In Exercises 31 to 40, find a polynomial function of lowest degree with integer coefficients that has the given zeros.

31. $4, -3, 2$

32. $-1, 1, -5$

33. $3, 2i, -2i$

34. $0, i, -i$

35. $3 + i, 3 - i, 2 + 5i, 2 - 5i$

36. $2 + 3i, 2 - 3i, -5, 2$

37. $6 + 5i, 6 - 5i, 2, 3, 5$

38. $\dfrac{1}{2}, 4 - i, 4 + i$

39. $\dfrac{3}{4}, 2 + 7i, 2 - 7i$

40. $\dfrac{1}{4}, -\dfrac{1}{5}, i, -i$

In Exercises 41 to 46, find a polynomial function $P(x)$ that has the indicated zeros.

41. Zeros: $2 - 5i, -4$; degree 3

▶ 42. Zeros: $3 + 2i, 7$; degree 3

43. Zeros: $4 + 3i, 5 - i$; degree 4

44. Zeros: $i, 3 - 5i$; degree 4

45. Zeros: $-2, 1, 3, 1 + 4i, 1 - 4i$; degree 5

46. Zeros: $-5, 3$ (multiplicity 2), $2 + i, 2 - i$; degree 5

CONNECTING CONCEPTS

In Exercises 47 to 50, find a polynomial function $P(x)$ with real coefficients that has the indicated zeros and satisfies the given conditions.

47. Zeros: $-1, 2, 3$; degree 3; $P(1) = 12$

48. Zeros: $3i, 2$; degree 3; $P(3) = 27$

49. Zeros: $3, -5, 2 + i$; degree 4; $P(1) = 48$

50. Zeros: $\dfrac{1}{2}, 1 - i$; degree 3; $P(4) = 140$

51. Verify that $P(x) = x^3 - x^2 - ix^2 - 9x + 9 + 9i$ has $1 + i$ as a zero and that its conjugate $1 - i$ is not a zero. Explain why this does not contradict the Conjugate Pair Theorem.

52. Verify that $P(x) = x^3 - x^2 - ix^2 - 20x + ix + 20i$ has a zero of i and that its conjugate $-i$ is not a zero. Explain why this does not contradict the Conjugate Pair Theorem.

PREPARE FOR SECTION 3.5

53. Simplify: $\dfrac{x^2 - 9}{x^2 - 2x - 15}$ [P.4]

54. Evaluate $\dfrac{x + 4}{x^2 - 2x - 5}$ for $x = -1$. [P.1]

55. Evaluate $\dfrac{2x^2 + 4x - 5}{x + 6}$ for $x = -3$. [P.1]

56. For what values of x does the denominator of $\dfrac{x^2 - x - 5}{2x^3 + x^2 - 15x}$ equal zero? [1.4]

57. Determine the degree of the numerator and the degree of the denominator of $\dfrac{x^3 + 3x^2 - 5}{x^2 - 4}$. [P.3]

58. Write $\dfrac{x^3 + 2x^2 - x - 11}{x^2 - 2x}$ in $Q(x) + \dfrac{R(x)}{x^2 - 2x}$ form. [3.1]

PROJECTS

1. **INVESTIGATE THE ROOTS OF A CUBIC EQUATION** Hieronimo Cardano, using a technique he learned from Nicolo Tartaglia, was able to solve some cubic equations.

a. Show that the cubic equation $x^3 + bx^2 + cx + d = 0$ can be transformed into the "reduced" cubic $y^3 + my = n$, where m and n are constants, depending on $b, c,$ and $d,$ by using the substitution $x = y - \dfrac{b}{3}$.

b. Cardano then showed that a solution of the reduced cubic is given by

$$\sqrt[3]{\dfrac{n}{2} + \sqrt{\dfrac{n^2}{4} + \dfrac{m^3}{27}}} - \sqrt[3]{-\dfrac{n}{2} + \sqrt{\dfrac{n^2}{4} + \dfrac{m^3}{27}}}$$

Use Cardano's procedure to solve the equation $x^3 - 6x^2 + 20x - 33 = 0$.

GRAPHS OF RATIONAL FUNCTIONS AND THEIR APPLICATIONS

● VERTICAL AND HORIZONTAL ASYMPTOTES

If $P(x)$ and $Q(x)$ are polynomials, then the function F given by

$$F(x) = \frac{P(x)}{Q(x)}$$

is called a **rational function.** The domain of F is the set of all real numbers except those for which $Q(x) = 0$. For example, let

$$F(x) = \frac{x^2 - x - 5}{2x^3 + x^2 - 15x}$$

Setting the denominator equal to zero, we have

$$2x^3 + x^2 - 15x = 0$$
$$x(2x - 5)(x + 3) = 0$$

The denominator is 0 for $x = 0$, $x = \dfrac{5}{2}$, and $x = -3$. Thus the domain of F is the set of all real numbers except 0, $\dfrac{5}{2}$, and -3.

The graph of $G(x) = \dfrac{x + 1}{x - 2}$ is given in **Figure 3.24.** The graph shows that G has the following properties:

● The graph has an x-intercept at $(-1, 0)$ and a y-intercept at $\left(0, -\dfrac{1}{2}\right)$.

● The graph does not exist when $x = 2$.

Note the behavior of the graph as x takes on values that are close to 2 but *less* than 2. Mathematically, we say that "x approaches 2 from the left."

x	1.9	1.95	1.99	1.995	1.999
$G(x)$	-29	-59	-299	-599	-2999

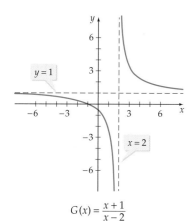

$$G(x) = \frac{x + 1}{x - 2}$$

FIGURE 3.24

From this table and the graph, it appears that as x approaches 2 from the left, the functional values $G(x)$ decrease without bound.

● In this case, we say that "$G(x)$ approaches negative infinity."

Now observe the behavior of the graph as x takes on values that are close to 2 but *greater* than 2. Mathematically, we say that "x approaches 2 from the right."

x	2.1	2.05	2.01	2.005	2.001
$G(x)$	31	61	301	601	3001

From this table and the graph, it appears that as x approaches 2 from the right, the functional values $G(x)$ increase without bound.

• In this case, we say that "$G(x)$ approaches positive infinity."

Now consider the values of $G(x)$ as x *increases* without bound. The following table gives values of $G(x)$ for selected values of x.

x	1000	5000	10,000	50,000	100,000
$G(x)$	1.00301	1.00060	1.00030	1.00006	1.00003

• As x increases without bound, the values of $G(x)$ become closer to 1.

Now let the values of x *decrease* without bound. The table below gives the values of $G(x)$ for selected values of x.

x	-1000	-5000	$-10,000$	$-50,000$	$-100,000$
$G(x)$	0.997006	0.999400	0.999700	0.999940	0.999970

• As x decreases without bound, the values of $G(x)$ become closer to 1.

When we are discussing graphs that increase or decrease without bound, it is convenient to use mathematical notation. The notation

$$f(x) \rightarrow \infty \quad \text{as} \quad x \rightarrow a^+$$

means that the functional values $f(x)$ increase without bound as x approaches a from the right. Recall that the symbol ∞ does not represent a real number but is used merely to describe the concept of a variable taking on larger and larger values without bound. See **Figure 3.25a.**

The notation

$$f(x) \rightarrow \infty \quad \text{as} \quad x \rightarrow a^-$$

means that the function values $f(x)$ increase without bound as x approaches a from the left. See **Figure 3.25b.**

The notation

$$f(x) \rightarrow -\infty \quad \text{as} \quad x \rightarrow a^+$$

means that the functional values $f(x)$ decrease without bound as x approaches a from the right. See **Figure 3.25c.**

The notation

$$f(x) \rightarrow -\infty \quad \text{as} \quad x \rightarrow a^-$$

means that the functional values $f(x)$ decrease without bound as x approaches a from the left. See **Figure 3.25d.**

Each graph in **Figure 3.25** approaches a vertical line through $(a, 0)$ as $x \rightarrow a^+$ or a^-. The line is said to be a *vertical asymptote* of the graph.

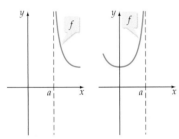

a. $f(x) \rightarrow \infty$
as $x \rightarrow a^+$

b. $f(x) \rightarrow \infty$
as $x \rightarrow a^-$

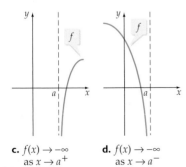

c. $f(x) \rightarrow -\infty$
as $x \rightarrow a^+$

d. $f(x) \rightarrow -\infty$
as $x \rightarrow a^-$

FIGURE 3.25

Definition of a Vertical Asymptote

The line $x = a$ is a **vertical asymptote** of the graph of a function F provided

$$F(x) \to \infty \quad \text{or} \quad F(x) \to -\infty$$

as x approaches a from either the left or right.

In **Figure 3.24,** the line $x = 2$ is a vertical asymptote of the graph of G. Note that the graph of G in **Figure 3.24** also approaches the horizontal line $y = 1$ as $x \to \infty$ and as $x \to -\infty$. The line $y = 1$ is a *horizontal asymptote* of the graph of G.

Definition of a Horizontal Asymptote

The line $y = b$ is a **horizontal asymptote** of the graph of a function F provided

$$F(x) \to b \quad \text{as} \quad x \to \infty \quad \text{or} \quad x \to -\infty$$

Figure 3.26 illustrates some of the ways in which the graph of a rational function may approach its horizontal asymptote. It is common practice to display the asymptotes of the graph of a rational function by using dashed lines. Although a rational function may have several vertical asymptotes, it can have at most one horizontal asymptote. The graph may intersect its horizontal asymptote.

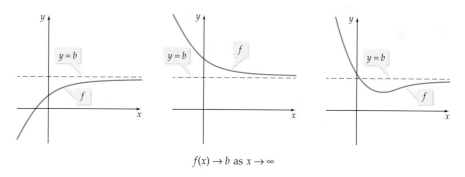

$f(x) \to b$ as $x \to \infty$

FIGURE 3.26

❓ QUESTION Can a graph of a rational function cross its vertical asymptote? Why or why not?

Geometrically, a line is an asymptote of a curve if the distance between the line and a point $P(x, y)$ on the curve approaches zero as the distance between the origin and the point P increases without bound.

❓ ANSWER No. If $x = a$ is a vertical asymptote of a rational function R, then $R(a)$ is undefined.

Vertical asymptotes of the graph of a rational function can be found by using the following theorem.

Theorem on Vertical Asymptotes

If the real number a is a zero of the denominator $Q(x)$, then the graph of $F(x) = P(x)/Q(x)$, where $P(x)$ and $Q(x)$ have no common factors, has the vertical asymptote $x = a$.

EXAMPLE 1 **Find the Vertical Asymptotes of a Rational Function**

Find the vertical asymptotes of each rational function.

a. $f(x) = \dfrac{x^3}{x^2 + 1}$ b. $g(x) = \dfrac{x}{x^2 - x - 6}$

Solution

a. To find the vertical asymptotes, determine the real zeros of the denominator. The denominator $x^2 + 1$ has no real zeros, so the graph of f has no vertical asymptotes. See **Figure 3.27**.

b. The denominator $x^2 - x - 6 = (x - 3)(x + 2)$ has zeros of 3 and -2. The numerator has no common factors with the denominator, so $x = 3$ and $x = -2$ are both vertical asymptotes of the graph of g, as shown in **Figure 3.28**.

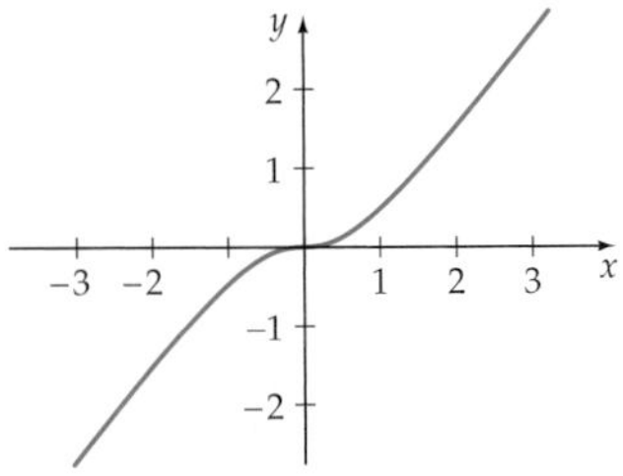

▶ **TRY EXERCISE 2, PAGE 311**

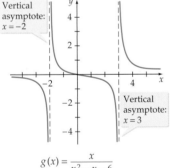

$f(x) = \dfrac{x^3}{x^2 + 1}$

FIGURE 3.27

Vertical asymptote: $x = -2$

Vertical asymptote: $x = 3$

$g(x) = \dfrac{x}{x^2 - x - 6}$

FIGURE 3.28

The following theorem indicates that a horizontal asymptote can be determined by examining the leading terms of the numerator and the denominator of a rational function.

Theorem on Horizontal Asymptotes

Let
$$F(x) = \frac{a_n x^n + a_{n-1} x^{n-1} + \cdots + a_1 x + a_0}{b_m x^m + b_{m-1} x^{m-1} + \cdots + b_1 x + b_0}$$

be a rational function with numerator of degree n and denominator of degree m.

1. If $n < m$, then the x-axis, which is the line given by $y = 0$, is the horizontal asymptote of the graph of F.

2. If $n = m$, then the line given by $y = a_n/b_m$ is the horizontal asymptote of the graph of F.

3. If $n > m$, the graph of F has no horizontal asymptote.

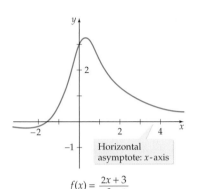

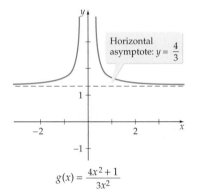

EXAMPLE 2   **Find the Horizontal Asymptote of a Rational Function**

Find the horizontal asymptote of each rational function.

a. $f(x) = \dfrac{2x + 3}{x^2 + 1}$ b. $g(x) = \dfrac{4x^2 + 1}{3x^2}$ c. $h(x) = \dfrac{x^3 + 1}{x - 2}$

Solution

a. The degree of the numerator $2x + 3$ is less than the degree of the denominator $x^2 + 1$. By the Theorem on Horizontal Asymptotes, the x-axis is the horizontal asymptote of f. See the graph of f in **Figure 3.29.**

b. The numerator $4x^2 + 1$ and the denominator $3x^2$ of g are both of degree 2. By the Theorem on Horizontal Asymptotes, the line $y = \dfrac{4}{3}$ is the horizontal asymptote of g. See the graph of g in **Figure 3.30.**

c. The degree of the numerator $x^3 + 1$ is larger than the degree of the denominator $x - 2$, so by the Theorem on Horizontal Asymptotes, the graph of h has no horizontal asymptotes.

▶ **TRY EXERCISE 6, PAGE 311**

The proof of the Theorem on Horizontal Asymptotes makes use of the technique employed in the following verification. To verify that

$$y = \frac{5x^2 + 4}{3x^2 + 8x + 7}$$

has a horizontal asymptote of $y = \dfrac{5}{3}$, divide the numerator and the denominator by the largest power of the variable x (x^2 in this case).

$$y = \frac{\dfrac{5x^2 + 4}{x^2}}{\dfrac{3x^2 + 8x + 7}{x^2}} = \frac{5 + \dfrac{4}{x^2}}{3 + \dfrac{8}{x} + \dfrac{7}{x^2}}, \quad x \neq 0$$

As x increases without bound or decreases without bound, the fractions $\dfrac{4}{x^2}, \dfrac{8}{x},$ and $\dfrac{7}{x^2}$ approach zero. Thus

$$y \rightarrow \frac{5 + 0}{3 + 0 + 0} = \frac{5}{3} \quad \text{as} \quad x \rightarrow \pm\infty$$

and hence the line $y = \dfrac{5}{3}$ is a horizontal asymptote of the graph.

• A Sign Property of Rational Functions

The zeros and vertical asymptotes of a rational function F divide the x-axis into intervals. In each interval, $F(x)$ is positive for all x in the interval or $F(x)$ is negative for all x in the interval. For example, consider the rational function

$$g(x) = \frac{x+1}{x^2 + 2x - 3}$$

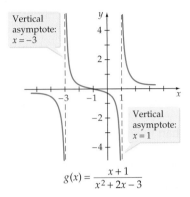

Vertical asymptote: $x = -3$

Vertical asymptote: $x = 1$

$$g(x) = \frac{x+1}{x^2 + 2x - 3}$$

FIGURE 3.31

which has vertical asymptotes of $x = -3$ and $x = 1$ and a zero of -1. These three numbers divide the x-axis into the four intervals $(-\infty, -3)$, $(-3, -1)$, $(-1, 1)$, and $(1, \infty)$. Note in **Figure 3.31** that the graph of g is negative for all x such that $x < -3$, positive for all x such that $-3 < x < -1$, negative for all x such that $-1 < x < 1$, and positive for all x such that $x > 1$.

• A General Graphing Procedure

If $F(x) = P(x)/Q(x)$, where $P(x)$ and $Q(x)$ are polynomials that have no common factors, then the following general procedure offers useful guidelines for graphing F.

General Procedure for Graphing Rational Functions That Have No Common Factors

1. *Asymptotes*   Find the real zeros of the denominator $Q(x)$. For each zero a, draw the dashed line $x = a$. Each line is a vertical asymptote of the graph of F. Also graph any horizontal asymptotes.

2. *Intercepts*   Find the real zeros of the numerator $P(x)$. For each real zero c, plot the point $(c, 0)$. Each such point is an x-intercept of the graph of F. For each x-intercept use the even and odd powers of $(x - c)$ to determine if the graph crosses the x-axis at the intercept or if the graph intersects but does not cross the x-axis. Also evaluate $F(0)$. Plot $(0, F(0))$, the y-intercept of the graph of F.

3. *Symmetry*   Use the tests for symmetry to determine whether the graph of the function has symmetry with respect to the y-axis or symmetry with respect to the origin.

4. *Additional points*   Plot some points that lie in the intervals between and beyond the vertical asymptotes and the x-intercepts.

5. *Behavior near asymptotes*   If $x = a$ is a vertical asymptote, determine whether $F(x) \to \infty$ or $F(x) \to -\infty$ as $x \to a^-$ and also as $x \to a^+$.

6. *Complete the sketch*   Use all the information obtained above to sketch the graph of F.

EXAMPLE 3   **Graph a Rational Function**

Sketch a graph of $f(x) = \dfrac{2x^2 - 18}{x^2 + 3}$.

Solution

Asymptotes   The denominator $x^2 + 3$ has no real zeros, so the graph of f has no vertical asymptotes. The numerator and denominator both are of degree 2. The leading coefficients are 2 and 1, respectively. By the Theorem on Horizontal Asymptotes, the graph of f has a horizontal asymptote of

$$y = \frac{2}{1} = 2.$$

Intercepts   The zeros of the numerator occur when $2x^2 - 18 = 0$ or, solving for x, when $x = -3$ and $x = 3$. Therefore, the x-intercepts are $(-3, 0)$ and $(3, 0)$. The factored numerator is $2(x + 3)(x - 3)$. Each linear factor has an exponent of 1, an odd number. Thus the graph crosses the x-axis at its x-intercepts. To find the y-intercept, evaluate f when $x = 0$. This gives $y = -6$. Therefore, the y-intercept is $(0, -6)$.

Symmetry   Below we show that $f(-x) = f(x)$, which means that f is an even function and therefore its graph is symmetric with respect to the y-axis.

$$f(-x) = \frac{2(-x)^2 - 18}{(-x)^2 + 3} = \frac{2x^2 - 18}{x^2 + 3} = f(x)$$

Additional Points   The intervals determined by the x-intercepts are $x < -3$, $-3 < x < 3$, and $x > 3$. Generally, it is necessary to determine points in all intervals. However, because f is an even function, its graph is symmetric with respect to the y-axis. The following table lists a few points for $x > 0$. Symmetry can be used to locate corresponding points for $x < 0$.

x	1	2	6
$f(x)$	-4	$-\dfrac{10}{7} \approx -1.43$	$\dfrac{18}{13} \approx 1.38$

Behavior Near Asymptotes   As x increases or decreases without bound, $f(x)$ approaches the horizontal asymptote $y = 2$.

To determine whether the graph of f intersects the horizontal asymptote at any point, solve the equation $f(x) = 2$.

There are no solutions of $f(x) = 2$ because

$$\frac{2x^2 - 18}{x^2 + 3} = 2 \quad \text{implies} \quad 2x^2 - 18 = 2x^2 + 6 \quad \text{implies} \quad -18 = 6$$

This is not possible. Thus the graph of f does not intersect the horizontal asymptote but approaches it from below as x increases or decreases without bound.

Continued ▶

Complete the Sketch   Use the summary in **Table 3.3,** to the left, to finish the sketch. The completed graph is shown in **Figure 3.32.**

TABLE 3.3

Vertical Asymptote	None
Horizontal Asymptote	$y = 2$
x-Intercepts	crosses at $(-3, 0)$, crosses at $(3, 0)$
y-Intercept	$(0, -6)$
Additional Points	$(1, -4)$, $(2, -1.43)$, $(6, 1.38)$

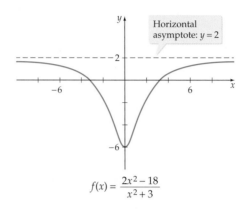

$$f(x) = \frac{2x^2 - 18}{x^2 + 3}$$

FIGURE 3.32

▶ **TRY EXERCISE 10, PAGE 311**

EXAMPLE 4   **Graph a Rational Function**

Sketch a graph of $h(x) = \dfrac{x^2 + 1}{x^2 + x - 2}$.

Solution

Asymptotes   The denominator $x^2 + x - 2 = (x + 2)(x - 1)$ has zeros -2 and 1; because there are no common factors of the numerator and the denominator, the lines $x = -2$ and $x = 1$ are vertical asymptotes.

   The numerator and denominator both are of degree 2. The leading coefficients of the numerator and denominator are both 1. Thus h has the horizontal asymptote $y = \dfrac{1}{1} = 1$.

Intercepts   The numerator $x^2 + 1$ has no real zeros, so the graph of h has no x-intercepts. Because $h(0) = -0.5$, h has the y-intercept $(0, -0.5)$.

Symmetry   By applying the tests for symmetry, we can determine that the graph of h is not symmetric with respect to the origin or to the y-axis.

Additional Points   The intervals determined by the vertical asymptotes are $(-\infty, -2)$, $(-2, 1)$, and $(1, \infty)$. Plot a few points from each interval.

x	−5	−3	−1	0.5	2	3	4
h(x)	$\dfrac{13}{9}$	$\dfrac{5}{2}$	−1	−1	$\dfrac{5}{4}$	1	$\dfrac{17}{18}$

The graph of h will intersect the horizontal asymptote $y = 1$ exactly once. This can be determined by solving the equation $h(x) = 1$.

$$\frac{x^2 + 1}{x^2 + x - 2} = 1$$

$$x^2 + 1 = x^2 + x - 2 \qquad \text{• Multiply both sides by } x^2 + x - 2.$$

$$1 = x - 2$$

$$3 = x$$

The only solution is $x = 3$. Therefore, the graph of h intersects the horizontal asymptote at $(3, 1)$.

Behavior Near Asymptotes As x approaches -2 from the left, the denominator $(x + 2)(x - 1)$ approaches 0 but remains positive. The numerator $x^2 + 1$ approaches 5, which is positive, so the quotient $h(x)$ increases without bound. Stated in mathematical notation,

$$h(x) \to \infty \quad \text{as} \quad x \to -2^-$$

Similarly, it can be determined that

$$h(x) \to -\infty \quad \text{as} \quad x \to -2^+$$
$$h(x) \to -\infty \quad \text{as} \quad x \to 1^-$$
$$h(x) \to \infty \quad \text{as} \quad x \to 1^+$$

Complete the Sketch Use the summary in **Table 3.4** to obtain the graph sketched in **Figure 3.33**.

TABLE 3.4

Vertical Asymptote	$x = -2, x = 1$
Horizontal Asymptote	$y = 1$
x-Intercepts	None
y-Intercept	$(0, -0.5)$
Additional Points	$(-5, 1.\overline{4}), (-3, 2.5),$ $(-1, -1), (0.5, -1),$ $(2, 1.25), (3, 1),$ $(4, 0.9\overline{4})$

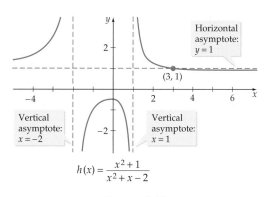

$$h(x) = \frac{x^2 + 1}{x^2 + x - 2}$$

FIGURE 3.33

▶ **TRY EXERCISE 26, PAGE 311**

● SLANT ASYMPTOTES

Some rational functions have an asymptote that is neither vertical nor horizontal, but slanted.

Theorem on Slant Asymptotes

The rational function given by $F(x) = P(x)/Q(x)$, where $P(x)$ and $Q(x)$ have no common factors, has a **slant asymptote** if the degree of the polynomial $P(x)$ in the numerator is one greater than the degree of the polynomial $Q(x)$ in the denominator.

To find the slant asymptote, divide $P(x)$ by $Q(x)$ and write $F(x)$ in the form

$$F(x) = \frac{P(x)}{Q(x)} = (mx + b) + \frac{r(x)}{Q(x)}$$

where the degree of $r(x)$ is less than the degree of $Q(x)$. Because

$$\frac{r(x)}{Q(x)} \rightarrow 0 \quad \text{as} \quad x \rightarrow \pm\infty$$

we know that $F(x) \rightarrow mx + b$ as $x \rightarrow \pm\infty$.

The line represented by $y = mx + b$ is the slant asymptote of the graph of F.

EXAMPLE 5 **Find the Slant Asymptote of a Rational Function**

Find the slant asymptote of $f(x) = \dfrac{2x^3 + 5x^2 + 1}{x^2 + x + 3}$.

Solution

Because the degree of the numerator $2x^3 + 5x^2 + 1$ is exactly one larger than the degree of the denominator $x^2 + x + 3$ and f is in simplest form, f has a slant asymptote. To find the asymptote, divide $2x^3 + 5x^2 + 1$ by $x^2 + x + 3$.

$$
\begin{array}{r}
2x + 3 \\
x^2 + x + 3 \overline{)2x^3 + 5x^2 + 0x + 1} \\
\underline{2x^3 + 2x^2 + 6x} \\
3x^2 - 6x + 1 \\
\underline{3x^2 + 3x + 9} \\
-9x - 8
\end{array}
$$

Therefore,

$$f(x) = \frac{2x^3 + 5x^2 + 1}{x^2 + x + 3} = 2x + 3 + \frac{-9x - 8}{x^2 + x + 3}$$

and the line given by $y = 2x + 3$ is the slant asymptote for the graph of f. **Figure 3.34** shows the graph of f and its slant asymptote.

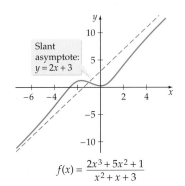

Slant asymptote: $y = 2x + 3$

$f(x) = \dfrac{2x^3 + 5x^2 + 1}{x^2 + x + 3}$

FIGURE 3.34

▶ **TRY EXERCISE 32, PAGE 311**

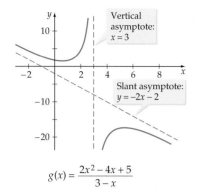

$$g(x) = \frac{2x^2 - 4x + 5}{3 - x}$$

FIGURE 3.35

The function f in Example 5 does not have a vertical asymptote because the denominator $x^2 + x + 3$ does not have any real zeros. However, the function

$$g(x) = \frac{2x^2 - 4x + 5}{3 - x}$$

has both a slant asymptote and a vertical asymptote. The vertical asymptote is $x = 3$, and the slant asymptote is $y = -2x - 2$. **Figure 3.35** shows the graph of g and its asymptotes.

● GRAPH RATIONAL FUNCTIONS THAT HAVE A COMMON FACTOR

If a rational function has a numerator and denominator that have a common factor, then you should reduce the rational function to lowest terms before you apply the general procedure for sketching the graph of a rational function.

EXAMPLE 6 | **Graph a Rational Function That Has a Common Factor**

Sketch the graph of $f(x) = \dfrac{x^2 - 3x - 4}{x^2 - 6x + 8}$.

Solution

Factor the numerator and denominator to obtain

$$f(x) = \frac{x^2 - 3x - 4}{x^2 - 6x + 8} = \frac{(x + 1)(x - 4)}{(x - 2)(x - 4)}, \quad x \neq 2, x \neq 4$$

Thus for all x values other than $x = 4$, the graph of f is the same as the graph of

$$G(x) = \frac{x + 1}{x - 2}$$

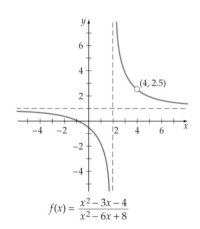

$$f(x) = \frac{x^2 - 3x - 4}{x^2 - 6x + 8}$$

FIGURE 3.36

Figure 3.24 on page 299 shows a graph of G. The graph of f will be the same as this graph, except that it will have an open circle at $(4, 2.5)$ to indicate that it is undefined at $x = 4$. See the graph of f in **Figure 3.36**. The height of the open circle was found by evaluating the resulting reduced rational function $G(x) = \dfrac{x + 1}{x - 2}$ at $x = 4$.

▶ **TRY EXERCISE 48, PAGE 312**

❓ **QUESTION** Does $F(x) = \dfrac{x^2 - x - 6}{x^2 - 9}$ have a vertical asymptote at $x = 3$?

❓ **ANSWER** No. $F(x) = \dfrac{x^2 - x - 6}{x^2 - 9} = \dfrac{(x - 3)(x + 2)}{(x - 3)(x + 3)} = \dfrac{x + 2}{x + 3}$, $x \neq 3$. As $x \to 3$,

$$F(x) \to \frac{5}{6}.$$

• APPLICATIONS OF RATIONAL FUNCTIONS

EXAMPLE 7   Solve an Application

 A cylindrical soft drink can is to be constructed so that it will have a volume of 21.6 cubic inches. See **Figure 3.37.**

a. Write the total surface area A of the can as a function of r, where r is the radius of the can in inches.

b. Use a graphing utility to estimate the value of r (to the nearest tenth of an inch) that produces the minimum surface area.

FIGURE 3.37

Solution

a. The formula for the volume of a cylinder is $V = \pi r^2 h$, where r is the radius and h is the height. Because we are given that the volume is 21.6 cubic inches, we have

$$21.6 = \pi r^2 h$$

$$\frac{21.6}{\pi r^2} = h \qquad \text{• Solve for } h.$$

The surface area of the cylinder is given by

$$A = 2\pi r^2 + 2\pi r h$$

$$A = 2\pi r^2 + 2\pi r\left(\frac{21.6}{\pi r^2}\right) \qquad \text{• Substitute for } h.$$

$$A = 2\pi r^2 + \frac{2(21.6)}{r} \qquad \text{• Simplify.}$$

$$A = \frac{2\pi r^3 + 43.2}{r} \qquad\qquad (1)$$

b. Use Equation (1) with $y = A$ and $x = r$ and a graphing utility to determine that A is a minimum when $r \approx 1.5$ inches. See **Figure 3.38.**

▶ **TRY EXERCISE 56, PAGE 312**

INTEGRATING TECHNOLOGY

A Web applet is available to explore the relationship between the radius of a cylinder with a given volume and the surface area of the cylinder. This applet, CYLINDER, can be found on our website at

math.college.hmco.com

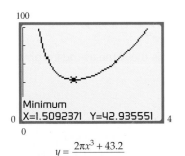

$$y = \frac{2\pi x^3 + 43.2}{x}$$

FIGURE 3.38

 TOPICS FOR DISCUSSION

1. What is a rational function? Give examples of functions that are rational functions and of functions that are not rational functions.

2. Does the graph of every rational function have at least one vertical asymptote? If so, explain why. If not, give an example of a rational function without a vertical asymptote.

3. Does the graph of every rational function have a horizontal asymptote? If so, explain why. If not, give an example of a rational function without a horizontal asymptote.

4. Can the graph of a polynomial function have a vertical asymptote? a horizontal asymptote?

EXERCISE SET 3.5

In Exercises 1 to 4, find all vertical asymptotes of each rational function.

1. $F(x) = \dfrac{2x - 1}{x^2 + 3x}$

▶ **2.** $F(x) = \dfrac{3x^2 + 5}{x^2 - 4}$

3. $F(x) = \dfrac{x^2 + 11}{6x^2 - 5x - 4}$

4. $F(x) = \dfrac{3x - 5}{x^3 - 8}$

In Exercises 5 to 8, find the horizontal asymptote of each rational function.

5. $F(x) = \dfrac{4x^2 + 1}{x^2 + x + 1}$

▶ **6.** $F(x) = \dfrac{3x^3 - 27x^2 + 5x - 11}{x^5 - 2x^3 + 7}$

7. $F(x) = \dfrac{15{,}000x^3 + 500x - 2000}{700 + 500x^3}$

8. $F(x) = 6000\left(1 - \dfrac{25}{(x + 5)^2}\right)$

In Exercises 9 to 30, determine the vertical and horizontal asymptotes and sketch the graph of the rational function F. Label all intercepts and asymptotes.

9. $F(x) = \dfrac{1}{x + 4}$

▶ **10.** $F(x) = \dfrac{1}{x - 2}$

11. $F(x) = \dfrac{-4}{x - 3}$

12. $F(x) = \dfrac{-3}{x + 2}$

13. $F(x) = \dfrac{4}{x}$

14. $F(x) = \dfrac{-4}{x}$

15. $F(x) = \dfrac{x}{x + 4}$

16. $F(x) = \dfrac{x}{x - 2}$

17. $F(x) = \dfrac{x + 4}{2 - x}$

18. $F(x) = \dfrac{x + 3}{1 - x}$

19. $F(x) = \dfrac{1}{x^2 - 9}$

20. $F(x) = \dfrac{-2}{x^2 - 4}$

21. $F(x) = \dfrac{1}{x^2 + 2x - 3}$

22. $F(x) = \dfrac{1}{x^2 - 2x - 8}$

23. $F(x) = \dfrac{x^2}{x^2 + 4x + 4}$

24. $F(x) = \dfrac{2x^2}{x^2 - 1}$

25. $F(x) = \dfrac{10}{x^2 + 2}$

▶ **26.** $F(x) = \dfrac{x^2}{x^2 - 6x + 9}$

27. $F(x) = \dfrac{2x^2 - 2}{x^2 - 9}$

28. $F(x) = \dfrac{6x^2 - 5}{2x^2 + 6}$

29. $F(x) = \dfrac{x^2 + x + 4}{x^2 + 2x - 1}$

30. $F(x) = \dfrac{2x^2 - 14}{x^2 - 6x + 5}$

In Exercises 31 to 34, find the slant asymptote of each rational function.

31. $F(x) = \dfrac{3x^2 + 5x - 1}{x + 4}$

▶ **32.** $F(x) = \dfrac{x^3 - 2x^2 + 3x + 4}{x^2 - 3x + 5}$

33. $F(x) = \dfrac{x^3 - 1}{x^2}$

34. $F(x) = \dfrac{4000 + 20x + 0.0001x^2}{x}$

In Exercises 35 to 44, determine the vertical and slant asymptotes and sketch the graph of the rational function *F*.

35. $F(x) = \dfrac{x^2 - 4}{x}$

36. $F(x) = \dfrac{x^2 + 10}{2x}$

37. $F(x) = \dfrac{x^2 - 3x - 4}{x + 3}$

38. $F(x) = \dfrac{x^2 - 4x - 5}{2x + 5}$

39. $F(x) = \dfrac{2x^2 + 5x + 3}{x - 4}$

40. $F(x) = \dfrac{4x^2 - 9}{x + 3}$

41. $F(x) = \dfrac{x^2 - x}{x + 2}$

42. $F(x) = \dfrac{x^2 + x}{x - 1}$

43. $F(x) = \dfrac{x^3 + 1}{x^2 - 4}$

44. $F(x) = \dfrac{x^3 - 1}{3x^2}$

In Exercises 45 to 52, sketch the graph of the rational function *F*. (*Hint:* First examine the numerator and denominator to determine whether there are any common factors.)

45. $F(x) = \dfrac{x^2 + x}{x + 1}$

46. $F(x) = \dfrac{x^2 - 3x}{x - 3}$

47. $F(x) = \dfrac{2x^3 + 4x^2}{2x + 4}$

▶ **48.** $F(x) = \dfrac{x^2 - x - 12}{x^2 - 2x - 8}$

49. $F(x) = \dfrac{-2x^3 + 6x}{2x^2 - 6x}$

50. $F(x) = \dfrac{x^3 + 3x^2}{x(x + 3)(x - 1)}$

51. $F(x) = \dfrac{x^2 - 3x - 10}{x^2 + 4x + 4}$

52. $F(x) = \dfrac{2x^2 + x - 3}{x^2 - 2x + 1}$

53.   **AVERAGE COST OF GOLF BALLS** The cost, in dollars, of producing *x* golf balls is given by

$$C(x) = 0.43x + 76,000$$

The average cost per golf ball is given by

$$\overline{C}(x) = \dfrac{C(x)}{x} = \dfrac{0.43x + 76,000}{x}$$

a. Find the average cost of producing 1000, 10,000, and 100,000 golf balls.

b. What is the equation of the horizontal asymptote of the graph of $\overline{C}$? Explain the significance of the horizontal asymptote as it relates to this application.

54. **AVERAGE COST OF CD PLAYERS** The cost, in dollars, of producing *x* CD players is given by

$$C(x) = 0.001x^2 + 54x + 175,000$$

The average cost per CD player is given by

$$\overline{C}(x) = \dfrac{C(x)}{x} = \dfrac{0.001x^2 + 54x + 175,000}{x}$$

a. Find the average cost of producing 1000, 10,000, and 100,000 CD players.

b. What is the minimum average cost per CD player? How many CD players should be produced to minimize the average cost per CD player?

55. **DESALINIZATION** The cost *C*, in dollars, to remove *p*% of the salt in a tank of seawater is given by

$$C(p) = \dfrac{2000p}{100 - p}, \quad 0 \le p < 100$$

a. Find the cost of removing 40% of the salt.

b. Find the cost of removing 80% of the salt.

c. Sketch the graph of *C*.

▶ **56.** **PRODUCTION COSTS** The cost, in dollars, of producing *x* cellular telephones is given by

$$C(x) = 0.0006x^2 + 9x + 401,000$$

The average cost per telephone is

$$\overline{C}(x) = \dfrac{C(x)}{x} = \dfrac{0.0006x^2 + 9x + 401,000}{x}$$

a. Find the average cost per telephone when 1000, 10,000, and 100,000 telephones are produced.

b. What is the minimum average cost per telephone? How many cellular telephones should be produced to minimize the average cost per telephone?

57.   **WEDDING EXPENSES** The function $C(t) = 17t^2 + 128t + 5900$ models the average cost of a wedding reception, and the function $W(t) = 38t^2 + 291t + 15,208$ models the  average cost of a wedding, where $t = 0$ represents the year 1990 and $0 \le t \le 12$. The rational function

$$R(t) = \dfrac{C(t)}{W(t)} = \dfrac{17t^2 + 128t + 5900}{38t^2 + 291t + 15,208}$$

gives the relative cost of the reception compared to the cost of a wedding.

a. Use $R(t)$ to estimate the relative cost of the reception compared to the cost of a wedding for the years $t = 0$, $t = 7$, and $t = 12$. Round your results to the nearest tenth of a percent.

b. According to the function $R(t)$, what percent of the total cost of a wedding, to the nearest tenth of a percent, will the cost of the reception approach as the years go by?

58. **INCOME TAX THEORY** The economist Arthur Laffer conjectured that if taxes were increased starting from very low levels, then the tax revenue received by the government would increase. But as tax rates continued to increase, there would be a point at which the tax revenue would start to decrease. The underlying concept was that if taxes were increased too much, people would not work as hard because much of their additional income would be taken from them by the increase in taxes. Laffer illustrated his concept by drawing a curve similar to the following.

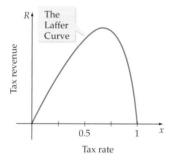

Laffer's curve shows that if the tax rate is 0%, the tax revenue will be \$0, and if the tax rate is 100%, the tax revenue also will be \$0. Laffer assumed that most people would not work if all their income went for taxes.

Most economists agree with Laffer's basic concept, but there is much disagreement about the equation of the actual tax revenue curve R and the tax rate x that will maximize the government's tax revenues.

a. Assume that Laffer's curve is given by

$$R(x) = \frac{-6.5(x^3 + 2x^2 - 3x)}{x^2 + x + 1}$$

where R is measured in trillions of dollars. Use a graphing utility to determine the tax rate x, to the nearest tenth of a percent, that would produce the maximum tax revenue. (*Hint:* Use a domain of $[0, 1]$ and a range of $[0, 4]$.)

b. Assume that Laffer's curve is given by

$$R(x) = \frac{-1500(x^3 + 2x^2 - 3x)}{x^2 + x + 400}$$

where R is measured in trillions of dollars. Use a graphing utility to determine the tax rate x, to the nearest tenth of a percent, that would produce the maximum tax revenue.

59. **A POPULATION MODEL** The population of a suburb, in thousands, is given by

$$P(t) = \frac{420t}{0.6t^2 + 15}$$

where t is the time in years after June 1, 1996.

a. Find the population of the suburb for $t = 1, 4$, and 10 years.

b. In what year will the population of the suburb reach its maximum?

c. What will happen to the population as $t \to \infty$?

60. **A MEDICATION MODEL** The rational function

$$M(t) = \frac{0.5t + 400}{0.04t^2 + 1}$$

models the number of milligrams of medication in the bloodstream of a patient t hours after 400 milligrams of the medication have been injected into the patient's bloodstream.

a. Find $M(5)$ and $M(10)$. Round to the nearest milligram.

b. What will M approach as $t \to \infty$?

61. **MINIMIZING SURFACE AREA** A cylindrical soft drink can is to be made so that it will have a volume of 354 milliliters. If r is the radius of the can in centimeters, then the total surface area A of the can is given by the rational function

$$A(r) = \frac{2\pi r^3 + 708}{r}$$

a. Graph A and use the graph to estimate (to the nearest tenth of a centimeter) the value of r that produces the minimum value of A.

b. Does the graph of A have a slant asymptote?

c. Explain the meaning of the following statement as it applies to the graph of A.

$$\text{As } r \to \infty, A \to 2\pi r^2.$$

62. **RESISTORS IN PARALLEL** The electronic circuit at the right shows two resistors connected in parallel.

One resistor has a resistance of R_1 ohms and the other has a resistance of R_2 ohms. The total resistance for the circuit, measured in ohms, is given by the formula

$$R_T = \frac{R_1 R_2}{R_1 + R_2}$$

Assume R_1 has a fixed resistance of 10 ohms.

a. Compute R_T for $R_2 = 2$ ohms and for $R_2 = 20$ ohms.

b. What happens to R_T as $R_2 \to \infty$?

CONNECTING CONCEPTS

63. Determine the point at which the graph of

$$F(x) = \frac{2x^2 + 3x + 4}{x^2 + 4x + 7}$$

intersects its horizontal asymptote.

64. Determine the point at which the graph of

$$F(x) = \frac{3x^3 + 2x^2 - 8x - 12}{x^2 + 4}$$

intersects its slant asymptote.

65. Determine the two points at which the graph of

$$F(x) = \frac{x^3 + x^2 + 4x + 1}{x^3 + 1}$$

intersects its horizontal asymptote.

66. Give an example of a rational function that intersects its slant asymptote at two points.

67. Write a rational function that has vertical asymptotes at $x = -2$ and $x = 3$ and a horizontal asymptote at $y = 1$.

68. Write a rational function that has vertical asymptotes at $x = -3$ and $x = 1$ and a horizontal asymptote at $y = 0$.

PROJECTS

1. **PARABOLIC ASYMPTOTES** It can be shown that the rational function $F(x) = R(x)/S(x)$, where $R(x)$ and $S(x)$ have no common factors, has a parabolic asymptote provided the degree of $R(x)$ is *two* greater than the degree of $S(x)$. For instance, the rational function

$$F(x) = \frac{x^3 + 2}{x + 1}$$

has a parabolic asymptote given by $y = x^2 - x + 1$.

a. Use a graphing utility to graph $F(x)$ and the parabola given by $y = x^2 - x + 1$ in the same viewing window. Does the parabola appear to be an asymptote for the graph of F? Explain.

b. Write a paragraph that explains how to determine the equation of the parabolic asymptote for a rational func-

tion $F(x) = R(x)/S(x)$, where $R(x)$ and $S(x)$ have no common factors and the degree of $R(x)$ is two greater than the degree of $S(x)$.

c. What is the equation of the parabolic asymptote for the rational function $G(x) = \dfrac{x^4 + x^2 + 2}{x^2 - 1}$? Use a graphing utility to graph $G(x)$ and the parabolic asymptote in the same viewing window. Does the parabola appear to be an asymptote for the graph of G?

d. Create a rational function that has $y = x^2 + x + 2$ as its parabolic asymptote. Explain the procedure you used to create your rational function.

EXPLORING CONCEPTS WITH TECHNOLOGY

Finding Zeros of a Polynomial Using *Mathematica*

Computer algebra systems (CAS) are computer programs that are used to solve equations, graph functions, simplify algebraic expressions, and help us perform many other mathematical tasks. In this exploration, we will demonstrate how to use one of these programs, *Mathematica*, to find zeros of a polynomial function.

Recall that a zero of a function P is a number x for which $P(x) = 0$. The idea behind finding a zero of a polynomial function by using a CAS is to solve the polynomial equation $P(x) = 0$ for x.

Two commands in *Mathematica* that can be used to solve an equation are **Solve** and **NSolve**. [*Mathematica* is sensitive about syntax (the way in which an expression is typed). You *must* use upper-case and lower-case letters as we indicate.] **Solve** will attempt to find an *exact* solution of the equation; **NSolve** attempts to find *approximate* solutions. Here are some examples.

To find the exact values of the zeros of $P(x) = x^3 + 5x^2 + 11x + 15$, input the following. *Note:* The two equals signs are necessary.

$$\text{Solve[x\^3+5x\^2+11x+15==0]}$$

Press Enter . The result should be

$$\{\{x->-3\}, \{x->-1-2\ I\}, \{x->-1+2\ I\}\}$$

Thus the three zeros of P are -3, $-1 - 2i$, and $-1 + 2i$.

To find the approximate values of the zeros of $P(x) = x^4 - 3x^3 + 4x^2 + x - 4$, input the following.

$$\text{NSolve[x\^4-3x\^3+4x\^2+x-4==0]}$$

Press Enter . The result should be

$$\{\{x->-0.821746\}, \{x->1.2326\}, \{x->1.29457-1.50771\ I\},$$
$$\{x->1.29457+1.50771\ I\}\}$$

The four zeros are (approximately) -0.821746, 1.2326, $1.29457 - 1.50771i$, and $1.29457 + 1.50771i$.

Not all polynomial equations can be solved exactly. This means that **Solve** will not always give solutions with *Mathematica*. Consider the two examples below.

Input	NSolve[x^5-3x^3+2x^2-5==0]
Output	{{x->-1.80492}, {x->-1.12491}, {x->0.620319-1.03589 I}, {x->0.620319+1.03589 I}, {x->1.68919}}

These are the approximate zeros of the polynomial.

Input	Solve[x^5-3x^3+2x^2-5==0]
Output	{ToRules[Roots[2x²-3x³+x⁵==5]]}

In this case, no exact solution could be found. In general, there are no formulas like the quadratic formula, for instance, that yield exact solutions for fifth- or higher-degree polynomial equations.

Use *Mathematica* (or another CAS) to find the zeros of each of the following polynomial functions.

1. $P(x) = x^4 - 3x^3 + x - 5$ **2.** $P(x) = 3x^3 - 4x^2 + x - 3$

3. $P(x) = 4x^5 - 3x^3 + 2x^2 - x + 2$ **4.** $P(x) = -3x^4 - 6x^3 + 2x - 8$

CHAPTER 3 SUMMARY

3.1 The Remainder Theorem and the Factor Theorem

- *The Remainder Theorem*   If a polynomial function $P(x)$ is divided by $(x - c)$, then the remainder equals $P(c)$.

- *The Factor Theorem*   A polynomial function $P(x)$ has a factor $(x - c)$ if and only if $P(c) = 0$.

3.2 Polynomial Functions of Higher Degree

- Characteristics and properties used in graphing polynomial functions include:

  1. Continuity—Polynomial functions are smooth continuous curves.

  2. Leading term test—Determines the behavior of the graph of a polynomial function at the far right and at the far left.

  3. The real zeros of the function determine the x-intercepts.

- *Relative Minimum and Relative Maximum*   If there is an open interval I containing c on which

  $f(c) \leq f(x)$ for all x in I, then $f(c)$ is a relative minimum of f.

  $f(c) \geq f(x)$ for all x in I, then $f(c)$ is a relative maximum of f.

- *The Zero Location Theorem*   Let $P(x)$ be a polynomial function. If $a < b$, and if $P(a)$ and $P(b)$ have opposite signs, then there is at least one real number c between a and b such that $P(c) = 0$.

3.3 Zeros of Polynomial Functions

- Values of x that satisfy $P(x) = 0$ are called zeros of P.

- *Definition of Multiple Zeros of a Polynomial*   If a polynomial function $P(x)$ has $(x - r)$ as a factor exactly k times, then r is said to be a zero of multiplicity k of the polynomial function $P(x)$.

- *The Rational Zero Theorem*   If

  $$P(x) = a_n x^n + a_{n-1} x^{n-1} + \cdots + a_1 x + a_0, a_n \neq 0$$

has integer coefficients, and $\dfrac{p}{q}$ (where p and q have no common factors) is a rational zero of P, then p is a factor of a_0 and q is a factor of a_n.

- *Upper- and Lower-Bound Theorem*
Let $P(x)$ be a polynomial function with real coefficients. Use synthetic division to divide $P(x)$ by $x - b$, where b is a nonzero real number.

  Upper Bound
  a. If $b > 0$ and the leading coefficient of P is positive, then b is an upper bound for the real zeros of P provided none of the numbers in the bottom row of the synthetic division are negative.

  b. If $b > 0$ and the leading coefficient of P is negative, then b is an upper bound for the real zeros of P provided none of the numbers in the bottom row of the synthetic division are positive.

  Lower Bound   If $b < 0$ and the numbers in the bottom row of the synthetic division of P by $x - b$ alternate in sign, then b is a lower bound for the real zeros of P.

- *Descartes' Rule of Signs*   Let $P(x)$ be a polynomial function with real coefficients and with terms arranged in order of decreasing powers of x.

  1. The number of positive real zeros of $P(x)$ is equal to the number of variations in sign of $P(x)$, or is equal to that number decreased by an even integer.

  2. The number of negative real zeros of $P(x)$ is equal to the number of variations in sign of $P(-x)$ or is equal to that number decreased by an even integer.

- The zeros of some polynomial functions with integer coefficients can be found by using the guidelines stated on page 281.

3.4 The Fundamental Theorem of Algebra

- *The Fundamental Theorem of Algebra*   If $P(x)$ is a polynomial function of degree $n \geq 1$ with complex coefficients, then $P(x)$ has at least one complex zero.

- *The Conjugate Pair Theorem*   If $a + bi$ ($b \neq 0$) is a complex zero of the polynomial function $P(x)$, with real coefficients, then the conjugate $a - bi$ is also a complex zero of the polynomial function.

3.5 Graphs of Rational Functions and Their Applications

• If $P(x)$ and $Q(x)$ are polynomials, then the function F given by

$$F(x) = \frac{P(x)}{Q(x)}$$

is called a rational function.

• *General Procedure for Graphing Rational Functions That Have No Common Factors*

  1. Find the real zeros of the denominator. For each zero a, the vertical line $x = a$ will be a vertical asymptote. Use the Theorem on Horizontal Asymptotes to determine if the function has a horizontal asymptote. Graph the horizontal asymptote.

  2. Find the real zeros of the numerator. For each real zero a, plot $(a, 0)$. These points are the x-intercepts. The y-intercept of the graph of $F(x)$ is the point $(0, F(0))$.

  3. Use the tests for symmetry to determine whether the graph has symmetry with respect to the y-axis or to the origin.

  4. Find additional points that lie in the intervals between the x-intercepts and the vertical asymptotes.

  5. Determine the behavior of the graph near the asymptotes.

  6. Use the information obtained in the above steps to sketch the graph.

• *Theorem on Slant Asymptotes* The rational function given by $F(x) = P(x)/Q(x)$, where $P(x)$ and $Q(x)$ have no common factors, has a slant asymptote if the degree of the polynomial $P(x)$ in the numerator is one greater than the degree of the polynomial $Q(x)$ in the denominator.

CHAPTER 3 TRUE/FALSE EXERCISES

In Exercises 1 to 12, answer true or false. If the statement is false, explain why the statement is false or give an example to show that the statement is false.

1. The complex zeros of a polynomial function with complex coefficients always occur in conjugate pairs.

2. Descartes' Rule of Signs indicates that the polynomial function $P(x) = x^3 - x^2 + x - 1$ must have three positive zeros.

3. The polynomial $2x^5 + x^4 - 7x^3 - 5x^2 + 4x + 10$ has two variations in sign.

4. If 4 is an upper bound of the zeros of the polynomial function P, then 5 is also an upper bound of the zeros of P.

5. The graph of every rational function has a vertical asymptote.

6. The graph of the rational function $F(x) = \dfrac{x^2 - 4x + 4}{x^2 - 5x + 6}$ has a vertical asymptote of $x = 2$.

7. If 7 is a zero of the polynomial function P, then $x - 7$ is a factor of P.

8. According to the Zero Location Theorem, the polynomial function $P(x) = x^3 + 6x - 2$ has a real zero between 0 and 1.

9. Every fourth-degree polynomial function with complex coefficients has exactly four complex zeros, provided each zero is counted according to its multiplicity.

10. The graph of a rational function can have at most one horizontal asymptote.

11. Descartes' Rule of Signs indicates that the polynomial function $P(x) = x^3 + 2x^2 + 4x - 7$ does have a positive zero.

12. Every polynomial function has at least one real zero.

CHAPTER 3 REVIEW EXERCISES

In Exercises 1 to 6, use synthetic division to divide the first polynomial by the second.

1. $4x^3 - 11x^2 + 5x - 2, x - 3$

2. $5x^3 - 18x + 2, x - 1$

3. $3x^3 - 5x + 1, x + 2$

4. $2x^3 + 7x^2 + 16x - 10, x - \dfrac{1}{2}$

5. $3x^3 - 10x^2 - 36x + 55, x - 5$

6. $x^4 + 9x^3 + 6x^2 - 65x - 63, x + 7$

In Exercises 7 to 10, use the Remainder Theorem to find $P(c)$.

7. $P(x) = x^3 + 2x^2 - 5x + 1, c = 4$

8. $P(x) = -4x^3 - 10x + 8, c = -1$

9. $P(x) = 6x^4 - 12x^2 + 8x + 1, c = -2$

10. $P(x) = 5x^5 - 8x^4 + 2x^3 - 6x^2 - 9, c = 3$

In Exercises 11 to 14, use synthetic division to show that c is a zero of the given polynomial function.

11. $P(x) = x^3 + 2x^2 - 26x + 33, c = 3$

12. $P(x) = 2x^4 + 8x^3 - 8x^2 - 31x + 4, c = -4$

13. $P(x) = x^5 - x^4 - 2x^2 + x + 1, c = 1$

14. $P(x) = 2x^3 + 3x^2 - 8x + 3, c = \dfrac{1}{2}$

In Exercises 15 to 20, graph the polynomial function.

15. $P(x) = x^3 - x$

16. $P(x) = -x^3 - x^2 + 8x + 12$

17. $P(x) = x^4 - 6$ **18.** $P(x) = x^5 - x$

19. $P(x) = x^4 - 10x^2 + 9$ **20.** $P(x) = x^5 - 5x^3$

In Exercises 21 to 26, use the Rational Zero Theorem to list all possible rational zeros for each polynomial function.

21. $P(x) = x^3 - 7x - 6$

22. $P(x) = 2x^3 + 3x^2 - 29x - 30$

23. $P(x) = 15x^3 - 91x^2 + 4x + 12$

24. $P(x) = x^4 - 12x^3 + 52x^2 - 96x + 64$

25. $P(x) = x^3 + x^2 - x - 1$

26. $P(x) = 6x^5 + 3x - 2$

In Exercises 27 to 30, use Descartes' Rule of Signs to state the number of possible positive and negative real zeros of each polynomial function.

27. $P(x) = x^3 + 3x^2 + x + 3$

28. $P(x) = x^4 - 6x^3 - 5x^2 + 74x - 120$

29. $P(x) = x^4 - x - 1$

30. $P(x) = x^5 - 4x^4 + 2x^3 - x^2 + x - 8$

In Exercises 31 to 36, find the zeros of the polynomial function.

31. $P(x) = x^3 + 6x^2 + 3x - 10$

32. $P(x) = x^3 - 10x^2 + 31x - 30$

33. $P(x) = 6x^4 + 35x^3 + 72x^2 + 60x + 16$

34. $P(x) = 2x^4 + 7x^3 + 5x^2 + 7x + 3$

35. $P(x) = x^4 - 4x^3 + 6x^2 - 4x + 1$

36. $P(x) = 2x^3 - 7x^2 + 22x + 13$

In Exercises 37 and 38, use the given zero to find the remaining zeros of each polynomial function.

37. $P(x) = x^4 - 4x^3 + 6x^2 - 4x - 15; 1 - 2i$

38. $P(x) = x^4 - x^3 - 17x^2 + 55x - 50; 2 + i$

39. Find a third-degree polynomial function with integer coefficients and zeros of 4, -3, and $\dfrac{1}{2}$.

40. Find a fourth-degree polynomial function with zeros of 2, -3, i, and $-i$.

41. Find a fourth-degree polynomial function with real coefficients that has zeros of 1, 2, and $5i$.

42. Find a fourth-degree polynomial function with real coefficients that has -2 as a zero of multiplicity 2 and also has $1 + 3i$ as a zero.

In Exercises 43 to 46, find the vertical, horizontal, and slant asymptotes for each rational function.

43. $f(x) = \dfrac{3x + 5}{x + 2}$

44. $f(x) = \dfrac{2x^2 + 12x + 2}{x^2 + 2x - 3}$

45. $f(x) = \dfrac{2x^2 + 5x + 11}{x + 1}$

46. $f(x) = \dfrac{6x^2 - 1}{2x^2 + x + 7}$

In Exercises 47 to 54, graph each rational function.

47. $f(x) = \dfrac{3x - 2}{x}$

48. $f(x) = \dfrac{x + 4}{x - 2}$

49. $f(x) = \dfrac{6}{x^2 + 2}$

50. $f(x) = \dfrac{4x^2}{x^2 + 1}$

51. $f(x) = \dfrac{2x^3 - 4x + 6}{x^2 - 4}$

52. $f(x) = \dfrac{x}{x^3 - 1}$

53. $f(x) = \dfrac{3x^2 - 6}{x^2 - 9}$

54. $f(x) = \dfrac{-x^3 + 6}{x^2}$

55. AVERAGE COST OF SKATEBOARDS The cost, in dollars, of producing x skateboards is given by

$$C(x) = 5.75x + 34{,}200$$

The average cost per skateboard is given by

$$\overline{C}(x) = \frac{C(x)}{x} = \frac{5.75x + 34{,}200}{x}$$

a. Find the average cost per skateboard, to the nearest cent, of producing 5000 and 50,000 skateboards.

b. What is the equation of the horizontal asymptote of the graph of $\overline{C}$? Explain the significance of the horizontal asymptote as it relates to this application.

56. FOOD TEMPERATURE The temperature F, in degrees Fahrenheit, of a dessert placed in a freezer for t hours is given by the rational function

$$F(t) = \frac{60}{t^2 + 2t + 1}, \quad t \ge 0$$

a. Find the temperature of the dessert after it has been in the freezer for 1 hour.

b. Find the temperature of the dessert after 4 hours.

c. What temperature will the dessert approach as $t \to \infty$?

57. **PHYSIOLOGY** One of Poiseuille's Laws states that the resistance R encountered by blood flowing through a blood vessel is given by

$$R(r) = C\frac{L}{r^4}$$

where C is a positive constant determined by the viscosity of the blood, L is the length of the blood vessel, and r is its radius.

a. Explain the meaning of $R(r) \to \infty$ as $r \to 0$.

b. Explain the meaning of $R(r) \to 0$ as $r \to \infty$.

CHAPTER 3 TEST

1. Use synthetic division to divide:

$$(3x^3 + 5x^2 + 4x - 1) \div (x + 2)$$

2. Use the Remainder Theorem to find $P(-2)$ if

$$P(x) = -3x^3 + 7x^2 + 2x - 5$$

3. Show that $x - 1$ is a factor of

$$x^4 - 4x^3 + 7x^2 - 6x + 2$$

4. Examine the leading term of the function given by the equation $P(x) = -3x^3 + 2x^2 - 5x + 2$ and determine the far-left and far-right behavior of the graph of P.

5. Find the real solutions of $3x^3 + 7x^2 - 6x = 0$.

6. Use the Zero Location Theorem to verify that
$$P(x) = 2x^3 - 3x^2 - x + 1$$
has a zero between 1 and 2.

7. Find the zeros of
$$P(x) = (x^2 - 4)^2(2x - 3)(x + 1)^3$$
and state the multiplicity of each.

8. Use the Rational Zero Theorem to list the possible rational zeros of
$$P(x) = 6x^3 - 3x^2 + 2x - 3$$

9. Find, by using the Upper- and Lower-Bound Theorem, the smallest positive integer and the largest negative integer that are upper and lower bounds for the real zeros of the polynomial function
$$P(x) = 2x^4 + 5x^3 - 23x^2 - 38x + 24$$

10. Use Descartes' Rule of Signs to state the number of possible positive and negative real zeros of
$$P(x) = x^4 - 3x^3 + 2x^2 - 5x + 1$$

11. Find the zeros of $P(x) = 2x^3 - 3x^2 - 11x + 6$

12. Given that $2 + 3i$ is a zero of
$$P(x) = 6x^4 - 5x^3 + 12x^2 + 207x + 130$$
find the remaining zeros.

13. Find all the zeros of
$$P(x) = x^5 - 6x^4 + 14x^3 - 14x^2 + 5x$$

14. Find a polynomial of smallest degree that has real coefficients and zeros $1 + i$, 3, and 0.

15. Find all vertical asymptotes of the graph of
$$f(x) = \frac{3x^2 - 2x + 1}{x^2 - 5x + 6}$$

16. Find the horizontal asymptote of the graph of
$$f(x) = \frac{3x^2 - 2x + 1}{2x^2 - 1}$$

17. Graph $f(x) = \dfrac{x^2 - 1}{x^2 - 2x - 3}$. Use an open circle to show the hole in the graph of f.

18. Graph $f(x) = \dfrac{2x^2 + 2x + 1}{x + 1}$ and label the slant asymptote with its equation.

19. The rational function
$$w(t) = \frac{70t + 120}{t + 40}, \quad t \geq 0$$
models Rene's typing speed, in words per minute, after t hours of typing lessons.

   a. Find $w(1)$, $w(10)$, and $w(20)$. Round to the nearest word per minute.

   b. What will Rene's typing speed approach as $t \to \infty$?

20. **MAXIMIZING VOLUME** You are to construct an open box from a rectangular sheet of cardboard that measures 18 inches by 25 inches. To assemble the box, make the four cuts shown in the figure below and then fold on the dashed lines. What value of x (to the nearest 0.01 inch) will produce a box with maximum volume? What is the maximum volume (to the nearest 0.1 cubic inch)?

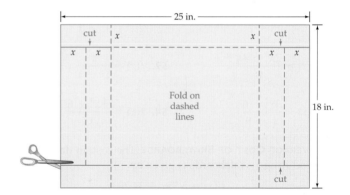

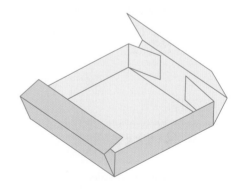

CUMULATIVE REVIEW EXERCISES

1. Write $\dfrac{3 + 4i}{1 - 2i}$ in $a + bi$ form.

2. Use the quadratic formula to solve $x^2 - x - 1 = 0$.

3. Solve: $\sqrt{2x + 5} - \sqrt{x - 1} = 2$

4. Solve: $|x - 3| \leq 11$

5. Find the distance between the points $(2, 5)$ and $(7, -11)$.

6. Explain how to use the graph of $y = x^2$ to produce the graph of $y = (x - 2)^2 + 4$.

7. Find the difference quotient for the function $P(x) = x^2 - 2x - 3$.

8. Given $f(x) = 2x^2 + 5x - 3$ and $g(x) = 4x - 7$, find $(f \circ g)(x)$.

9. Given $f(x) = x^3 - 2x + 7$ and $g(x) = x^2 - 3x - 4$, find $(f - g)(x)$.

10. Use synthetic division to divide $(4x^4 - 2x^2 - 4x - 5)$ by $(x + 2)$.

11. Use the Remainder Theorem to find $P(3)$ for $P(x) = 2x^4 - 3x^2 + 4x - 6$.

12. Determine the far-right behavior of the graph of $P(x) = -3x^4 - x^2 + 7x - 6$.

13. Determine the relative maximum of the polynomial function $P(x) = -3x^3 - x^2 + 4x - 1$. Round to the nearest ten thousandth.

14. Use the Rational Zero Theorem to list all possible rational zeros of $P(x) = 3x^4 - 4x^3 - 11x^2 + 16x - 4$.

15. Use Descartes' Rule of Signs to state the number of possible positive and negative real zeros of $P(x) = x^3 + x^2 + 2x + 4$.

16. Find all zeros of $P(x) = x^3 + x + 10$.

17. Find a polynomial function of smallest degree that has real coefficients and -2 and $3 + i$ as zeros.

18. Write $P(x) = x^3 - 2x^2 + 9x - 18$ as a product of linear factors.

19. Determine the vertical and horizontal asymptotes of the graph of $F(x) = \dfrac{4x^2}{x^2 + x - 6}$.

20. Find the equation of the slant asymptote for the graph of $F(x) = \dfrac{x^3 + 4x^2 + 1}{x^2 + 4}$.

EXPONENTIAL AND LOGARITHMIC FUNCTIONS

Modeling Data with an Exponential Function

The following table shows the time, in hours, before the body of a scuba diver, wearing a 5-millimeter-thick wet suit, reaches hypothermia (95°F) for various water temperatures.

Water Temperature, °F	Time, hours
36	1.5
41	1.8
46	2.6
50	3.1
55	4.9

Source: Data extracted from the *American Journal of Physics,* vol. 71, no. 4 (April 2003), Fig. 3, p. 336.

The following function, which is an example of an exponential function, closely models the data in the table:

$$T(F) = 0.1509(1.0639)^F$$

In this function F represents the Fahrenheit temperature of the water, and T represents the time in hours. A diver can use the function to determine the time it takes to reach hypothermia for water temperatures that are not included in the table.

Exponential functions can be used to model many other situations. Exercise 43 on page 348 uses an exponential function to estimate the growth of broadband Internet connections.

  VIDEO & DVD SSG

Use Two Methods to Solve and Compare Results

Sometimes it is possible to solve a problem in two or more ways. In such situations it is recommended that you use at least two methods to solve the problem, and compare your results. Here is an example of an application that can be solved in more than one way.

Example

In a league of eight basketball teams, each team plays every other team in the league exactly once. How many league games will take place?

Solution

Method 1:   *Use an analytic approach.* Each of the eight teams must play the other seven teams. Using this information, you might be tempted to conclude that there will be $8 \cdot 7 = 56$ games, but this result is too large because it counts each game between two individual teams as two different games. Thus the number of league games will be

$$\frac{8 \cdot 7}{2} = \frac{56}{2} = 28$$

Method 2:   *Make an organized list.* Use the letters A, B, C, D, E, F, G, and H to represent the eight teams. Use the notation AB to represent the game between team A and team B. Do not include BA in your list because it represents the same game between team A and team B.

AB AC AD AE AF AG AH

BC BD BE BF BG BH

CD CE CF CG CH

DE DF DG DH

EF EG EH

FG FH

GH

The list shows that there will be 28 league games.

The procedure of using two different solution methods and comparing results is employed often in this chapter. For instance, see **Example 2, page 379.** In this example, a solution is found by applying algebraic procedures and also by graphing. Notice that both methods produce the same result.

INVERSE FUNCTIONS

• INTRODUCTION TO INVERSE FUNCTIONS

Consider the "doubling" function $f(x) = 2x$ that doubles every input. Some of the ordered pairs of this function are

$$\left\{ (-4, -8), (-1.5, -3), (1, 2), \left(\frac{5}{3}, \frac{10}{3}\right), (7, 14) \right\}$$

Now consider the "halving" function $g(x) = \frac{1}{2}x$ that takes one-half of every input. Some of the ordered pairs of this function are

$$\left\{ (-8, -4), (-3, -1.5), (2, 1), \left(\frac{10}{3}, \frac{5}{3}\right), (14, 7) \right\}$$

Observe that the coordinates of the ordered pairs of g are the reverse of the coordinates of the ordered pairs of f. This is always the case for f and g. Here are two more examples.

$$f(5) = 2(5) = 10 \qquad\qquad g(10) = \frac{1}{2}(10) = 5$$

Ordered pair: (5, 10)   **Ordered pair: (10, 5)**

$$f(a) = 2(a) = 2a \qquad\qquad g(2a) = \frac{1}{2}(2a) = a$$

Ordered pair: (a, 2a)   **Ordered pair: (2a, a)**

For these functions, f and g are called *inverse functions* of one another.

Inverse Function

If the coordinates of the ordered pairs of a function g are the reverse of the coordinates of the ordered pairs of a function f, then g is said to be the inverse function of f.

take note

It is important to remember the information in the paragraph at the right. If f is a function and g is the inverse of f, then

Domain of g = range of f

and

Range of g = domain of f

Because the coordinates of the ordered pairs of the inverse function g are the reverse of the coordinates of the ordered pairs of the function f, the domain of g is the range of f, and the range of g is the domain of f.

Not all functions have an inverse that is a function. Consider, for instance, the "square" function $S(x) = x^2$. Some of the ordered pairs of S are

$$\{(-3, 9), (-1, 1), (0, 0), (1, 1), (3, 9), (5, 25)\}$$

If we reverse the coordinates of the ordered pairs, we have

$$\{(9, -3), (1, -1), (0, 0), (1, 1), (9, 3), (25, 5)\}$$

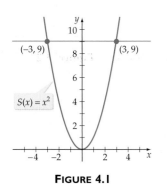

FIGURE 4.1

This set of ordered pairs is not a function because there are ordered pairs, for instance $(9, -3)$ and $(9, 3)$, with the same first coordinate and different second coordinates. In this case, S has an inverse *relation* but not an inverse *function*.

A graph of S is shown in **Figure 4.1.** Note that $x = -3$ and $x = 3$ produce the same value of y. Thus the graph of S fails the horizontal line test, and therefore S is not a one-to-one function. This observation is used in the following theorem.

Condition for an Inverse Function

A function f has an inverse function if and only if f is a one-to-one function.

Recall that increasing functions or decreasing functions are one-to-one functions. Thus we can state the following theorem.

Alternative Condition for an Inverse Function

If f is an increasing function or a decreasing function, then f has an inverse function.

? QUESTION Which of the functions graphed below has an inverse function?

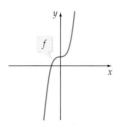
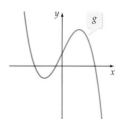
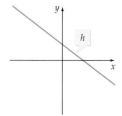

If a function g is the inverse of a function f, we usually denote the inverse function by f^{-1} rather than g. For the doubling and halving functions f and g discussed on page 325, we write

$$f(x) = 2x \qquad f^{-1}(x) = \frac{1}{2}x$$

• GRAPHS OF INVERSE FUNCTIONS

Because the coordinates of the ordered pairs of the inverse of a function f are the reverse of the coordinates of f, we can use them to create a graph of f^{-1}.

---

take note

$f^{-1}(x)$ does not mean $\dfrac{1}{f(x)}$. For

$f(x) = 2x$, $f^{-1}(x) = \dfrac{1}{2}x$ but

$\dfrac{1}{f(x)} = \dfrac{1}{2x}$.

? ANSWER The graph of f is the graph of an increasing function. Therefore, f is a one-to-one function and has an inverse function. The graph of h is the graph of a decreasing function. Therefore, h is a one-to-one function and has an inverse function. The graph of g is not the graph of a one-to-one function. g does not have an inverse function.

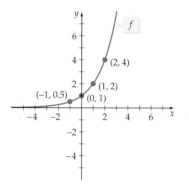

FIGURE 4.2

EXAMPLE 1   **Sketch the Graph of the Inverse of a Function**

Sketch the graph of f^{-1} given that f is the function shown in **Figure 4.2.**

Solution

Because the graph of f passes through $(-1, 0.5)$, $(0, 1)$, $(1, 2)$, and $(2, 4)$, the graph of f^{-1} must pass through $(0.5, -1)$, $(1, 0)$, $(2, 1)$, and $(4, 2)$. Plot the points and then draw a smooth graph through the points, as shown in **Figure 4.3.**

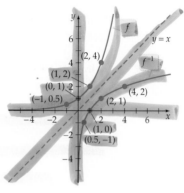

FIGURE 4.3

▶ **TRY EXERCISE 10, PAGE 334**

The graph from the solution to Example 1 is shown again in **Figure 4.4.** Note that the graph of f^{-1} is symmetric to the graph of f with respect to the graph of $y = x$. If the graph were folded along the dashed line, the graph of f would lie on top of the graph of f^{-1}. This is a characteristic of all graphs of functions and their inverses. In **Figure 4.5,** although S does not have an inverse that is a function, the graph of the inverse relation S^{-1} is symmetric to S with respect to the graph of $y = x$.

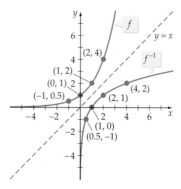

FIGURE 4.4

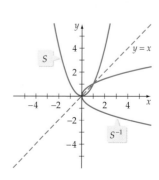

FIGURE 4.5

● COMPOSITION OF A FUNCTION AND ITS INVERSE

Observe the effect, as shown below, of taking the composition of functions that are inverses of one another.

$$f(x) = 2x \qquad\qquad\qquad g(x) = \frac{1}{2}x$$

$$f[g(x)] = 2\left[\frac{1}{2}x\right] \quad \bullet \text{ Replace } x \qquad g[f(x)] = \frac{1}{2}[2x] \quad \bullet \text{ Replace } x$$
$$\text{by } g(x). \qquad\qquad\qquad\qquad\qquad \text{by } f(x).$$
$$f[g(x)] = x \qquad\qquad\qquad\qquad g[f(x)] = x$$

This property of the composition of inverse functions always holds true. When taking the composition of inverse functions, the inverse function reverses the effect of the original function. For the two functions above, f doubles a number, and g halves a number. If you double a number and then take one-half of the result, you are back to the original number.

take note

If we think of a function as a machine, then the Composition of Inverse Functions Property can be represented as shown below. Take any input x for f. Use the output of f as the input for f^{-1}. The result is the original input, x.

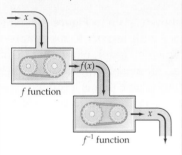

→ x

→ f(x)

f function

→ x

f^{-1} function

Composition of Inverse Functions Property

If f is a one-to-one function, then f^{-1} is the inverse function of f if and only if

$$(f \circ f^{-1})(x) = f[f^{-1}(x)] = x \qquad \text{for all } x \text{ in the domain of } f^{-1}$$

and

$$(f^{-1} \circ f)(x) = f^{-1}[f(x)] = x \qquad \text{for all } x \text{ in the domain of } f.$$

EXAMPLE 2   **Use the Composition of Inverse Functions Property**

Use composition of functions to show that $f^{-1}(x) = 3x - 6$ is the inverse function of $f(x) = \frac{1}{3}x + 2$.

Solution

We must show that $f[f^{-1}(x)] = x$ and $f^{-1}[f(x)] = x$.

$$f(x) = \frac{1}{3}x + 2 \qquad\qquad\qquad f^{-1}(x) = 3x - 6$$

$$f[f^{-1}(x)] = \frac{1}{3}[3x - 6] + 2 \qquad f^{-1}[f(x)] = 3\left[\frac{1}{3}x + 2\right] - 6$$

$$f[f^{-1}(x)] = x \qquad\qquad\qquad f^{-1}[f(x)] = x$$

▶ **TRY EXERCISE 20, PAGE 335**

⌐▱⌐ **INTEGRATING**
▱▱▱ **TECHNOLOGY**

In the standard viewing window of a calculator, the distance between two tic marks on the x-axis is not equal to the distance between two tic marks on the y-axis. As a result, the graph of $y = x$ does not appear to bisect the first and third quadrants. See **Figure 4.6.** This anomaly is important if a graphing calculator is being used to check whether two functions are inverses of one another. Because the graph of $y = x$ does not appear to bisect the first and third quadrants, the graphs of f and f^{-1} will not appear to be symmetric about the graph of $y = x$. The graphs of $f(x) = \dfrac{1}{3}x + 2$ and $f^{-1}(x) = 3x - 6$ from Example 2 are shown in **Figure 4.7.** Notice that the graphs do not appear to be quite symmetric about the graph of $y = x$.

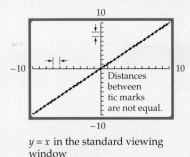

$y = x$ in the standard viewing window

FIGURE 4.6

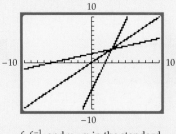

f, f^{-1}, and $y = x$ in the standard viewing window

FIGURE 4.7

To get a better view of a function and its inverse, it is necessary to use the SQUARE viewing window, as in **Figure 4.8.** In this window, the distance between two tic marks on the x-axis is equal to the distance between two tic marks on the y-axis.

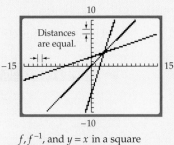

f, f^{-1}, and $y = x$ in a square viewing window

FIGURE 4.8

• FIND AN INVERSE FUNCTION

If a one-to-one function f is defined by an equation, then we can use the following method to find the equation for f^{-1}.

Steps for Finding the Inverse of a Function

To find the equation of the inverse f^{-1} of the one-to-one function f:

1. Substitute y for $f(x)$.
2. Interchange x and y.
3. Solve, if possible, for y in terms of x.
4. Substitute $f^{-1}(x)$ for y.

EXAMPLE 3   Find the Inverse of a Function

Find the inverse of $f(x) = 3x + 8$.

Solution

$$f(x) = 3x + 8$$
$$y = 3x + 8 \qquad \text{• Replace } f(x) \text{ by } y.$$
$$x = 3y + 8 \qquad \text{• Interchange } x \text{ and } y.$$
$$x - 8 = 3y \qquad \text{• Solve for } y.$$
$$\frac{x - 8}{3} = y$$
$$\frac{1}{3}x - \frac{8}{3} = f^{-1}(x) \qquad \text{• Replace } y \text{ by } f^{-1}(x).$$

The inverse function is given by $f^{-1}(x) = \dfrac{1}{3}x - \dfrac{8}{3}$.

▶ **TRY EXERCISE 28, PAGE 335**

EXAMPLE 4   Find the Inverse of a Function

Find the inverse of $f(x) = \dfrac{2x + 1}{x}$, $x \neq 0$.

Solution

$$f(x) = \frac{2x + 1}{x}$$

$$y = \frac{2x + 1}{x}$$

• Replace $f(x)$ by y.

$$x = \frac{2y + 1}{y}$$

• Interchange x and y.

$$xy = 2y + 1$$

• Solve for y.

$$xy - 2y = 1$$

$$y(x - 2) = 1$$

• Factor the left side.

$$y = \frac{1}{x - 2}$$

$$f^{-1}(x) = \frac{1}{x - 2}, x \neq 2$$

• Replace y by $f^{-1}(x)$.

▶ **TRY EXERCISE 34, PAGE 335**

? QUESTION If f is a one-to-one function and $f(4) = 5$, what is $f^{-1}(5)$?

The graph of $f(x) = x^2 + 4x + 3$ is shown in **Figure 4.9a.** The function f is not a one-to-one function and therefore does not have an inverse function. However, the function given by $G(x) = x^2 + 4x + 3$, shown in **Figure 4.9b,** for which the domain is restricted to $\{x \mid x \geq -2\}$, is a one-to-one function and has an inverse function G^{-1}. This is shown in Example 5.

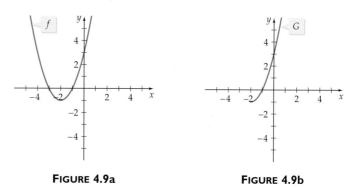

FIGURE 4.9a   **FIGURE 4.9b**

? ANSWER Because f^{-1} is the inverse function of f, the coordinates of the ordered pairs of f^{-1} are the reverse of the coordinates of the ordered pairs of f. Therefore, $f^{-1}(5) = 4$.

EXAMPLE 5   **Find the Inverse of a Function with a Restricted Domain**

Find the inverse of $G(x) = x^2 + 4x + 3$, where the domain of G is $\{x \mid x \geq -2\}$.

Solution

$$G(x) = x^2 + 4x + 3$$
$$y = x^2 + 4x + 3$$ • Replace $G(x)$ by y.
$$x = y^2 + 4y + 3$$ • Interchange x and y.
$$x = (y^2 + 4y + 4) - 4 + 3$$ • Solve for y by completing the square of $y^2 + 4y$.
$$x = (y + 2)^2 - 1$$ • Factor.
$$x + 1 = (y + 2)^2$$ • Add 1 to each side of the equation.
$$\sqrt{x + 1} = \sqrt{(y + 2)^2}$$ • Take the square root of each side of the equation.
$$\pm\sqrt{x + 1} = y + 2$$ • Recall that if $a^2 = b$, then $a = \pm\sqrt{b}$.
$$\pm\sqrt{x + 1} - 2 = y$$

Because the domain of G is $\{x \mid x \geq -2\}$, the range of G^{-1} is $\{y \mid y \geq -2\}$. This means that we must choose the positive value of $\pm\sqrt{x + 1}$. Thus $G^{-1}(x) = \sqrt{x + 1} - 2$. See **Figure 4.10.**

▶ TRY EXERCISE 40, PAGE 335

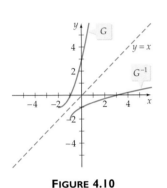

take note

Recall that the range of a function f is the domain of f^{-1}, and the domain of f is the range of f^{-1}.

FIGURE 4.10

● **APPLICATION**

There are practical applications of finding the inverse of a function. Here is one in which a shirt size in the United States is converted to a shirt size in Italy. Finding the inverse function gives the function that converts a shirt size in Italy to a shirt size in the United States.

EXAMPLE 6   **Solve an Application**

   The function $IT(x) = 2x + 8$ converts a men's shirt size x in the United States to the equivalent shirt size in Italy.

a.   Use IT to determine the equivalent Italian shirt size for a size 16.5 U.S. shirt.

b.   Find IT^{-1} and use IT^{-1} to determine the U.S. men's shirt size that is equivalent to an Italian shirt size of 36.

Solution

a. $IT(16.5) = 2(16.5) + 8 = 33 + 8 = 41$

A size 16.5 U.S. shirt is equivalent to a size 41 Italian shirt.

b. To find the inverse function, begin by substituting y for $IT(x)$.

$$IT(x) = 2x + 8$$
$$y = 2x + 8$$
$$x = 2y + 8 \qquad \text{• Interchange } x \text{ and } y.$$
$$x - 8 = 2y \qquad \text{• Solve for } y.$$
$$\frac{x - 8}{2} = y$$

In inverse notation, the above equation can be written as

$$IT^{-1}(x) = \frac{x - 8}{2} \qquad \text{or} \qquad IT^{-1}(x) = \frac{1}{2}x - 4$$

Substitute 36 for x to find the equivalent U.S. shirt size.

$$IT^{-1}(36) = \frac{1}{2}(36) - 4 = 18 - 4 = 14$$

A size 36 Italian shirt is equivalent to a size 14 U.S. shirt.

▶ **TRY EXERCISE 50, PAGE 336**

 **INTEGRATING TECHNOLOGY**

Some graphing utilities can be used to draw the graph of the inverse of a function without the user having to find the inverse function. For instance, **Figure 4.11** shows the graph of $f(x) = 0.1x^3 - 4$. The graphs of f and f^{-1} are both shown in **Figure 4.12,** along with the graph of $y = x$. Note that the graph of f^{-1} is the reflection of the graph of f with respect to the graph of $y = x$. The display shown in **Figure 4.12** was produced on a TI-83 graphing calculator by using the DrawInv command, which is in the DRAW menu.

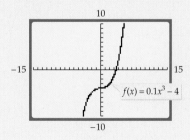

FIGURE 4.11

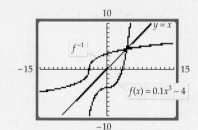

FIGURE 4.12

 TOPICS FOR DISCUSSION

1. If $f(x) = 3x + 1$, what are the values of $f^{-1}(2)$ and $[f(2)]^{-1}$?

2. How are the domain and range of a one-to-one function f related to the domain and range of the inverse function of f?

3. How is the graph of the inverse of a function f related to the graph of f?

4. The function $f(x) = -x$ is its own inverse. Find at least two other functions that are their own inverses.

5. What are the steps in finding the inverse of a one-to-one function?

EXERCISE SET 4.1

In Exercises 1 to 4, assume that the given function has an inverse function.

1. Given $f(3) = 7$, find $f^{-1}(7)$.

2. Given $g(-3) = 5$, find $g^{-1}(5)$.

3. Given $h^{-1}(-3) = -4$, find $h(-4)$.

4. Given $f^{-1}(7) = 0$, find $f(0)$.

5. If 3 is in the domain of f^{-1}, find $f[f^{-1}(3)]$.

6. If f is a one-to-one function and $f(0) = 5$, $f(1) = 2$, and $f(2) = 7$, find:

    a. $f^{-1}(5)$ **b.** $f^{-1}(2)$

7. The domain of the inverse function f^{-1} is the _____ of f.

8. The range of the inverse function f^{-1} is the _____ of f.

In Exercises 9 to 16, draw the graph of the inverse relation. Is the inverse relation a function?

9.

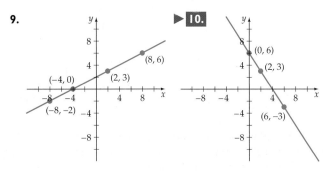

10.

11.

12.

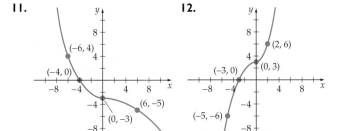

13.

14.

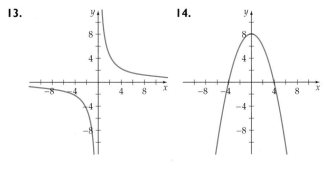

15.

16.

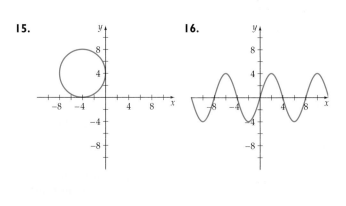

In Exercises 17 to 22, use composition of functions to determine whether f and g are inverses of one another.

17. $f(x) = 4x$; $g(x) = \dfrac{x}{4}$

18. $f(x) = 3x$; $g(x) = \dfrac{1}{3x}$

19. $f(x) = 4x - 1$; $g(x) = \dfrac{1}{4}x + \dfrac{1}{4}$

▶ **20.** $f(x) = \dfrac{1}{2}x - \dfrac{3}{2}$; $g(x) = 2x + 3$

21. $f(x) = -\dfrac{1}{2}x - \dfrac{1}{2}$; $g(x) = -2x + 1$

22. $f(x) = 3x + 2$; $g(x) = \dfrac{1}{3}x - \dfrac{2}{3}$

In Exercises 23 to 26, find the inverse of the function. If the function does not have an inverse function, write "no inverse function."

23. $\{(-3, 1), (-2, 2), (1, 5), (4, -7)\}$

24. $\{(-5, 4), (-2, 3), (0, 1), (3, 2), (7, 11)\}$

25. $\{(0, 1), (1, 2), (2, 4), (3, 8), (4, 16)\}$

26. $\{(1, 0), (10, 1), (100, 2), (1000, 3), (10{,}000, 4)\}$

In Exercises 27 to 44, find $f^{-1}(x)$. State any restrictions on the domain of $f^{-1}(x)$.

27. $f(x) = 2x + 4$

▶ **28.** $f(x) = 4x - 8$

29. $f(x) = 3x - 7$

30. $f(x) = -3x - 8$

31. $f(x) = -2x + 5$

32. $f(x) = -x + 3$

33. $f(x) = \dfrac{2x}{x - 1}$, $x \neq 1$

▶ **34.** $f(x) = \dfrac{x}{x - 2}$, $x \neq 2$

35. $f(x) = \dfrac{x - 1}{x + 1}$, $x \neq -1$

36. $f(x) = \dfrac{2x - 1}{x + 3}$, $x \neq -3$

37. $f(x) = x^2 + 1$, $x \geq 0$

38. $f(x) = x^2 - 4$, $x \geq 0$

39. $f(x) = \sqrt{x - 2}$, $x \geq 2$

▶ **40.** $f(x) = \sqrt{4 - x}$, $x \leq 4$

41. $f(x) = x^2 + 4x$, $x \geq -2$

42. $f(x) = x^2 - 6x$, $x \leq 3$

43. $f(x) = x^2 + 4x - 1$, $x \leq -2$

44. $f(x) = x^2 - 6x + 1$, $x \geq 3$

45.   **GEOMETRY** The volume of a cube is given by $V(x) = x^3$, where x is the measure of the length of a side of the cube. Find $V^{-1}(x)$ and explain what it represents.

46.   **UNIT CONVERSIONS** The function $f(x) = 12x$ converts feet, x, into inches, $f(x)$. Find $f^{-1}(x)$ and explain what it determines.

47.   **UNIT CONVERSIONS** A conversion function such as the one in Exercise 46 converts a measurement in one unit into another unit. Is a conversion function always a one-to-one function? Does a conversion function always have an inverse function? Explain your answer.

48.   **GRADING SCALE** Does the grading scale function given below have an inverse function? Explain your answer.

Score	Grade
90–100	A
80–89	B
70–79	C
60–69	D
0–59	F

49.   **FASHION** The function $s(x) = 2x + 24$ can be used to convert a U.S. women's shoe size into an Italian women's shoe size. Determine the function $s^{-1}(x)$ that can be used to convert an Italian women's shoe size to its equivalent U.S. shoe size.

▶ **50.** FASHION The function $K(x) = 1.3x - 4.7$ converts a men's shoe size in the United States to the equivalent shoe size in the United Kingdom. Determine the function $K^{-1}(x)$ that can be used to convert a United Kingdom men's shoe size to its equivalent U.S. shoe size.

51. COMPENSATION The monthly earnings $E(s)$, in dollars, of a software sales executive is given by $E(s) = 0.05s + 2500$, where s is the value, in dollars, of the software sold by the executive during the month. Find $E^{-1}(s)$, and explain how the executive could use this function.

52. POSTAGE Does the first-class postage rate function given below have an inverse function? Explain your answer.

Weight (in ounces)	Cost
$0 < w \le 1$	\$.37
$1 < w \le 2$	\$.60
$2 < w \le 3$	\$.83
$3 < w \le 4$	\$1.06

53. INTERNET COMMERCE Functions and their inverses can be used to create secret codes that are used to secure business transactions made over the Internet. Let A = 10, B = 11, ..., and Z = 35. Let $f(x) = 2x - 1$ define a coding function. Code the word MATH (M—22, A—10, T—29, H—17), which is 22102917, by finding $f(22102917)$. Now find the inverse of f and show that applying f^{-1} to the output of f returns the original word.

54. CRYPTOGRAPHY A friend is using the letter-number correspondence in Exercise 53 and the coding function $f(x) = 2x + 3$. Suppose this friend sends you the coded message 5658602671. Decode this message.

In Exercises 55 to 60, answer the question without finding the equation of the linear function.

55. Suppose that f is a linear function, $f(2) = 7$, and $f(5) = 12$. If $f(4) = c$, then is c less than 7, between 7 and 12, or greater than 12? Explain your answer.

56. Suppose that f is a linear function, $f(1) = 13$, and $f(4) = 9$. If $f(3) = c$, then is c less than 9, between 9 and 13, or greater than 13? Explain your answer.

57. Suppose that f is a linear function, $f(2) = 3$, and $f(5) = 9$. Between which two numbers is $f^{-1}(6)$?

58. Suppose that f is a linear function, $f(5) = -1$, and $f(9) = -3$. Between which two numbers is $f^{-1}(-2)$?

59. Suppose that g is a linear function, $g^{-1}(3) = 4$, and $g^{-1}(7) = 8$. Between which two numbers is $g(5)$?

60. Suppose that g is a linear function, $g^{-1}(-2) = 5$, and $g^{-1}(0) = -3$. Between which two numbers is $g(0)$?

───── *CONNECTING CONCEPTS* ─────

In Exercises 61 and 62, find the inverse of the given function.

61. $f(x) = ax + b, \quad a \ne 0$

62. $f(x) = ax^2 + bx + c, \quad a \ne 0, \quad x \ge -\dfrac{b}{2a}$

63. Use a graph of $f(x) = -x + 3$ to explain why f is its own inverse.

64. Use a graph of $f(x) = \sqrt{16 - x^2}$, with $0 \le x \le 4$, to explain why f is its own inverse.

Only one-to-one functions have inverses that are functions. **In Exercises 65 to 68, determine if the given function is a one-to-one function.**

65. $p(t) = \sqrt{9 - t}$

66. $v(t) = \sqrt{16 + t}$

67. $F(x) = |x| + x$

68. $T(x) = |x^2 - 6|, \quad x \ge 0$

PREPARE FOR SECTION 4.2

69. Evaluate: 2^3 [P.2]

70. Evaluate: 3^{-4} [P.2]

71. Evaluate: $\dfrac{2^2 + 2^{-2}}{2}$ [P.2/P.4]

72. Evaluate: $\dfrac{3^2 - 3^{-2}}{2}$ [P.2/P.4]

73. Evaluate $f(x) = 10^x$ for $x = -1, 0, 1,$ and 2. [2.2]

74. Evaluate $f(x) = \left(\dfrac{1}{2}\right)^x$ for $x = -1, 0, 1,$ and 2. [2.2]

PROJECTS

1. **INTERSECTION POINTS FOR THE GRAPHS OF f AND f^{-1}** For each of the following, graph f and its inverse.

i. $f(x) = 2x - 4$

ii. $f(x) = -x + 2$

iii. $f(x) = x^3 + 1$

iv. $f(x) = x - 3$

v. $f(x) = -3x + 2$

vi. $f(x) = \dfrac{1}{x}$

a. Do the graphs of a function and its inverse always intersect?

b. If the graphs of a function and its inverse intersect at one point, what is true about the coordinates of the point of intersection?

c. Can the graphs of a function and its inverse intersect at more than one point?

EXPONENTIAL FUNCTIONS AND THEIR APPLICATIONS

SECTION 4.2

- EXPONENTIAL FUNCTIONS
- GRAPHS OF EXPONENTIAL FUNCTIONS
- THE NATURAL EXPONENTIAL FUNCTION
- APPLICATIONS OF EXPONENTIAL FUNCTIONS

● EXPONENTIAL FUNCTIONS

In 1965, Gordon Moore, one of the cofounders of Intel Corporation, observed that the maximum number of transistors that could be placed on a microprocessor seemed to be doubling every 18 to 24 months. **Table 4.1** below shows how the maximum number of transistors on various Intel processors has changed over time. (*Source:* Intel Museum home page.)

TABLE 4.1

Year	1971	1979	1983	1985	1990	1993	1995	1998	2000
Number of transistors per microprocessor (in thousands)	2.3	31	110	280	1200	3100	5500	14,000	42,000

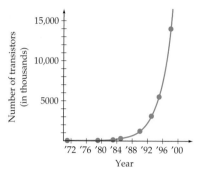

FIGURE 4.13

Moore's Law

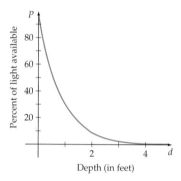

FIGURE 4.14

The curve that approximately passes through the points is a mathematical model of the data. See **Figure 4.13.** The model is based on an *exponential* function.

When light enters water, the intensity of the light decreases with the depth of the water. The graph in **Figure 4.14** shows a model, for Lake Michigan, of the decrease in the percentage of available light as the depth of the water increases. This model is also based on an exponential function.

> **Definition of an Exponential Function**
>
> The exponential function with base b is defined by
>
> $$f(x) = b^x$$
>
> where $b > 0$, $b \neq 1$, and x is a real number.

The base b of $f(x) = b^x$ is required to be positive. If the base were a negative number, the value of the function would be a complex number for some values of x. For instance, if $b = -4$ and $x = \dfrac{1}{2}$, then $f\left(\dfrac{1}{2}\right) = (-4)^{1/2} = 2i$. To avoid complex number values of a function, the base of any exponential function must be a nonnegative number. Also, b is defined such that $b \neq 1$ because $f(x) = 1^x = 1$ is a constant function.

In the following examples we evaluate $f(x) = 2^x$ at $x = 3$ and $x = -2$.

$$f(3) = 2^3 = 8 \qquad f(-2) = 2^{-2} = \frac{1}{2^2} = \frac{1}{4}$$

To evaluate the exponential function $f(x) = 2^x$ at an irrational number such as $x = \sqrt{2}$, we use a rational approximation of $\sqrt{2}$, such as 1.4142, and a calculator to obtain an approximation of the function. For instance, if $f(x) = 2^x$, then $f\left(\sqrt{2}\right) = 2^{\sqrt{2}} \approx 2^{1.4142} \approx 2.6651$.

EXAMPLE 1   **Evaluate an Exponential Function**

Evaluate $f(x) = 3^x$ at $x = 2$, $x = -4$, and $x = \pi$.

Solution

$$f(2) = 3^2 = 9$$

$$f(-4) = 3^{-4} = \frac{1}{3^4} = \frac{1}{81}$$

$$f(\pi) = 3^\pi \approx 3^{3.1415927} \approx 31.54428 \qquad \bullet \text{ Evaluate with the aid of a calculator.}$$

▶ **TRY EXERCISE 2, PAGE 346**

● **GRAPHS OF EXPONENTIAL FUNCTIONS**

The graph of $f(x) = 2^x$ is shown in **Figure 4.15.** The coordinates of some of the points on the curve are given in **Table 4.2.**

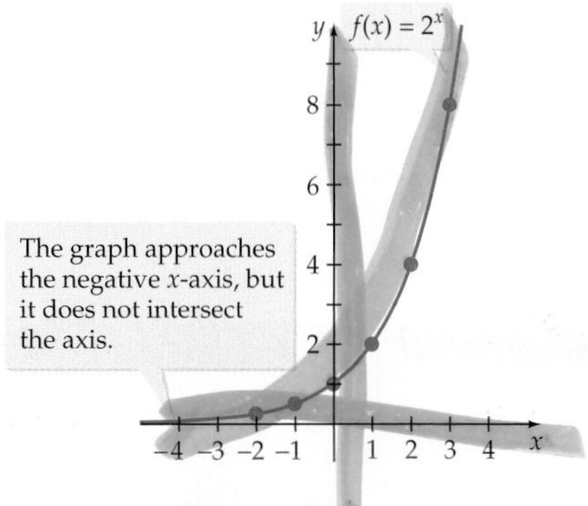

FIGURE 4.15

The graph approaches the negative x-axis, but it does not intersect the axis.

TABLE 4.2

x	$y = f(x) = 2^x$	(x, y)
−2	$f(-2) = 2^{-2} = \dfrac{1}{4}$	$\left(-2, \dfrac{1}{4}\right)$
−1	$f(-1) = 2^{-1} = \dfrac{1}{2}$	$\left(-1, \dfrac{1}{2}\right)$
0	$f(0) = 2^0 = 1$	$(0, 1)$
1	$f(1) = 2^1 = 2$	$(1, 2)$
2	$f(2) = 2^2 = 4$	$(2, 4)$
3	$f(3) = 2^3 = 8$	$(3, 8)$

Note the following properties of the graph of the exponential function $f(x) = 2^x$.

- The y-intercept is $(0, 1)$.
- The graph passes through $(1, 2)$.
- As x decreases without bound (that is, as $x \to -\infty$), $f(x) \to 0$.
- The graph is a smooth continuous increasing curve.

Now consider the graph of an exponential function for which the base is between 0 and 1. The graph of $f(x) = \left(\dfrac{1}{2}\right)^x$ is shown in **Figure 4.16.** The coordinates of some of the points on the curve are given in **Table 4.3.**

TABLE 4.3

x	$y = f(x) = \left(\dfrac{1}{2}\right)^x$	(x, y)
−3	$f(-3) = \left(\dfrac{1}{2}\right)^{-3} = 8$	$(-3, 8)$
−2	$f(-2) = \left(\dfrac{1}{2}\right)^{-2} = 4$	$(-2, 4)$
−1	$f(-1) = \left(\dfrac{1}{2}\right)^{-1} = 2$	$(-1, 2)$
0	$f(0) = \left(\dfrac{1}{2}\right)^0 = 1$	$(0, 1)$
1	$f(1) = \left(\dfrac{1}{2}\right)^1 = \dfrac{1}{2}$	$\left(1, \dfrac{1}{2}\right)$
2	$f(2) = \left(\dfrac{1}{2}\right)^2 = \dfrac{1}{4}$	$\left(2, \dfrac{1}{4}\right)$

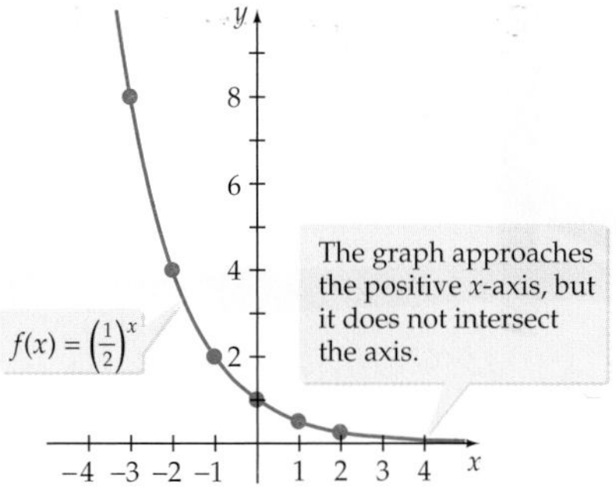

The graph approaches the positive x-axis, but it does not intersect the axis.

$f(x) = \left(\dfrac{1}{2}\right)^x$

FIGURE 4.16

Note the following properties of the graph of $f(x) = \left(\dfrac{1}{2}\right)^x$ in **Figure 4.16.**

- The y-intercept is $(0, 1)$.
- The graph passes through $\left(1, \dfrac{1}{2}\right)$.

- As x increases without bound, the y-values decrease toward 0. That is, as $x \to \infty$, $f(x) \to 0$.

- The graph is a smooth continuous decreasing curve.

The basic properties of exponential functions are provided in the following summary.

Properties of $f(x) = b^x$

For positive real numbers b, $b \neq 1$, the exponential function defined by $f(x) = b^x$ has the following properties:

1. The function f is a one-to-one function. It has the set of real numbers as its domain and the set of positive real numbers as its range.

2. The graph of f is a smooth continuous curve with a y-intercept of $(0, 1)$, and the graph passes through $(1, b)$.

3. If $b > 1$, f is an increasing function and the graph of f is asymptotic to the negative x-axis. [As $x \to \infty$, $f(x) \to \infty$, and as $x \to -\infty$, $f(x) \to 0$.] See **Figure 4.17a.**

4. If $0 < b < 1$, f is a decreasing function and the graph of f is asymptotic to the positive x-axis. [As $x \to -\infty$, $f(x) \to \infty$, and as $x \to \infty$, $f(x) \to 0$.] See **Figure 4.17b.**

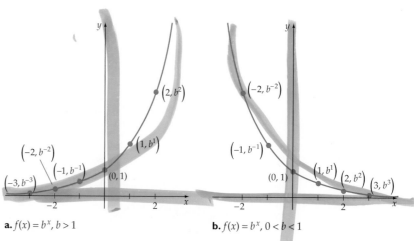

a. $f(x) = b^x$, $b > 1$

b. $f(x) = b^x$, $0 < b < 1$

FIGURE 4.17

❓ QUESTION What is the x-intercept of the graph of $f(x) = \left(\dfrac{1}{3}\right)^x$?

❓ ANSWER The graph does not have an x-intercept. As x increases, the graph approaches the x-axis, but it does not intersect the x-axis.

EXAMPLE 2   **Graph an Exponential Function**

Graph $g(x) = \left(\dfrac{3}{4}\right)^x$.

Solution

Because the base $\dfrac{3}{4}$ is less than 1, we know that the graph of g is a decreasing function that is asymptotic to the positive x-axis. The y-intercept of the graph is the point $(0, 1)$, and the graph also passes through $\left(1, \dfrac{3}{4}\right)$.

Plot a few additional points (see **Table 4.4**), and then draw a smooth curve through the points as in **Figure 4.18**.

TABLE 4.4

x	$y = g(x) = \left(\dfrac{3}{4}\right)^x$	(x, y)
-3	$\left(\dfrac{3}{4}\right)^{-3} = \dfrac{64}{27}$	$\left(-3, \dfrac{64}{27}\right)$
-2	$\left(\dfrac{3}{4}\right)^{-2} = \dfrac{16}{9}$	$\left(-2, \dfrac{16}{9}\right)$
-1	$\left(\dfrac{3}{4}\right)^{-1} = \dfrac{4}{3}$	$\left(-1, \dfrac{4}{3}\right)$
2	$\left(\dfrac{3}{4}\right)^{2} = \dfrac{9}{16}$	$\left(2, \dfrac{9}{16}\right)$
3	$\left(\dfrac{3}{4}\right)^{3} = \dfrac{27}{64}$	$\left(3, \dfrac{27}{64}\right)$

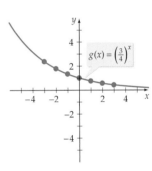

FIGURE 4.18

▶ **TRY EXERCISE 22, PAGE 347**

Consider the functions $F(x) = 2^x - 3$ and $G(x) = 2^{x-3}$. You can construct the graphs of these functions by plotting points; however, it is easier to construct their graphs by using translations of the graph of $f(x) = 2^x$, as shown in Example 3.

EXAMPLE 3   **Use a Translation to Produce a Graph**

a. Explain how to use the graph of $f(x) = 2^x$ to produce the graph of $F(x) = 2^x - 3$.

b. Explain how to use the graph of $f(x) = 2^x$ to produce the graph of $G(x) = 2^{x-3}$.

Solution

a. $F(x) = 2^x - 3 = f(x) - 3$. The graph of F is a vertical translation of f down 3 units, as shown in **Figure 4.19**.

Continued ▶

b. $G(x) = 2^{x-3} = f(x - 3)$. The graph of G is a horizontal translation of f to the right 3 units, as shown in **Figure 4.20.**

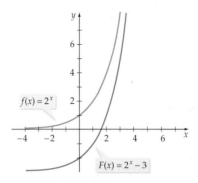

FIGURE 4.19

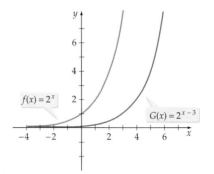

FIGURE 4.20

▶ **TRY EXERCISE 28, PAGE 347**

The graphs of some functions can be constructed by stretching, compressing, or reflecting the graph of an exponential function.

| **EXAMPLE 4** | **Use Stretching or Reflecting Procedures to Produce a Graph** |

a. Explain how to use the graph of $f(x) = 2^x$ to produce the graph of $M(x) = 2(2^x)$.

b. Explain how to use the graph of $f(x) = 2^x$ to produce the graph of $N(x) = 2^{-x}$.

Solution

a. $M(x) = 2(2^x) = 2f(x)$. The graph of M is a vertical stretching of f, as shown in **Figure 4.21.** If (x, y) is a point on the graph of $f(x) = 2^x$, then $(x, 2y)$ is a point on the graph of M.

b. $N(x) = 2^{-x} = f(-x)$. The graph of N is the graph of f reflected across the y-axis, as shown in **Figure 4.22.** If (x, y) is a point on the graph of $f(x) = 2^x$, then $(-x, y)$ is a point on the graph of N.

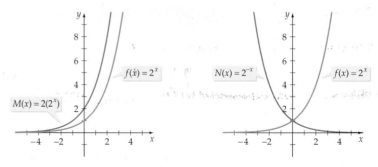

FIGURE 4.21

FIGURE 4.22

▶ **TRY EXERCISE 30, PAGE 347**

• THE NATURAL EXPONENTIAL FUNCTION

The irrational number π is often used in applications that involve circles. Another irrational number, denoted by the letter e, is useful in applications that involve growth or decay.

Definition of e

The **number e** is defined as the number that

$$\left(1 + \frac{1}{n}\right)^n$$

approaches as n increases without bound.

The letter e was chosen in honor of the Swiss mathematician Leonhard Euler. He was able to compute the value of e to several decimal places by evaluating $\left(1 + \dfrac{1}{n}\right)^n$ for large values of n, as shown in **Table 4.5**.

TABLE 4.5

Value of n	Value of $\left(1 + \dfrac{1}{n}\right)^n$
1	2
10	2.59374246
100	2.704813829
1000	2.716923932
10,000	2.718145927
100,000	2.718268237
1,000,000	2.718280469
10,000,000	2.718281693

The value of e accurate to eight decimal places is 2.71828183.

The Natural Exponential Function

For all real numbers x, the function defined by

$$f(x) = e^x$$

is called the **natural exponential function.**

A calculator can be used to evaluate e^x for specific values of x. For instance,

$$e^2 \approx 7.389056, \quad e^{3.5} \approx 33.115452, \quad \text{and} \quad e^{-1.4} \approx 0.246597$$

On a TI-83 calculator the e^x function is located above the $\boxed{\text{LN}}$ key.

MATH MATTERS

Leonhard Euler (1707–1783)

Some mathematicians consider Euler to be the greatest mathematician of all time. He certainly was the most prolific writer of mathematics of all time. He made substantial contributions in the areas of number theory, geometry, calculus, differential equations, differential geometry, topology, complex variables, and analysis, to name but a few. Euler was the first to introduce many of the mathematical notations that we use today. For instance, he introduced the symbol i for the square root of -1, the symbol π for pi, the functional notation $f(x)$, and the letter e for the base of the natural exponential function. Euler's computational skills were truly amazing. The mathematician François Arago remarked, "Euler calculated without apparent effort, as men breathe, or as eagles sustain themselves in the wind."

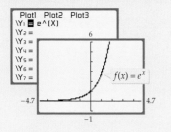
To graph $f(x) = e^x$, use a calculator to find the range values for a few domain values. The range values in **Table 4.6** have been rounded to the nearest tenth.

TABLE 4.6

x	-2	-1	0	1	2
$f(x) = e^x$	0.1	0.4	1.0	2.7	7.4

Plot the points given in **Table 4.6,** and then connect the points with a smooth curve. Because $e > 1$, we know that the graph is an increasing function. To the far left, the graph will approach the x-axis. The y-intercept is $(0, 1)$. See **Figure 4.23.** Note in **Figure 4.24** how the graph of $f(x) = e^x$ compares with the graphs of $g(x) = 2^x$ and $h(x) = 3^x$. You may have anticipated that the graph of $f(x) = e^x$ would lie between the two other graphs because e is between 2 and 3.

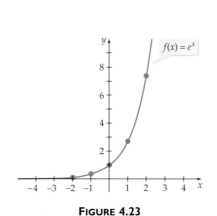

FIGURE 4.23

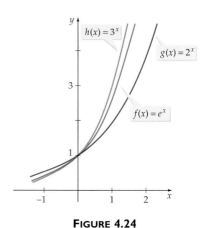

FIGURE 4.24

● APPLICATIONS OF EXPONENTIAL FUNCTIONS

Many applications can be effectively modeled by functions that involve an exponential function. For instance, in Example 5 we make use of a function that involves an exponential function to model the temperature of a cup of coffee.

EXAMPLE 5 **Use a Mathematical Model**

A cup of coffee is heated to 160°F and placed in a room that maintains a temperature of 70°F. The temperature T of the coffee, in degrees Fahrenheit, after t minutes is given by

$$T = 70 + 90e^{-0.0485t}$$

a. Find the temperature of the coffee, to the nearest degree, 20 minutes after it is placed in the room.

b. Use a graphing utility to determine when the temperature of the coffee will reach 90°F.

take note

In Example 5**b.**, we use a graphing
utility to solve the equation
$90 = 70 + 90e^{-0.0485t}$. Analytic
methods of solving this type of
equation without the use of a
graphing utility will be developed in
Section 4.5.

Solution

a. $T = 70 + 90e^{-0.0485t}$

$= 70 + 90e^{-0.0485 \cdot (20)}$ • **Substitute 20 for t.**

$\approx 70 + 34.1$

≈ 104.1

After 20 minutes the temperature of the coffee is about 104°F.

b.   Graph $T = 70 + 90e^{-0.0485t}$ and $T = 90$. See the following figure.

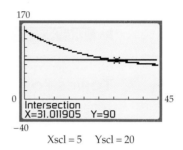

Xscl = 5   Yscl = 20

The graphs intersect at about $(31.01, 90)$. It takes the coffee about
31 minutes to cool to 90°F.

▶ **TRY EXERCISE 44, PAGE 348**

EXAMPLE 6   **Use a Mathematical Model**

 The weekly revenue R, in dollars, from the sale of a product varies with
time according to the function

$$R(x) = \frac{1760}{8 + 14e^{-0.03x}}$$

where x is the number of weeks that have passed since the product was put
on the market. What will the weekly revenue approach as time goes by?

Solution

Method 1   Use a graphing utility to graph $R(x)$, and use the TRACE feature
to see what happens to the revenue as the time increases. The following
graph shows that as the weeks go by, the
weekly revenue will increase and approach
$220.00 per week.

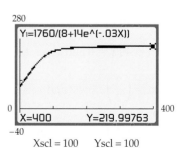

Xscl = 100   Yscl = 100

Continued ▶

Method 2 Write the revenue function in the following form.

$$R(x) = \frac{1760}{8 + \dfrac{14}{e^{0.03x}}} \qquad \bullet\ 14e^{-0.03x} = \frac{14}{e^{0.03x}}$$

As x increases without bound, $e^{0.03x}$ increases without bound, and the fraction $\dfrac{14}{e^{0.03x}}$ approaches 0. Therefore, as $x \to \infty$, $R(x) \to \dfrac{1760}{8 + 0} = 220$. Both methods indicate that as the number of weeks increases, the revenue approaches \$220 per week.

▶ **TRY EXERCISE 54, PAGE 349**

TOPICS FOR DISCUSSION

1. Explain how to use the graph of $f(x) = 2^x$ to produce the graph of $g(x) = 2^{(x-3)} + 4$.

2. At what point does the function $g(x) = e^{-x^2/2}$ take on its maximum value?

3. Without using a graphing utility, determine whether the revenue function $R(t) = 10 + e^{-0.05t}$ is an increasing function or a decreasing function.

4. Discuss the properties of the graph of $f(x) = b^x$ when $b > 1$.

5. What is the base of the natural exponential function? How is it calculated? What is its approximate value?

EXERCISE SET 4.2

In Exercises 1 to 8, evaluate the exponential function for the given x-values.

1. $f(x) = 3^x$; $x = 0$ and $x = 4$

▶ **2.** $f(x) = 5^x$; $x = 3$ and $x = -2$

3. $g(x) = 10^x$; $x = -2$ and $x = 3$

4. $g(x) = 4^x$; $x = 0$ and $x = -1$

5. $h(x) = \left(\dfrac{3}{2}\right)^x$; $x = 2$ and $x = -3$

6. $h(x) = \left(\dfrac{2}{5}\right)^x$; $x = -1$ and $x = 3$

7. $j(x) = \left(\dfrac{1}{2}\right)^x$; $x = -2$ and $x = 4$

8. $j(x) = \left(\dfrac{1}{4}\right)^x$; $x = -1$ and $x = 5$

 In Exercises 9 to 14, use a calculator to evaluate the exponential function for the given x-value. Round to the nearest hundredth.

9. $f(x) = 2^x$, $x = 3.2$

10. $f(x) = 3^x$, $x = -1.5$

11. $g(x) = e^x$, $x = 2.2$

12. $g(x) = e^x$, $x = -1.3$

13. $h(x) = 5^x$, $x = \sqrt{2}$

14. $h(x) = 0.5^x$, $x = \pi$

15. Examine the following four functions and the graphs labeled **a, b, c,** and **d.** For each graph, determine which function has been graphed.

$$f(x) = 5^x \qquad g(x) = 1 + 5^{-x}$$
$$h(x) = 5^{x+3} \qquad k(x) = 5^x + 3$$

a.
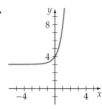

b.
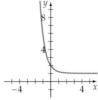

c.
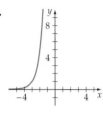

d.
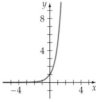

16. Examine the following four functions and the graphs labeled **a, b, c,** and **d.** For each graph, determine which function has been graphed.

$$f(x) = \left(\frac{1}{4}\right)^x \qquad g(x) = \left(\frac{1}{4}\right)^{-x}$$
$$h(x) = \left(\frac{1}{4}\right)^{x-2} \qquad k(x) = 3\left(\frac{1}{4}\right)^x$$

a.
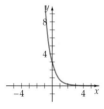

b.
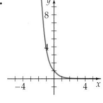

c.
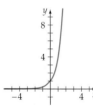

d.

In Exercises 17 to 24, sketch the graph of each function.

17. $f(x) = 3^x$

18. $f(x) = 4^x$

19. $f(x) = 10^x$

20. $f(x) = 6^x$

21. $f(x) = \left(\dfrac{3}{2}\right)^x$

22. $f(x) = \left(\dfrac{5}{2}\right)^x$

23. $f(x) = \left(\dfrac{1}{3}\right)^x$

24. $f(x) = \left(\dfrac{2}{3}\right)^x$

In Exercises 25 to 34, explain how to use the graph of the first function f to produce the graph of the second function F.

25. $f(x) = 3^x,\ F(x) = 3^x + 2$

26. $f(x) = 4^x,\ F(x) = 4^x - 3$

27. $f(x) = 10^x,\ F(x) = 10^{x-2}$

▶ 28. $f(x) = 6^x,\ F(x) = 6^{x+5}$

29. $f(x) = \left(\dfrac{3}{2}\right)^x,\ F(x) = \left(\dfrac{3}{2}\right)^{-x}$

▶ 30. $f(x) = \left(\dfrac{5}{2}\right)^x,\ F(x) = -\left[\left(\dfrac{5}{2}\right)^x\right]$

31. $f(x) = \left(\dfrac{1}{3}\right)^x,\ F(x) = 2\left[\left(\dfrac{1}{3}\right)^x\right]$

32. $f(x) = \left(\dfrac{2}{3}\right)^x,\ F(x) = \dfrac{1}{2}\left[\left(\dfrac{2}{3}\right)^x\right]$

33. $f(x) = e^x,\ F(x) = e^{-x} + 2$

34. $f(x) = e^x,\ F(x) = e^{x-3} + 1$

In Exercises 35 to 42, use a graphing utility to graph each function. If the function has a horizontal asymptote, state the equation of the horizontal asymptote.

35. $f(x) = \dfrac{3^x + 3^{-x}}{2}$

36. $f(x) = 4 \cdot 3^{-x^2}$

37. $f(x) = \dfrac{e^x - e^{-x}}{2}$

38. $f(x) = \dfrac{e^x + e^{-x}}{2}$

39. $f(x) = -e^{(x-4)}$

40. $f(x) = 0.5e^{-x}$

41. $f(x) = \dfrac{10}{1 + 0.4e^{-0.5x}},$
$x \geq 0$

42. $f(x) = \dfrac{10}{1 + 1.5e^{-0.5x}},$
$x \geq 0$

43. **INTERNET CONNECTIONS** Data from Forrester Research suggest that the number of broadband [cable and digital subscriber line (DSL)] connections to the Internet can be modeled by $f(x) = 1.353(1.9025)^x$, where x is the number of years after January 1, 1998, and $f(x)$ is the number of connections in millions.

a. How many broadband Internet connections, to the nearest million, does this model predict will exist on January 1, 2005?

b. According to the model, in what year will the number of broadband connections first reach 300 million? [*Hint:* Use the intersect feature of a graphing utility to determine the x-coordinate of the point of intersection of the graphs of $f(x)$ and $y = 300$.]

▶ **44.** **MEDICATION IN BLOODSTREAM** The function $A(t) = 200e^{-0.014t}$ gives the amount of medication, in milligrams, in a patient's bloodstream t minutes after the medication has been injected into the patient's bloodstream.

a. Find the amount of medication, to the nearest milligram, in the patient's bloodstream after 45 minutes.

b. Use a graphing utility to determine how long it will take, to the nearest minute, for the amount of medication in the patient's bloodstream to reach 50 milligrams.

45. **DEMAND FOR A PRODUCT** The demand d for a specific product, in items per month, is given by

$$d(p) = 25 + 880e^{-0.18p}$$

where p is the price, in dollars, of the product.

a. What will be the monthly demand, to the nearest unit, when the price of the product is $8 and when the price is $18?

b. What will happen to the demand as the price increases without bound?

46. **SALES** The monthly income I, in dollars, from a new product is given by

$$I(t) = 24,000 - 22,000e^{-0.005t}$$

where t is the time, in months, since the product was first put on the market.

a. What was the monthly income after the 10th month and after the 100th month?

b. What will the monthly income from the product approach as the time increases without bound?

47. **A PROBABILITY FUNCTION** The manager of a home improvement store finds that between 10 A.M. and 11 A.M., customers enter the store at the average rate of 45 customers per hour. The following function gives the probability that a customer will arrive within t minutes of 10 A.M. (*Note:* A probability of 0.6 means there is a 60% chance that a customer will arrive during a given time period.)

$$P(t) = 1 - e^{-0.75t}$$

a. Find the probability, to the nearest hundredth, that a customer will arrive within 1 minute of 10 A.M.

b. Find the probability, to the nearest hundredth, that a customer will arrive within 3 minutes of 10 A.M.

c. Use a graph of $P(t)$ to determine how many minutes, to the nearest tenth of a minute, it takes for $P(t)$ to equal 98%.

d. Write a sentence that explains the meaning of the answer in part **c.**

48. **A PROBABILITY FUNCTION** The owner of a sporting goods store finds that between 9 A.M. and 10 A.M., customers enter the store at the average rate of 12 customers per hour. The following function gives the probability that a customer will arrive within t minutes of 9 A.M.

$$P(t) = 1 - e^{-0.2t}$$

a. Find the probability, to the nearest hundredth, that a customer will arrive within 5 minutes of 9 A.M.

b. Find the probability, to the nearest hundredth, that a customer will arrive within 15 minutes of 9 A.M.

c. Use a graph of $P(t)$ to determine how many minutes, to the nearest 0.1 minute, it takes for $P(t)$ to equal 90%.

d. Write a sentence that explains the meaning of the answer in part **c.**

Exercises 49 and 50 involve the factorial function $x!$, which is defined for whole numbers x as

$$x! = \begin{cases} 1, & \text{if } x = 0 \\ x \cdot (x - 1) \cdot (x - 2) \cdot \cdots \cdot 3 \cdot 2 \cdot 1, & \text{if } x \geq 1 \end{cases}$$

For example, $3! = 3 \cdot 2 \cdot 1 = 6$ and $5! = 5 \cdot 4 \cdot 3 \cdot 2 \cdot 1 = 120$.

49. **QUEUING THEORY** During the 30-minute period before a Broadway play begins, the members of the audience arrive at the theater at the average rate of 12 people per minute. The probability that x people will arrive during a particular minute is given by $P(x) = \dfrac{12^x e^{-12}}{x!}$. Find the probability, to the nearest 0.1%, that

a. 9 people will arrive during a given minute.

b. 18 people will arrive during a given minute.

50. QUEUING THEORY During the period from 2:00 P.M. to 3:00 P.M., a bank finds that an average of seven people enter the bank every minute. The probability that x people will enter the bank during a particular minute is given by $P(x) = \dfrac{7^x e^{-7}}{x!}$. Find the probability, to the nearest 0.1%, that

a. only two people will enter the bank during a given minute.

b. 11 people will enter the bank during a given minute.

51. 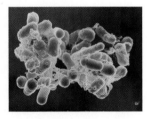 E. COLI INFEC-
TION *Escherichia coli (E. coli)* is a bacterium that can reproduce at an exponential rate. The *E. coli* reproduce by dividing. A small number of *E. coli* bacteria in the large intestine of a human can trigger a serious infection within a few hours. Consider a particular *E. coli* infection that starts with 100 *E. coli* bacteria. Each bacterium splits into two parts every half hour. Assuming none of the bacteria die, the size of the *E. coli* population after t hours is given by $P(t) = 100 \cdot 2^{2t}$, where $0 \le t \le 16$.

a. Find $P(3)$ and $P(6)$.

b. Use a graphing utility to find the time, to the nearest tenth of an hour, it takes for the *E. coli* population to number 1 billion.

52. RADIATION Lead shielding is used to contain radiation. The percentage of a certain radiation that can penetrate x millimeters of lead shielding is given by $I(x) = 100e^{-1.5x}$.

a. What percentage of radiation, to the nearest tenth of a percent, will penetrate a lead shield that is 1 millimeter thick?

b. How many millimeters of lead shielding are required so that less than 0.05% of the radiation penetrates the shielding? Round to the nearest millimeter.

53. AIDS An exponential function that approximates the number of people in the United States who have been infected with AIDS is given by $N(t) = 138,000(1.39)^t$, where t is the number of years after January 1, 1990.

a. According to this function, how many people had been infected with AIDS as of January 1, 1994? Round to the nearest thousand.

b. Use a graph to estimate during what year the number of people in the United States who had been infected with AIDS first reached 1.5 million.

▶ **54.** FISH POPULATION The number of bass in a lake is given by

$$P(t) = \frac{3600}{1 + 7e^{-0.05t}}$$

where t is the number of months that have passed since the lake was stocked with bass.

a. How many bass were in the lake immediately after it was stocked?

b. How many bass were in the lake 1 year after the lake was stocked?

c. What will happen to the bass population as t increases without bound?

55. THE PAY IT FORWARD MODEL In the movie *Pay It Forward*, Trevor McKinney, played by Haley Joel Osment, is given a school assignment to "think of an idea to change the world—and then put it into action." In response to this assignment, Trevor develops a *pay it forward* project. In this project, anyone who benefits from another person's good deed must do a good deed for

three additional people. Each of these three people is then obligated to do a good deed for another three people, and so on.

The following diagram shows the number of people who have been a beneficiary of a good deed after 1 round and after 2 rounds of this project.

Three beneficiaries after one round.

A total of 12 beneficiaries after two rounds (3 + 9 = 12).

A mathematical model for the number of pay it forward beneficiaries after n rounds is given by $B(n) = \dfrac{3^{n+1} - 3}{2}$.

Use this model to determine

a. the number of beneficiaries after 5 rounds and after 10 rounds. Assume that no person is a beneficiary of more than one good deed.

b. how many rounds are required to produce at least 2 million beneficiaries.

56. **INTENSITY OF LIGHT** The percent $I(x)$ of the original intensity of light striking the surface of a lake that is available x feet below the surface of the lake is given by $I(x) = 100e^{-0.95x}$.

a. What percentage of the light, to the nearest tenth of a percent, is available 2 feet below the surface of the lake?

b. At what depth, to the nearest hundredth of a foot, is the intensity of the light one-half the intensity at the surface?

57. **A TEMPERATURE MODEL** A cup of coffee is heated to 180°F and placed in a room that maintains a temperature of 65°F. The temperature of the coffee after t minutes is given by $T(t) = 65 + 115e^{-0.042t}$.

a. Find the temperature, to the nearest degree, of the coffee 10 minutes after it is placed in the room.

b. Use a graphing utility to determine when, to the nearest tenth of a minute, the temperature of the coffee will reach 100°F.

58. **A TEMPERATURE MODEL** Soup that is at a temperature of 170°F is poured into a bowl in a room that maintains a constant temperature. The temperature of the soup decreases according to the model given by $T(t) = 75 + 95e^{-0.12t}$, where t is time in minutes after the soup is poured.

a. What is the temperature, to the nearest tenth of a degree, of the soup after 2 minutes?

b. A certain customer prefers soup at a temperature of 110°F. How many minutes, to the nearest 0.1 minute, after the soup is poured does the soup reach that temperature?

c. What is the temperature of the room?

59. **MUSICAL SCALES** Starting on the left side of a standard 88-key piano, the frequency, in vibrations per second, of the nth note is given by $f(n) = (27.5)2^{(n-1)/12}$.

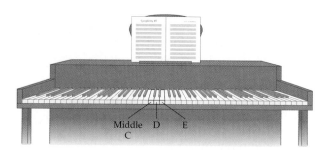

Middle C   D   E

a. Using this formula, determine the frequency, to the nearest hundredth of a vibration per second, of middle C, key number 40 on an 88-key piano.

b. Is the difference in frequency between middle C (key number 40) and D (key number 42) the same as the difference in frequency between D (key number 42) and E (key number 44)? Explain.

CONNECTING CONCEPTS

60. Verify that the hyperbolic cosine function
$\cosh(x) = \dfrac{e^x + e^{-x}}{2}$ is an even function.

61. Verify that the hyperbolic sine function $\sinh(x) = \dfrac{e^x - e^{-x}}{2}$
is an odd function.

62. Graph $g(x) = 10^x$, and then sketch the graph of g reflected across the line given by $y = x$.

63. Graph $f(x) = e^x$, and then sketch the graph of f reflected across the line given by $y = x$.

In Exercises 64 to 67, determine the domain of the given function. Write the domain using interval notation.

64. $f(x) = \dfrac{e^x - e^{-x}}{e^x + e^{-x}}$

65. $f(x) = \dfrac{e^{|x|}}{1 + e^x}$

66. $f(x) = \sqrt{1 - e^x}$

67. $f(x) = \sqrt{e^x - e^{-x}}$

PREPARE FOR SECTION 4.3

68. If $2^x = 16$, determine the value of x. [4.2]

69. If $3^{-x} = \dfrac{1}{27}$, determine the value of x. [4.2]

70. If $x^4 = 625$, determine the value of x. [4.2]

71. Find the inverse of $f(x) = \dfrac{2x}{x + 3}$. [4.1]

72. State the domain of $g(x) = \sqrt{x - 2}$. [2.2]

73. If the range of $h(x)$ is the set of all positive real numbers, then what is the domain of $h^{-1}(x)$? [4.2]

PROJECTS

1.   **THE SAINT LOUIS GATEWAY ARCH** The Gateway Arch in Saint Louis was designed in the shape of an inverted **catenary,** as shown by the red curve in the drawing at the right. The Gateway Arch is one of the largest optical illusions ever created. As you look at the arch (and its basic shape defined by the catenary curve), it appears to be much taller than it is wide. However, this is not the case. The height of the catenary is given by

$$h(x) = 693.8597 - 68.7672\left(\dfrac{e^{0.0100333x} + e^{-0.0100333x}}{2}\right)$$

where x and $h(x)$ are measured in feet and $x = 0$ represents the position at ground level that is directly below the highest point of the catenary.

a. Use a graphing utility to graph $h(x)$.

b. Use your graph to find the height of the catenary for $x = 0$, 100, 200, and 299 feet. Round each result to the nearest tenth of a foot.

c. What is the width of the catenary at ground level and what is the maximum height of the catenary? Round each result to the nearest tenth of a foot.

d. By how much does the maximum height of the catenary exceed its width at ground level? Round to the nearest tenth of a foot.

2. 　　**AN EXPONENTIAL REWARD** According to legend, when Sissa Ben Dahir of India invented the game of chess, King Shirham was so impressed with the game that he summoned the game's inventor and offered him the reward of his choosing. The inventor pointed to the chessboard and requested, for his reward, one grain of wheat on the first square, two grains of wheat on the second square, four grains on the third square, eight grains on the fourth square, and so on for all 64 squares on the chessboard. The King considered this a very modest reward and said he would grant the inventor's wish. The following table shows how many grains of wheat are on each of the first six squares and the total number of grains of wheat needed to cover squares 1 to n for $n \leq 6$.

a. If all 64 squares of the chessboard are piled with wheat as requested by Sissa Ben Dahir, how many grains of wheat are on the board?

b. A grain of wheat weighs approximately 0.000008 kilogram. Find the total weight of the wheat requested by Sissa Ben Dahir.

c. In a recent year, a total of 6.5×10^8 metric tons of wheat were produced in the world. At this level, how many years, to the nearest year, of wheat production would be required to fill the request of Sissa Ben Dahir? One metric ton equals 1000 kilograms.

Square number, n	Number of grains of wheat on square n	Total number of grains of wheat on squares 1 through n
1	1	1
2	2	3
3	4	7
4	8	15
5	16	31
6	32	63

LOGARITHMIC FUNCTIONS AND THEIR APPLICATIONS

● LOGARITHMIC FUNCTIONS

Every exponential function of the form $g(x) = b^x$ is a one-to-one function and therefore has an inverse function. Sometimes we can determine the inverse of a function represented by an equation by interchanging the variables of its equation and then solving for the dependent variable. If we attempt to use this procedure for $g(x) = b^x$, we obtain

$$g(x) = b^x$$
$$y = b^x$$
$$x = b^y \qquad \text{• Interchange the variables.}$$

None of our previous methods can be used to solve the equation $x = b^y$ for the exponent y. Thus we need to develop a new procedure. One method would be to merely write

$$y = \text{the power of } b \text{ that produces } x$$

Although this would work, it is not very concise. We need a compact notation to represent "y is the power of b that produces x." This more compact notation is given in the following definition.

Logarithms were developed by John Napier (1550–1617) as a means of simplifying the calculations of astronomers. One of his ideas was to devise a method by which the product of two numbers could be determined by performing an addition.

Definition of a Logarithm and a Logarithmic Function

If $x > 0$ and b is a positive constant ($b \neq 1$), then,

$$y = \log_b x \qquad \text{if and only if} \qquad b^y = x$$

The notation $\log_b x$ is read "the logarithm (or log) base b of x." The function defined by $f(x) = \log_b x$ is a logarithmic function with base b. This function is the inverse of the exponential function $g(x) = b^x$.

It is essential to remember that $f(x) = \log_b x$ is the inverse function of $g(x) = b^x$. Because these functions are inverses and because functions that are inverses have the property that $f(g(x)) = x$ and $g(f(x)) = x$, we have the following important relationships.

Composition of Logarithmic and Exponential Functions

Let $g(x) = b^x$ and $f(x) = \log_b x$ ($x > 0, b > 0, b \neq 1$). Then

$$g(f(x)) = b^{\log_b x} = x \qquad \text{and} \qquad f(g(x)) = \log_b b^x = x$$

take note

The notation $\log_b x$ replaces the phrase "the power of b that produces x." For instance, "3 is the power of 2 that produces 8" is abbreviated $3 = \log_2 8$. In your work with logarithms, remember that a logarithm is an exponent.

As an example of these relationships, let $g(x) = 2^x$ and $f(x) = \log_2 x$. Then

$$2^{\log_2 x} = x \qquad \text{and} \qquad \log_2 2^x = x$$

The equations

$$y = \log_b x \qquad \text{and} \qquad b^y = x$$

are different ways of expressing the same concept.

Exponential Form and Logarithmic Form

The exponential form of $y = \log_b x$ is $b^y = x$.

The logarithmic form of $b^y = x$ is $y = \log_b x$.

These concepts are illustrated in the next two examples.

> **EXAMPLE 1** **Change from Logarithmic to Exponential Form**
>
> Write each equation in its exponential form.
>
> **a.** $3 = \log_2 8$ **b.** $2 = \log_{10}(x + 5)$ **c.** $\log_e x = 4$ **d.** $\log_b b^3 = 3$
>
> **Solution**
>
> Use the definition $y = \log_b x$ if and only if $b^y = x$.
>
> ┌─── Logarithms are exponents. ───┐
> **a.** $3 = \log_2 8$ if and only if $2^3 = 8$
> └────────── Base ──────────┘
>
> **b.** $2 = \log_{10}(x + 5)$ if and only if $10^2 = x + 5$.
>
> **c.** $\log_e x = 4$ if and only if $e^4 = x$.
>
> **d.** $\log_b b^3 = 3$ if and only if $b^3 = b^3$.
>
> ▶ **TRY EXERCISE 4, PAGE 361**

> **EXAMPLE 2** **Change from Exponential to Logarithmic Form**
>
> Write each equation in its logarithmic form.
>
> **a.** $3^2 = 9$ **b.** $5^3 = x$ **c.** $a^b = c$ **d.** $b^{\log_b 5} = 5$
>
> **Solution**
>
> The logarithmic form of $b^y = x$ is $y = \log_b x$.
>
> ┌─── Exponent ───┐
> **a.** $3^2 = 9$ if and only if $2 = \log_3 9$
> └────────── Base ──────────┘
>
> **b.** $5^3 = x$ if and only if $3 = \log_5 x$.
>
> **c.** $a^b = c$ if and only if $b = \log_a c$.
>
> **d.** $b^{\log_b 5} = 5$ if and only if $\log_b 5 = \log_b 5$.
>
> ▶ **TRY EXERCISE 12, PAGE 361**

The definition of a logarithm and the definition of inverse functions can be used to establish many properties of logarithms. For instance:

- $\log_b b = 1$ because $b = b^1$.

- $\log_b 1 = 0$ because $1 = b^0$.

- $\log_b(b^x) = x$ because $b^x = b^x$.

- $b^{\log_b x} = x$ because $f(x) = \log_b x$ and $g(x) = b^x$ are inverse functions. Thus $g[f(x)] = x$.

We will refer to the preceding properties as the *basic logarithmic properties.*

Basic Logarithmic Properties

1. $\log_b b = 1$ **2.** $\log_b 1 = 0$ **3.** $\log_b(b^x) = x$ **4.** $b^{\log_b x} = x$

EXAMPLE 3   **Apply the Basic Logarithmic Properties**

Evaluate each of the following logarithms.

a. $\log_8 1$ b. $\log_5 5$ c. $\log_2(2^4)$ d. $3^{\log_3 7}$

Solution

a. By Property 2, $\log_8 1 = 0$.

b. By Property 1, $\log_5 5 = 1$.

c. By Property 3, $\log_2(2^4) = 4$.

d. By Property 4, $3^{\log_3 7} = 7$.

▶ **TRY EXERCISE 28, PAGE 361**

Some logarithms can be evaluated just by remembering that a logarithm is an exponent. For instance, $\log_5 25$ equals 2 because the base 5 raised to the second power equals 25.

- $\log_{10} 100 = 2$ because $10^2 = 100$.

- $\log_4 64 = 3$ because $4^3 = 64$.

- $\log_7 \dfrac{1}{49} = -2$ because $7^{-2} = \dfrac{1}{7^2} = \dfrac{1}{49}$.

❓ **QUESTION**   What is the value of $\log_5 625$?

● **GRAPHS OF LOGARITHMIC FUNCTIONS**

Because $f(x) = \log_b x$ is the inverse function of $g(x) = b^x$, the graph of f is a reflection of the graph of g across the line given by $y = x$. The graph of $g(x) = 2^x$ is shown in **Figure 4.25**. **Table 4.7** below shows some of the ordered pairs on the graph of g.

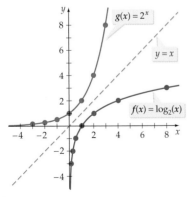

FIGURE 4.25

TABLE 4.7

x	-3	-2	-1	0	1	2	3
$g(x) = 2^x$	$\dfrac{1}{8}$	$\dfrac{1}{4}$	$\dfrac{1}{2}$	1	2	4	8

❓ **ANSWER** $\log_5 625 = 4$ because $5^4 = 625$.

The graph of the inverse of g, which is $f(x) = \log_2 x$, is also shown in **Figure 4.25**. Some of the ordered pairs of f are shown in **Table 4.8**. Note that if (x, y) is a point on the graph of g, then (y, x) is a point on the graph of f. Also notice that the graph of f is a reflection of the graph of g across the line given by $y = x$.

TABLE 4.8

x	$\dfrac{1}{8}$	$\dfrac{1}{4}$	$\dfrac{1}{2}$	1	2	4	8
$f(x) = \log_2 x$	-3	-2	-1	0	1	2	3

The graph of a logarithmic function can be drawn by first rewriting the function in its exponential form. This procedure is illustrated in Example 4.

EXAMPLE 4   **Graph a Logarithmic Function**

Graph $f(x) = \log_3 x$.

Solution

To graph $f(x) = \log_3 x$, consider the equivalent exponential equation $x = 3^y$. Because this equation is solved for x, choose values of y and calculate the corresponding values of x, as shown in **Table 4.9**.

TABLE 4.9

$x = 3^y$	$\dfrac{1}{9}$	$\dfrac{1}{3}$	1	3	9
y	-2	-1	0	1	2

Now plot the ordered pairs and connect the points with a smooth curve, as shown in **Figure 4.26**.

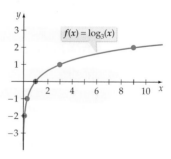

FIGURE 4.26

▶ **TRY EXERCISE 32, PAGE 361**

We can use a similar procedure to draw the graph of a logarithmic function with a fractional base. For instance, consider $y = \log_{2/3} x$. Rewriting this in expo-

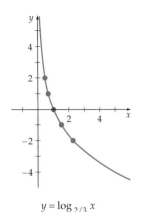

$y = \log_{2/3} x$

FIGURE 4.27

nential form gives us $\left(\dfrac{2}{3}\right)^y = x$. Choose values of y and calculate the corresponding x-values. See **Table 4.10.** Plot the points corresponding to the ordered pairs (x, y), and then draw a smooth curve through the points, as shown in **Figure 4.27.**

TABLE 4.10

$x = \left(\dfrac{2}{3}\right)^y$	$\left(\dfrac{2}{3}\right)^{-2} = \dfrac{9}{4}$	$\left(\dfrac{2}{3}\right)^{-1} = \dfrac{3}{2}$	$\left(\dfrac{2}{3}\right)^{0} = 1$	$\left(\dfrac{2}{3}\right)^{1} = \dfrac{2}{3}$	$\left(\dfrac{2}{3}\right)^{2} = \dfrac{4}{9}$
y	-2	-1	0	1	2

Properties of $f(x) = \log_b x$

For all positive real numbers b, $b \neq 1$, the function $f(x) = \log_b x$ has the following properties:

1. The domain of f consists of the set of positive real numbers and its range consists of the set of all real numbers.

2. The graph of f has an x-intercept of $(1, 0)$ and passes through $(b, 1)$.

3. If $b > 1$, f is an increasing function and its graph is asymptotic to the negative y-axis. [As $x \to \infty$, $f(x) \to \infty$, and as $x \to 0$ from the right, $f(x) \to -\infty$.] See **Figure 4.28a.**

4. If $0 < b < 1$, f is a decreasing function and its graph is asymptotic to the positive y-axis. [As $x \to \infty$, $f(x) \to -\infty$, and as $x \to 0$ from the right, $f(x) \to \infty$.] See **Figure 4.28b.**

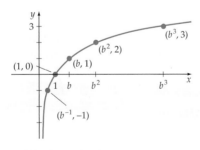

a. $f(x) = \log_b x$, $b > 1$

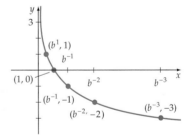

b. $f(x) = \log_b x$, $0 < b < 1$

FIGURE 4.28

● DOMAINS OF LOGARITHMIC FUNCTIONS

The function $f(x) = \log_b x$ has as its domain the set of positive real numbers. The function $f(x) = \log_b(g(x))$ has as its domain the set of all x for which $g(x) > 0$. To determine the domain of a function such as $f(x) = \log_b(g(x))$, we must determine the values of x that make $g(x)$ positive. This process is illustrated in Example 5.

> **EXAMPLE 5** **Find the Domain of a Logarithmic Function**

Find the domain of each of the following logarithmic functions.

a. $f(x) = \log_6(x - 3)$ **b.** $F(x) = \log_2|x + 2|$ **c.** $R(x) = \log_5\left(\dfrac{x}{8 - x}\right)$

Solution

a. Solving $(x - 3) > 0$ for x gives us $x > 3$. The domain of f consists of all real numbers greater than 3. In interval notation the domain is $(3, \infty)$.

b. The solution set of $|x + 2| > 0$ consists of all real numbers x except $x = -2$. The domain of F consists of all real numbers $x \neq -2$. In interval notation the domain is $(-\infty, -2) \cup (-2, \infty)$.

c. Solving $\left(\dfrac{x}{8 - x}\right) > 0$ yields the set of all real numbers x between 0 and 8. The domain of R is all real numbers x such that $0 < x < 8$. In interval notation the domain is $(0, 8)$.

▶ **TRY EXERCISE 40, PAGE 361**

Some logarithmic functions can be graphed by using horizontal and/or vertical translations of a previously drawn graph.

> **EXAMPLE 6** **Use Translations to Graph Logarithmic Functions**

Graph: **a.** $f(x) = \log_4(x + 3)$ **b.** $f(x) = \log_4 x + 3$

Solution

a. The graph of $f(x) = \log_4(x + 3)$ can be obtained by shifting the graph of $g(x) = \log_4 x$ to the left 3 units. See **Figure 4.29**. Note that the domain of f consists of all real numbers x greater than -3 because $x + 3 > 0$ for $x > -3$. The graph of f is asymptotic to the vertical line $x = -3$.

b. The graph of $f(x) = \log_4 x + 3$ can be obtained by shifting the graph of $g(x) = \log_4 x$ upward 3 units. See **Figure 4.30**.

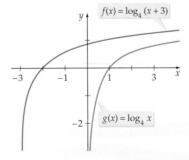

FIGURE 4.29

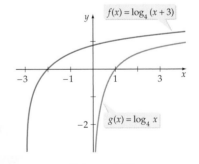

FIGURE 4.30

▶ **TRY EXERCISE 50, PAGE 361**

• COMMON AND NATURAL LOGARITHMS

Two of the most frequently used logarithmic functions are *common logarithms,* which have base 10, and *natural logarithms,* which have base e (the base of the natural exponential function).

Definition of Common and Natural Logarithms

The function defined by $f(x) = \log_{10} x$ is called the **common logarithmic function**. It is customarily written without stating the base as $f(x) = \log x$.

The function defined by $f(x) = \log_e x$ is called the **natural logarithmic function**. It is customarily written as $f(x) = \ln x$.

Most scientific or graphing calculators have a $\boxed{\text{LOG}}$ key for evaluating common logarithms and an $\boxed{\text{LN}}$ key to evaluate natural logarithms. For instance, using a graphing calculator,

$$\log 24 \approx 1.3802112 \qquad \text{and} \qquad \ln 81 \approx 4.3944492$$

The graphs of $f(x) = \log x$ and $f(x) = \ln x$ can be drawn using the same techniques we used to draw the graphs in the preceding examples. However, these graphs also can be produced with a graphing calculator by entering $\log x$ and $\ln x$ into the Y= menu. See **Figure 4.31** and **Figure 4.32.**

FIGURE 4.31

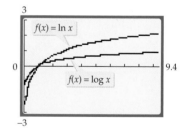

FIGURE 4.32

Observe that each graph passes through $(1, 0)$. Also note that as $x \to 0$ from the right, the functional values $f(x) \to -\infty$. Thus the y-axis is a vertical asymptote for each of the graphs. The domain of both $f(x) = \log x$ and $f(x) = \ln x$ is the set of positive real numbers. Each of these functions has a range consisting of the set of real numbers.

• APPLICATIONS OF LOGARITHMIC FUNCTIONS

Many applications can be modeled by logarithmic functions.

Average time of a baseball game (in minutes)

$T(x) = 149.57 + 7.63 \ln(x)$

Year (1 represents 1981)

EXAMPLE 7   *Average Time of a Major League Baseball Game*

From 1981 to 1999, the average time of a major league baseball game tended to increase each year. If the year 1981 is represented by $x = 1$, then the function

$$T(x) = 149.57 + 7.63 \ln x$$

approximates the average time T, in minutes, of a major league baseball game for the years 1981 to 1999—that is, for $x = 1$ to $x = 19$.

a. Use the function T to determine the average time of a major league baseball game during the 1981 season and during the 1999 season.

b. By how much did the average time of a major league baseball game increase during the years 1981 to 1999?

Solution

a. The year 1981 is represented by $x = 1$ and the year 1999 by $x = 19$.

$$T(1) = 149.57 + 7.63 \ln(1) = 149.57$$

In 1981 the average time of a baseball game was about 149.57 minutes.

$$T(19) = 149.57 + 7.63 \ln(19) \approx 172.04$$

In 1999 the average time of a baseball game was about 172.04 minutes.

b. $T(19) - T(1) \approx 172.04 - 149.57 = 22.47$. During the years 1981 to 1999, the average time of a baseball game increased by about 22.47 minutes.

▶ **TRY EXERCISE 70, PAGE 362**

TOPICS FOR DISCUSSION

1. If $m > n$, must $\log_b m > \log_b n$?

2. For what values of x is $\ln x > \log x$?

3. What is the domain of $f(x) = \log(x^2 + 1)$? Explain why the graph of f does not have a vertical asymptote.

4. The subtraction $3 - 5$ does not have an answer if we require that the answer be positive. Keep this idea in mind as you work the rest of this exercise.

   Press the ⎡MODE⎤ key of a TI-83 graphing calculator, and choose "Real" from the menu. Now use the calculator to evaluate $\log(-2)$. What output is given by the calculator? Press the ⎡MODE⎤ key, and choose "a + bi" from the menu. Now use the calculator to evaluate $\log(-2)$. What output is given by the calculator? Write a sentence or two that explain why the output is different for these two evaluations.

EXERCISE SET 4.3

In Exercises 1 to 10, change each equation to its exponential form.

1. $\log 10 = 1$

2. $\log 10{,}000 = 4$

3. $\log_8 64 = 2$

▶ **4.** $\log_4 64 = 3$

5. $\log_7 x = 0$

6. $\log_3 \dfrac{1}{81} = -4$

7. $\ln x = 4$

8. $\ln e^2 = 2$

9. $\ln 1 = 0$

10. $\ln x = -3$

In Exercises 11 to 20, change each equation to its logarithmic form. Assume $y > 0$ and $b > 0$.

11. $3^2 = 9$

▶ **12.** $5^3 = 125$

13. $4^{-2} = \dfrac{1}{16}$

14. $10^0 = 1$

15. $b^x = y$

16. $2^x = y$

17. $y = e^x$

18. $5^1 = 5$

19. $100 = 10^2$

20. $2^{-4} = \dfrac{1}{16}$

In Exercises 21 to 30, evaluate each logarithm. Do not use a calculator.

21. $\log_4 16$

22. $\log_{3/2} \dfrac{8}{27}$

23. $\log_3 \dfrac{1}{243}$

24. $\log_b 1$

25. $\ln e^3$

26. $\log_b b$

27. $\log \dfrac{1}{100}$

▶ **28.** $\log 1{,}000{,}000$

29. $\log_{0.5} 16$

30. $\log_{0.3} \dfrac{100}{9}$

In Exercises 31 to 38, graph each function by using its exponential form.

31. $f(x) = \log_4 x$

▶ **32.** $f(x) = \log_6 x$

33. $f(x) = \log_{12} x$

34. $f(x) = \log_8 x$

35. $f(x) = \log_{1/2} x$

36. $f(x) = \log_{1/4} x$

37. $f(x) = \log_{5/2} x$

38. $f(x) = \log_{7/3} x$

In Exercises 39 to 48, find the domain of the function. Write the domains using interval notation.

39. $f(x) = \log_5(x - 3)$

▶ **40.** $k(x) = \log_4(5 - x)$

41. $k(x) = \log_{2/3}(11 - x)$

42. $H(x) = \log_{1/4}(x^2 + 1)$

43. $P(x) = \ln(x^2 - 4)$

44. $J(x) = \ln\left(\dfrac{x - 3}{x}\right)$

45. $h(x) = \ln\left(\dfrac{x^2}{x - 4}\right)$

46. $R(x) = \ln(x^4 - x^2)$

47. $N(x) = \log_2(x^3 - x)$

48. $s(x) = \log_7(x^2 + 7x + 10)$

In Exercises 49 to 56, use translations of the graphs in Exercises 31 to 38 to produce the graph of the given function.

49. $f(x) = \log_4(x - 3)$

▶ **50.** $f(x) = \log_6(x + 3)$

51. $f(x) = \log_{12} x + 2$

52. $f(x) = \log_8 x - 4$

53. $f(x) = 3 + \log_{1/2} x$

54. $f(x) = 2 + \log_{1/4} x$

55. $f(x) = 1 + \log_{5/2}(x - 4)$

56. $f(x) = \log_{7/3}(x - 3) - 1$

57. Examine the following four functions and the graphs labeled **a, b, c,** and **d.** Determine which graph is the graph of each function.

$$f(x) = \log_5(x - 2) \qquad g(x) = 2 + \log_5 x$$
$$h(x) = \log_5(-x) \qquad k(x) = -\log_5(x + 3)$$

a.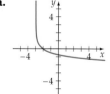

b.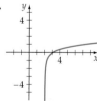

c.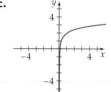

d.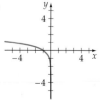

58. Examine the following four functions and the graphs labeled **a, b, c,** and **d.** Determine which graph is the graph of each function.

$$f(x) = \ln x + 3 \qquad g(x) = \ln(x - 3)$$
$$h(x) = \ln(3 - x) \qquad k(x) = -\ln(-x)$$

a.

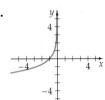

b.

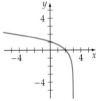

c.

d.

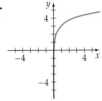

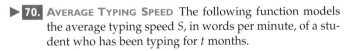 **In Exercises 59 to 68, use a graphing utility to graph the function.**

59. $f(x) = -2 \ln x$

60. $f(x) = -\log x$

61. $f(x) = |\ln x|$

62. $f(x) = \ln |x|$

63. $f(x) = \log \sqrt[3]{x}$

64. $f(x) = \ln \sqrt{x}$

65. $f(x) = \log(x + 10)$

66. $f(x) = \ln(x + 3)$

67. $f(x) = 3 \log |2x + 10|$

68. $f(x) = \dfrac{1}{2} \ln |x - 4|$

69. **MONEY MARKET RATES** The function

$$r(t) = 0.69607 + 0.60781 \ln t$$

gives the annual interest rate r, as a percent, a bank will pay on its money market accounts, where t is the term (the time the money is invested) in months.

a. What interest rate, to the nearest tenth of a percent, will the bank pay on a money market account with a term of 9 months?

b. What is the minimum number of complete months during which a person must invest to receive an interest rate of at least 3%?

▶ **70.** **AVERAGE TYPING SPEED** The following function models the average typing speed S, in words per minute, of a student who has been typing for t months.

$$S(t) = 5 + 29 \ln(t + 1), \quad 0 \le t \le 16$$

a. What was the student's average typing speed, to the nearest word per minute, when the student first started to type? What was the student's average typing speed, to the nearest word per minute, after 3 months?

b. Use a graph of S to determine how long, to the nearest tenth of a month, it will take the student to achieve an average typing speed of 65 words per minute.

71. **ADVERTISING COSTS AND SALES** The function

$$N(x) = 2750 + 180 \ln\left(\dfrac{x}{1000} + 1\right)$$

models the relationship between the dollar amount x spent on advertising a product and the number of units N that a company can sell.

a. Find the number of units that will be sold with advertising expenditures of \$20,000, \$40,000, and \$60,000.

b. How many units will be sold if the company does not pay to advertise the product?

In anesthesiology it is necessary to accurately estimate the body surface area of a patient. One formula for estimating body surface area (*BSA*) was developed by Edith Boyd (University of Minnesota Press, 1935). Her formula for the *BSA* (in square meters) of a patient of height H (in centimeters) and weight W (in grams) is

$$BSA = 0.0003207 \cdot H^{0.3} \cdot W^{(0.7285 - 0.0188 \log W)}$$

 MEDICINE In Exercises 72 and 73, use Boyd's formula to estimate the body surface area of a patient with the given weight and height. Round to the nearest hundredth of a square meter.

72. $W = 110$ pounds (49,895.2 grams); $H = 5$ feet 4 inches (162.56 centimeters)

73. $W = 180$ pounds (81,646.6 grams); $H = 6$ feet 1 inch (185.42 centimeters)

74. **ASTRONOMY** Astronomers measure the apparent brightness of a star by a unit called the **apparent magnitude.** This unit was created in the second century B.C. when the Greek astronomer Hipparchus classified the relative brightness of several stars. In his list he assigned the number 1 to the stars that appeared to be the brightest (Sirius, Vega, and Deneb). They are first-magnitude stars. Hipparchus assigned the number 2 to all the stars in the Big Dipper. They are second-magnitude stars. The following table shows the relationship between a star's brightness relative to a first-magnitude star and the star's apparent magnitude. Notice from the table that a first-magnitude star appears in the sky to be about 2.51 times as bright as a second-magnitude star.

Brightness relative to a first-magnitude star x	Apparent magnitude $M(x)$
1	1
$\dfrac{1}{2.51}$	2
$\dfrac{1}{6.31} \approx \dfrac{1}{2.51^2}$	3
$\dfrac{1}{15.85} \approx \dfrac{1}{2.51^3}$	4
$\dfrac{1}{39.82} \approx \dfrac{1}{2.51^4}$	5
$\dfrac{1}{100} \approx \dfrac{1}{2.51^5}$	6

The following logarithmic function gives the apparent magnitude $M(x)$ of a star as a function of its brightness x.

$$M(x) = -2.51 \log x + 1, \quad 0 < x \le 1$$

a. Use $M(x)$ to find the apparent magnitude of a star that is $\dfrac{1}{10}$ as bright as a first-magnitude star. Round to the nearest hundredth.

b. Find the approximate apparent magnitude of a star that is $\dfrac{1}{400}$ as bright as a first-magnitude star. Round to the nearest hundredth.

c. Which star appears brighter: a star with an apparent magnitude of 12 or a star with an apparent magnitude of 15?

d. Is $M(x)$ an increasing function or a decreasing function?

75. **NUMBER OF DIGITS IN** b^x An engineer has determined that the number of digits N in the expansion of b^x, where both b and x are positive integers, is $N = \text{int}(x \log b) + 1$, where $\text{int}(x \log b)$ denotes the greatest integer of $x \log b$. (*Note:* The greatest integer of the real number x is x if x is an integer and is the largest integer less than x if x is not an integer. For example, the greatest integer of 5 is 5 and the greatest integer of 7.8 is 7.)

a. Because $2^{10} = 1024$, we know that 2^{10} has four digits. Use the equation $N = \text{int}(x \log b) + 1$ to verify this result.

b. Find the number of digits in 3^{200}.

c. Find the number of digits in 7^{4005}.

d. The largest known prime number as of November 17, 2003 was $2^{20996011} - 1$. Find the number of digits in this prime number. (*Hint:* Because $2^{20996011}$ is not a power of 10, both $2^{20996011}$ and $2^{20996011} - 1$ have the same number of digits.)

76. **NUMBER OF DIGITS IN** $9^{(9^9)}$ A science teacher has offered 10 points extra credit to any student who will write out all the digits in the expansion of $9^{(9^9)}$.

a. Use the formula from Exercise 75 to determine the number of digits in this number.

b. Assume that you can write 1000 digits per page and that 500 pages of paper are in a ream of paper. How many reams of paper, to the nearest tenth of a ream, are required to write out the expansion of $9^{(9^9)}$? Assume that you write on only one side of each page.

CONNECTING CONCEPTS

77. Use a graphing utility to graph $f(x) = \dfrac{e^x - e^{-x}}{2}$ and $g(x) = \ln(x + \sqrt{x^2 + 1})$ on the same screen. Use a square viewing window. What appears to be the relationship between f and g?

78. Use a graphing utility to graph $f(x) = \dfrac{e^x + e^{-x}}{2}$, for $x \ge 0$, and $g(x) = \ln(x + \sqrt{x^2 - 1})$, for $x \ge 1$, on the same screen. Use a square viewing window. What appears to be the relationship between f and g?

79. The functions $f(x) = \dfrac{e^x - e^{-x}}{e^x + e^{-x}}$ and $g(x) = \dfrac{1}{2} \ln \dfrac{1 + x}{1 - x}$ are inverse functions. The domain of f is the set of all real numbers. The domain of g is $\{x \mid -1 < x < 1\}$. Use this information to determine the range of f and the range of g.

80. Use a graph of $f(x) = \dfrac{2}{e^x + e^{-x}}$ to determine the domain and the range of f.

PREPARE FOR SECTION 4.4

In Exercises 81 to 86, use a calculator to compare each of the given expressions.

81. $\log 3 + \log 2$; $\log 6$ [4.2]

82. $\ln 8 - \ln 3$; $\ln\left(\dfrac{8}{3}\right)$ [4.2]

83. $3 \log 4$; $\log(4^3)$ [4.2]

84. $2 \ln 5$; $\ln(5^2)$ [4.2]

85. $\ln 5$; $\dfrac{\log 5}{\log e}$ [4.2]

86. $\log 8$; $\dfrac{\ln 8}{\ln 10}$ [4.2]

PROJECTS

1. **BENFORD'S LAW** The authors of this text know some interesting details about your finances. For instance, of the last 150 checks you have written, about 30% are for amounts that start with the number 1. Also, you have written about 3 times as many checks for amounts that start with the number 2 as you have for amounts that start with the number 7.

We are sure of these results because of a mathematical formula known as **Benford's Law.** This law was first discovered by the mathematician Simon Newcomb in 1881 and then rediscovered by the physicist Frank Benford in 1938. Benford's Law states that the probability P that the first digit of a number selected from a wide range of numbers is d is given by

$$P(d) = \log\left(1 + \frac{1}{d}\right)$$

a. Use Benford's Law to complete the table below and the bar graph at the top of the next column.

d	$P(d) = \log\left(1 + \dfrac{1}{d}\right)$
1	0.301
2	0.176
3	0.125
4	
5	
6	
7	
8	
9	

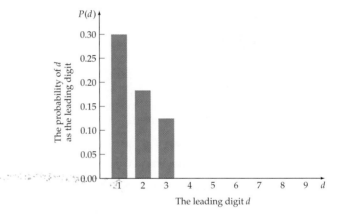

Benford's Law applies to most data with a wide range. For instance, it applies to

- the populations of the cities in the U.S.

- the numbers of dollars in the savings accounts at your local bank.

- the number of miles driven during a month by each person in a state.

b. Use the table in part **a.** to find the probability that in a U.S. city selected at random, the number of telephones in that city will be a number starting with 6.

c. Use the table in part **a.** to estimate how many times as many purchases you have made for dollar amounts that start with a 1 as you have for dollar amounts that start with a 9.

d. Explain why Benford's Law would not apply to the set of telephone numbers of the people living in a small city such as Le Mars, Iowa.

e. ✏ Explain why Benford's Law would not apply to the set of all the ages, in years, of students at a local high school.

AN APPLICATION OF BENFORD'S LAW Benford's Law has been used to identify fraudulent accountants. In most cases these accountants are unaware of Benford's Law and have replaced valid numbers with numbers selected at random. Their numbers do not conform to Benford's Law. Hence an audit is warranted.

LOGARITHMS AND LOGARITHMIC SCALES

● PROPERTIES OF LOGARITHMS

In Section 4.3 we introduced the following basic properties of logarithms.

$$\log_b b = 1 \quad \text{and} \quad \log_b 1 = 0$$

Also, because exponential functions and logarithmic functions are inverses of each other, we observed the relationships

$$\log_b(b^x) = x \quad \text{and} \quad b^{\log_b x} = x$$

We can use the properties of exponents to establish the following additional logarithmic properties.

Properties of Logarithms

In the following properties, b, M, and N are positive real numbers ($b \neq 1$).

Product property	$\log_b(MN) = \log_b M + \log_b N$
Quotient property	$\log_b \dfrac{M}{N} = \log_b M - \log_b N$
Power property	$\log_b(M^p) = p \log_b M$
Logarithm-of-each-side property	$M = N$ implies $\log_b M = \log_b N$
One-to-one property	$\log_b M = \log_b N$ implies $M = N$

take note

Pay close attention to these properties. Note that
$$\log_b(MN) \neq \log_b M \cdot \log_b N$$
and
$$\log_b \frac{M}{N} \neq \frac{\log_b M}{\log_b N}$$
Also,
$$\log_b(M + N) \neq \log_b M + \log_b N$$
In fact, the expression $\log_b(M + N)$ cannot be expanded at all.

❓ QUESTION Is it true that $\ln 5 + \ln 10 = \ln 50$?

The above properties of logarithms are often used to rewrite logarithmic expressions in an equivalent form.

❓ ANSWER Yes. By the product property, $\ln 5 + \ln 10 = \ln(5 \cdot 10)$.

> **EXAMPLE 1**   **Rewrite Logarithmic Expressions**
>
> Use the properties of logarithms to express the following logarithms in terms of logarithms of x, y, and z.
>
> a. $\log_5(xy^2)$ b. $\log_b \dfrac{2\sqrt{y}}{z^5}$
>
> **Solution**
>
> a. $\log_5(xy^2) = \log_5 x + \log_5 y^2$ • **Product property**
>
> $\qquad\qquad = \log_5 x + 2\log_5 y$ • **Power property**
>
> b. $\log_b \dfrac{2\sqrt{y}}{z^5} = \log_b(2\sqrt{y}) - \log_b z^5$ • **Quotient property**
>
> $\qquad\qquad = \log_b 2 + \log_b \sqrt{y} - \log_b z^5$ • **Product property**
>
> $\qquad\qquad = \log_b 2 + \log_b y^{1/2} - \log_b z^5$ • **Replace $\sqrt{y}$ with $y^{1/2}$.**
>
> $\qquad\qquad = \log_b 2 + \dfrac{1}{2}\log_b y - 5\log_b z$ • **Power property**
>
> ▶ **TRY EXERCISE 2, PAGE 373**

The properties of logarithms are also used to rewrite expressions that involve several logarithms as a single logarithm.

> **EXAMPLE 2**   **Rewrite Logarithmic Expressions**
>
> Use the properties of logarithms to rewrite each expression as a single logarithm with a coefficient of 1.
>
> a. $2\log_b x + \dfrac{1}{2}\log_b(x + 4)$ b. $4\log_3(x + 2) - 3\log_3(x - 5)$
>
> **Solution**
>
> a. $2\log_b x + \dfrac{1}{2}\log_b(x + 4)$
>
> $\qquad = \log_b x^2 + \log_b(x + 4)^{1/2}$ • **Power property**
>
> $\qquad = \log_b[x^2(x + 4)^{1/2}]$ • **Product property**
>
> $\qquad = \log_b(x^2\sqrt{x + 4})$
>
> b. $4\log_3(x + 2) - 3\log_3(x - 5)$
>
> $\qquad = \log_3(x + 2)^4 - \log_3(x - 5)^3$ • **Power property**
>
> $\qquad = \log_3 \dfrac{(x + 2)^4}{(x - 5)^3}$ • **Quotient property**
>
> ▶ **TRY EXERCISE 10, PAGE 373**

● CHANGE-OF-BASE FORMULA

Recall that to determine the value of y in $\log_3 81 = y$, we are basically asking, "What power of 3 is equal to 81?" Because $3^4 = 81$, we have $\log_3 81 = 4$. Now suppose that we need to determine the value of $\log_3 50$. In this case we need to find the power of 3 that produces 50. Because $3^3 = 27$ and $3^4 = 81$, the value we are seeking is somewhere between 3 and 4. The following procedure can be used to produce an estimate of $\log_3 50$.

The exponential form of $\log_3 50 = y$ is $3^y = 50$. Applying logarithmic properties gives us

$$3^y = 50$$

$$\ln 3^y = \ln 50 \qquad \bullet \text{ Logarithm-of-each-side property}$$

$$y \ln 3 = \ln 50 \qquad \bullet \text{ Power property}$$

$$y = \frac{\ln 50}{\ln 3} \approx 3.56088 \qquad \bullet \text{ Solve for y.}$$

Thus $\log_3 50 \approx 3.56088$. In the above procedure we could just as well have used logarithms of any base and arrived at the same value. Thus any logarithm can be expressed in terms of logarithms of any base we wish. This general result is summarized in the following formula.

Change-of-Base Formula

If x, a, and b are positive real numbers with $a \neq 1$ and $b \neq 1$, then

$$\log_b x = \frac{\log_a x}{\log_a b}$$

Because most calculators use only common logarithms ($a = 10$) or natural logarithms ($a = e$), the change-of-base formula is used most often in the following form.

If x and b are positive real numbers and $b \neq 1$, then

$$\log_b x = \frac{\log x}{\log b} = \frac{\ln x}{\ln b}$$

EXAMPLE 3   Use the Change-of-Base Formula

Evaluate each logarithm. Round to the nearest hundred thousandth.

a. $\log_3 18$ **b.** $\log_{12} 400$

Solution

To approximate these logarithms, we may use the change-of-base formula with $a = 10$ or $a = e$. For this example we choose to use the change-of-base formula with $a = e$. That is, we will evaluate these logarithms by using the $\boxed{\text{LN}}$ key on a scientific or graphing calculator.

a. $\log_3 18 = \dfrac{\ln 18}{\ln 3} \approx 2.63093$ **b.** $\log_{12} 400 = \dfrac{\ln 400}{\ln 12} \approx 2.41114$

---

take note

If common logarithms had been used for the calculation in Example **3a.**, the final result would be the same.

$$\log_3 18 = \frac{\log 18}{\log 3} \approx 2.63093$$

▶ TRY EXERCISE 16, PAGE 374

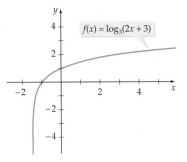

FIGURE 4.33

The change-of-base formula and a graphing calculator can be used to graph logarithmic functions that have a base other than 10 or e. For instance, to graph $f(x) = \log_3(2x + 3)$, we rewrite the function in terms of base 10 or base e. Using base 10 logarithms, we have $f(x) = \log_3(2x + 3) = \dfrac{\log(2x + 3)}{\log 3}$. The graph is shown in **Figure 4.33**.

EXAMPLE 4   **Use the Change-of-Base Formula to Graph a Logarithmic Function**

Graph $f(x) = \log_2|x - 3|$.

Solution

Rewrite f using the change-of-base formula. We will use the natural logarithm function; however, the common logarithm function could be used instead.

$$f(x) = \log_2|x - 3| = \frac{\ln|x - 3|}{\ln 2}$$

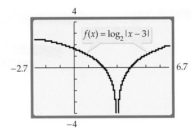

- Enter $\dfrac{\ln|x - 3|}{\ln 2}$ into Y1. Note that the domain of $f(x) = \log_2|x - 3|$ is all real numbers except 3, because $|x - 3| = 0$ when $x = 3$ and $|x - 3|$ is positive for all other values of x.

▶ **TRY EXERCISE 24, PAGE 374**

● LOGARITHMIC SCALES

Logarithmic functions are often used to scale very large (or very small) numbers into numbers that are easier to comprehend. For instance, the *Richter scale* magnitude of an earthquake uses a logarithmic function to convert the intensity of the earthquake's shock waves I into a number M, which for most earthquakes is in the range of 0 to 10. The intensity I of an earthquake is often given in terms of the constant I_0, where I_0 is the intensity of the smallest earthquake (called a **zero-level earthquake**) that can be measured on a seismograph near the earthquake's epicenter. The following formula is used to compute the Richter scale magnitude of an earthquake.

The Richter Scale Magnitude of an Earthquake

An earthquake with an intensity of I has a Richter scale magnitude of

$$M = \log\left(\frac{I}{I_0}\right)$$

where I_0 is the measure of the intensity of a zero-level earthquake.

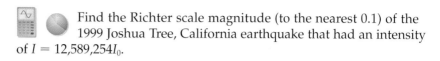

Determine the Magnitude of an Earthquake

Find the Richter scale magnitude (to the nearest 0.1) of the 1999 Joshua Tree, California earthquake that had an intensity of $I = 12{,}589{,}254I_0$.

Solution

$$M = \log\left(\frac{I}{I_0}\right) = \log\left(\frac{12{,}589{,}254I_0}{I_0}\right) = \log(12{,}589{,}254) \approx 7.1$$

The 1999 Joshua Tree earthquake had a Richter scale magnitude of 7.1.

▶ **TRY EXERCISE 56, PAGE 375**

take note

Notice in Example 5 that we didn't need to know the value of I_0 to determine the Richter scale magnitude of the quake.

If you know the Richter scale magnitude of an earthquake, you can determine the intensity of the earthquake.

EXAMPLE 6 **Determine the Intensity of an Earthquake**

Find the intensity of the 1999 Taiwan earthquake, which measured 7.6 on the Richter scale.

Solution

$$\log\left(\frac{I}{I_0}\right) = 7.6$$

$$\frac{I}{I_0} = 10^{7.6} \qquad \text{• Write in exponential form.}$$

$$I = 10^{7.6}I_0 \qquad \text{• Solve for } I.$$

$$I \approx 39{,}810{,}717I_0$$

The 1999 Taiwan earthquake had an intensity that was approximately 39,811,000 times the intensity of a zero-level earthquake.

▶ **TRY EXERCISE 58, PAGE 375**

In Example 7 we make use of the Richter scale magnitudes of two earthquakes to compare the intensities of the earthquakes.

EXAMPLE 7 **Compare Earthquakes**

The 1960 Chile earthquake had a Richter scale magnitude of 9.5. The 1989 San Francisco earthquake had a Richter scale magnitude of 7.1. Compare the intensities of the earthquakes.

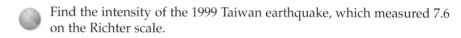

Continued ▶

take note

The results of Example 7 show that if an earthquake has a Richter scale magnitude of M_1 and a smaller earthquake has a Richter scale magnitude of M_2, then the larger earthquake is $10^{M_1 - M_2}$ times as intense as the smaller earthquake.

Solution

Let I_1 be the intensity of the Chilean earthquake and I_2 the intensity of the San Francisco earthquake. Then

$$\log\left(\frac{I_1}{I_0}\right) = 9.5 \qquad \text{and} \qquad \log\left(\frac{I_2}{I_0}\right) = 7.1$$

$$\frac{I_1}{I_0} = 10^{9.5} \qquad\qquad \frac{I_2}{I_0} = 10^{7.1}$$

$$I_1 = 10^{9.5} I_0 \qquad\qquad I_2 = 10^{7.1} I_0$$

To compare the intensities of the earthquakes, we compute the ratio I_1/I_2.

$$\frac{I_1}{I_2} = \frac{10^{9.5} I_0}{10^{7.1} I_0} = \frac{10^{9.5}}{10^{7.1}} = 10^{9.5-7.1} = 10^{2.4} \approx 251$$

The earthquake in Chile was approximately 251 times as intense as the San Francisco earthquake.

▶ **TRY EXERCISE 60, PAGE 375**

Seismologists generally determine the Richter scale magnitude of an earthquake by examining a *seismogram*. See **Figure 4.34.**

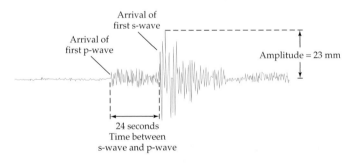

FIGURE 4.34

The magnitude of an earthquake cannot be determined just by examining the amplitude of a seismogram because this amplitude decreases as the distance between the epicenter of the earthquake and the observation station increases. To account for the distance between the epicenter and the observation station, a seismologist examines a seismogram for both small waves called **p-waves** and larger waves called **s-waves.** The Richter scale magnitude M of the earthquake is a function of both the amplitude A of the s-waves and the difference in time t between the occurrence of the s-waves and the p-waves. In the 1950s, Charles Richter developed the following formula to determine the magnitude of an earthquake from the data in a seismogram.

Amplitude-Time-Difference Formula

The Richter scale magnitude M of an earthquake is given by

$$M = \log A + 3 \log 8t - 2.92$$

where A is the amplitude, in millimeters, of the s-waves on a seismogram and t is the difference in time, in seconds, between the s-waves and the p-waves.

EXAMPLE 8   **Determine the Magnitude of an Earthquake from Its Seismogram**

Find the Richter scale magnitude of the earthquake that produced the seismogram in **Figure 4.34.**

Solution

$$M = \log A + 3 \log 8t - 2.92$$
$$= \log 23 + 3 \log[8 \cdot 24] - 2.92 \qquad \bullet \text{ Substitute 23 for } A \text{ and 24 for } t.$$
$$\approx 1.36173 + 6.84990 - 2.92$$
$$\approx 5.3$$

The earthquake had a magnitude of about 5.3 on the Richter scale.

▶ **TRY EXERCISE 64, PAGE 375**

> **take note**
>
> The Richter scale magnitude is usually rounded to the nearest tenth.

Logarithmic scales are also used in chemistry. One example concerns the pH of a liquid, which is a measure of the liquid's **acidity** or **alkalinity.** (You may have tested the pH of a swimming pool or an aquarium.) Pure water, which is considered neutral, has a pH of 7.0. The pH scale ranges from 0 to 14, with 0 corresponding to the most acidic solutions and 14 to the most alkaline. Lemon juice has a pH of about 2, whereas household ammonia measures about 11.

Specifically, the pH of a solution is a function of the hydronium-ion concentration of the solution. Because the hydronium-ion concentration of a solution can be very small (with values such as 0.00000001), pH uses a logarithmic scale.

> **take note**
>
> One mole is equivalent to 6.022×10^{23} ions.

The pH of a Solution

The **pH of a solution** with a hydronium-ion concentration of H^+ moles per liter is given by

$$pH = -\log[H^+]$$

EXAMPLE 9   **Find the pH of a Solution**

Find the pH of each liquid. Round to the nearest tenth.

a.   Orange juice with $H^+ = 2.8 \times 10^{-4}$ mole per liter

b.   Milk with $H^+ = 3.97 \times 10^{-7}$ mole per liter

c.   Rainwater with $H^+ = 6.31 \times 10^{-5}$ mole per liter

d.   A baking soda solution with $H^+ = 3.98 \times 10^{-9}$ mole per liter

Solution

a. $pH = -\log[H^+] = -\log(2.8 \times 10^{-4}) \approx 3.6$
The orange juice has a pH of 3.6.

b. $pH = -\log[H^+] = -\log(3.97 \times 10^{-7}) \approx 6.4$
The milk has a pH of 6.4.

c. $pH = -\log[H^+] = -\log(6.31 \times 10^{-5}) \approx 4.2$
The rainwater has a pH of 4.2.

d. $pH = -\log[H^+] = -\log(3.98 \times 10^{-9}) \approx 8.4$
The baking soda solution has a pH of 8.4.

▶ **TRY EXERCISE 48, PAGE 374**

MATH MATTERS

The pH scale was created by the Danish biochemist Søren Sørensen in 1909 to measure the acidity of water used in the brewing of beer. pH is an abbreviation for *pondus hydrogenii*, which translates as "potential hydrogen."

Figure 4.35 illustrates the pH scale, along with the corresponding hydronium-ion concentrations. A solution on the left half of the scale, with a pH of less than 7, is an **acid,** and a solution on the right half of the scale is an **alkaline solution** or a **base.** Because the scale is logarithmic, a solution with a pH of 5 is 10 times more acidic than a solution with a pH of 6. From Example 9 we see that the orange juice, rainwater, and milk are acids, whereas the baking soda solution is a base.

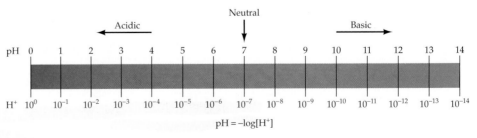

FIGURE 4.35

EXAMPLE 10 **Find the Hydronium-Ion Concentration**

A sample of blood has a pH of 7.3. Find the hydronium-ion concentration of the blood.

Solution

$$pH = -\log[H^+]$$
$$7.3 = -\log[H^+] \qquad \bullet \text{ Substitute 7.3 for pH.}$$
$$-7.3 = \log[H^+] \qquad \bullet \text{ Multiply both sides by } -1.$$
$$10^{-7.3} = H^+ \qquad \bullet \text{ Change to exponential form.}$$
$$5.0 \times 10^{-8} \approx H^+$$

The hydronium-ion concentration of the blood is about 5.0×10^{-8} mole per liter.

▶ **TRY EXERCISE 50, PAGE 375**

TOPICS FOR DISCUSSION

1. The function $f(x) = \log_b x$ is defined only for $x > 0$. Explain why this condition is imposed.

2. If p and q are positive numbers, explain why $\ln(p + q)$ isn't normally equal to $\ln p + \ln q$.

3. If $f(x) = \log_b x$ and $f(c) = f(d)$, can we conclude that $c = d$?

4. Give examples of situations in which it is advantageous to use logarithmic scales.

EXERCISE SET 4.4

In Exercises 1 to 8, write the given logarithm in terms of logarithms of x, y, and z.

1. $\log_b(xyz)$

▶ **2.** $\ln \dfrac{z^3}{\sqrt{xy}}$

3. $\ln \dfrac{x}{z^4}$

4. $\log_5 \dfrac{xy^2}{z^4}$

5. $\log_2 \dfrac{\sqrt{x}}{y^3}$

6. $\log_b\left(x\sqrt[3]{y}\right)$

7. $\log_7 \dfrac{\sqrt{xz}}{y^2}$

8. $\ln \sqrt[3]{x^2\sqrt{y}}$

In Exercises 9 to 14, write each logarithmic expression as a single logarithm with a coefficient of 1. Simplify when possible.

9. $\log(x + 5) + 2\log x$

▶ **10.** $3\log_2 t - \dfrac{1}{3}\log_2 u + 4\log_2 v$

11. $\ln(x^2 - y^2) - \ln(x - y)$

12. $\dfrac{1}{2}\log_8(x + 5) - 3\log_8 y$

13. $3 \log x + \dfrac{1}{3} \log y + \log(x + 1)$

14. $\ln(xz) - \ln\!\left(x\sqrt{y}\right) + 2 \ln \dfrac{y}{z}$

In Exercises 15 to 22, use the change-of-base formula to approximate the logarithm accurate to the nearest ten thousandth.

15. $\log_7 20$ ▶ **16.** $\log_5 37$

17. $\log_{11} 8$ **18.** $\log_{50} 22$

19. $\log_6 \dfrac{1}{3}$ **20.** $\log_3 \dfrac{7}{8}$

21. $\log_9 \sqrt{17}$ **22.** $\log_4 \sqrt{7}$

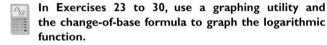

In Exercises 23 to 30, use a graphing utility and the change-of-base formula to graph the logarithmic function.

23. $f(x) = \log_4 x$ ▶ **24.** $g(x) = \log_8 (5 - x)$

25. $g(x) = \log_8(x - 3)$ **26.** $t(x) = \log_9(5 - x)$

27. $h(x) = \log_3(x - 3)^2$ **28.** $J(x) = \log_{12}(-x)$

29. $F(x) = -\log_5|x - 2|$ **30.** $n(x) = \log_2\sqrt{x - 8}$

In Exercises 31 to 40, determine if the statement is true or false for all $x > 0, y > 0$. If it is false, write an example that disproves the statement.

31. $\log_b(x + y) = \log_b x + \log_b y$

32. $\log_b(xy) = \log_b x \cdot \log_b y$

33. $\log_b(xy) = \log_b x + \log_b y$

34. $\log_b x \cdot \log_b y = \log_b x + \log_b y$

35. $\log_b x - \log_b y = \log_b(x - y), \quad x > y$

36. $\log_b \dfrac{x}{y} = \dfrac{\log_b x}{\log_b y}$

37. $\dfrac{\log_b x}{\log_b y} = \log_b x - \log_b y$

38. $\log_b(x^n) = n \log_b x$

39. $(\log_b x)^n = n \log_b x$

40. $\log_b \sqrt{x} = \dfrac{1}{2} \log_b x$

41. Evaluate the following *without* using a calculator.

$$\log_3 5 \cdot \log_5 7 \cdot \log_7 9$$

42. Evaluate the following *without* using a calculator.

$$\log_5 20 \cdot \log_{20} 60 \cdot \log_{60} 100 \cdot \log_{100} 125$$

43. Which is larger, 500^{501} or 506^{500}? These numbers are too large for most calculators to handle. (They each have 1353 digits!) (*Hint:* Let $x = 500^{501}$ and $y = 506^{500}$ and then compare $\ln x$ with $\ln y$.)

44. Which number is smaller, $\dfrac{1}{50^{300}}$ or $\dfrac{1}{151^{233}}$?

45. **ANIMATED MAPS** A software company that creates interactive maps for websites has designed an animated zooming feature so that when a user selects the zoom-in option, the map appears to expand on a location. This is accomplished by displaying several intermediate maps to give the illusion of motion. The company has determined that zooming in on a location is more informative and pleasing to observe when the scale of each step of the animation is determined using the equation

$$S_n = S_0 \cdot 10^{\frac{n}{N}(\log S_f - \log S_0)}$$

where S_n represents the scale of the current step n ($n = 0$ corresponds to the initial scale), S_0 is the starting scale of the map, S_f is the final scale, and N is the number of steps in the animation following the initial scale. (If the initial scale of the map is $1:200$, then $S_0 = 200$.) Determine the scales to be used at each intermediate step if a map is to start with a scale of $1:1,000,000$ and proceed through five intermediate steps to end with a scale of $1:500,000$.

46. **ANIMATED MAPS** Use the equation in Exercise 45 to determine the scales for each stage of an animated map zoom that goes from a scale of $1:250,000$ to a scale of $1:100,000$ in four steps (following the initial scale).

47. **pH** Milk of magnesia has a hydronium-ion concentration of about 3.97×10^{-11} mole per liter. Determine the pH of milk of magnesia and state whether it is an acid or a base.

▶ **48.** **pH** Vinegar has a hydronium-ion concentration of 1.26×10^{-3} mole per liter. Determine the pH of vinegar and state whether it is an acid or a base.

49. **HYDRONIUM-ION CONCENTRATION** A morphine solution has a pH of 9.5. Determine the hydronium-ion concentration of the morphine solution.

▶ **50.** **HYDRONIUM-ION CONCENTRATION** A rainstorm in New York City produced rainwater with a pH of 5.6. Determine the hydronium-ion concentration of the rainwater.

51. **DECIBEL LEVEL** The range of sound intensities that the human ear can detect is so large that a special decibel scale (named after Alexander Graham Bell) is used to measure and compare sound intensities. The **decibel level** dB of a sound is given by

$$dB(I) = 10 \log \left(\frac{I}{I_0} \right)$$

where I_0 is the intensity of sound that is barely audible to the human ear. Find the decibel level for the following sounds. Round to the nearest tenth of a decibel.

Sound	Intensity
a. Automobile traffic	$I = 1.58 \times 10^8 \cdot I_0$
b. Quiet conversation	$I = 10,800 \cdot I_0$
c. Fender guitar	$I = 3.16 \times 10^{11} \cdot I_0$
d. Jet engine	$I = 1.58 \times 10^{15} \cdot I_0$

52. **COMPARISON OF SOUND INTENSITIES** A team in Arizona installed a 48,000-watt sound system in a Ford Bronco that it claims can output 175-decibel sound. The human pain threshold for sound is 125 decibels. How many times more intense is the sound from the Bronco than the human pain threshold?

53. **COMPARISON OF SOUND INTENSITIES** How many times more intense is a sound that measures 120 decibels than a sound that measures 110 decibels?

54. **DECIBEL LEVEL** If the intensity of a sound is doubled, what is the increase in the decibel level? [*Hint:* Find $dB(2I) - dB(I)$.]

55. **EARTHQUAKE MAGNITUDE** What is the Richter scale magnitude of an earthquake with an intensity of $I = 100,000I_0$?

▶ **56.** **EARTHQUAKE MAGNITUDE** The Colombia earthquake of 1906 had an intensity of $I = 398,107,000I_0$. What did it measure on the Richter scale?

57. **EARTHQUAKE INTENSITY** The Coalinga, California, earthquake of 1983 had a Richter scale magnitude of 6.5. Find the intensity of this earthquake.

▶ **58.** **EARTHQUAKE INTENSITY** The earthquake that occurred just south of Concepción, Chile, in 1960 had a Richter scale magnitude of 9.5. Find the intensity of this earthquake.

59. **COMPARISON OF EARTHQUAKES** Compare the intensity of an earthquake that measures 5.0 on the Richter scale to the intensity of an earthquake that measures 3.0 on the Richter scale by finding the ratio of the larger intensity to the smaller intensity.

▶ **60.** **COMPARISON OF EARTHQUAKES** How many times more intense was the 1960 earthquake in Chile, which measured 9.5 on the Richter scale, than the San Francisco earthquake of 1906, which measured 8.3 on the Richter scale?

61. **COMPARISON OF EARTHQUAKES** On March 2, 1933, an earthquake of magnitude 8.9 on the Richter scale struck Japan. In October 1989, an earthquake of magnitude 7.1 on the Richter scale struck San Francisco, California. Compare the intensity of the larger earthquake to the intensity of the smaller earthquake by finding the ratio of the larger intensity to the smaller intensity.

62. **COMPARISON OF EARTHQUAKES** An earthquake that occurred in China in 1978 measured 8.2 on the Richter scale. In 1988, an earthquake in California measured 6.9 on the Richter scale. Compare the intensity of the larger earthquake to the intensity of the smaller earthquake by finding the ratio of the larger intensity to the smaller intensity.

63. **EARTHQUAKE MAGNITUDE** Find the Richter scale magnitude of the earthquake that produced the seismogram in the following figure.

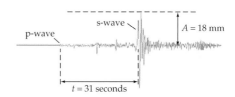

64. **EARTHQUAKE MAGNITUDE** Find the Richter scale magnitude of the earthquake that produced the seismogram in the following figure.

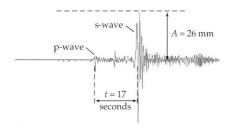

━━━━ *CONNECTING CONCEPTS* ━━━━

65. **NOMOGRAMS AND LOGARITHMIC SCALES** A **nomogram** is a diagram used to determine a numerical result by drawing a line across numerical scales. The following nomogram, used by Richter, determines the magnitude of an earthquake from its seismogram. To use the nomogram, mark the amplitude of a seismogram on the amplitude scale and mark the time between the s-wave and the p-wave on the S-P scale. Draw a line between these marks. The Richter scale magnitude of the earthquake that produced the seismogram is shown by the intersection of the line and the center scale. The example below shows that an earthquake with a seismogram amplitude of 23 millimeters and an S-P time of 24 seconds has a Richter scale magnitude of about 5.

The amplitude and the S-P time are shown on logarithmic scales. On the amplitude scale, the distance from 1 to 10 is the same as the distance from 10 to 100, because $\log 100 - \log 10 = \log 10 - \log 1$.

Use the nomogram at the left to determine the Richter scale magnitude of an earthquake with a seismogram

a. amplitude of 50 millimeters and S-P time of 40 seconds.

b. amplitude of 1 millimeter and S-P time of 30 seconds.

c. How do the results in parts **a.** and **b.** compare with the Richter scale magnitudes produced by using the amplitude-time-difference formula?

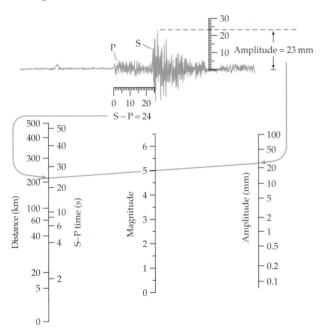

Richter's earthquake nomogram

━━━━ *PREPARE FOR SECTION 4.5* ━━━━

66. Use the definition of a logarithm to write the exponential equation $3^6 = 729$ in logarithmic form. [4.2]

67. Use the definition of a logarithm to write the logarithmic equation $\log_5 625 = 4$ in exponential form. [4.2]

68. Use the definition of a logarithm to write the exponential equation $a^{x+2} = b$ in logarithmic form. [4.2]

69. Solve for x: $4a = 7bx + 2cx$. [1.2]

70. Solve for x: $165 = \dfrac{300}{1 + 12x}$. [1.4]

71. Solve for x: $A = \dfrac{100 + x}{100 - x}$. [1.4]

PROJECTS

1. LOGARITHMIC SCALES Sometimes **logarithmic scales** are used to better view a collection of data that span a wide range of values. For instance, consider the table below, which lists the approximate masses of various marine creatures in grams. Next we have attempted to plot the masses on a number line.

Animal	Mass (g)
Rotifer	0.000000006
Dwarf goby	0.30
Lobster	15,900
Leatherback turtle	851,000
Giant squid	1,820,000
Whale shark	4,700,000
Blue whale	120,000,000

Mass (in millions of grams)

As you can see, we had to use such a large span of numbers that the data for most of the animals are bunched up at the left. Visually, this number line isn't very helpful for any comparisons.

a. Make a new number line, this time plotting the logarithm (base 10) of each of the masses.

b. Which number line is more helpful to compare the masses of the different animals?

c. If the data points for two animals on the logarithmic number line are 1 unit apart, how do the animals' masses compare? What if the points are 2 units apart?

2. LOGARITHMIC SCALES The distances of the planets in our solar system from the sun are given in the table at the top of the next column.

a. Draw a number line with an appropriate scale to plot the distances.

b. Draw a second number line, this time plotting the logarithm (base 10) of each distance.

c. Which number line do you find more helpful to compare the different distances?

d. If two distances are 3 units apart on the logarithmic number line, how do the distances of the corresponding planets compare?

Planet	Distance (million km)
Mercury	58
Venus	108
Earth	150
Mars	228
Jupiter	778
Saturn	1427
Uranus	2871
Neptune	4497
Pluto	5913

3. BIOLOGIC DIVERSITY To discuss the variety of species that live in a certain environment, a biologist needs a precise definition of *diversity*. Let $p_1, p_2, \ldots, p_n$ be the proportions of n species that live in an environment. The biologic diversity D of this system is

$$D = -(p_1 \log_2 p_1 + p_2 \log_2 p_2 + \cdots + p_n \log_2 p_n)$$

Suppose that an ecosystem has exactly five different varieties of grass: rye (R), bermuda (B), blue (L), fescue (F), and St. Augustine (A).

a. Calculate the diversity of this ecosystem if the proportions of these grasses are as shown in Table 1. Round to the nearest hundredth.

Table 1

R	B	L	F	A
$\frac{1}{5}$	$\frac{1}{5}$	$\frac{1}{5}$	$\frac{1}{5}$	$\frac{1}{5}$

b. Because bermuda and St. Augustine are virulent grasses, after a time the proportions will be as shown in Table 2. Calculate the diversity of this system. Does this system have more or less diversity than the system given in Table 1?

Table 2

R	B	L	F	A
$\frac{1}{8}$	$\frac{3}{8}$	$\frac{1}{16}$	$\frac{1}{8}$	$\frac{5}{16}$

c. After an even longer time period, the bermuda and St. Augustine grasses completely overrun the environment and the proportions are as shown in Table 3. Calculate the diversity of this system. (*Note:* Although the equation is not technically correct, for purposes of the diversity definition, we may say that $0 \log_2 0 = 0$. By using very small values of p_i, we can demonstrate that this definition makes sense.) Does this system have more or less diversity than the system given in Table 2?

d. Finally, the St. Augustine grasses overrun the bermuda grasses and the proportions are as shown in Table 4. Calculate the diversity of this system. Write a sentence that explains the meaning of the value you obtained.

Table 3

R	B	L	F	A
0	$\frac{1}{4}$	0	0	$\frac{3}{4}$

Table 4

R	B	L	F	A
0	0	0	0	1

SECTION 4.5

EXPONENTIAL AND LOGARITHMIC EQUATIONS

- SOLVE EXPONENTIAL EQUATIONS
- SOLVE LOGARITHMIC EQUATIONS
- APPLICATION

SOLVE EXPONENTIAL EQUATIONS

If a variable appears in an exponent of a term of an equation, such as $2^{x+1} = 32$, then the equation is called an **exponential equation**. Example 1 uses the following Equality-of-Exponents Theorem to solve $2^{x+1} = 32$.

> **Equality of Exponents Theorem**
>
> If $b^x = b^y$, then $x = y$, provided $b > 0$ and $b \neq 1$.

EXAMPLE 1 **Solve an Exponential Equation**

Use the Equality of Exponents Theorem to solve $2^{x+1} = 32$.

Solution

$$2^{x+1} = 32$$
$$2^{x+1} = 2^5 \qquad \bullet \text{ Write each side as a power of 2.}$$
$$x + 1 = 5 \qquad \bullet \text{ Equate the exponents.}$$
$$x = 4$$

Check: Let $x = 4$, then $2^{x+1} = 2^{4+1}$
$$= 2^5$$
$$= 32$$

▶ **TRY EXERCISE 2, PAGE 385**

A graphing utility can also be used to find the solutions of an equation of the form $f(x) = g(x)$. Either of the following two methods can be employed.

> ### Using a Graphing Utility to Find the Solutions of $f(x) = g(x)$
>
> *Intersection Method*   Graph $y_1 = f(x)$ and $y_2 = g(x)$ on the same screen. The solutions of $f(x) = g(x)$ are the x-coordinates of the points of intersection of the graphs.
>
> *Intercept Method*   The solutions of $f(x) = g(x)$ are the x-coordinates of the x-intercepts of the graph of $y = f(x) - g(x)$.
>
> **Figures 4.36** and **4.37** illustrate the graphical methods for solving $2^{x+1} = 32$.

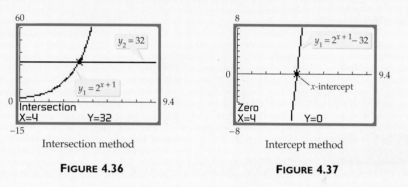

Intersection method

FIGURE 4.36

Intercept method

FIGURE 4.37

In Example 1 we were able to write both sides of the equation as a power of the same base. If you find it difficult to write both sides of an exponential equation in terms of the same base, then try the procedure of taking the logarithm of each side of the equation. This procedure is used in Example 2.

EXAMPLE 2   Solve an Exponential Equation

Solve: $5^x = 40$

Algebraic Solution

$$5^x = 40$$
$$\log(5^x) = \log 40 \quad \text{• Take the logarithm of each side.}$$
$$x \log 5 = \log 40 \quad \text{• Power property}$$
$$x = \frac{\log 40}{\log 5} \quad \text{• Exact solution}$$
$$x \approx 2.3 \quad \text{• Decimal approximation}$$

To the nearest tenth, the solution is 2.3.

Visualize the Solution

Intersection Method   The solution of $5^x = 40$ is the x-coordinate of the point of intersection of $y = 5^x$ and $y = 40$ (see **Figure 4.38**).

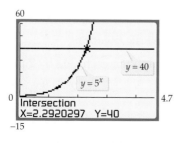

FIGURE 4.38

▶ **TRY EXERCISE 10, PAGE 385**

An alternative approach to solving the equation in Example 2 is to rewrite the exponential equation in logarithmic form: $5^x = 40$ is equivalent to the logarithmic equation $\log_5 40 = x$. Using the change-of-base formula, we find that $x = \log_5 40 = \dfrac{\log 40}{\log 5}$. In the following example, however, we must take logarithms of both sides to reach a solution.

EXAMPLE 3   Solve an Exponential Equation

Solve: $3^{2x-1} = 5^{x+2}$

Algebraic Solution

$$3^{2x-1} = 5^{x+2}$$
$$\ln 3^{2x-1} = \ln 5^{x+2}$$

• Take the natural logarithm of each side.

$$(2x - 1)\ln 3 = (x + 2)\ln 5$$

• Power property

$$2x \ln 3 - \ln 3 = x \ln 5 + 2 \ln 5$$

• Distributive property

$$2x \ln 3 - x \ln 5 = 2 \ln 5 + \ln 3$$

• Solve for x.

$$x(2 \ln 3 - \ln 5) = 2 \ln 5 + \ln 3$$

$$x = \frac{2 \ln 5 + \ln 3}{2 \ln 3 - \ln 5}$$

• Exact solution

$$x \approx 7.3$$

• Decimal approximation

To the nearest tenth, the solution is 7.3.

Visualize the Solution

Intercept Method   The solution of $3^{2x-1} = 5^{x+2}$ is the x-coordinate of the x-intercept of $y = 3^{2x-1} - 5^{x+2}$ (see **Figure 4.39**).

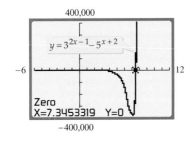

FIGURE 4.39

▶ **TRY EXERCISE 18, PAGE 385**

In Example 4 we solve an exponential equation that has two solutions.

EXAMPLE 4   Solve an Exponential Equation Involving $b^x + b^{-x}$

Solve: $\dfrac{2^x + 2^{-x}}{2} = 3$

Algebraic Solution

Multiplying each side by 2 produces

$$2^x + 2^{-x} = 6$$
$$2^{2x} + 2^0 = 6(2^x)$$

• Multiply each side by 2^x to clear negative exponents.

$$(2^x)^2 - 6(2^x) + 1 = 0$$

• Write in quadratic form.

$$(u)^2 - 6(u) + 1 = 0$$

• Substitute u for 2^x.

Visualize the Solution

Intersection Method   The solutions of $\dfrac{2^x + 2^{-x}}{2} = 3$ are the x-coordinates of the points of intersection of

By the quadratic formula,

$$u = \frac{6 \pm \sqrt{36 - 4}}{2} = \frac{6 \pm 4\sqrt{2}}{2} = 3 \pm 2\sqrt{2}$$

$$2^x = 3 \pm 2\sqrt{2}$$ • Replace u with 2^x.

$$\log 2^x = \log(3 \pm 2\sqrt{2})$$ • Take the common
logarithm of each
side.

$$x \log 2 = \log(3 \pm 2\sqrt{2})$$ • Power property

$$x = \frac{\log(3 \pm 2\sqrt{2})}{\log 2} \approx \pm 2.54$$

The approximate solutions are -2.54 and 2.54.

▶ **TRY EXERCISE 40, PAGE 385**

$y = \dfrac{2^x + 2^{-x}}{2}$ and $y = 3$ (see

Figure 4.40).

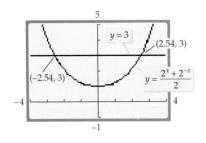

FIGURE 4.40

● SOLVE LOGARITHMIC EQUATIONS

Equations that involve logarithms are called logarithmic equations. The properties of logarithms, along with the definition of a logarithm, are often used to find the solutions of a logarithmic equation.

EXAMPLE 5   **Solve a Logarithmic Equation**

Solve: $\log(3x - 5) = 2$

Solution

$$\log(3x - 5) = 2$$

$$3x - 5 = 10^2$$ • Definition of a logarithm

$$3x = 105$$ • Solve for x.

$$x = 35$$

Check: $\log[3(35) - 5] = \log 100 = 2$

▶ **TRY EXERCISE 22, PAGE 385**

❓ QUESTION Can a negative number be a solution of a logarithmic equation?

❓ ANSWER   Yes. For instance, -10 is a solution of $\log(-x) = 1$.

| **EXAMPLE 6** | **Solve a Logarithmic Equation** |

Solve: $\log 2x - \log(x - 3) = 1$

Solution

$$\log 2x - \log(x - 3) = 1$$

$$\log \frac{2x}{x - 3} = 1 \qquad \text{• Quotient property}$$

$$\frac{2x}{x - 3} = 10^1 \qquad \text{• Definition of logarithm}$$

$$2x = 10x - 30 \qquad \text{• Solve for } x.$$

$$-8x = -30$$

$$x = \frac{15}{4}$$

Check the solution by substituting $\frac{15}{4}$ into the original equation.

▶ **TRY EXERCISE 26, PAGE 385**

In Example 7 we make use of the one-to-one property of logarithms to find the solution of a logarithmic equation. This example illustrates that the process of solving a logarithmic equation by using logarithmic properties may introduce an extraneous solution.

| **EXAMPLE 7** | **Solve a Logarithmic Equation** |

Solve: $\ln(3x + 8) = \ln(2x + 2) + \ln(x - 2)$

Algebraic Solution

$$\ln(3x + 8) = \ln(2x + 2) + \ln(x - 2)$$

$$\ln(3x + 8) = \ln[(2x + 2)(x - 2)] \qquad \text{• Product property}$$

$$\ln(3x + 8) = \ln(2x^2 - 2x - 4)$$

$$3x + 8 = 2x^2 - 2x - 4 \qquad \text{• One-to-one property of logarithms}$$

$$0 = 2x^2 - 5x - 12$$

$$0 = (2x + 3)(x - 4) \qquad \text{• Solve for } x.$$

$$x = -\frac{3}{2} \quad \text{or} \quad x = 4$$

Thus $-\frac{3}{2}$ and 4 are possible solutions. A check will show that 4 is a

solution, but $-\frac{3}{2}$ is not a solution.

Visualize the Solution

The graph of $y = \ln(3x + 8) - \ln(2x + 2) - \ln(x - 2)$ has only one x-intercept (see **Figure 4.41**). Thus there is only one real solution.

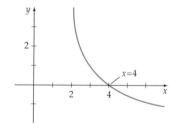

$y = \ln(3x + 8) - \ln(2x + 2) - \ln(x - 2)$

FIGURE 4.41

▶ **TRY EXERCISE 36, PAGE 385**

? QUESTION Why does $x = -\dfrac{3}{2}$ not check in Example 7?

• APPLICATION

EXAMPLE 8   **Velocity of a Sky Diver Experiencing Air Resistance**

During the free-fall portion of a jump, the time t in seconds required for a sky diver to reach a velocity of v feet per second is given by

$$t = -\frac{175}{32} \ln\left(1 - \frac{v}{175}\right)$$

a. Determine the velocity of the diver after 5 seconds.

b. The graph of the above function has a vertical asymptote at $v = 175$. Explain the meaning of the vertical asymptote in the context of this example.

Solution

a. Substitute 5 for t and solve for v.

$$t = -\frac{175}{32} \ln\left(1 - \frac{v}{175}\right)$$

$$5 = -\frac{175}{32} \ln\left(1 - \frac{v}{175}\right) \qquad \text{• Replace } t \text{ with 5.}$$

$$\left(-\frac{32}{175}\right)5 = \ln\left(1 - \frac{v}{175}\right) \qquad \text{• Solve for } v.$$

$$-\frac{32}{35} = \ln\left(1 - \frac{v}{175}\right)$$

$$e^{-32/35} = 1 - \frac{v}{175} \qquad \text{• Write in exponential form.}$$

$$e^{-32/35} - 1 = -\frac{v}{175}$$

$$v = 175(1 - e^{-32/35})$$

$$v \approx 104.86$$

Continued ▶

take note

If air resistance is not considered, then the time in seconds required for a sky diver to reach a given velocity (in feet per second) is

$t = \dfrac{v}{32}$. The function in Example 8

is a more realistic model of the time required to reach a given velocity during the free-fall of a sky diver who is experiencing air resistance.

? ANSWER If $x = -\dfrac{3}{2}$, the original equation becomes $\ln\left(\dfrac{7}{2}\right) = \ln(-1) + \ln\left(-\dfrac{7}{2}\right)$.
This cannot be true, because the function $f(x) = \ln x$ is not defined for negative values of x.

After 5 seconds the velocity of the sky diver will be about 104.9 feet per second. See **Figure 4.42.**

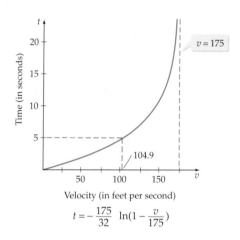

$$t = -\frac{175}{32} \ln\left(1 - \frac{v}{175}\right)$$

FIGURE 4.42

b. The vertical asymptote $v = 175$ indicates that the sky diver will not attain a velocity greater than 175 feet per second. In **Figure 4.42,** note that as $v \to 175$ from the left, $t \to \infty$.

▶ **TRY EXERCISE 68, PAGE 387**

 TOPICS FOR DISCUSSION

1. Discuss how to solve the equation $a = \log_b x$ for x.

2. What is the domain of $y = \log_4(2x - 5)$? Explain why this means that the equation $\log_4(x - 3) = \log_4(2x - 5)$ has no real number solution.

3. -8 is not a solution of the equation $\log_2 x + \log_2(x + 6) = 4$. Discuss at which step in the following solution the extraneous solution -8 was introduced.

$$\log_2 x + \log_2(x + 6) = 4$$
$$\log_2 x(x + 6) = 4$$
$$x(x + 6) = 2^4$$
$$x^2 + 6x = 16$$
$$x^2 + 6x - 16 = 0$$
$$(x + 8)(x - 2) = 0$$
$$x = -8 \quad \text{or} \quad x = 2$$

EXERCISE SET 4.5

In Exercises 1 to 46, solve for x algebraically.

1. $2^x = 64$

2. $3^x = 243$

3. $49^x = \dfrac{1}{343}$

4. $9^x = \dfrac{1}{243}$

5. $2^{5x+3} = \dfrac{1}{8}$

6. $3^{4x-7} = \dfrac{1}{9}$

7. $\left(\dfrac{2}{5}\right)^x = \dfrac{8}{125}$

8. $\left(\dfrac{2}{5}\right)^x = \dfrac{25}{4}$

9. $5^x = 70$

10. $6^x = 50$

11. $3^{-x} = 120$

12. $7^{-x} = 63$

13. $10^{2x+3} = 315$

14. $10^{6-x} = 550$

15. $e^x = 10$

16. $e^{x+1} = 20$

17. $2^{1-x} = 3^{x+1}$

18. $3^{x-2} = 4^{2x+1}$

19. $2^{2x-3} = 5^{-x-1}$

20. $5^{3x} = 3^{x+4}$

21. $\log(4x - 18) = 1$

22. $\log(x^2 + 19) = 2$

23. $\ln(x^2 - 12) = \ln x$

24. $\log(2x^2 + 3x) = \log(10x + 30)$

25. $\log_2 x + \log_2(x - 4) = 2$

26. $\log_3 x + \log_3(x + 6) = 3$

27. $\log(5x - 1) = 2 + \log(x - 2)$

28. $1 + \log(3x - 1) = \log(2x + 1)$

29. $\ln(1 - x) + \ln(3 - x) = \ln 8$

30. $\log(4 - x) = \log(x + 8) + \log(2x + 13)$

31. $\log \sqrt{x^3 - 17} = \dfrac{1}{2}$

32. $\log(x^3) = (\log x)^2$

33. $\log(\log x) = 1$

34. $\ln(\ln x) = 2$

35. $\ln(e^{3x}) = 6$

36. $\ln x = \dfrac{1}{2}\ln\left(2x + \dfrac{5}{2}\right) + \dfrac{1}{2}\ln 2$

37. $e^{\ln(x-1)} = 4$

38. $10^{\log(2x+7)} = 8$

39. $\dfrac{10^x - 10^{-x}}{2} = 20$

40. $\dfrac{10^x + 10^{-x}}{2} = 8$

41. $\dfrac{10^x + 10^{-x}}{10^x - 10^{-x}} = 5$

42. $\dfrac{10^x - 10^{-x}}{10^x + 10^{-x}} = \dfrac{1}{2}$

43. $\dfrac{e^x + e^{-x}}{2} = 15$

44. $\dfrac{e^x - e^{-x}}{2} = 15$

45. $\dfrac{1}{e^x - e^{-x}} = 4$

46. $\dfrac{e^x + e^{-x}}{e^x - e^{-x}} = 3$

 In Exercises 47 to 56, use a graphing utility to approximate the solutions of the equation to the nearest hundredth.

47. $2^{-x+3} = x + 1$

48. $3^{x-2} = -2x - 1$

49. $e^{3-2x} - 2x = 1$

50. $2e^{x+2} + 3x = 2$

51. $3\log_2(x - 1) = -x + 3$

52. $2\log_3(2 - 3x) = 2x - 1$

53. $\ln(2x + 4) + \dfrac{1}{2}x = -3$

54. $2\ln(3 - x) + 3x = 4$

55. $2^{x+1} = x^2 - 1$

56. $\ln x = -x^2 + 4$

57. POPULATION GROWTH The population P of a city grows exponentially according to the function

$$P(t) = 8500(1.1)^t, \quad 0 \le t \le 8$$

where t is measured in years.

a. Find the population at time $t = 0$ and also at time $t = 2$.

b. When, to the nearest year, will the population reach 15,000?

58. PHYSICAL FITNESS After a race, a runner's pulse rate R in beats per minute decreases according to the function

$$R(t) = 145e^{-0.092t}, \quad 0 \le t \le 15$$

where t is measured in minutes.

a. Find the runner's pulse rate at the end of the race and also 1 minute after the end of the race.

b. How long, to the nearest minute, after the end of the race will the runner's pulse rate be 80 beats per minute?

59. RATE OF COOLING A can of soda at 79°F is placed in a refrigerator that maintains a constant temperature of 36°F. The temperature T of the soda t minutes after it is placed in the refrigerator is given by

$$T(t) = 36 + 43e^{-0.058t}$$

a. Find the temperature, to the nearest degree, of the soda 10 minutes after it is placed in the refrigerator.

b. When, to the nearest minute, will the temperature of the soda be 45°F?

60. MEDICINE During surgery, a patient's circulatory system requires at least 50 milligrams of an anesthetic. The amount of anesthetic present t hours after 80 milligrams of anesthetic is administered is given by

$$T(t) = 80(0.727)^t$$

a. How much, to the nearest milligram, of the anesthetic is present in the patient's circulatory system 30 minutes after the anesthetic is administered?

b. How long, to the nearest minute, can the operation last if the patient does not receive additional anesthetic?

61. PSYCHOLOGY Industrial psychologists study employee training programs to assess the effectiveness of the instruction. In one study, the percent score P on a test for a person who had completed t hours of training was given by

$$P = \frac{100}{1 + 30e^{-0.088t}}$$

a. Use a graphing utility to graph the equation for $t \geq 0$.

b. Use the graph to estimate (to the nearest hour) the number of hours of training necessary to achieve a 70% score on the test.

c. From the graph, determine the horizontal asymptote.

d. Write a sentence that explains the meaning of the horizontal asymptote.

62. PSYCHOLOGY An industrial psychologist has determined that the average percent score for an employee on a test of the employee's knowledge of the company's product is given by

$$P = \frac{100}{1 + 40e^{-0.1t}}$$

where t is the number of weeks on the job and P is the percent score.

a. Use a graphing utility to graph the equation for $t \geq 0$.

b. Use the graph to estimate (to the nearest week) the number of weeks of employment that are necessary for the average employee to earn a 70% score on the test.

c. Determine the horizontal asymptote of the graph.

d. Write a sentence that explains the meaning of the horizontal asymptote.

63. ECOLOGY A herd of bison was placed in a wildlife preserve that can support a maximum of 1000 bison. A population model for the bison is given by

$$B = \frac{1000}{1 + 30e^{-0.127t}}$$

where B is the number of bison in the preserve and t is time in years, with the year 1999 represented by $t = 0$.

a. Use a graphing utility to graph the equation for $t \geq 0$.

b. Use the graph to estimate (to the nearest year) the number of years before the bison population reaches 500.

c. Determine the horizontal asymptote of the graph.

d. Write a sentence that explains the meaning of the horizontal asymptote.

64. POPULATION GROWTH A yeast culture grows according to the equation

$$Y = \frac{50,000}{1 + 250e^{-0.305t}}$$

where Y is the number of yeast and t is time in hours.

a. Use a graphing utility to graph the equation for $t \geq 0$.

b. Use the graph to estimate (to the nearest hour) the number of hours before the yeast population reaches 35,000.

c. From the graph, estimate the horizontal asymptote.

d. Write a sentence that explains the meaning of the horizontal asymptote.

65. CONSUMPTION OF NATURAL RESOURCES A model for how long our coal resources will last is given by

$$T = \frac{\ln(300r + 1)}{\ln(r + 1)}$$

where r is the percent increase in consumption from current levels of use and T is the time (in years) before the resource is depleted.

a. Use a graphing utility to graph this equation.

b. If our consumption of coal increases by 3% per year, in how many years will we deplete our coal resources?

c. What percent increase in consumption of coal will deplete the resource in 100 years? Round to the nearest tenth of a percent.

66. CONSUMPTION OF NATURAL RESOURCES A model for how long our aluminum resources will last is given by

$$T = \frac{\ln(20{,}500r + 1)}{\ln(r + 1)}$$

where r is the percent increase in consumption from current levels of use and T is the time (in years) before the resource is depleted.

a. Use a graphing utility to graph this equation.

b. If our consumption of aluminum increases by 5% per year, in how many years (to the nearest year) will we deplete our aluminum resources?

c. What percent increase in consumption of aluminum will deplete the resource in 100 years? Round to the nearest tenth of a percent.

67. VELOCITY OF A MEDICAL CARE PACKAGE A medical care package is air lifted and dropped to a disaster area. During the free-fall portion of the drop, the time, in seconds, required for the package to obtain a velocity of v feet per second is given by the function

$$t = 2.43 \ln \frac{150 + v}{150 - v}, \quad 0 \le v < 150$$

a. Determine the velocity of the package 5 seconds after it is dropped. Round to the nearest foot per second.

b. Determine the vertical asymptote of the function.

c. Write a sentence that explains the meaning of the vertical asymptote in the context of this application.

▶ **68.** EFFECTS OF AIR RESISTANCE ON VELOCITY If we assume that air resistance is proportional to the square of the velocity, then the time t in seconds required for an object to reach a velocity of v feet per second is given by

$$t = \frac{9}{24} \ln \frac{24 + v}{24 - v}, 0 \le v < 24$$

a. Determine the velocity, to the nearest hundredth of a foot per second, of the object after 1.5 seconds.

b. Determine the vertical asymptote for the graph of this function.

c. Write a sentence that describes the meaning of the vertical asymptote in the context of this problem.

69. TERMINAL VELOCITY WITH AIR RESISTANCE The velocity v of an object t seconds after it has been dropped from a height above the surface of the earth is given by the equation $v = 32t$ feet per second, assuming no air resistance. If we assume that air resistance is proportional to the square of the velocity, then the velocity after t seconds is given by

$$v = 100\left(\frac{e^{0.64t} - 1}{e^{0.64t} + 1}\right)$$

a. In how many seconds will the velocity be 50 feet per second?

b. Determine the horizontal asymptote for the graph of this function.

c. Write a sentence that describes the meaning of the horizontal asymptote in the context of this problem.

70. TERMINAL VELOCITY WITH AIR RESISTANCE If we assume that air resistance is proportional to the square of the velocity, then the velocity v in feet per second of an object t seconds after it has been dropped is given by

$$v = 50\left(\frac{e^{1.6t} - 1}{e^{1.6t} + 1}\right)$$

(See Exercise 69. The reason for the difference in the equations is that the proportionality constants are different.)

a. In how many seconds will the velocity be 20 feet per second?

b. Determine the horizontal asymptote for the graph of this function.

c. Write a sentence that describes the meaning of the horizontal asymptote in the context of this problem.

71. EFFECTS OF AIR RESISTANCE ON DISTANCE The distance s, in feet, that the object in Exercise 69 will fall in t seconds is given by

$$s = \frac{100^2}{32} \ln\left(\frac{e^{0.32t} + e^{-0.32t}}{2}\right)$$

a. Use a graphing utility to graph this equation for $t \ge 0$.

b. How long does it take for the object to fall 100 feet? Round to the nearest tenth of a second.

72. 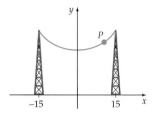 **EFFECTS OF AIR RESISTANCE ON DISTANCE** The distance s, in feet, that the object in Exercise 70 will fall in t seconds is given by

$$s = \frac{50^2}{40} \ln\left(\frac{e^{0.8t} + e^{-0.8t}}{2}\right)$$

a. Use a graphing utility to graph this equation for $t \geq 0$.

b. How long does it take for the object to fall 100 feet? Round to the nearest tenth of a second.

73. **RETIREMENT PLANNING** The retirement account for a graphic designer contains $250,000 on January 1, 2002, and earns interest at a rate of 0.5% per month. On February 1, 2002, the designer withdraws $2000 and plans to continue these withdrawals as retirement income each month. The value V of the account after x months is

$$V = 400{,}000 - 150{,}000(1.005)^x$$

If the designer wishes to leave $100,000 to a scholarship foundation, what is the maximum number of withdrawals (to the nearest month) the designer can make from this account and still have $100,000 to donate?

74. **HANGING CABLE** The height h, in feet, of any point P on the cable shown is given by

$$h(x) = 10(e^{x/20} + e^{-x/20}), \quad -15 \leq x \leq 15$$

where $|x|$ is the horizontal distance in feet between P and the y-axis.

a. What is the lowest height of the cable?

b. What is the height of the cable 10 feet to the right of the y-axis? Round to the nearest tenth of a foot.

c. How far to the right of the y-axis is the cable 24 feet in height? Round to the nearest tenth of a foot.

CONNECTING CONCEPTS

75. The following argument seems to indicate that $0.125 > 0.25$. Find the first incorrect statement in the argument.

$$3 > 2$$
$$3(\log 0.5) > 2(\log 0.5)$$
$$\log 0.5^3 > \log 0.5^2$$
$$0.5^3 > 0.5^2$$
$$0.125 > 0.25$$

76. The following argument seems to indicate that $4 = 6$. Find the first incorrect statement in the argument.

$$4 = \log_2 16$$
$$4 = \log_2(8 + 8)$$
$$4 = \log_2 8 + \log_2 8$$
$$4 = 3 + 3$$
$$4 = 6$$

77. A common mistake that students make is to write $\log(x + y)$ as $\log x + \log y$. For what values of x and y does $\log(x + y) = \log x + \log y$? (*Hint:* Solve for x in terms of y.)

78. Let $f(x) = 2 \ln x$ and $g(x) = \ln x^2$. Does $f(x) = g(x)$ for all real numbers x?

79. Explain why the functions $F(x) = 1.4^x$ and $G(x) = e^{0.336x}$ represent essentially the same function.

80. Find the constant k that will make $f(t) = 2.2^t$ and $g(t) = e^{-kt}$ represent essentially the same function.

PREPARE FOR SECTION 4.6

81. Evaluate $A = 1000\left(1 + \dfrac{0.1}{12}\right)^{12t}$ for $t = 2$. Round to the nearest hundredth. [4.2]

82. Evaluate $A = 600\left(1 + \dfrac{0.04}{4}\right)^{4t}$ for $t = 8$. Round to the nearest hundredth. [4.2]

83. Solve $0.5 = e^{14k}$ for k. Round to the nearest ten-thousandth. [4.5]

84. Solve $0.85 = 0.5^{t/5730}$ for t. Round to the nearest ten. [4.5]

85. Solve $6 = \dfrac{70}{5 + 9e^{-k \cdot 12}}$ for k. Round to the nearest thousandth. [4.5]

86. Solve $2{,}000{,}000 = \dfrac{3^{n+1} - 3}{2}$ for n. Round to the nearest tenth. [4.5]

PROJECTS

1.   NAVIGATING   The pilot of a boat is trying to cross a river to a point O two miles due west of the boat's starting position by always pointing the nose of the boat toward O. Suppose the speed of the current is w miles per hour and the speed of the boat is v miles per hour. If point O is the origin and the boat's starting position is $(2, 0)$ (see the diagram at the right), then the equation of the boat's path is given by

$$y = \left(\frac{x}{2}\right)^{1-(w/v)} - \left(\frac{x}{2}\right)^{1+(w/v)}$$

a. If the speed of the current and the speed of the boat are the same, can the pilot reach point O by always having the nose of the boat pointed toward O? If not, at what point will the pilot arrive? Explain your answer.

b. If the speed of the current is greater than the speed of the boat, can the pilot reach point O by always pointing the nose of the boat toward O? If not, where will the pilot arrive? Explain.

c. If the speed of the current is less than the speed of the boat, can the pilot reach point O by always pointing the nose of the boat toward O? If not, where will the pilot arrive? Explain.

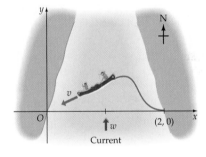

EXPONENTIAL GROWTH AND DECAY

In many applications, a quantity N grows or decays according to the function $N(t) = N_0 e^{kt}$. In this function, N is a function of time t, and N_0 is the value of N at time $t = 0$. If k is a *positive* constant, then $N(t) = N_0 e^{kt}$ is called an **exponential growth function**. If k is a *negative* constant, then $N(t) = N_0 e^{kt}$ is called an **exponential decay function**. The following examples illustrate how growth and decay functions arise naturally in the investigation of certain phenomena.

Interest is money paid for the use of money. The interest I is called **simple interest** if it is a fixed percent r, per time period t, of the amount of money invested. The amount of money invested is called the **principal** P. Simple interest is computed using the formula $I = Prt$. For example, if $1000 is invested at 12% for 3 years, the simple interest is

$$I = Prt = \$1000(0.12)(3) = \$360$$

The balance after t years is $A = P + I = P + Prt$. In the previous example, the $1000 invested for 3 years produced $360 interest. Thus the balance after 3 years is $1000 + $360 = $1360.

• COMPOUND INTEREST

In many financial transactions, interest is added to the principal at regular intervals so that interest is paid on interest as well as on the principal. Interest earned in this manner is called **compound interest**. For example, if $1000 is invested at 12% annual interest compounded annually for 3 years, then the total interest after 3 years is

First-year interest	$1000(0.12) = \$120.00
Second-year interest	$1120(0.12) = \$134.40
Third-year interest	$1254.40(0.12) \approx \underline{\$150.53}
	$404.93 • Total interest

This method of computing the balance can be tedious and time-consuming. A *compound interest formula* that can be used to determine the balance due after t years of compounding can be developed as follows.

Note that if P dollars is invested at an interest rate of r per year, then the balance after one year is $A_1 = P + Pr = P(1 + r)$, where Pr represents the interest earned for the year. Observe that A_1 is the product of the original principal P and $(1 + r)$. If the amount A_1 is reinvested for another year, then the balance after the second year is

$$A_2 = (A_1)(1 + r) = P(1 + r)(1 + r) = P(1 + r)^2$$

Successive reinvestments lead to the results shown in **Table 4.11.** The equation $A_t = P(1 + r)^t$ is valid if r is the annual interest rate paid during each of the t years.

TABLE 4.11

Number of Years	Balance
3	$A_3 = P(1 + r)^3$
4	$A_4 = P(1 + r)^4$
⋮	⋮
t	$A_t = P(1 + r)^t$

If r is an annual interest rate and n is the number of compounding periods per year, then the interest rate each period is r/n and the number of compounding periods after t years is nt. Thus the compound interest formula is expressed as follows:

The Compound Interest Formula

A principal P invested at an annual interest rate r, expressed as a decimal and compounded n times per year for t years, produces the balance

$$A = P\left(1 + \frac{r}{n}\right)^{nt}$$

EXAMPLE 1 Solve a Compound Interest Application

Find the balance if $1000 is invested at an annual interest rate of 10% for 2 years compounded

a. annually   b. monthly   c. daily

Solution

a. Use the compound interest formula with $P = 1000$, $r = 0.1$, $t = 2$, and $n = 1$.

$$A = \$1000\left(1 + \frac{0.1}{1}\right)^{1 \cdot 2} = \$1000(1.1)^2 = \$1210.00$$

b. Because there are 12 months in a year, use $n = 12$.

$$A = \$1000\left(1 + \frac{0.1}{12}\right)^{12 \cdot 2} \approx \$1000(1.008333333)^{24} \approx \$1220.39$$

c. Because there are 365 days in a year, use $n = 365$.

$$A = \$1000\left(1 + \frac{0.1}{365}\right)^{365 \cdot 2} \approx \$1000(1.000273973)^{730} \approx \$1221.37$$

▶ **TRY EXERCISE 4, PAGE 400**

To **compound continuously** means to increase the number of compounding periods without bound.

To derive a continuous compounding interest formula, substitute $\frac{1}{m}$ for $\frac{r}{n}$ in the compound interest formula

$$A = P\left(1 + \frac{r}{n}\right)^{nt} \tag{1}$$

to produce

$$A = P\left(1 + \frac{1}{m}\right)^{nt} \tag{2}$$

This substitution is motivated by the desire to express $\left(1 + \dfrac{r}{n}\right)^n$ as $\left[\left(1 + \dfrac{1}{m}\right)^m\right]^r$, which approaches e^r as m gets larger without bound.

Solving the equation $\dfrac{1}{m} = \dfrac{r}{n}$ for n yields $n = mr$, so the exponent nt can be written as mrt. Therefore Equation (2) can be expressed as

$$A = P\left(1 + \frac{1}{m}\right)^{mrt} = P\left[\left(1 + \frac{1}{m}\right)^m\right]^{rt} \qquad (3)$$

By the definition of e, we know that as m increases without bound,

$$\left(1 + \frac{1}{m}\right)^m \qquad \text{approaches} \qquad e$$

Thus, using continuous compounding, Equation (3) simplifies to $A = Pe^{rt}$.

Continuous Compounding Interest Formula

If an account with principal P and annual interest rate r is compounded continuously for t years, then the balance is $A = Pe^{rt}$.

EXAMPLE 2 | **Solve a Continuous Compound Interest Application**

Find the balance after 4 years on $800 invested at an annual rate of 6% compounded continuously.

Algebraic Solution

Use the continuous compounding formula with $P = 800$, $r = 0.06$, and $t = 4$.

$$A = Pe^{rt}$$
$$= 800e^{0.06(4)}$$
$$= 800e^{0.24}$$
$$\approx 800(1.27124915)$$
$$\approx 1017.00 \qquad \bullet \text{ To the nearest cent}$$

The balance after 4 years will be $1017.00.

Visualize the Solution

Figure 4.43, a graph of $A = 800e^{0.06t}$, shows that the balance is about $1017.00 when $t = 4$.

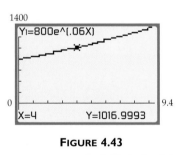

FIGURE 4.43

▶ **TRY EXERCISE 6, PAGE 400**

You have probably heard it said that time is money. In fact, many investors ask the question "How long will it take to double my money?" The following example answers this question for two different investments.

EXAMPLE 3   **Double Your Money**

Find the time required for money invested at an annual rate of 6% to double in value if the investment is compounded

a.   semiannually

b.   continuously

Solution

a.   Use $A = P\left(1 + \dfrac{r}{n}\right)^{nt}$ with $r = 0.06$, $n = 2$, and the balance A equal to twice the principal $(A = 2P)$.

$$2P = P\left(1 + \frac{0.06}{2}\right)^{2t}$$

$$2 = \left(1 + \frac{0.06}{2}\right)^{2t} \qquad \bullet \textbf{ Divide each side by } P.$$

$$\ln 2 = \ln\left(1 + \frac{0.06}{2}\right)^{2t} \qquad \bullet \textbf{Take the natural logarithm of each side.}$$

$$\ln 2 = 2t \ln\left(1 + \frac{0.06}{2}\right) \qquad \bullet \textbf{Apply the power property.}$$

$$2t = \frac{\ln 2}{\ln\left(1 + \dfrac{0.06}{2}\right)} \qquad \bullet \textbf{Solve for } t.$$

$$t = \frac{1}{2} \cdot \frac{\ln 2}{\ln\left(1 + \dfrac{0.06}{2}\right)}$$

$$t \approx 11.72$$

If the investment is compounded semiannually, it will double in value in about 11.72 years.

b.   Use $A = Pe^{rt}$ with $r = 0.06$ and $A = 2P$.

$$2P = Pe^{0.06t}$$

$$2 = e^{0.06t} \qquad \bullet \textbf{ Divide each side by } P.$$

$$\ln 2 = 0.06t \qquad \bullet \textbf{Write in logarithmic form.}$$

$$t = \frac{\ln 2}{0.06} \qquad \bullet \textbf{Solve for } t.$$

$$t \approx 11.55$$

If the investment is compounded continuously, it will double in value in about 11.55 years.

▶   **TRY EXERCISE 10, PAGE 400**

• EXPONENTIAL GROWTH

Given any two points on the graph of $N(t) = N_0 e^{kt}$, you can use the given data to solve for the constants N_0 and k.

EXAMPLE 4 | **Find the Exponential Growth Function That Models Given Data**

a. Find the exponential growth function for a town whose population was 16,400 in 1990 and 20,200 in 2000.

b. Use the function from part **a.** to predict, to the nearest 100, the population of the town in 2005.

Solution

a. We need to determine N_0 and k in $N(t) = N_0 e^{kt}$. If we represent the year 1990 by $t = 0$, then our given data are $N(0) = 16{,}400$ and $N(10) = 20{,}200$. Because N_0 is defined to be $N(0)$, we know that $N_0 = 16{,}400$. To determine k, substitute $t = 10$ and $N_0 = 16{,}400$ into $N(t) = N_0 e^{kt}$ to produce

$$N(10) = 16{,}400e^{k \cdot 10}$$

$$20{,}200 = 16{,}400e^{10k} \qquad \text{• Substitute 20,200 for } N(10).$$

$$\frac{20{,}200}{16{,}400} = e^{10k} \qquad \text{• Solve for } e^{10k}.$$

$$\ln \frac{20{,}200}{16{,}400} = 10k \qquad \text{• Write in logarithmic form.}$$

$$\frac{1}{10}\ln \frac{20{,}200}{16{,}400} = k \qquad \text{• Solve for } k.$$

$$0.0208 \approx k$$

The exponential growth function is $N(t) \approx 16{,}400e^{0.0208t}$.

b. The year 1990 was represented by $t = 0$, so we will use $t = 15$ to represent the year 2005.

$$N(t) \approx 16{,}400e^{0.0208t}$$

$$N(15) \approx 16{,}400e^{0.0208 \cdot 15}$$

$$\approx 22{,}400 \quad \text{(nearest 100)}$$

The exponential growth function yields 22,400 as the approximate population of the town in 2005.

> ▶ **TRY EXERCISE 18, PAGE 401**

take note

Because $e^{0.0208} \approx 1.021$, the growth equation can also be written as

$$N(t) \approx 16{,}400(1.021)^t$$

In this form we see that the population is growing by 2.1% $(1.021 - 1 = 0.021 = 2.1\%)$ per year.

• EXPONENTIAL DECAY

Many radioactive materials *decrease* in mass exponentially over time. This decrease, called radioactive decay, is measured in terms of half-life, which is defined as the time required for the disintegration of half the atoms in a sample of a radioactive substance. **Table 4.12** shows the half-lives of selected radioactive isotopes.

TABLE 4.12

Isotope	Half-Life
Carbon (^{14}C)	5730 years
Radium (^{226}Ra)	1660 years
Polonium (^{210}Po)	138 days
Phosphorus (^{32}P)	14 days
Polonium (^{214}Po)	1/10,000th of a second

EXAMPLE 5   Find the Exponential Decay Function That Models Given Data

Find the exponential decay function for the amount of phosphorus (^{32}P) that remains in a sample after t days.

Solution

When $t = 0$, $N(0) = N_0 e^{k(0)} = N_0$. Thus $N(0) = N_0$. Also, because the phosphorus has a half-life of 14 days (from **Table 4.12**), $N(14) = 0.5N_0$. To find k, substitute $t = 14$ into $N(t) = N_0 e^{kt}$ and solve for k.

$$N(14) = N_0 \cdot e^{k \cdot 14}$$

$$0.5N_0 = N_0 e^{14k} \qquad \text{• Substitute } 0.5N_0 \text{ for } N(14).$$

$$0.5 = e^{14k} \qquad \text{• Divide each side by } N_0.$$

$$\ln 0.5 = 14k \qquad \text{• Write in logarithmic form.}$$

$$\frac{1}{14} \ln 0.5 = k \qquad \text{• Solve for } k.$$

$$-0.0495 \approx k$$

The exponential decay function is $N(t) = N_0 e^{-0.0495t}$.

take note

Because $e^{-0.0495} \approx (0.5)^{1/14}$, the decay function $N(t) = N_0 e^{-0.0495t}$ can also be written as $N(t) = N_0(0.5)^{t/14}$. In this form it is easy to see that if t is increased by 14, then N will decrease by a factor of 0.5.

▶ **TRY EXERCISE 20, PAGE 401**

EXAMPLE 6 Application to Air Resistance

Assuming that air resistance is proportional to the velocity of a falling object, the velocity (in feet per second) of the object t seconds after it has been dropped is given by $v = 82(1 - e^{-0.39t})$.

a. Determine when the velocity will be 70 feet per second.

b. Write a sentence that explains the meaning of the horizontal asymptote, which is $v = 82$, in the context of this example.

Algebraic Solution

a.
$$v = 82(1 - e^{-0.39t})$$

$$70 = 82(1 - e^{-0.39t})$$ • Replace v by 70.

$$\frac{70}{82} = 1 - e^{-0.39t}$$ • Divide each side by 82.

$$e^{-0.39t} = 1 - \frac{70}{82}$$ • Solve for $e^{-0.39t}$.

$$-0.39t = \ln \frac{6}{41}$$ • Write in logarithmic form.

$$t = \frac{\ln(6/41)}{-0.39} \approx 4.9277246$$ • Solve for t.

The time is approximately 4.9 seconds.

b. The horizontal asymptote $v = 82$ means that as time increases, the velocity of the object will approach but never reach or exceed 82 feet per second.

Visualize the Solution

a. A graph of $y = 82(1 - e^{-0.39x})$ and $y = 70$ shows that the x-coordinate of the point of intersection is about 4.9.

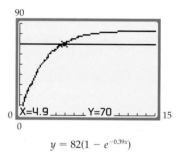

$$y = 82(1 - e^{-0.39x})$$

FIGURE 4.44

Note: The x value shown is rounded to the nearest tenth.

▶ **TRY EXERCISE 32, PAGE 402**

● **CARBON DATING**

The bone tissue in all living animals contains both carbon-12, which is nonradioactive, and carbon-14, which is radioactive with a half-life of approximately 5730 years. See **Figure 4.45.** As long as the animal is alive, the ratio of carbon-14 to carbon-12 remains constant. When the animal dies ($t = 0$), the carbon-14 begins to decay. Thus a bone that has a smaller ratio of carbon-14 to carbon-12 is older than a bone that has a larger ratio. The percent of carbon-14 present at time t is

$$P(t) = 0.5^{t/5730}$$

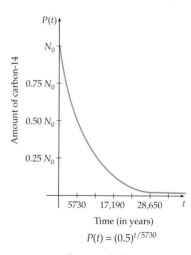

$$P(t) = (0.5)^{t/5730}$$

FIGURE 4.45

EXAMPLE 7   **Application to Archeology**

Find the age of a bone if it now has 85% of the carbon-14 it had when $t = 0$.

Solution

Let t be the time at which $P(t) = 0.85$.

$$0.85 = 0.5^{t/5730}$$

$$\ln 0.85 = \ln 0.5^{t/5730} \qquad \bullet \text{ Take the natural logarithm of each side.}$$

$$\ln 0.85 = \frac{t}{5730} \ln 0.5 \qquad \bullet \text{ Power property}$$

$$5730 \left(\frac{\ln 0.85}{\ln 0.5} \right) = t \qquad \bullet \text{ Solve for } t.$$

$$1340 \approx t$$

The bone is about 1340 years old.

▶ **TRY EXERCISE 24, PAGE 401**

MATH MATTERS

The chemist Willard Frank Libby developed the carbon-14 dating technique in 1947. In 1960 he was awarded the Nobel Prize in chemistry for this achievement.

● THE LOGISTIC MODEL

The population growth function $P(t) = P_0 e^{kt}$ is called the **Malthusian growth model.** It was developed by Robert Malthus (1766–1834) in *An Essay on the Principle of Population Growth,* which was published in 1798. The Malthusian growth model is an unrestricted growth model that does not consider any limited resources that eventually will curb population growth.

The **logistic model** is a restricted growth model that takes into consideration the effects of limited resources. The logistic model was developed by Pierre Verhulst in 1836.

The Logistic Model (A Restricted Growth Model)

The magnitude of a population at time $t \geq 0$ is given by

$$P(t) = \frac{c}{1 + ae^{-bt}}$$

where c is the **carrying capacity** (the maximum population that can be supported by available resources as $t \to \infty$) and b is a positive constant called the **growth rate constant.**

The **initial population** is $P_0 = P(0)$. The constant a is related to the initial population P_0 and the carrying capacity c by the formula

$$a = \frac{c - P_0}{P_0}$$

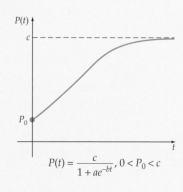

$$P(t) = \frac{c}{1 + ae^{-bt}}, \, 0 < P_0 < c$$

In the following example we determine a logistic growth model for a coyote population.

EXAMPLE 8 Find and Use a Logistic Model

At the beginning of 2002, the coyote population in a wilderness area was estimated at 200. By the beginning of 2004, the coyote population had increased to 250. A park ranger estimates that the carrying capacity of the wilderness area is 500 coyotes.

a. Use the given data to determine the growth rate constant for the logistic model of this coyote population.

b. Use the logistic model determined in part **a.** to predict the year in which the coyote population will first reach 400.

Solution

a. If we represent the beginning of the year 2002 by $t = 0$, then the beginning of the year 2004 will be represented by $t = 2$. In the logistic model, make the following substitutions: $P(2) = 250$, $c = 500$, and

$$a = \frac{c - P_0}{P_0} = \frac{500 - 200}{200} = 1.5.$$

$$P(t) = \frac{c}{1 + ae^{-bt}}$$

$$P(2) = \frac{500}{1 + 1.5e^{-b\cdot2}} \qquad \bullet \text{ Substitute the given values.}$$

$$250 = \frac{500}{1 + 1.5e^{-b\cdot2}}$$

$$250(1 + 1.5e^{-b\cdot2}) = 500 \qquad \bullet \text{ Solve for the growth rate constant } b.$$

$$1 + 1.5e^{-b\cdot2} = \frac{500}{250}$$

$$1.5e^{-b\cdot2} = 2 - 1$$

$$e^{-b\cdot2} = \frac{1}{1.5}$$

$$-2b = \ln\left(\frac{1}{1.5}\right)$$

$$b = -\frac{1}{2}\ln\left(\frac{1}{1.5}\right)$$

$$b \approx 0.20273255$$

Using $a = 1.5$, $b = 0.20273255$, and $c = 500$ gives us the following logistic model.

$$P(t) = \frac{500}{1 + 1.5e^{-0.20273255t}}$$

b. To determine during what year the logistic model predicts the coyote population will first reach 400, replace $P(t)$ with 400 and solve for t.

$$400 = \frac{500}{1 + 1.5e^{-0.20273255t}}$$

$$400(1 + 1.5e^{-0.20273255t}) = 500$$

$$1 + 1.5e^{-0.20273255t} = \frac{500}{400}$$

$$1.5e^{-0.20273255t} = 1.25 - 1$$

$$e^{-0.20273255t} = \frac{0.25}{1.5}$$

$$-0.20273255t = \ln\left(\frac{0.25}{1.5}\right) \qquad \bullet \textbf{ Write in logarithmic form.}$$

$$t = \frac{1}{-0.20273255}\ln\left(\frac{0.25}{1.5}\right) \qquad \bullet \textbf{ Solve for } t.$$

$$\approx 8.8$$

According to the logistic model, the coyote population will reach 400 about 8.8 years after the beginning of 2002, which is during the year 2010. The graph of the logistic model is shown in **Figure 4.46.** Note that $P(8.8) \approx 400$ and that as $t \to \infty$, $P(t) \to 500$.

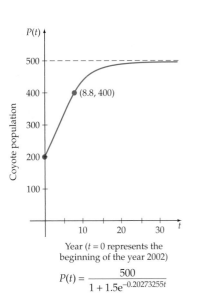

$$P(t) = \frac{500}{1 + 1.5e^{-0.20273255t}}$$

FIGURE 4.46

▶ **TRY EXERCISE 48, PAGE 403**

 TOPICS FOR DISCUSSION

1. Explain the difference between compound interest and simple interest.

2. What is an exponential growth model? Give an example of an application for which the exponential growth model might be appropriate.

3. What is an exponential decay model? Give an example of an application for which the exponential decay model might be appropriate.

4. Consider the exponential model $P(t) = P_0 e^{kt}$ and the logistic model $P(t) = \dfrac{c}{1 + ae^{-bt}}$. Explain the similarities and differences between the two models.

EXERCISE SET 4.6

1. **COMPOUND INTEREST** If $8000 is invested at an annual interest rate of 5% and compounded annually, find the balance after

   a. 4 years   **b.** 7 years

2. **COMPOUND INTEREST** If $22,000 is invested at an annual interest rate of 4.5% and compounded annually, find the balance after

   a. 2 years   **b.** 10 years

3. **COMPOUND INTEREST** If $38,000 is invested at an annual interest rate of 6.5% for 4 years, find the balance if the interest is compounded

   a. annually   **b.** daily   **c.** hourly

▶ **4.** **COMPOUND INTEREST** If $12,500 is invested at an annual interest rate of 8% for 10 years, find the balance if the interest is compounded

   a. annually   **b.** daily   **c.** hourly

5. **COMPOUND INTEREST** Find the balance if $15,000 is invested at an annual rate of 10% for 5 years, compounded continuously.

▶ **6.** **COMPOUND INTEREST** Find the balance if $32,000 is invested at an annual rate of 8% for 3 years, compounded continuously.

7. **COMPOUND INTEREST** How long will it take $4000 to double if it is invested in a certificate of deposit that pays 7.84% annual interest compounded continuously? Round to the nearest tenth of a year.

8. **COMPOUND INTEREST** How long will it take $25,000 to double if it is invested in a savings account that pays 5.88% annual interest compounded continuously? Round to the nearest tenth of a year.

9. **CONTINUOUS COMPOUNDING INTEREST** Use the Continuous Compounding Interest Formula to derive an expression for the time it will take money to triple when invested at an annual interest rate of r compounded continuously.

▶ **10.** **CONTINUOUS COMPOUNDING INTEREST** How long will it take $1000 to triple if it is invested at an annual interest rate of 5.5% compounded continuously? Round to the nearest year.

11. **CONTINUOUS COMPOUNDING INTEREST** How long will it take $6000 to triple if it is invested in a savings account that pays 7.6% annual interest compounded continuously? Round to the nearest year.

12. **CONTINUOUS COMPOUNDING INTEREST** How long will it take $10,000 to triple if it is invested in a savings account that pays 5.5% annual interest compounded continuously? Round to the nearest year.

13. **POPULATION GROWTH** The number of bacteria $N(t)$ present in a culture at time t hours is given by

   $$N(t) = 2200(2)^t$$

   Find the number of bacteria present when

   a. $t = 0$ hours   **b.** $t = 3$ hours

14. **POPULATION GROWTH** The population of a town grows exponentially according to the function

   $$f(t) = 12,400(1.14)^t$$

   for $0 \le t \le 5$ years. Find, to the nearest hundred, the population of the town when t is

   a. 3 years   **b.** 4.25 years

15. **POPULATION GROWTH** A town had a population of 22,600 in 1990 and a population of 24,200 in 1995.

   a. Find the exponential growth function for the town. Use $t = 0$ to represent the year 1990.

   b. Use the growth function to predict the population of the town in 2005. Round to the nearest hundred.

16. **POPULATION GROWTH** A town had a population of 53,700 in 1996 and a population of 58,100 in 2000.

   a. Find the exponential growth function for the town. Use $t = 0$ to represent the year 1996.

   b. Use the growth function to predict the population of the town in 2008. Round to the nearest hundred.

17. **POPULATION GROWTH** The growth of the population of Los Angeles, California, for the years 1992 through 1996 can be approximated by the equation

   $$P = 10,130(1.005)^t$$

   where $t = 0$ corresponds to January 1, 1992 and P is in thousands.

   a. Assuming this growth rate continues, what will be the population of Los Angeles on January 1 in the year 2004?

b. In what year will the population of Los Angeles first exceed 13,000,000?

▶ **18.** 🔵 **POPULATION GROWTH** The growth of the population of Mexico City, Mexico, for the years 1991 through 1998 can be approximated by the equation

$$P = 20{,}899(1.027)^t$$

where $t = 0$ corresponds to 1991 and P is in thousands.

a. Assuming this growth rate continues, what will be the population of Mexico City in the year 2003?

b. Assuming this growth rate continues, in what year will the population of Mexico City first exceed 35,000,000?

19. **MEDICINE** Sodium-24 is a radioactive isotope of sodium that is used to study circulatory dysfunction. Assuming that 4 micrograms of sodium-24 is injected into a person, the amount A in micrograms remaining in that person after t hours is given by the equation $A = 4e^{-0.046t}$.

a. Graph this equation.

b. What amount of sodium-24 remains after 5 hours?

c. What is the half-life of sodium-24?

d. In how many hours will the amount of sodium-24 be 1 microgram?

▶ **20.** 🔵 **RADIOACTIVE DECAY** Polonium (^{210}Po) has a half-life of 138 days. Find the decay function for the amount of polonium (^{210}Po) that remains in a sample after t days.

21. 🔵 **GEOLOGY** Geologists have determined that Crater Lake in Oregon was formed by a volcanic eruption. Chemical analysis of a wood chip that is assumed to be from a tree that died during the eruption has shown that it contains approximately 45% of its original carbon-14. Determine how long ago the volcanic eruption occurred. Use 5730 years as the half-life of carbon-14.

22. 🔵 **RADIOACTIVE DECAY** Use $N(t) = N_0(0.5)^{t/138}$, where t is measured in days, to estimate the percentage of polonium (^{210}Po) that remains in a sample after 2 years. Round to the nearest hundredth of a percent.

23. 🔵 **ARCHEOLOGY** The Rhind papyrus, named after A. Henry Rhind, contains most of what we know today of ancient Egyptian mathematics. A chemical analysis of a sample from the papyrus has shown that it contains approximately 75% of its original carbon-14. What is the age of the Rhind papyrus? Use 5730 years as the half-life of carbon-14.

▶ **24.** **ARCHEOLOGY** Determine the age of a bone if it now contains 65% of its original amount of carbon-14. Round to the nearest 100 years.

25. **PHYSICS** Newton's Law of Cooling states that if an object at temperature T_0 is placed into an environment at constant temperature A, then the temperature of the object, $T(t)$ (in degrees Fahrenheit), after t minutes is given by $T(t) = A + (T_0 - A)e^{-kt}$, where k is a constant that depends on the object.

a. Determine the constant k (to the nearest thousandth) for a canned soda drink that takes 5 minutes to cool from 75°F to 65°F after being placed in a refrigerator that maintains a constant temperature of 34°F.

b. What will be the temperature (to the nearest degree) of the soda drink after 30 minutes?

c. When (to the nearest minute) will the temperature of the soda drink be 36°F?

26. **PSYCHOLOGY** According to a software company, the users of its typing tutorial can expect to type $N(t)$ words per minute after t hours of practice with the product, according to the function $N(t) = 100(1.04 - 0.99^t)$.

a. How many words per minute can a student expect to type after 2 hours of practice?

b. How many words per minute can a student expect to type after 40 hours of practice?

c. According to the function N, how many hours (to the nearest hour) of practice will be required before a student can expect to type 60 words per minute?

27. **PSYCHOLOGY** In the city of Whispering Palms, which has a population of 80,000 people, the number of people $P(t)$ exposed to a rumor in t hours is given by the function $P(t) = 80{,}000(1 - e^{-0.0005t})$.

a. Find the number of hours until 10% of the population has heard the rumor.

b. Find the number of hours until 50% of the population has heard the rumor.

28. **LAW** A lawyer has determined that the number of people $P(t)$ in a city of 1,200,000 people who have been exposed to a news item after t days is given by the function

$$P(t) = 1{,}200{,}000(1 - e^{-0.03t})$$

a. How many days after a major crime has been reported has 40% of the population heard of the crime?

b. A defense lawyer knows it will be very difficult to pick an unbiased jury after 80% of the population has heard of the crime. After how many days will 80% of the population have heard of the crime?

29. DEPRECIATION An automobile depreciates according to the function $V(t) = V_0(1 - r)^t$, where $V(t)$ is the value in dollars after t years, V_0 is the original value, and r is the yearly depreciation rate. A car has a yearly depreciation rate of 20%. Determine, to the nearest 0.1 year, in how many years the car will depreciate to half its original value.

30. PHYSICS The current $I(t)$ (measured in amperes) of a circuit is given by the function $I(t) = 6(1 - e^{-2.5t})$, where t is the number of seconds after the switch is closed.

a. Find the current when $t = 0$.

b. Find the current when $t = 0.5$.

c. Solve the equation for t.

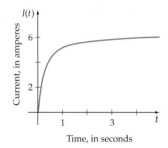

Time, in seconds

31. AIR RESISTANCE Assuming that air resistance is proportional to velocity, the velocity v, in feet per second, of a falling object after t seconds is given by $v = 32(1 - e^{-t})$.

a. Graph this equation for $t \geq 0$.

b. Determine algebraically, to the nearest 0.01 second, when the velocity is 20 feet per second.

c. Determine the horizontal asymptote of the graph of v.

d. Write a sentence that explains the meaning of the horizontal asymptote in the context of this application.

32. AIR RESISTANCE Assuming that air resistance is proportional to velocity, the velocity v, in feet per second, of a falling object after t seconds is given by $v = 64(1 - e^{-t/2})$.

a. Graph this equation for $t \geq 0$.

b. Determine algebraically, to the nearest 0.1 second, when the velocity is 50 feet per second.

c. Determine the horizontal asymptote of the graph of v.

d. Write a sentence that explains the meaning of the horizontal asymptote in the context of this application.

33. The distance s (in feet) that the object in Exercise 31 will fall in t seconds is given by $s = 32t + 32(e^{-t} - 1)$.

a. Use a graphing utility to graph this equation for $t \geq 0$.

b. Determine, to the nearest 0.1 second, the time it takes the object to fall 50 feet.

c. Calculate the slope of the secant line through $(1, s(1))$ and $(2, s(2))$.

d. Write a sentence that explains the meaning of the slope of the secant line you calculated in **c.**

34. The distance s (in feet) that the object in Exercise 32 will fall in t seconds is given by $s = 64t + 128(e^{-t/2} - 1)$.

a. Use a graphing utility to graph this equation for $t \geq 0$.

b. Determine, to the nearest 0.1 second, the time it takes the object to fall 50 feet.

c. Calculate the slope of the secant line through $(1, s(1))$ and $(2, s(2))$.

d. Write a sentence that explains the meaning of the slope of the secant line you calculated in **c.**

In Exercises 35 to 40, determine the following constants for the given logistic growth model.
a. The carrying capacity
b. The growth rate constant
c. The initial population P_0

35. $P(t) = \dfrac{1900}{1 + 8.5e^{-0.16t}}$

36. $P(t) = \dfrac{32{,}550}{1 + 0.75e^{-0.08t}}$

37. $P(t) = \dfrac{157{,}500}{1 + 2.5e^{-0.04t}}$

38. $P(t) = \dfrac{51}{1 + 1.04e^{-0.03t}}$

39. $P(t) = \dfrac{2400}{1 + 7e^{-0.12t}}$

40. $P(t) = \dfrac{320}{1 + 15e^{-0.12t}}$

In Exercises 41 to 44, use algebraic procedures to find the logistic growth model for the data.

41. $P_0 = 400$, $P(2) = 780$, and the carrying capacity is 5500.

42. $P_0 = 6200$, $P(8) = 7100$, and the carrying capacity is 9500.

43. $P_0 = 18$, $P(3) = 30$, and the carrying capacity is 100.

44. $P_0 = 3200$, $P(22) \approx 5565$, and the growth rate constant is 0.056.

45. REVENUE The annual revenue R, in dollars, of a new company can be closely modeled by the logistic growth function

$$R(t) = \frac{625{,}000}{1 + 3.1e^{-0.045t}}$$

where the *natural* number t is the time, in years, since the company was founded.

a. According to the model, what will be the company's annual revenue for its first year and its second year ($t = 1$ and $t = 2$) of operation? Round to the nearest $1000.

b. According to the model, what will the company's annual revenue approach in the long-term future?

46. NEW CAR SALES The number of cars A sold annually by an automobile dealership can be closely modeled by the logistic growth function

$$A(t) = \frac{1650}{1 + 2.4e^{-0.055t}}$$

where the *natural* number t is the time, in years, since the dealership was founded.

a. According to the model, what number of cars will the dealership sell during its first year and its second year ($t = 1$ and $t = 2$) of operation? Round to the nearest unit.

b. According to the model, what will the dealership's annual car sales approach in the long-term future?

47. POPULATION GROWTH The population of wolves in a preserve satisfies a logistic growth model in which $P_0 = 312$ in the year 2002, $c = 1600$, and $P(6) = 416$.

a. Determine the logistic growth model for this population, where t is the number of years after 2002.

b. Use the logistic growth model from part **a.** to predict the size of the wolf population in 2012.

▶ 48. POPULATION GROWTH The population of groundhogs on a ranch satisfies a logistic growth model in which $P_0 = 240$ in the year 2001, $c = 3400$, and $P(1) = 310$.

a. Determine the logistic growth model for this population, where t is the number of years after 2001.

b. Use the logistic growth model from part **a.** to predict the size of the groundhog population in 2008.

49. POPULATION GROWTH The population of squirrels in a nature preserve satisfies a logistic growth model in which $P_0 = 1500$ in the year 2001. The carrying capacity of the preserve is estimated at 8500 squirrels and $P(2) = 1900$.

a. Determine the logistic growth model for this population, where t is the number of years after 2001.

b. Use the logistic growth model from part **a.** to predict the year in which the squirrel population will first exceed 4000.

50. POPULATION GROWTH The population of walruses on an island satisfies a logistic growth model in which $P_0 = 800$ in the year 2000. The carrying capacity of the island is estimated at 5500 walruses, and $P(1) = 900$.

a. Determine the logistic growth model for this population, where t is the number of years after 2000.

b. Use the logistic growth model from part **a.** to predict the year in which the walrus population will first exceed 2000.

51. LEARNING THEORY The logistic model is also used in learning theory. Suppose that historical records from employee training at a company show that the percent score on a product information test is given by

$$P = \frac{100}{1 + 25e^{-0.095t}}$$

where t is the number of hours of training. What is the number of hours (to the nearest hour) of training needed before a new employee will answer 75% of the questions correctly?

52. LEARNING THEORY A company provides training in the assembly of a computer circuit to new employees. Past experience has shown that the number of correctly assembled circuits per week can be modeled by

$$N = \frac{250}{1 + 249e^{-0.503t}}$$

where t is the number of weeks of training. What is the number of weeks (to the nearest week) of training needed before a new employee will correctly make 140 circuits?

CONNECTING CONCEPTS

53. MEDICATION LEVEL A patient is given three dosages of aspirin. Each dosage contains 1 gram of aspirin. The second and third dosages are each taken 3 hours after the previous dosage is administered. The half-life of the aspirin is 2 hours. The amount of aspirin, A, in the patient's body t hours after the first dosage is administered is

$$A(t) = \begin{cases} 0.5^{t/2} & 0 \le t < 3 \\ 0.5^{t/2} + 0.5^{(t-3)/2} & 3 \le t < 6 \\ 0.5^{t/2} + 0.5^{(t-3)/2} + 0.5^{(t-6)/2} & t \ge 6 \end{cases}$$

Find, to the nearest hundredth of a gram, the amount of aspirin in the patient's body when

a. $t = 1$ **b.** $t = 4$ **c.** $t = 9$

54. MEDICATION LEVEL Use a graphing calculator and the dosage formula in Exercise 53 to determine when, to the nearest tenth of an hour, the amount of aspirin in the patient's body first reaches 0.25 gram.

Exercises 55 to 57 make use of the factorial function, which is defined as follows. For whole numbers n, the number $n!$ (which is read "n factorial") is given by

$$n! = \begin{cases} n(n-1)(n-2)\cdots 1, & \text{if } n \ge 1 \\ 1, & \text{if } n = 0 \end{cases}$$

Thus, $0! = 1$ and $4! = 4 \cdot 3 \cdot 2 \cdot 1 = 24$.

55. QUEUEING THEORY A study shows that the number of people who arrive at a bank teller's window averages 4.1 people every 10 minutes. The probability P that exactly x people will arrive at the teller's window in a given 10-minute period is

$$P(x) = \frac{4.1^x e^{-4.1}}{x!}$$

Find, to the nearest 0.1%, the probability that in a given 10-minute period, exactly

a. 0 people arrive at the window.

b. 2 people arrive at the window.

c. 3 people arrive at the window.

d. 4 people arrive at the window.

e. 9 people arrive at the window.

As $x \to \infty$, what does P approach?

56. STIRLING'S FORMULA Stirling's Formula (after James Stirling, 1692–1770),

$$n! \approx \left(\frac{n}{e}\right)^n \sqrt{2\pi n}$$

is often used to approximate very large factorials. Use Stirling's Formula to approximate 10!, and then compute the ratio of Stirling's approximation of 10! divided by the actual value of 10!, which is 3,628,800.

57. RUBIK'S CUBE The Rubik's cube shown here was invented by Erno Rubik in 1975. The small outer cubes are held together in such a way that they can be rotated around three axes. The total number of positions in which the Rubik's cube can be arranged is

$$\frac{3^8 2^{12} 8! \, 12!}{2 \cdot 3 \cdot 2}$$

If you can arrange a Rubik's cube into a new arrangement every second, how many centuries would it take to place the cube into each of its arrangements? Assume that there are 365 days in a year.

58. OIL SPILLS Crude oil leaks from a tank at a rate that depends on the amount of oil that remains in the tank. Because $\frac{1}{8}$ of the oil in the tank leaks out every 2 hours, the volume of oil $V(t)$ in the tank after t hours is given by $V(t) = V_0(0.875)^{t/2}$, where $V_0 = 350,000$ gallons is the number of gallons in the tank at the time the tank started to leak ($t = 0$).

a. How many gallons does the tank hold after 3 hours?

b. How many gallons does the tank hold after 5 hours?

c. How long, to the nearest hour, will it take until 90% of the oil has leaked from the tank?

PROJECTS

A DECLINING LOGISTIC MODEL If $P_0 > c$ (which implies that $-1 < a < 0$), then the logistic function $P(t) = \dfrac{c}{1 + ae^{-bt}}$ decreases as t increases. Biologists often use this type of logistic function to model populations that decrease over time. See the following figure.

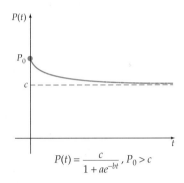

$$P(t) = \frac{c}{1 + ae^{-bt}}, \; P_0 > c$$

1. **A DECLINING FISH POPULATION** A biologist finds that the fish population in a small lake can be closely modeled by the logistic function

$$P(t) = \frac{1000}{1 + (-0.3333)e^{-0.05t}}$$

where t is the time, in years, since the lake was first stocked with fish.

a. What was the fish population when the lake was first stocked with fish?

b. According to the logistic model, what will the fish population approach in the long-term future?

2. **A DECLINING DEER POPULATION** The deer population in a reserve is given by the logistic function

$$P(t) = \frac{1800}{1 + (-0.25)e^{-0.07t}}$$

where t is the time, in years, since July 1, 2001.

a. What was the deer population on July 1, 2001? What was the deer population on July 1, 2003?

b. According to the logistic model, what will the deer population approach in the long-term future?

3.  **MODELING WORLD RECORD TIMES IN THE MEN'S MILE RACE** In the early 1950s, many people speculated that no runner would ever run a mile race in under 4 minutes. During the period from 1913 to 1945, the world record in the mile event had been reduced from 4.14.4 (4 minutes, 14.4 seconds) to 4.01.4, but no one seemed capable of running a sub-four-minute mile. Then, in 1954, Roger Bannister broke through the four-minute barrier by running a mile in 3.59.6. In 1999, the current record of 3.43.13 was established. It is fun to think about future record times in the mile race. Will they ever go below 3 minutes, 30 seconds? Below 3 minutes, 20 seconds? What about a sub-three-minute mile?

A declining logistic function that closely models the world record times WR, in seconds, in the men's mile run from 1913 ($t = 0$) to 1999 ($t = 86$) is given by

$$WR(t) = \frac{199.13}{1 + (-0.21726)e^{-0.0079889t}}$$

a. Use the above logistic model to predict the world record time for the men's mile run in the year 2020 and the year 2050.

b. According to the logistic function, what time will the world record in the men's mile event approach but never break through?

EXPLORING CONCEPTS WITH TECHNOLOGY

TABLE 4.13

T	V
90	700
100	500
110	350
120	250
130	190
140	150
150	120

Using a Semilog Graph to Model Exponential Decay

Consider the data in **Table 4.13,** which shows the viscosity V of SAE 40 motor oil at various temperatures T. The graph of these data is shown below, along with a curve that passes through the points. The graph in **Figure 4.47** appears to have the shape of an exponential decay model.

One way to determine whether the graph in **Figure 4.47** is the graph of an exponential function is to plot the data on *semilog* graph paper. On this graph paper, the horizontal axis remains the same, but the vertical axis uses a logarithmic scale.

The data in **Table 4.13** are graphed again in **Figure 4.48,** but this time the vertical axis is a natural logarithm axis. This graph is approximately a straight line.

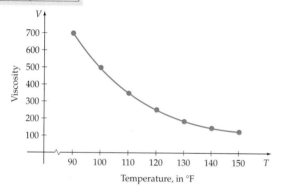

FIGURE 4.47

FIGURE 4.48

The slope of the line in **Figure 4.48,** to the nearest ten-thousandth, is

$$m = \frac{\ln 500 - \ln 120}{100 - 150} \approx -0.0285$$

Using this slope and the point-slope formula with V replaced by $\ln V$, we have

$$\ln V - \ln 120 = -0.0285(T - 150)$$
$$\ln V \approx -0.0285T + 9.062 \qquad (1)$$

Equation (1) is the equation of the line on a semilog coordinate grid.

Now solve Equation (1) for V.

$$e^{\ln V} = e^{-0.0285T + 9.062}$$
$$V = e^{-0.0285T}e^{9.062}$$
$$V \approx 8621e^{-0.0285T} \qquad (2)$$

TABLE 4.14

t	A
1	91.77
4	70.92
8	50.30
15	27.57
20	17.95
30	7.60

Equation (2) is a model of the data in the rectangular coordinate system shown in **Figure 4.47.**

1. A chemist wishes to determine the decay characteristics of iodine-131. A 100-mg sample of iodine-131 is observed over a 30-day period. **Table 4.14** shows the amount A (in milligrams) of iodine-131 remaining after t days.

   a. Graph the ordered pairs (t, A) on semilog paper. (*Note:* Semilog paper comes in different varieties. Our calculations are based on semilog paper that has a natural logarithm scale on the vertical axis.)

b. Use the points $(4, 4.3)$ and $(15, 3.3)$ to approximate the slope of the line that passes through the points.

c. Using the slope calculated in part **b.** and the point $(4, 4.3)$, determine the equation of the line.

d. Solve the equation you derived in part **c.** for A.

e. Graph the equation you derived in part **d.** in a rectangular coordinate system.

f. What is the half-life of iodine-131?

2. The live birth rates B per thousand births in the United States are given in **Table 4.15** for the years 1986 through 1990 ($t = 0$ corresponds to 1986).

TABLE 4.15

t	B
0	15.5
1	15.7
2	15.9
3	16.2
4	16.7

a. Graph the ordered pairs $(t, \ln B)$. (You will need to adjust the scale so that you can discriminate between plotted points. A suggestion is given in **Figure 4.49.**)

b. Use the points $(1, 2.754)$ and $(3, 2.785)$ to approximate the slope of the line that passes through the points.

c. Using the slope calculated in part **b.** and the point $(1, 2.754)$, determine the equation of the line.

d. Solve the equation you derived in part **c.** for B.

e. Graph the equation you derived in part **d.** in a rectangular coordinate system.

f. If the birth rate continues as predicted by your model, in what year will the birth rate be 17.5 per thousand?

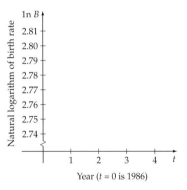

FIGURE 4.49

The difference in graphing strategies between Exercise 1 and Exercise 2 is that in Exercise 1, semilog paper was used. When a point is graphed on this coordinate paper, the y-coordinate is $\ln y$. In Exercise 2, graphing a point $(x, \ln y)$ in a rectangular coordinate system has the same effect as graphing (x, y) in a semilog coordinate system.

CHAPTER 4 SUMMARY

4.1 Inverse Functions

- If f is a one-to-one function with domain X and range Y, and g is a function with domain Y and range X, then g is the inverse function of f if and only if $(f \circ g)(x) = x$ for all x in the domain of g and $(g \circ f)(x) = x$ for all x in the domain of f.

- A function f has an inverse function if and only if it is a one-to-one function. The graph of a function f and the graph of the inverse function f^{-1} are symmetric with respect to the line given by $y = x$.

4.2 Exponential Functions and Their Applications

- For all positive real numbers b, $b \neq 1$, the exponential function defined by $f(x) = b^x$ has the following properties:

  1. f has the set of real numbers as its domain.

  2. f has the set of positive real numbers as its range.

  3. f has a graph with a y-intercept of $(0, 1)$.

  4. f has a graph asymptotic to the x-axis.

  5. f is a one-to-one function.

6. f is an increasing function if $b > 1$.

7. f is a decreasing function if $0 < b < 1$.

- As n increases without bound, $(1 + 1/n)^n$ approaches an irrational number denoted by e. The value of e accurate to eight decimal places is 2.71828183.

- The function defined by $f(x) = e^x$ is called the natural exponential function.

4.3 Logarithmic Functions and Their Applications

- *Definition of a Logarithm* If $x > 0$ and b is a positive constant $(b \neq 1)$, then

$$y = \log_b x \quad \text{if and only if} \quad b^y = x$$

- For all positive real numbers b, $b \neq 1$, the function defined by $f(x) = \log_b x$ has the following properties:

  1. f has the set of positive real numbers as its domain.

  2. f has the set of real numbers as its range.

  3. f has a graph with an x-intercept of $(1, 0)$.

  4. f has a graph asymptotic to the y-axis.

  5. f is a one-to-one function.

  6. f is an increasing function if $b > 1$.

  7. f is a decreasing function if $0 < b < 1$.

- The exponential form of $y = \log_b x$ is $b^y = x$.

- The logarithmic form of $b^y = x$ is $y = \log_b x$.

- *Basic Logarithmic Properties*

  1. $\log_b b = 1$ **2.** $\log_b 1 = 0$ **3.** $\log_b (b^p) = p$

- The function $f(x) = \log_{10} x$ is the common logarithmic function. It is customarily written as $f(x) = \log x$.

- The function $f(x) = \log_e x$ is the natural logarithmic function. It is customarily written as $f(x) = \ln x$.

4.4 Logarithms and Logarithmic Scales

- If b, M, and N are positive real numbers $(b \neq 1)$, and p is any real number, then

$\log_b(MN) = \log_b M + \log_b N$

$\log_b \dfrac{M}{N} = \log_b M - \log_b N$

$\log_b(M^p) = p \log_b M$

$\log_b M = \log_b N \quad \text{implies} \quad M = N$

$M = N \quad \text{implies} \quad \log_b M = \log_b N$

$b^{\log_b p} = p \quad (\text{for } p > 0)$

- *Change-of-Base Formula* If x, a, and b are positive real numbers with $a \neq 1$ and $b \neq 1$, then

$$\log_b x = \frac{\log_a x}{\log_a b}$$

- An earthquake with an intensity of I has a Richter scale magnitude of $M = \log\left(\dfrac{I}{I_0}\right)$, where I_0 is the measure of the intensity of a zero-level earthquake.

- The pH of a solution with a hydronium-ion concentration of H^+ mole per liter is given by $\text{pH} = -\log[H^+]$.

4.5 Exponential and Logarithmic Equations

- *Equality of Exponents Theorem* If b is a positive real number $(b \neq 1)$ such that $b^x = b^y$, then $x = y$.

- Exponential equations of the form $b^x = b^y$ can be solved by using the Equality of Exponents Theorem.

- Exponential equations of the form $b^x = c$ can be solved by taking either the common logarithm or the natural logarithm of each side of the equation.

- Logarithmic equations can often be solved by using the properties of logarithms and the definition of a logarithm.

4.6 Exponential Growth and Decay

- The function defined by $N(t) = N_0 e^{kt}$ is called an exponential growth function if k is a positive constant, and it is called an exponential decay function if k is a negative constant.

- *The Compound Interest Formula* A principal P invested at an annual interest rate r, expressed as a decimal and compounded n times per year for t years, produces the balance

$$A = P\left(1 + \frac{r}{n}\right)^{nt}$$

- *Continuous Compounding Interest Formula* If an account with principal P and annual interest rate r is compounded continuously for t years, then the balance is $A = Pe^{rt}$.

- *The Logistic Model* The magnitude of a population at time t is given by

$$P(t) = \frac{c}{1 + ae^{-bt}}$$

where $P_0 = P(0)$ is the population at time $t = 0$, c is the carrying capacity of the population, and b is a constant called the growth rate constant.

CHAPTER 4 TRUE/FALSE EXERCISES

In Exercises 1 to 16, answer true or false. If the statement is false, give an example or state a reason to demonstrate that the statement is false.

1. Every function has an inverse function.

2. If $(f \circ g)(a) = a$ and $(g \circ f)(a) = a$ for some constant a, then f and g are inverse functions.

3. If $7^x = 40$, then $\log_7 40 = x$.

4. If $\log_4 x = 3.1$, then $4^{3.1} = x$.

5. If $f(x) = \log x$ and $g(x) = 10^x$, then $f[g(x)] = x$ for all real numbers x.

6. If $f(x) = \log x$ and $g(x) = 10^x$, then $g[f(x)] = x$ for all real numbers x.

7. The exponential function $h(x) = b^x$ is an increasing function.

8. The logarithmic function $j(x) = \log_b x$ is an increasing function.

9. The exponential function $h(x) = b^x$ is a one-to-one function.

10. The logarithmic function $j(x) = \log_b x$ is a one-to-one function.

11. The graph of $f(x) = \dfrac{2^x + 2^{-x}}{2}$ is symmetric with respect to the y-axis.

12. The graph of $f(x) = \dfrac{2^x - 2^{-x}}{2}$ is symmetric with respect to the origin.

13. If $x > 0$ and $y > 0$, then $\log(x + y) = \log x + \log y$.

14. If $x > 0$, then $\log x^2 = 2 \log x$.

15. If M and N are positive real numbers, then
$$\ln \frac{M}{N} = \ln M - \ln N$$

16. For all $p > 0$, $e^{\ln p} = p$.

CHAPTER 4 REVIEW EXERCISES

In Exercises 1 to 4, determine whether the given functions are inverses.

1. $F(x) = 2x - 5 \qquad G(x) = \dfrac{x + 5}{2}$

2. $h(x) = \sqrt{x} \qquad k(x) = x^2, \quad x \geq 0$

3. $l(x) = \dfrac{x + 3}{x} \qquad m(x) = \dfrac{3}{x - 1}$

4. $p(x) = \dfrac{x - 5}{2x} \qquad q(x) = \dfrac{2x}{x - 5}$

In Exercises 5 to 8, find the inverse of the function. Sketch the graph of the function and its inverse on the same set of coordinate axes.

5. $f(x) = 3x - 4$

6. $g(x) = -2x + 3$

7. $h(x) = -\dfrac{1}{2}x - 2$

8. $k(x) = \dfrac{1}{x}$

In Exercises 9 to 20, solve each equation. Do not use a calculator.

9. $\log_5 25 = x$

10. $\log_3 81 = x$

11. $\ln e^3 = x$

12. $\ln e^\pi = x$

13. $3^{2x+7} = 27$

14. $5^{x-4} = 625$

15. $2^x = \dfrac{1}{8}$

16. $27(3^x) = 3^{-1}$

17. $\log x^2 = 6$

18. $\dfrac{1}{2} \log |x| = 5$

19. $10^{\log 2x} = 14$

20. $e^{\ln x^2} = 64$

In Exercises 21 to 30, sketch the graph of each function.

21. $f(x) = (2.5)^x$

22. $f(x) = \left(\dfrac{1}{4}\right)^x$

23. $f(x) = 3^{|x|}$

24. $f(x) = 4^{-|x|}$

25. $f(x) = 2^x - 3$

26. $f(x) = 2^{(x-3)}$

27. $f(x) = \dfrac{1}{3} \log x$

28. $f(x) = 3 \log x^{1/3}$

29. $f(x) = -\dfrac{1}{2} \ln x$

30. $f(x) = -\ln |x|$

 In Exercises 31 and 32, use a graphing utility to graph each function.

31. $f(x) = \dfrac{4^x + 4^{-x}}{2}$

32. $f(x) = \dfrac{3^x - 3^{-x}}{2}$

In Exercises 33 to 36, change each logarithmic equation to its exponential form.

33. $\log_4 64 = 3$

34. $\log_{1/2} 8 = -3$

35. $\log_{\sqrt{2}} 4 = 4$

36. $\ln 1 = 0$

In Exercises 37 to 40, change each exponential equation to its logarithmic form.

37. $5^3 = 125$

38. $2^{10} = 1024$

39. $10^0 = 1$

40. $8^{1/2} = 2\sqrt{2}$

In Exercises 41 to 44, write the given logarithm in terms of logarithms of x, y, and z.

41. $\log_b \dfrac{x^2 y^3}{z}$

42. $\log_b \dfrac{\sqrt{x}}{y^2 z}$

43. $\ln xy^3$

44. $\ln \dfrac{\sqrt{xy}}{z^4}$

In Exercises 45 to 48, write each logarithmic expression as a single logarithm with a coefficient of 1.

45. $2 \log x + \dfrac{1}{3} \log (x + 1)$

46. $5 \log x - 2 \log (x + 5)$

47. $\dfrac{1}{2} \ln 2xy - 3 \ln z$

48. $\ln x - (\ln y - \ln z)$

In Exercises 49 to 52, use the change-of-base formula and a calculator to approximate each logarithm accurate to six significant digits.

49. $\log_5 101$

50. $\log_3 40$

51. $\log_4 0.85$

52. $\log_8 0.3$

In Exercises 53 to 68, solve each equation for x. Give exact answers. Do not use a calculator.

53. $4^x = 30$

54. $5^{x+1} = 41$

55. $\ln 3x - \ln(x - 1) = \ln 4$

56. $\ln 3x + \ln 2 = 1$

57. $e^{\ln(x+2)} = 6$

58. $10^{\log(2x+1)} = 31$

59. $\dfrac{4^x + 4^{-x}}{4^x - 4^{-x}} = 2$

60. $\dfrac{5^x + 5^{-x}}{2} = 8$

61. $\log(\log x) = 3$

62. $\ln(\ln x) = 2$

63. $\log \sqrt{x - 5} = 3$

64. $\log x + \log(x - 15) = 1$

65. $\log_4(\log_3 x) = 1$

66. $\log_7(\log_5 x^2) = 0$

67. $\log_5 x^3 = \log_5 16x$

68. $25 = 16^{\log_4 x}$

69. **EARTHQUAKE MAGNITUDE** Determine, to the nearest 0.1, the Richter scale magnitude of an earthquake with an intensity of $I = 51{,}782{,}000 I_0$.

70. **EARTHQUAKE MAGNITUDE** A seismogram has an amplitude of 18 millimeters and a time delay of 21 seconds. Find, to the nearest tenth, the Richter scale magnitude of the earthquake that produced the seismogram.

71. **COMPARISON OF EARTHQUAKES** An earthquake had a Richter scale magnitude of 7.2. Its aftershock had a Richter scale magnitude of 3.7. Compare the intensity of the earthquake to the intensity of the aftershock by finding, to the nearest unit, the ratio of the larger intensity to the smaller intensity.

72. **COMPARISON OF EARTHQUAKES** An earthquake has an intensity 600 times the intensity of a second earthquake. Find, to the nearest tenth, the difference between the Richter scale magnitudes of the earthquakes.

73. **CHEMISTRY** Find the pH of tomatoes that have a hydronium-ion concentration of 6.28×10^{-5}. Round to the nearest tenth.

74. **CHEMISTRY** Find the hydronium-ion concentration of rainwater that has a pH of 5.4.

75. **COMPOUND INTEREST** Find the balance when $16,000 is invested at an annual rate of 8% for 3 years if the interest is compounded

  a. monthly   **b.** continuously

76. **COMPOUND INTEREST** Find the balance when $19,000 is invested at an annual rate of 6% for 5 years if the interest is compounded

  a. daily   **b.** continuously

77. DEPRECIATION The scrap value S of a product with an expected life span of n years is given by $S(n) = P(1 - r)^n$, where P is the original purchase price of the product and r is the annual rate of depreciation. A taxicab is purchased for $12,400 and is expected to last 3 years. What is its scrap value if it depreciates at a rate of 29% per year?

78. MEDICINE A skin wound heals according to the function given by $N(t) = N_0 e^{-0.12t}$, where N is the number of square centimeters of unhealed skin t days after the injury, and N_0 is the number of square centimeters covered by the original wound.

a. What percentage of the wound will be healed after 10 days?

b. How many days, to the nearest day, will it take for 50% of the wound to heal?

c. How long, to the nearest day, will it take for 90% of the wound to heal?

In Exercises 79 to 82, find the exponential growth/decay function $N(t) = N_0 e^{kt}$ that satisfies the given conditions.

79. $N(0) = 1, N(2) = 5$ **80.** $N(0) = 2, N(3) = 11$

81. $N(1) = 4, N(5) = 5$ **82.** $N(-1) = 2, N(0) = 1$

83. POPULATION GROWTH

a. Find the exponential growth function for a city whose population was 25,200 in 2002 and 26,800 in 2003. Use $t = 0$ to represent the year 2002.

b. Use the growth function to predict, to the nearest hundred, the population of the city in 2009.

84. CARBON DATING Determine, to the nearest ten years, the age of a bone if it now contains 96% of its original amount of carbon-14. The half-life of carbon-14 is 5730 years.

85. LOGISTIC GROWTH The population of coyotes in a national park satisfies the logistic model with $P_0 = 210$ in 1992, $c = 1400$, and $P(3) = 360$ (the population in 1995).

a. Determine the logistic model.

b. Use the model to predict, to the nearest 10, the coyote population in 2005.

86. Consider the logistic function

$$P(t) = \frac{128}{1 + 5e^{-0.27t}}$$

a. Find P_0.

b. What does $P(t)$ approach as $t \to \infty$?

CHAPTER 4 TEST

1. Find the inverse of $f(x) = 2x - 3$. Graph f and f^{-1} on the same coordinate axes.

2. Find the inverse of $f(x) = \dfrac{x}{4x - 8}$. State the domain and the range of f^{-1}.

3. a. Write $\log_b(5x - 3) = c$ in exponential form.

b. Write $3^{x/2} = y$ in logarithmic form.

4. Write $\log_b \dfrac{z^2}{y^3 \sqrt{x}}$ in terms of logarithms of x, y, and z.

5. Write $\log(2x + 3) - 3\log(x - 2)$ as a single logarithm with a coefficient of 1.

6. Use the change-of-base formula and a calculator to approximate $\log_4 12$. Round your result to the nearest ten thousandth.

7. Graph: $f(x) = 3^{-x/2}$

8. Graph: $f(x) = -\ln(x + 1)$

9. Solve: $5^x = 22$. Round your solution to the nearest ten thousandth.

10. Find the *exact* solution of $4^{5-x} = 7^x$.

11. Solve: $\log(x + 99) - \log(3x - 2) = 2$

12. Solve: $\ln(2 - x) + \ln(5 - x) = \ln(37 - x)$

13. Find the balance on $20,000 invested at an annual interest rate of 7.8% for 5 years:

  a. compounded monthly.

  b. compounded continuously.

14. Find the time required for money invested at an annual rate of 4% to double in value if the investment is compounded monthly. Round to the nearest hundredth of a year.

15. a. What, to the nearest tenth, will an earthquake measure on the Richter scale if it has an intensity of $I = 42,304,000I_0$?

  b. Compare the intensity of an earthquake that measures 6.3 on the Richter scale to the intensity of an earthquake that measures 4.5 on the Richter scale by finding the ratio of the larger intensity to the smaller intensity. Round to the nearest whole number.

16. a. Find the exponential growth function for a city whose population was 34,600 in 1996 and 39,800 in 1999. Use $t = 0$ to represent the year 1996.

  b. Use the growth function to predict the population of the city in 2006. Round to the nearest thousand.

17. Determine, to the nearest ten years, the age of a bone if it now contains 92% of its original amount of carbon-14. The half-life of carbon-14 is 5730 years.

18. A heated metal ball is placed in a room and allowed to cool. The temperature T of the ball in degrees Fahrenheit t minutes after it is placed in the room can be approximated by $T = 75 + 100e^{-0.05t}$. In how many minutes will the temperature of the ball be 95°F? Round to the nearest tenth of a minute.

19. During the free-fall portion of a jump, the time t in seconds required for a sky diver to reach a velocity of v feet per second is given by $t = -5\ln\left(1 - \dfrac{v}{160}\right)$. Determine the velocity of the sky diver 2 seconds after the free-fall portion of the jump begins. Round to the nearest foot per second.

20. The population of raccoons in a state park satisfies a logistic growth model with $P_0 = 160$ in 1999 and $P(1) = 190$. A park ranger has estimated the carrying capacity of the park to be 1100 raccoons. Use these data to

  a. find the logistic growth model for the raccoon population.

  b. predict the raccoon population in 2006.

CUMULATIVE REVIEW EXERCISES

1. Solve $|x - 4| \le 2$. Write the solution set using interval notation.

2. Solve $\dfrac{x}{2x - 6} \ge 1$. Write the solution set using set-builder notation.

3. Find, to the nearest tenth, the distance between the points $(5, 2)$ and $(11, 7)$.

4. The height, in feet, of a ball released with an initial upward velocity of 44 feet per second and at an initial height of 8 feet is given by $h(t) = -16t^2 + 44t + 8$, where t is the time in seconds after the ball is released. Find the maximum height the ball will reach.

5. Given $f(x) = 2x + 1$ and $g(x) = x^2 - 5$, find $(g \circ f)$.

6. Find the inverse of $f(x) = 3x - 5$.

7. The length, L, of a rectangle is twice its width, W. Express the area of the rectangle as a function of its width.

8. Use Descartes' Rule of Signs to determine the number of possible real zeros of $P(x) = x^4 - 3x^3 + x^2 - x - 6$.

9. Find the zeros of $P(x) = x^4 - 5x^3 + x^2 + 15x - 12$.

10. Find a polynomial function of lowest degree that has 2, $1 - i$, and $1 + i$ as zeros.

11. Find the equations of the vertical and horizontal asymptotes of the graph of $r(x) = \dfrac{3x - 5}{x - 4}$.

12. Determine the domain and the range of the rational function $R(x) = \dfrac{4}{x^2 + 1}$.

13. State whether $f(x) = 0.4^x$ is an increasing function or a decreasing function.

14. Write $\log_4 x = y$ in exponential form.

15. Write $5^3 = 125$ in logarithmic form.

16. Find $\log_3 4$. Round to the nearest thousandth.

17. Find, to the nearest tenth, the Richter scale magnitude of an earthquake with an intensity of $I = 11,650,600I_0$.

18. Solve $2e^x = 15$. Round to the nearest ten thousandth.

19. Find the age of a bone if it now has 94% of the carbon-14 it had at time $t = 0$. Round to the nearest ten years.

20. The wolf population in a national park satisfies a logistic growth model with $P_0 = 160$ in 1998 and $P(3) = 205$ (the population in 2001). It has been determined that the maximum population the park can support is 450 wolves.

   a. Determine the logistic growth model for the data.

   b. Use the model to predict, to the nearest 10, the wolf population in 2008.

SYSTEMS OF EQUATIONS

Tax Brackets Versus a Flat Tax

A perennial issue in Congress is the equity of the income tax laws. Currently, there are six tax brackets. As a person's income increases, the percent of adjusted gross income (income after allowable deductions) that the person pays in income tax increases. In 2003, Congress passed the Tax Relief Reconciliation Act of 2003, which changed the laws relating to income taxes. An alternative proposal before Congress that year was to replace these brackets with one bracket so that all taxpayers paid 20% of their income as tax. This type of tax structure is called a *flat tax* and is used in some states and countries. The flat-tax proposal was not passed by Congress.

The graph below shows the amount of tax a single person would pay using tax brackets and a flat tax. (Only the first three tax brackets are shown.)

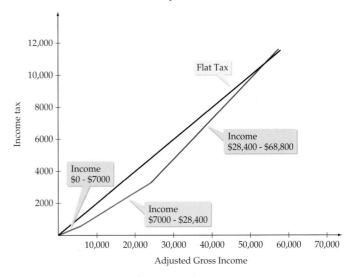

Source: www.irs.gov

The *x*-coordinate of the point at which the graphs intersect is the income at which a taxpayer pays the same amount of tax whether brackets are used or a flat tax is used. The coordinates of that point can be found by solving a system of equations. See **Exercise 59 on page 427.**

Implication, Converse, and Contrapositive

In the movie *Star Wars*, Episode IV, *A New Hope*, Obi-Wan Kenobi says, "If you strike me down, I shall become more powerful than you can possibly imagine." The sentence uttered by Obi-Wan is called an *implication*. Statements such as these can be symbolized by *if p, then q* or *if p, q*. The statement *p* is called the *antecedent*, and *q* is called the *consequent*. For Obi-Wan Kenobi, we have

     p: You strike me down.

     q: I shall become more powerful than you can possibly imagine.

In mathematics, many theorems are stated as implications. For instance, consider the following theorem.

Theorem   If a number ends in 0, then the number is divisible by 5.

The antecedent is "a number ends in 0"; the consequent is "the number is divisible by 5."

Interchanging the order of the antecedent and the consequent results in the **converse** of the implication. For the theorem above, we have

Converse   If a number is divisible by 5, then the number ends in 0.

This is *not* a true statement because 15 is divisible by 5 but 15 does not end in 0. In this case, interchanging the antecedent and consequent turned a true statement into a false statement. This demonstrates that the converse of a theorem is not always true.

Although the converse of a theorem *may* not be a true statement, the **contrapositive** of a theorem is always a true statement. The contrapositive is formed by interchanging the antecedent and the consequent (as we did for the converse) and then negating each statement. For the theorem above, we have

Contrapositive   If a number is *not* divisible by 5, then the number does *not* end in 0.

Some theorems are stated in the form "*p* if and only if *q*." These are very powerful theorems because both the implication and its converse are true. Here is an example.

    A linear system of two equations in two variables has a unique solution if and only if the graphs of the equations are not parallel.

In this case, we have the implication, "If a linear system of equations in two variables has a unique solution, the graphs of the equations are not parallel"—a true statement. The converse of the implication is "If the graphs of the two equations of a linear system of equations in two variables are not parallel, the system has a unique solution"—also a true statement.

SYSTEMS OF LINEAR EQUATIONS IN TWO VARIABLES

SECTION **5.1**

- SUBSTITUTION METHOD FOR SOLVING A SYSTEM OF LINEAR EQUATIONS
- ELIMINATION METHOD FOR SOLVING A SYSTEM OF EQUATIONS
- APPLICATIONS OF SYSTEMS OF EQUATIONS

Recall that an equation of the form $Ax + By = C$ is a linear equation in two variables. A solution of a linear equation in two variables is an ordered pair (x, y) that makes the equation a true statement. For example, $(-2, 3)$ is a solution of the equation

$$2x + 3y = 5 \quad \text{since} \quad 2(-2) + 3(3) = 5$$

The graph of a linear equation in two variables, a straight line, is the set of points whose ordered pairs satisfy the equation. **Figure 5.1** is the graph of $2x + 3y = 5$.

A **system of equations** is two or more equations considered together. The following system of equations is a **linear system of equations** in two variables.

$$\begin{cases} 2x + 3y = 4 \\ 3x - 2y = -7 \end{cases}$$

A **solution** of a system of equations in two variables is an ordered pair that is a solution of both equations.

In **Figure 5.2,** the graphs of the two equations in the system of equations above intersect at the point $(-1, 2)$. Because that point lies on both lines, $(-1, 2)$ is a solution of both equations and thus is a solution of the system of equations. The point $(5, -2)$ is a solution of the first equation but not the second equation. Therefore, $(5, -2)$ is not a solution of the system of equations.

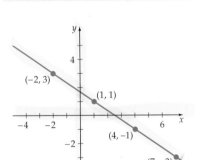

$2x + 3y = 5$

FIGURE 5.1

? QUESTION Is $(3, -4)$ a solution of $\begin{cases} 2x - 3y = 18 \\ x + 4y = -13 \end{cases}$?

The graphs of two linear equations in two variables can intersect at a single point, be the same line, or be parallel. When the graphs intersect at a single point or are the same line, the system is called a **consistent** system of equations. The system is called an **independent** system of equations when the lines intersect at exactly one point. The system is called a **dependent** system of equations when the equations represent the same line. In this case, the system has an infinite number of solutions. When the graphs of the two equations are parallel lines, the system is called **inconsistent** and has no solution. See **Figure 5.3.**

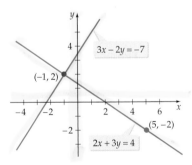

FIGURE 5.2

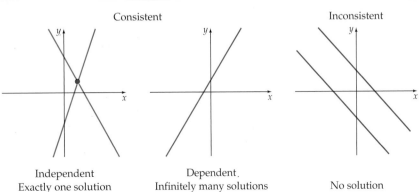

Consistent — Independent / Exactly one solution

Dependent / Infinitely many solutions

Inconsistent — No solution

FIGURE 5.3

? ANSWER Yes.

• SUBSTITUTION METHOD FOR SOLVING A SYSTEM OF LINEAR EQUATIONS

The **substitution method** is one procedure for solving a system of equations. This method is illustrated in Example 1.

EXAMPLE I   Solve a System of Equations by the Substitution Method

Solve: $\begin{cases} 3x - 5y = 7 & (1) \\ y = 2x & (2) \end{cases}$

Algebraic Solution

The solutions of $y = 2x$ are the ordered pairs $(x, 2x)$. For the system of equations to have a solution, ordered pairs of the form $(x, 2x)$ also must be solutions of $3x - 5y = 7$. To determine whether the ordered pairs $(x, 2x)$ are solutions of Equation (1), substitute $(x, 2x)$ into Equation (1) and solve for x. Think of this as *substituting* $2x$ for y.

$$3x - 5y = 7 \qquad \bullet \textbf{ Equation (1)}$$
$$3x - 5(2x) = 7 \qquad \bullet \textbf{ Substitute } 2x \textbf{ for y.}$$
$$3x - 10x = 7$$
$$-7x = 7$$
$$x = -1$$
$$y = 2x \qquad \bullet \textbf{ Equation (2)}$$
$$= 2(-1) = -2 \qquad \bullet \textbf{ Substitute } -1 \textbf{ for x in Equation 2.}$$

The only ordered-pair solution of the system of equations is $(-1, -2)$. When a system of equations has a unique solution, the system of equations is independent.

Visualize the Solution

Graphing $3x - 5y = 7$ and $y = 2x$ shows that the ordered pair $(-1, -2)$ belongs to both lines. Therefore, $(-1, -2)$ is a solution of the system of equations. See **Figure 5.4.**

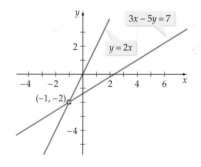

FIGURE 5.4

An independent system of equations

▶ **TRY EXERCISE 6, PAGE 425**

EXAMPLE 2   Identify an Inconsistent System of Equations

Solve: $\begin{cases} x + 3y = 6 & (1) \\ 2x + 6y = -18 & (2) \end{cases}$

Algebraic Solution

Solve Equation (1) for y:

$$x + 3y = 6$$
$$y = -\frac{1}{3}x + 2$$

The solutions of $y = -\frac{1}{3}x + 2$ are the ordered pairs $\left(x, -\frac{1}{3}x + 2\right)$. For the system of equations to have a solution, ordered pairs of this form must also be solutions of $2x + 6y = -18$. To determine whether the

Visualize the Solution

Solving Equations (1) and (2) for y gives $y = -\frac{1}{3}x + 2$ and $y = -\frac{1}{3}x - 3$. Note that these two equations have the same slope, $-\frac{1}{3}$, and different y-intercepts.

ordered pairs $\left(x, -\dfrac{1}{3}x + 2\right)$ are solutions of Equation (2), substitute $\left(x, -\dfrac{1}{3}x + 2\right)$ into Equation (2) and solve for x.

$$2x + 6y = -18 \qquad \text{• Equation (2)}$$

$$2x + 6\left(-\frac{1}{3}x + 2\right) = -18 \qquad \text{• Substitute } -\frac{1}{3}x + 2 \text{ for } y.$$

$$2x - 2x + 12 = -18$$

$$12 = -18 \qquad \text{• A false statement}$$

The false statement $12 = -18$ means that no ordered pair that is a solution of Equation (1) is also a solution of Equation (2). The equations have no ordered pairs in common and thus the system of equations has no solution. This is an inconsistent system of equations.

Therefore, the graphs of the two lines are parallel and never intersect. See **Figure 5.5**.

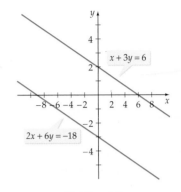

FIGURE 5.5

An inconsistent system of equations

▶ **TRY EXERCISE 18, PAGE 426**

EXAMPLE 3   Identify a Dependent System of Equations

Solve: $\begin{cases} 8x - 4y = 16 & (1) \\ 2x - y = 4 & (2) \end{cases}$

Algebraic Solution

Solve Equation (2) for y:

$$2x - y = 4$$

$$y = 2x - 4$$

The solutions of $y = 2x - 4$ are the ordered pairs $(x, 2x - 4)$. For the system of equations to have a solution, ordered pairs of the form $(x, 2x - 4)$ also must be solutions of $8x - 4y = 16$. To determine whether the ordered pairs $(x, 2x - 4)$ are solutions of Equation (1), substitute $(x, 2x - 4)$ into Equation (1) and solve for x.

$$8x - 4y = 16 \qquad \text{• Equation (1)}$$

$$8x - 4(2x - 4) = 16 \qquad \text{• Substitute } 2x - 4 \text{ for } y.$$

$$8x - 8x + 16 = 16$$

$$16 = 16 \qquad \text{• A true statement}$$

The true statement $16 = 16$ means that the ordered pairs $(x, 2x - 4)$ that are solutions of Equation (2) are also solutions of Equation (1). Because x can be replaced by any real number c, the solution of the system of equations is the set of ordered pairs $(c, 2c - 4)$. This is a dependent system of equations.

Visualize the Solution

Solving Equations (1) and (2) for y gives $y = 2x - 4$ and $y = 2x - 4$. Note that these two equations have the same slope, 2, and the same y-intercept, $(0, -4)$. Therefore, the graphs of the two lines are exactly the same. One graph intersects the second graph infinitely often. See **Figure 5.6**.

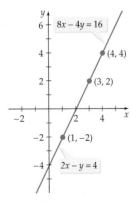

FIGURE 5.6

▶ **TRY EXERCISE 20, PAGE 426**

Some of the specific ordered-pair solutions in Example 3 can be found by choosing various values for c. The table below shows the ordered pairs that result from choosing 1, 3, and 4 as c. The ordered pairs $(1, -2)$, $(3, 2)$, and $(4, 4)$ are specific solutions of the system of equations. These points are on the graphs of Equation (1) and Equation (2), as shown in **Figure 5.6.**

c	$(c, 2c - 4)$	(x, y)
1	$(1, 2(1) - 4)$	$(1, -2)$
3	$(3, 2(3) - 4)$	$(3, 2)$
4	$(4, 2(4) - 4)$	$(4, 4)$

Before leaving Example 3, note that there is more than one way to represent the ordered-pair solutions. To illustrate this point, solve Equation (2) for x.

$$2x - y = 4 \qquad \bullet \text{ Equation (2)}$$

$$x = \frac{1}{2}y + 2 \qquad \bullet \text{ Solve for } x.$$

Because y can be replaced by any real number b, there are an infinite number of ordered pairs $\left(\frac{1}{2}b + 2, b\right)$ that are solutions of the system of equations. Choosing b as -2, 2, and 4 gives the same ordered pairs: $(1, -2)$, $(3, 2)$, and $(4, 4)$. There is always more than one way to describe the ordered pairs when writing the solution of a dependent system of equations. For Example 3, either the ordered pairs $(c, 2c - 4)$ or the ordered pairs $\left(\frac{1}{2}b + 2, b\right)$ would generate all the solutions of the system of equations.

• ELIMINATION METHOD FOR SOLVING A SYSTEM OF EQUATIONS

Two systems of equations are **equivalent** if each system has exactly the same solutions. The systems

$$\begin{cases} 3x + 5y = 9 \\ 2x - 3y = -13 \end{cases} \quad \text{and} \quad \begin{cases} x = -2 \\ y = 3 \end{cases}$$

are equivalent systems of equations. Each system has the solution $(-2, 3)$, as shown in **Figure 5.7.**

A second technique for solving a system of equations is similar to the strategy for solving first-degree equations in one variable. The system of equations is replaced by a series of equivalent systems until the solution is obvious.

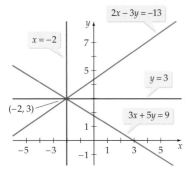

FIGURE 5.7

Operations That Produce Equivalent Systems of Equations

1. Interchange any two equations.

2. Replace an equation with a nonzero multiple of that equation.

3. Replace an equation with the sum of that equation and a nonzero constant multiple of another equation in the system.

Because the order in which the equations are written does not affect the system of equations, interchanging the equations does not affect its solution. The second operation restates the property that says that multiplying each side of an equation by the same nonzero constant does not change the solutions of the equation.

The third operation can be illustrated as follows. Consider the system of equations

$$\begin{cases} 3x + 2y = 10 & (1) \\ 2x - 3y = -2 & (2) \end{cases}$$

Multiply each side of Equation (2) by 2. (Any nonzero number would work.) Add the resulting equation to Equation (1).

$$\begin{array}{ll} 3x + 2y = 10 & \text{• Equation (1)} \\ \underline{4x - 6y = -4} & \text{• 2 times Equation (2)} \\ 7x - 4y = 6 \quad (3) & \text{• Add the equations.} \end{array}$$

Replace Equation (1) with the new Equation (3) to produce the following equivalent system of equations.

$$\begin{cases} 7x - 4y = 6 & (3) \\ 2x - 3y = -2 & (2) \end{cases}$$

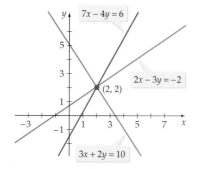

The third property states that the resulting system of equations has the same solutions as the original system and is therefore equivalent to the original system of equations. **Figure 5.8** shows the graph of $7x - 4y = 6$. Note that the line passes through the same point at which the lines of the original system of equations intersect, the point $(2, 2)$.

FIGURE 5.8

EXAMPLE 4   Solve a System of Equations by the Elimination Method

Solve: $\begin{cases} 3x - 4y = 10 & (1) \\ 2x + 5y = -1 & (2) \end{cases}$

Algebraic Solution

Use the operations that produce equivalent equations to eliminate a variable from one of the equations. We will eliminate x from Equation (2) by multiplying each equation by a different constant so as to have a new system of equations in which the coefficients of x are additive inverses.

$$\begin{array}{ll} 6x - 8y = 20 & \text{• 2 times Equation (1)} \\ \underline{-6x - 15y = 3} & \text{• −3 times Equation (2)} \\ -23y = 23 & \text{• Add the equations.} \\ y = -1 & \text{• Solve for y.} \end{array}$$

Solve Equation (1) for x by substituting -1 for y.

$$\begin{aligned} 3x - 4(-1) &= 10 \\ 3x &= 6 \\ x &= 2 \end{aligned}$$

The solution of the system of equations is $(2, -1)$.

Visualize the Solution

Graphing $3x - 4y = 10$ and $2x + 5y = -1$ shows that $(2, -1)$ belongs to both lines. Therefore, $(2, -1)$ is a solution of the system of equations. See **Figure 5.9.**

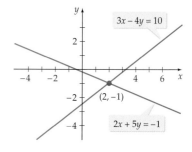

FIGURE 5.9

▶ **TRY EXERCISE 24, PAGE 426**

The method just described is called the **elimination method** for solving a system of equations, because it involves *eliminating* a variable from one of the equations.

INTEGRATING TECHNOLOGY

You can use a graphing calculator to solve a system of equations in two variables. First, algebraically solve each equation for y.

Solve for y.

$$3x - 4y = 10 \quad \rightarrow \quad y = 0.75x - 2.5$$
$$2x + 5y = -1 \quad \rightarrow \quad y = -0.4x - 0.2$$

Now graph the equations. Enter **0.75X-2.5** into Y₁ and **-0.4X-0.2** into Y₂ and graph the two equations in the standard viewing window. The sequence of steps shown in **Figure 5.10** can be used to find the point of intersection with a TI-83 calculator.

Press [2nd] CALC.
Select 5: intersect.
Press [ENTER].

The "First curve?" shown on the bottom of the screen means to select the first of the two graphs that intersect. Just press [ENTER].

The "Second curve?" shown on the bottom of the screen means to select the second of the two graphs that intersect. Just press [ENTER].

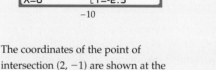

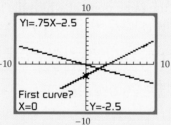

"Guess?" is shown on the bottom of the screen. Move the cursor until it is approximately on the point of intersection. Press [ENTER].

The coordinates of the point of intersection $(2, -1)$ are shown at the bottom of the screen.

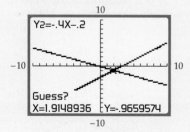

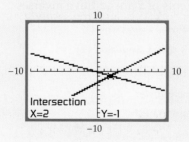

FIGURE 5.10

For the system of equations in Example 4, the intersection of the two graphs occurs at a point in the standard viewing window. If the point of intersection does not appear on the screen, you must adjust the viewing window so that the point of intersection is visible.

EXAMPLE 5   Solve a Dependent System of Equations

Solve: $\begin{cases} x - 2y = 2 & (1) \\ 3x - 6y = 6 & (2) \end{cases}$

Solution

Eliminate x by multiplying Equation (2) by $-\dfrac{1}{3}$ and then adding the result to Equation (1).

$$x - 2y = \quad 2 \qquad \text{• Equation (1)}$$
$$\underline{-x + 2y = -2} \qquad \text{• } -\dfrac{1}{3} \text{ times Equation (2)}$$
$$0 = \quad 0 \qquad \text{• Add the two equations.}$$

Replace Equation (2) by $0 = 0$.

$$\begin{cases} x - 2y = 2 \\ \qquad 0 = 0 \end{cases} \quad \text{• This is an equivalent system of equations.}$$

Because the equation $0 = 0$ is an identity, an ordered pair that is a solution of Equation (1) is also a solution of $0 = 0$. Thus the solutions are the solutions of $x - 2y = 2$. Solving for y, we find that $y = \dfrac{1}{2}x - 1$.

Because x can be replaced by any real number c, the solutions of the system of equations are the ordered pairs $\left(c, \dfrac{1}{2}c - 1 \right)$. See **Figure 5.11.**

▶ **TRY EXERCISE 28, PAGE 426**

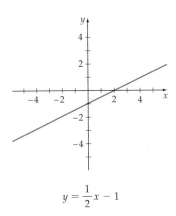

$y = \dfrac{1}{2}x - 1$

FIGURE 5.11

---

take note

Referring again to Example 5 and solving Equation (1) for x, we have x = 2y + 2. Because y can be any real number b, the ordered-pair solutions of the system of equations can also be written as (2b + 2, b).

If one equation of the system of equations is replaced by a false equation, the system of equations has no solution. For example, the system of equations

$$\begin{cases} x + y = 4 \\ \qquad 0 = 5 \end{cases}$$

has no solution because the second equation is false for any choice of x and y.

● APPLICATIONS OF SYSTEMS OF EQUATIONS

Consider the situation of a Corvette car dealership. If the dealership were willing to sell a Corvette for $10, there would be many consumers willing to buy a Corvette. The problem with this plan is that the dealership would soon be out of business. On the other hand, if the dealership tried to sell each Corvette for $1 million, the dealership would not sell any cars and would still go out of business. Between $10 and $1 million, there is a price at which a dealership can sell Corvettes (and stay in business) and at which consumers are willing to pay that price. This price is referred to as the **equilibrium price.**

Economists refer to these types of problems as **supply-demand problems.** Businesses are willing to *supply* a product at a certain price and there is consumer *demand* for the product at that price. To find the equilibrium point, a system of equations is created. One equation of the system is the demand model of the business. The second equation is the supply model of the consumer.

EXAMPLE 6   Solve a Supply-Demand Problem

Suppose that the number of bushels x of apples a farmer is willing to sell is given by $x = 100p - 25$, where p is the price, in dollars, per bushel of apples. The number of bushels x of apples a grocer is willing to purchase is given by $x = -150p + 655$, where p is the price per bushel of apples. Find the equilibrium price.

Solution

Using the supply and demand equations, we have the system of equations

$$\begin{cases} x = 100p - 25 \\ x = -150p + 655 \end{cases}$$

Solve the system of equations by substitution.

$$-150p + 655 = 100p - 25$$
$$-250p + 655 = -25 \qquad \text{• Subtract 100}p \text{ from each side.}$$
$$-250p = -680 \qquad \text{• Subtract 655 from each side.}$$
$$p = 2.72 \qquad \text{• Divide by } -250.$$

The equilibrium price is $2.72 per bushel.

▶ **TRY EXERCISE 42, PAGE 426**

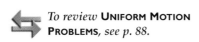

To review **UNIFORM MOTION PROBLEMS,** *see p. 88.*

As the application problems we studied earlier in the text become more complicated, a system of equations may be the method for solving these problems. The next example involves the distance-rate-time equation $d = rt$.

EXAMPLE 7   Solve an Application to River Currents

A rowing team rowing with the current traveled 18 miles in 2 hours. Against the current, the team rowed 10 miles in 2 hours. Find the rate of the boat in calm water and the rate of the current.

Solution

Let r_1 represent the rate of the boat in calm water, and let r_2 represent the rate of the current.

The rate of the boat *with the current* is $r_1 + r_2$.

The rate of the boat *against the current* is $r_1 - r_2$.

Because the rowing team traveled 18 miles in 2 hours with the current, we use the equation $d = rt$.

$$d = r \cdot t$$
$$18 = (r_1 + r_2) \cdot 2 \qquad \bullet\ d = 18, t = 2$$
$$9 = r_1 + r_2 \qquad \bullet\ \text{Divide each side by 2.}$$

Because the team rowed 10 miles in 2 hours against the current, we write

$$10 = (r_1 - r_2) \cdot 2 \qquad \bullet\ d = 10, t = 2$$
$$5 = r_1 - r_2 \qquad \bullet\ \text{Divide each side by 2.}$$

Thus we have a system of two linear equations in the variables r_1 and r_2.

$$\begin{cases} 9 = r_1 + r_2 \\ 5 = r_1 - r_2 \end{cases}$$

Solving the system by using the elimination method, we find that r_1 is 7 mph and r_2 is 2 mph. Thus the rate of the boat in calm water is 7 mph and the rate of the current is 2 mph. You should verify these solutions.

▶ **TRY EXERCISE 46, PAGE 426**

TOPICS FOR DISCUSSION

1. Explain how to use the substitution method to solve a system of equations.

2. Explain how to use the elimination method to solve a system of equations.

3. Give an example of a system of equations in two variables that is

   a. independent   **b.** dependent   **c.** inconsistent

4. If a linear system of equations in two variables has no solution, what does that mean about the graphs of the equations of the system?

5. If $A = \{(x, y) \mid x + y = 5\}$ and $B = \{(x, y) \mid x - y = 3\}$, explain the meaning of $A \cap B$.

EXERCISE SET 5.1

In Exercises 1 to 20, solve each system of equations by the substitution method.

1. $\begin{cases} 2x - 3y = 16 \\ x = 2 \end{cases}$

2. $\begin{cases} 3x - 2y = -11 \\ y = 1 \end{cases}$

3. $\begin{cases} 3x + 4y = 18 \\ y = -2x + 3 \end{cases}$

4. $\begin{cases} 5x - 4y = -22 \\ y = 5x - 2 \end{cases}$

5. $\begin{cases} -2x + 3y = 6 \\ x = 2y - 5 \end{cases}$

▶ **6.** $\begin{cases} 8x + 3y = -7 \\ x = 3y + 15 \end{cases}$

7. $\begin{cases} 6x + 5y = 1 \\ x - 3y = 4 \end{cases}$

8. $\begin{cases} -3x + 7y = 14 \\ 2x - y = -13 \end{cases}$

9. $\begin{cases} 7x + 6y = -3 \\ y = \dfrac{2}{3}x - 6 \end{cases}$

10. $\begin{cases} 9x - 4y = 3 \\ x = \dfrac{4}{3}y + 3 \end{cases}$

11. $\begin{cases} y = 4x - 3 \\ y = 3x - 1 \end{cases}$

12. $\begin{cases} y = 5x + 1 \\ y = 4x - 2 \end{cases}$

13. $\begin{cases} y = 5x + 4 \\ x = -3y - 4 \end{cases}$

14. $\begin{cases} y = -2x - 6 \\ x = -2y - 2 \end{cases}$

15. $\begin{cases} 3x - 4y = 2 \\ 4x + 3y = 14 \end{cases}$

16. $\begin{cases} 6x + 7y = -4 \\ 2x + 5y = 4 \end{cases}$

17. $\begin{cases} 3x - 3y = 5 \\ 4x - 4y = 9 \end{cases}$

▶ **18.** $\begin{cases} 3x - 4y = 8 \\ 6x - 8y = 9 \end{cases}$

19. $\begin{cases} 4x + 3y = 6 \\ \\ y = -\dfrac{4}{3}x + 2 \end{cases}$

▶ **20.** $\begin{cases} 5x + 2y = 2 \\ \\ y = -\dfrac{5}{2}x + 1 \end{cases}$

In Exercises 21 to 40, solve each system of equations by the elimination method.

21. $\begin{cases} 3x - y = 10 \\ 4x + 3y = -4 \end{cases}$

22. $\begin{cases} 3x + 4y = -5 \\ x - 5y = -8 \end{cases}$

23. $\begin{cases} 4x + 7y = 21 \\ 5x - 4y = -12 \end{cases}$

▶ **24.** $\begin{cases} 3x - 8y = -6 \\ -5x + 4y = 10 \end{cases}$

25. $\begin{cases} 5x - 3y = 0 \\ 10x - 6y = 0 \end{cases}$

26. $\begin{cases} 3x + 2y = 0 \\ 2x + 3y = 0 \end{cases}$

27. $\begin{cases} 6x + 6y = 1 \\ 4x + 9y = 4 \end{cases}$

▶ **28.** $\begin{cases} 4x + 5y = 2 \\ 8x - 15y = 9 \end{cases}$

29. $\begin{cases} 3x + 6y = 11 \\ 2x + 4y = 9 \end{cases}$

30. $\begin{cases} 4x - 2y = 9 \\ 2x - y = 3 \end{cases}$

31. $\begin{cases} \dfrac{5}{6}x - \dfrac{1}{3}y = -6 \\ \\ \dfrac{1}{6}x + \dfrac{2}{3}y = 1 \end{cases}$

32. $\begin{cases} \dfrac{3}{4}x + \dfrac{2}{5}y = 1 \\ \\ \dfrac{1}{2}x - \dfrac{3}{5}y = -1 \end{cases}$

33. $\begin{cases} \dfrac{3}{4}x + \dfrac{1}{3}y = 1 \\ \\ \dfrac{1}{2}x + \dfrac{2}{3}y = 0 \end{cases}$

34. $\begin{cases} \dfrac{3}{5}x - \dfrac{2}{3}y = 7 \\ \\ \dfrac{2}{5}x - \dfrac{5}{6}y = 7 \end{cases}$

35. $\begin{cases} 2\sqrt{3}x - 3y = 3 \\ 3\sqrt{3}x + 2y = 24 \end{cases}$

36. $\begin{cases} 4x - 3\sqrt{5}y = -19 \\ 3x + 4\sqrt{5}y = 17 \end{cases}$

37. $\begin{cases} 3\pi x - 4y = 6 \\ 2\pi x + 3y = 5 \end{cases}$

38. $\begin{cases} 2x - 5\pi y = 3 \\ 3x + 4\pi y = 2 \end{cases}$

39. $\begin{cases} 3\sqrt{2}x - 4\sqrt{3}y = -6 \\ 2\sqrt{2}x + 3\sqrt{3}y = 13 \end{cases}$

40. $\begin{cases} 2\sqrt{2}x + 3\sqrt{5}y = 7 \\ 3\sqrt{2}x - \sqrt{5}y = -17 \end{cases}$

In Exercises 41 to 61, solve by using a system of equations.

41. **SUPPLY/DEMAND** The number x of MP3 players a manufacturer is willing to sell is given by $x = 20p - 2000$, where p is the price, in dollars, per MP3 player. The number x of MP3 players a store is willing to purchase is given by $x = -4p + 1000$, where p is the price per MP3 player. Find the equilibrium price.

▶ **42.** **SUPPLY/DEMAND** The number x of digital cameras a manufacturer is willing to sell is given by $x = 25p - 500$, where p is the price, in dollars, per digital camera. The number x of digital cameras a store is willing to purchase is given by $x = -7p + 1100$, where p is the price per digital camera. Find the equilibrium price.

43. **RATE OF WIND** Flying with the wind, a plane traveled 450 miles in 3 hours. Flying against the wind, the plane traveled the same distance in 5 hours. Find the rate of the plane in calm air and the rate of the wind.

44. **RATE OF WIND** A plane flew 800 miles in 4 hours while flying with the wind. Against the wind, it took the plane 5 hours to travel 800 miles. Find the rate of the plane in calm air and the rate of the wind.

45. **RATE OF CURRENT** A motorboat traveled a distance of 120 miles in 4 hours while traveling with the current. Against the current, the same trip took 6 hours. Find the rate of the boat in calm water and the rate of the current.

▶ **46.** **RATE OF CURRENT** A canoeist can row 12 miles with the current in 2 hours. Rowing against the current, it takes the canoeist 4 hours to travel the same distance. Find the rate of the canoeist in calm water and the rate of the current.

47. **METALLURGY** A metallurgist made two purchases. The first purchase, which cost $1080, included 30 kilograms of an iron alloy and 45 kilograms of a lead alloy. The second purchase, at the same prices, cost $372 and included 15 kilograms of the iron alloy and 12 kilograms of the lead alloy. Find the cost per kilogram of the iron and lead alloys.

48. **CHEMISTRY** For $14.10, a chemist purchased 10 liters of hydrochloric acid and 15 liters of silver nitrate. A second purchase, at the same prices, cost $18.16 and included 12 liters of hydrochloric acid and 20 liters of silver nitrate. Find the cost per liter of each of the two chemicals.

49. **CHEMISTRY** A goldsmith has two gold alloys. The first alloy is 40% gold; the second alloy is 60% gold. How many grams of each should be mixed to produce 20 grams of an alloy that is 52% gold?

50. **CHEMISTRY** One acetic acid solution is 70% water and another is 30% water. How many liters of each solution

should be mixed to produce 20 liters of a solution that is 40% water?

51. GEOMETRY A right triangle in the first quadrant is bounded by the lines $y = 0$, $y = \frac{1}{2}x$, and $y = -2x + 6$. Find its area.

52. GEOMETRY The lines whose equations are $2x + 3y = 1$, $3x - 4y = 10$, and $4x + ky = 5$ all intersect at the same point. What is the value of k?

53. NUMBER THEORY Adding a three-digit number $5Z7$ to 256 gives $XY3$. If $XY3$ is divisible by 3, then what is the largest possible value of Z?

54. NUMBER THEORY Find the value of k if $2x + 5 = 6x + k = 4x - 7$.

55. NUMBER THEORY A *Pythagorean triple* is three positive integers a, b, and c for which $a^2 + b^2 = c^2$. Given $a = 42$, find all the values of b and c such that a, b, and c form a Pythagorean triple. (*Suggestion:* If $a = 42$, then $1764 + b^2 = c^2$ or $1764 = c^2 - b^2 = (c - b)(c + b)$. Because $(c - b)(c + b) = 1764$, $c - b$ and $b + c$ must be factors of 1764. For instance, one possibility is $2 = c - b$ and $882 = c + b$. Solving this system of equations yields one set of Pythagorean triples. Now repeat for other possible factors of 1764. Remember that answers must be positive integers.)

56. NUMBER THEORY Given $a = 30$, find all the values of b and c such that a, b, and c form a Pythagorean triple. (See the preceding exercise.)

57. MARKETING A marketing company asked 100 people whether they liked a new skin cream and lip balm. The company found that 80% of the people who liked the new skin cream also liked the new lip balm and that 50% of the people who did not like the skin cream liked the new lip balm. If 77 people liked the lip balm, how many people liked the skin cream?

58. FIRE SCIENCE An analysis of 200 scores on a firefighter qualifying exam found that 75% of those who passed the basic fire science exam also passed the exam on containing chemical fires. Of those who did not pass the basic fire science exam, 25% passed the exam on containing chemical fires. If 120 people passed the exam on containing chemical fires, how many people passed the basic fire science exam?

59. INCOME TAX The chapter opener on page 415 shows the graph of the income tax a taxpayer pays for various income levels for both a flat tax and variable tax brackets. Given that the equation of the line in red is $T = 0.25I - 3190$ for $28{,}400 < I \le 68{,}800$, find and interpret the point at which the flat tax graph crosses the variable tax bracket graph. (*Suggestion:* Use the fact that flat tax is 20% of adjusted gross income to find the equation of the line in black. Now solve a system of equations.)

60. INCOME TAX The chapter opener on page 415 shows the graph of the income tax a taxpayer pays for various income levels for both a flat tax and variable tax brackets. The equation of the line segment in green is $T = 0.10I$ for $0 \le I \le 7000$; the equation of the line segment in blue is $T = 0.15I - 350$ for $7000 < I \le 28{,}400$; the equation of the line segment in red is $T = 0.25I - 3190$ for $28{,}400 < I \le 68{,}800$. If the flat-tax proposal for each taxpayer is reduced to 14%, does the line representing the flat-tax proposal intersect the green, blue, or red line segment? What is the point of intersection?

61. INVESTMENT A broker invests $25,000 of a client's money in two different bond funds. The annual rate of return on one bond fund is 6%, and the annual rate of return on the second bond fund is 6.5%. The investor receives a total annual interest payment from the two bond funds of $1555. Find the amount invested in each fund.

CONNECTING CONCEPTS

In Exercises 62 to 71, solve for x and y. Use the fact that if $z_1 = a_1 + b_1 i$ and $z_2 = a_2 + b_2 i$ are two complex numbers, then $z_1 = z_2$ if and only if $a_1 = a_2$ and $b_1 = b_2$.

62. $(2 + i)x + (3 - i)y = 7$

63. $(3 + 2i)x + (4 - 3i)y = 2 - 16i$

64. $(4 - 3i)x + (5 + 2i)y = 11 + 9i$

65. $(2 + 6i)x + (4 - 5i)y = -8 - 7i$

66. $(-3 - i)x - (4 + 2i)y = 1 - i$

67. $(5 - 2i)x + (-3 - 4i)y = 12 - 35i$

68. $\begin{cases} 2x + 5y = 11 + 3i \\ 3x + \ y = 10 - 2i \end{cases}$

69. $\begin{cases} 4x + 3y = 11 + \ 6i \\ 3x - 5y = \ 1 + 19i \end{cases}$

70. $\begin{cases} 2x + 3y = 11 + \ 5i \\ 3x - 3y = \ 9 - 15i \end{cases}$

71. $\begin{cases} 5x - 4y = 15 - 41i \\ 3x + 5y = \ 9 + \ 5i \end{cases}$

PREPARE FOR SECTION 5.2

72. Solve $2x - 5y = 15$ for y. [1.1]

73. If $x = 2c + 1$, $y = -c + 3$, and $z = 2x + 5y - 4$, write z in terms of c. [P.1]

74. Solve: $\begin{cases} 5x - 2y = 10 \\ 2y = 8 \end{cases}$ [5.1]

75. Solve: $\begin{cases} 3x - y = 11 \\ 2x + 3y = -11 \end{cases}$ [5.1]

76. Solve: $\begin{cases} y = 3x - 4 \\ y = 4x - 2 \end{cases}$ [5.1]

77. Solve: $\begin{cases} 4x + y = 9 \\ -8x - 2y = -18 \end{cases}$ [5.1]

PROJECTS

1. INDEPENDENT AND DEPENDENT CONDITIONS Consider the following problem: "Maria and Michael drove from Los Angeles to New York in 60 hours. How long did Maria drive?" It is difficult to answer this question. She may have driven all 60 hours while Michael relaxed, or she may have relaxed while Michael drove all 60 hours. The difficulty is that there are two unknowns (how long each drove) and only one condition (the total driving time) relating the unknowns. If we added another condition, such as Michael drove 25 hours, then we could determine how long Maria drove, 35 hours. In most cases, an application problem will have a single answer only when there are as many *independent* conditions as there are variables. Conditions are independent if knowing one does *not* enable you to know the other.

Here is an example of conditions that are not independent. "The perimeter of a rectangle is 50 meters. The sum of the width and length is 25 meters." To see that these conditions are dependent, write the perimeter equation and divide each side by 2.

$$2w + 2l = 50$$
$$w + l = 25 \qquad \bullet \text{ Divide each side by 2.}$$

Note that the resulting equation is the second condition: the sum of the width and length is 25. Thus, knowing the first condition enables us to determine the second condition. The conditions are not independent, so there is no one solution to this problem.

For each of the following problems, determine whether the conditions are independent or dependent. For those problems that have independent conditions, find the solution (if possible). For those problems for which the conditions are dependent, find two solutions.

a. The sum of two numbers is 30. The difference between the two numbers is 10. Find the numbers.

b. The area of a square is 25 square meters. Find the length of each side.

c. The area of a rectangle is 25 square meters. Find the length of each side.

d. Emily spent $1000 for carpeting and tile. Carpeting cost $20 per square yard and tile cost $30 per square yard. How many square yards of each did she purchase?

e. The sum of two numbers is 20. Twice the smaller number is 10 minus twice the larger number. Find the two numbers.

f. Make up a word problem for which there are two independent conditions. Solve the problem.

g. Make up a word problem for which there are two dependent conditions. Find at least two solutions.

SYSTEMS OF LINEAR EQUATIONS IN MORE THAN TWO VARIABLES

• SYSTEMS OF EQUATIONS IN THREE VARIABLES

An equation of the form $Ax + By + Cz = D$, with A, B, and C not all zero, is a linear equation in three variables. A solution of an equation in three variables is an **ordered triple** (x, y, z).

The ordered triple $(2, -1, -3)$ is one of the solutions of the equation $2x - 3y + z = 4$. The ordered triple $(3, 1, 1)$ is another solution. In fact, an infinite number of ordered triples are solutions of the equation.

Graphing an equation in three variables requires a third coordinate axis perpendicular to the xy-plane. This third axis is commonly called the **z-axis.** The result is a three-dimensional coordinate system called the xyz-coordinate system (**Figure 5.12**). To help visualize a three-dimensional coordinate system, think of a corner of a room: the floor is the xy-plane, one wall is the yz-plane, and the other wall is the xz-plane.

Graphing an ordered triple requires three moves, the first along the x-axis, the second along the y-axis, and the third along the z-axis. **Figure 5.13** is the graph of the points $(-5, -4, 3)$ and $(4, 5, -2)$.

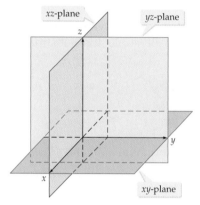

FIGURE 5.12

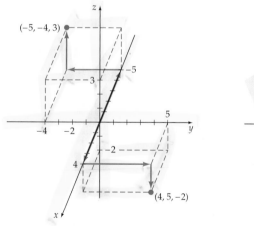

FIGURE 5.13

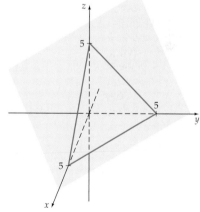

FIGURE 5.14

The graph of a linear equation in three variables is a plane. That is, if all the solutions of a linear equation in three variables were plotted in an xyz-coordinate system, the graph would look like a large, flat piece of paper with infinite extent. **Figure 5.14** is a portion of the graph of $x + y + z = 5$.

There are different ways in which three planes can be oriented in an xyz-coordinate system. **Figure 5.15** illustrates several ways.

For a linear system of equations in three variables to have a solution, the graphs of the planes must intersect at a single point, they must intersect along a common line, or all equations must have a graph that is the same plane. In **Figure 5.15,** the graphs in (a), (b), and (c) represent systems of equations that

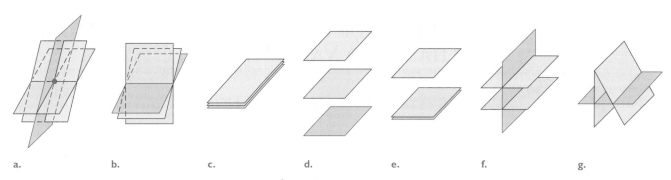

a.   b.   c.   d.   e.   f.   g.

FIGURE 5.15

have a solution. The system of equations represented in **Figure 5.15a** is a consistent system of equations. **Figures 5.15b** and **5.15c** are graphs of a dependent system of equations. The remaining graphs are examples of inconsistent systems of equations.

A system of equations in more than two variables can be solved by using the substitution method or the elimination method. To illustrate the substitution method, consider the system of equations

$$\begin{cases} x - 2y + z = 7 & (1) \\ 2x + y - z = 0 & (2) \\ 3x + 2y - 2z = -2 & (3) \end{cases}$$

Solve Equation (1) for x and substitute the result into Equations (2) and (3).

$$x = 2y - z + 7 \quad (4)$$

$$2(2y - z + 7) + y - z = 0$$ • Substitute **$2y - z + 7$ for x** in Equation (2).

$$4y - 2z + 14 + y - z = 0$$ • Simplify.

$$5y - 3z = -14 \quad (5)$$

$$3(2y - z + 7) + 2y - 2z = -2$$ • Substitute **$2y - z + 7$ for x** in Equation (3).

$$6y - 3z + 21 + 2y - 2z = -2$$ • Simplify.

$$8y - 5z = -23 \quad (6)$$

Now solve the system of equations formed from Equations (5) and (6).

$$\begin{cases} 5y - 3z = -14 \\ 8y - 5z = -23 \end{cases}$$
multiply by 8 $\rightarrow$ $40y - 24z = -112$
multiply by -5 $\rightarrow$ $\underline{-40y + 25z = 115}$

$$z = 3$$

Substitute 3 for z into Equation (5) and solve for y.

$$5y - 3z = -14 \qquad \text{• Equation (5)}$$
$$5y - 3(3) = -14$$
$$5y - 9 = -14$$
$$5y = -5$$
$$y = -1$$

Substitute -1 for y and 3 for z into Equation (4) and solve for x.

$$x = 2y - z + 7 = 2(-1) - (3) + 7 = 2$$

The ordered-triple solution is $(2, -1, 3)$. The graphs of the three planes intersect at a single point.

● TRIANGULAR FORM

There are many approaches one can take to determine the solution of a system of equations by the elimination method. For consistency, we will always follow a plan that produces an equivalent system of equations in **triangular form.** Three examples of systems of equations in triangular form are

$$\begin{cases} 2x - 3y + z = -4 \\ \quad\quad 2y + 3z = 9 \\ \quad\quad\quad\quad -2z = -2 \end{cases} \quad \begin{cases} w + 3x - 2y + 3z = 0 \\ \quad\quad 2x - y + 4z = 8 \\ \quad\quad\quad -3y - 2z = -1 \\ \quad\quad\quad\quad 3z = 9 \end{cases} \quad \begin{cases} 3x - 4y + z = 1 \\ \quad\quad 3y + 2z = 3 \end{cases}$$

Once a system of equations is written in triangular form, the solution can be found by *back substitution*—that is, by solving the last equation of the system and substituting *back* into the previous equation. This process is continued until the value of each variable has been found.

As an example of solving a system of equations by back substitution, consider the following system of equations in triangular form.

$$\begin{cases} 2x - 4y + z = -3 & (1) \\ \quad\quad 3y - 2z = 9 & (2) \\ \quad\quad\quad\quad 3z = -9 & (3) \end{cases}$$

Solve Equation (3) for z. Substitute the value of z into Equation (2) and solve for y.

$$3z = -9 \quad \text{• Equation (3)} \quad\quad 3y - 2z = 9 \quad \text{• Equation (2)}$$
$$z = -3 \quad\quad\quad\quad\quad\quad 3y - 2(-3) = 9 \quad \text{• } z = -3$$
$$\quad\quad\quad\quad\quad\quad\quad\quad\quad 3y = 3$$
$$\quad\quad\quad\quad\quad\quad\quad\quad\quad y = 1$$

Replace z by -3 and y by 1 in Equation (1) and then solve for x.

$$2x - 4y + z = -3 \quad \text{• Equation (1)}$$
$$2x - 4(1) + (-3) = -3$$
$$2x - 7 = -3$$
$$x = 2$$

The solution is the ordered triple $(2, 1, -3)$.

? **QUESTION** What is the solution of $\begin{cases} x + 2y + z = 2 \\ \quad\quad y - z = 3? \\ \quad\quad\quad z = 2 \end{cases}$

? **ANSWER** $(-10, 5, 2)$

EXAMPLE 1 **Solve an Independent System of Equations**

Solve: $\begin{cases} x + 2y - z = 1 & (1) \\ 2x - y + z = 6 & (2) \\ 2x - y - z = 0 & (3) \end{cases}$

Solution

Eliminate x from Equation (2) by multiplying Equation (1) by -2 and then adding it to Equation (2). Replace Equation (2) by the new equation.

$$\begin{array}{ll} -2x - 4y + 2z = -2 & \bullet\ -2 \text{ times Equation (1)} \\ \underline{2x - y + z = 6} & \bullet\ \text{Equation (2)} \\ -5y + 3z = 4 & \bullet\ \text{Add the equations.} \end{array}$$

$\begin{cases} x + 2y - z = 1 & (1) \\ -5y + 3z = 4 & (4) \\ 2x - y - z = 0 & (3) \end{cases}$ $\bullet$ Replace Equation (2).

Eliminate x from Equation (3) by multiplying Equation (1) by -2 and adding it to Equation (3). Replace Equation (3) by the new equation.

$$\begin{array}{ll} -2x - 4y + 2z = -2 & \bullet\ -2 \text{ times Equation (1)} \\ \underline{2x - y - z = 0} & \bullet\ \text{Equation (3)} \\ -5y + z = -2 & \bullet\ \text{Add the equations.} \end{array}$$

$\begin{cases} x + 2y - z = 1 & (1) \\ -5y + 3z = 4 & (4) \\ -5y + z = -2 & (5) \end{cases}$ $\bullet$ Replace Equation (3).

Eliminate y from Equation (5) by multiplying Equation (4) by -1 and then adding it to Equation (5). Replace Equation (5) by the new equation.

$$\begin{array}{ll} 5y - 3z = -4 & \bullet\ -1 \text{ times Equation (4)} \\ \underline{-5y + z = -2} & \bullet\ \text{Equation (5)} \\ -2z = -6 & \bullet\ \text{Add the equations.} \end{array}$$

$\begin{cases} x + 2y - z = 1 & (1) \\ -5y + 3z = 4 & (4) \\ -2z = -6 & (6) \end{cases}$ $\bullet$ Replace Equation (5).

The system of equations is now in triangular form. Solve the system of equations by back substitution.

Solve Equation (6) for z. Substitute the value into Equation (4) and then solve for y.

$$\begin{array}{ll} -2z = -6 & \bullet\ \text{Equation (6)} \\ z = 3 & \end{array} \qquad \begin{array}{ll} -5y + 3z = 4 & \bullet\ \text{Equation (4)} \\ -5y + 3(3) = 4 & \bullet\ \text{Replace z by 3.} \\ -5y = -5 & \bullet\ \text{Solve for y.} \\ y = 1 & \end{array}$$

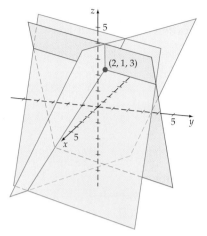

FIGURE 5.16

Replace z by 3 and y by 1 in Equation (1) and solve for x.

$$x + \ 2y - z = 1 \qquad \text{• Equation (1)}$$
$$x + 2(1) - 3 = 1 \qquad \text{• Replace y by 1; replace z by 3.}$$
$$x = 2$$

The system of equations is consistent. The solution is the ordered triple $(2, 1, 3)$. See **Figure 5.16.**

▶ **TRY EXERCISE 12, PAGE 440**

EXAMPLE 2 **Solve a Dependent System of Equations**

Solve: $\begin{cases} 2x - \ y - z = -1 & (1) \\ - \ x + 3y - z = -3 & (2) \\ -5x + 5y + z = -1 & (3) \end{cases}$

Solution

Eliminate x from Equation (2) by multiplying Equation (2) by 2 and then adding it to Equation (1). Replace Equation (2) by the new equation.

$$\begin{array}{ll} 2x - \ y - \ z = -1 & \text{• Equation (1)} \\ \underline{-2x + 6y - 2z = -6} & \text{• 2 times Equation (2)} \\ 5y - 3z = -7 & \text{• Add the equations.} \end{array}$$

$\begin{cases} 2x - \ y - \ z = -1 & (1) \\ 5y - 3z = -7 & (4) \\ -5x + 5y + \ z = -1 & (3) \end{cases}$ • Replace Equation (2).

Eliminate x from Equation (3) by multiplying Equation (1) by 5 and multiplying Equation (3) by 2. Then add. Replace Equation (3) by the new equation.

$$\begin{array}{ll} 10x - \ 5y - 5z = -5 & \text{• 5 times Equation (1)} \\ \underline{-10x + 10y + 2z = -2} & \text{• 2 times Equation (3)} \\ 5y - 3z = -7 & \text{• Add the equations.} \end{array}$$

$\begin{cases} 2x - \ y - \ z = -1 & (1) \\ 5y - 3z = -7 & (4) \\ 5y - 3z = -7 & (5) \end{cases}$ • Replace Equation (3).

Eliminate y from Equation (5) by multiplying Equation (4) by -1 and then adding it to Equation (5). Replace Equation (5) by the new equation.

$$\begin{array}{ll} -5y + 3z = \ \ 7 & \text{• } -1 \text{ times Equation (4)} \\ \underline{5y - 3z = -7} & \text{• Equation (5)} \\ 0 = \ \ 0 & \text{• Add the equations.} \end{array}$$

$\begin{cases} 2x - \ y - \ z = -1 & (1) \\ 5y - 3z = -7 & (4) \\ 0 = \ \ 0 & (6) \end{cases}$ • Replace Equation (5).

Continued ▶

Because any ordered triple (x, y, z) is a solution of Equation (6), the solutions of the system of equations will be the ordered triples that are solutions of Equations (1) and (4).

Solve Equation (4) for y.

$$5y - 3z = -7$$
$$5y = 3z - 7$$
$$y = \frac{3}{5}z - \frac{7}{5}$$

Substitute $\frac{3}{5}z - \frac{7}{5}$ for y in Equation (1) and solve for x.

$$2x - y - z = -1 \qquad \bullet \text{ Equation (1)}$$
$$2x - \left(\frac{3}{5}z - \frac{7}{5}\right) - z = -1 \qquad \bullet \text{ Replace } y \text{ by } \frac{3}{5}z - \frac{7}{5}.$$
$$2x - \frac{8}{5}z + \frac{7}{5} = -1 \qquad \bullet \text{ Simplify and solve for } x.$$
$$2x = \frac{8}{5}z - \frac{12}{5}$$
$$x = \frac{4}{5}z - \frac{6}{5}$$

By choosing any real number c for z, we have $y = \frac{3}{5}c - \frac{7}{5}$ and $x = \frac{4}{5}c - \frac{6}{5}$. For any real number c, the ordered-triple solutions of the system of equations are $\left(\frac{4}{5}c - \frac{6}{5}, \frac{3}{5}c - \frac{7}{5}, c\right)$. The solid red line shown in **Figure 5.17** is a graph of the solutions.

▶ **TRY EXERCISE 16, PAGE 440**

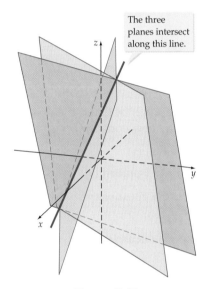

The three planes intersect along this line.

FIGURE 5.17

As in the case of a dependent system of equations in two variables, there is more than one way to represent the solutions of a dependent system of equations in three variables. For instance, from Example 2, let $a = \frac{4}{5}c - \frac{6}{5}$, the x-coordinate of the ordered triple $\left(\frac{4}{5}c - \frac{6}{5}, \frac{3}{5}c - \frac{7}{5}, c\right)$, and solve for c.

$$a = \frac{4}{5}c - \frac{6}{5} \quad \rightarrow \quad c = \frac{5}{4}a + \frac{3}{2}$$

Substitute this value of c into each component of the ordered triple.

$$\left(\frac{4}{5}\left(\frac{5}{4}a + \frac{3}{2}\right) - \frac{6}{5}, \frac{3}{5}\left(\frac{5}{4}a + \frac{3}{2}\right) - \frac{7}{5}, \frac{5}{4}a + \frac{3}{2}\right) = \left(a, \frac{3}{4}a - \frac{1}{2}, \frac{5}{4}a + \frac{3}{2}\right)$$

Thus the solutions of the system of equations can also be written as

$$\left(a, \frac{3}{4}a - \frac{1}{2}, \frac{5}{4}a + \frac{3}{2}\right)$$

> **EXAMPLE 3**   **Identify an Inconsistent System of Equations**

Solve: $\begin{cases} x + 2y + 3z = 4 & (1) \\ 2x - y - z = 3 & (2) \\ 3x + y + 2z = 5 & (3) \end{cases}$

Solution

Eliminate x from Equation (2) by multiplying Equation (1) by -2 and then adding it to Equation (2). Replace Equation (2). Eliminate x from Equation (3) by multiplying Equation (1) by -3 and adding it to Equation (3). Replace Equation (3). The equivalent system is

$\begin{cases} x + 2y + 3z = 4 & (1) \\ -5y - 7z = -5 & (4) \\ -5y - 7z = -7 & (5) \end{cases}$

Eliminate y from Equation (5) by multiplying Equation (4) by -1 and adding it to Equation (5). Replace Equation (5). The equivalent system is

$\begin{cases} x + 2y + 3z = 4 & (1) \\ -5y - 7z = -5 & (4) \\ 0 = -2 & (6) \end{cases}$

This system of equations contains a false equation. The system is inconsistent and has no solution. There is no point on all three planes, as shown in **Figure 5.18**.

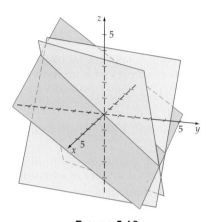

FIGURE 5.18

▶ **TRY EXERCISE 18, PAGE 440**

● NONSQUARE SYSTEMS OF EQUATIONS

The linear systems of equations that we have solved so far contain the same number of variables as equations. These are *square systems of equations*. If there are fewer equations than variables—a *nonsquare system of equations*—the system has either no solution or an infinite number of solutions.

> **EXAMPLE 4**   **Solve a Nonsquare System of Equations**

Solve: $\begin{cases} x - 2y + 2z = 3 & (1) \\ 2x - y - 2z = 15 & (2) \end{cases}$

Solution

Eliminate x from Equation (2) by multiplying Equation (1) by -2 and adding it to Equation (2). Replace Equation (2).

$\begin{cases} x - 2y + 2z = 3 & (1) \\ 3y - 6z = 9 & (3) \end{cases}$

Solve Equation (3) for y.

$$3y - 6z = 9$$
$$y = 2z + 3$$

Continued ▶

Substitute $2z + 3$ for y into Equation (1) and solve for x.

$$x - 2y + 2z = 3$$
$$x - 2(2z + 3) + 2z = 3 \qquad \bullet\ y = 2z + 3$$
$$x = 2z + 9$$

For each value of z selected, there correspond values for x and y. If z is any real number c, then the solutions of the system are the ordered triples $(2c + 9, 2c + 3, c)$.

▶ **TRY EXERCISE 20, PAGE 440**

● HOMOGENEOUS SYSTEMS OF EQUATIONS

A linear system of equations for which the constant term is zero for all equations is called a **homogeneous system of equations.** Two examples of homogeneous systems of equations are

$$\begin{cases} 3x + 4y = 0 \\ 2x + 3y = 0 \end{cases} \qquad \begin{cases} 2x - 3y + 5z = 0 \\ 3x + 2y + z = 0 \\ x - 4y + 5z = 0 \end{cases}$$

The solution $(0, 0)$ is always a solution of a homogeneous system of equations in two variables, and $(0, 0, 0)$ is always a solution of a homogeneous system of equations in three variables. This solution is called the **trivial solution.**

Sometimes a homogeneous system of equations may have solutions other than the trivial solution. For example, $(1, -1, -1)$ is a solution of the homogeneous system of three equations in three variables above.

If a homogeneous system of equations has a unique solution, the graphs intersect only at the origin. Solutions of a homogeneous system of equations can be found by using the substitution method or the elimination method.

EXAMPLE 5   **Solve a Homogeneous System of Equations**

Solve: $\begin{cases} x + 2y - 3z = 0 & (1) \\ 2x - y + z = 0 & (2) \\ 3x + y - 2z = 0 & (3) \end{cases}$

Solution

Eliminate x from Equations (2) and (3) and replace these equations by the new equations.

$$\begin{cases} x + 2y - 3z = 0 & (1) \\ -5y + 7z = 0 & (4) \\ -5y + 7z = 0 & (5) \end{cases}$$

Eliminate y from Equation (5). Replace Equation (5).

$$\begin{cases} x + 2y - 3z = 0 & (1) \\ -5y + 7z = 0 & (4) \\ 0 = 0 & (6) \end{cases}$$

Because Equation (6) is an identity, the solutions of the system are the solutions of Equations (1) and (4).

Solve Equation (4) for y.

$$y = \frac{7}{5}z$$

Substitute the expression for y into Equation (1) and solve for x.

$$x + 2y - 3z = 0 \qquad \text{• Equation (1)}$$

$$x + 2\left(\frac{7}{5}z\right) - 3z = 0 \qquad \text{• } y = \frac{7}{5}z$$

$$x = \frac{1}{5}z$$

Letting z be any real number c, we find that the solutions of the system are $\left(\frac{1}{5}c, \frac{7}{5}c, c\right)$.

▶ **TRY EXERCISE 32, PAGE 440**

● APPLICATIONS

One application of a system of equations is "curve fitting." Given a set of points in the plane, try to find an equation whose graph passes through those points, or "fits" those points.

EXAMPLE 6   **Solve an Application of a System of Equations to Curve Fitting**

Find an equation of the form $y = ax^2 + bx + c$ whose graph passes through the points whose coordinates are $(1, 4)$, $(-1, 6)$, and $(2, 9)$.

Solution

Substitute each of the given ordered pairs into the equation $y = ax^2 + bx + c$. Write the resulting system of equations.

$$\begin{cases} 4 = a(1)^2 + b(1) + c & \text{or} \\ 6 = a(-1)^2 + b(-1) + c & \text{or} \\ 9 = a(2)^2 + b(2) + c & \text{or} \end{cases} \quad \begin{cases} a + b + c = 4 & (1) \\ a - b + c = 6 & (2) \\ 4a + 2b + c = 9 & (3) \end{cases}$$

Solve the resulting system of equations for a, b, and c.

Eliminate a from Equation (2) by multiplying Equation (1) by -1 and then adding it to Equation (2). Now eliminate a from Equation (3) by multiplying Equation (1) by -4 and adding it to Equation (3). The result is

$$\begin{cases} a + b + c = 4 \\ -2b = 2 \\ -2b - 3c = -7 \end{cases}$$

Continued ▶

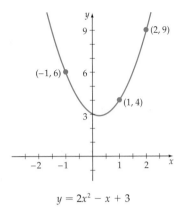

$$y = 2x^2 - x + 3$$

FIGURE 5.19

Although this system of equations is not in triangular form, we can solve the second equation for b and use this value to find a and c.

Solving by substitution, we obtain $a = 2$, $b = -1$, $c = 3$. The equation of the form $y = ax^2 + bx + c$ whose graph passes through $(1, 4)$, $(-1, 6)$, and $(2, 9)$ is $y = 2x^2 - x + 3$. See **Figure 5.19**.

▶ | TRY EXERCISE 36, PAGE 440 |

Traffic engineers use systems of equations to study the flow of traffic. The analysis of traffic flow is based on the principle that the numbers of cars that enter and leave an intersection must be equal.

EXAMPLE 7 \ **Traffic Flow**

Suppose the traffic flow for some one-way streets can be modeled by the diagram below, where the numbers and the variables represent the numbers of cars entering or leaving an intersection per hour.

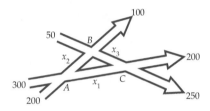

If the street connecting intersections A and C has an estimated traffic flow of between 100 and 200 cars per hour, what is the estimated traffic flow between A and B (which is x_2) and between B and C (which is x_3)?

Solution

Let x_1, x_2, and x_3 represent the numbers of cars per hour that are traveling on AC, AB, and BC, respectively. Now consider intersection A. There are $300 + 200 = 500$ cars per hour entering A and $x_1 + x_2$ cars leaving A. Therefore, $x_1 + x_2 = 500$. For intersection B, we have $50 + x_2$ cars per hour entering the intersection and $100 + x_3$ cars leaving the intersection. Thus $50 + x_2 = 100 + x_3$, or $x_2 - x_3 = 50$. Applying the same reasoning to C, we have $x_1 + x_3 = 450$. These equations result in the system of equations

$$\begin{cases} x_1 + x_2 = 500 & (1) \\ x_2 - x_3 = 50 & (2) \\ x_1 + x_3 = 450 & (3) \end{cases}$$

Subtracting Equation (2) from Equation (1) gives

$$\begin{aligned} x_1 + x_2 &= 500 & (1) \\ \underline{x_2 - x_3} &= \underline{50} & (2) \\ x_1 + x_3 &= 450 & (4) \end{aligned}$$

Subtracting Equation (4) from Equation (3) gives

$$x_1 + x_3 = 450 \qquad (3)$$
$$\underline{x_1 + x_3 = 450} \qquad (4)$$
$$0 = 0$$

This indicates that the system of equations is dependent. Because we are given that between 100 and 200 cars per hour flow between A and C (the value of x_1), we will solve each equation in terms of x_1. From Equation (1) we have $x_2 = -x_1 + 500$ and from Equation (3) we have $x_3 = -x_1 + 450$. Because $100 \le x_1 \le 200$, we have, by substituting for x_1, $300 \le x_2 \le 400$ and $250 \le x_3 \le 350$.

▶ **Try Exercise 42, page 441**

 Topics for Discussion

1. Can a system of equations contain more equations than variables? If not, explain why not. If so, give an example.

2. If a linear system of three equations in three variables is dependent, what does that mean about the graphs of the equations of the system?

3. If a linear system of three equations in three variables is inconsistent, what does that mean about the graphs of the equations of the system?

4. The equation of a circle centered at the origin with radius 5 is given by $x^2 + y^2 = 25$. Discuss the shape of $x^2 + y^2 + z^2 = 25$ in an xyz-coordinate system.

5. Consider the plane P given by $2x + 4y - 3z = 12$. The *trace* of the graph of P is obtained by letting one of the variables equal zero. For instance, the trace in the xy-plane is the graph of $2x + 4y = 12$ that is obtained by letting $z = 0$. Determine the traces of P in the xz- and yz-planes, and discuss how the traces can be used to visualize the graph of P.

Exercise Set 5.2

In Exercises 1 to 24, solve each system of equations.

1. $\begin{cases} 2x - y + z = 8 \\ 2y - 3z = -11 \\ 3y + 2z = 3 \end{cases}$

2. $\begin{cases} 3x + y + 2z = -4 \\ -3y - 2z = -5 \\ 2y + 5z = -4 \end{cases}$

3. $\begin{cases} x + 3y - 2z = 8 \\ 2x - y + z = 1 \\ 3x + 2y - 3z = 15 \end{cases}$

4. $\begin{cases} x - 2y + 3z = 5 \\ 3x - 3y + z = 9 \\ 5x + y - 3z = 3 \end{cases}$

5. $\begin{cases} 3x + 4y - z = -7 \\ x - 5y + 2z = 19 \\ 5x + y - 2z = 5 \end{cases}$

6. $\begin{cases} 2x - 3y - 2z = 12 \\ x + 4y + z = -9 \\ 4x + 2y - 3z = 6 \end{cases}$

7. $\begin{cases} 2x - 5y + 3z = -18 \\ 3x + 2y - z = -12 \\ x - 3y - 4z = -4 \end{cases}$

8. $\begin{cases} 4x - y + 2z = -1 \\ 2x + 3y - 3z = -13 \\ x + 5y + z = 7 \end{cases}$

9. $\begin{cases} x + 2y - 3z = -7 \\ 2x - y + 4z = 11 \\ 4x + 3y - 4z = -3 \end{cases}$

10. $\begin{cases} x - 3y + 2z = -11 \\ 3x + y + 4z = 4 \\ 5x - 5y + 8z = -18 \end{cases}$

11. $\begin{cases} 2x - 5y + 2z = -4 \\ 3x + 2y + 3z = 13 \\ 5x - 3y - 4z = -18 \end{cases}$

▶ **12.** $\begin{cases} 3x + 2y - 5z = 6 \\ 5x - 4y + 3z = -12 \\ 4x + 5y - 2z = 15 \end{cases}$

13. $\begin{cases} 2x + y - z = -2 \\ 3x + 2y + 3z = 21 \\ 7x + 4y + z = 17 \end{cases}$

14. $\begin{cases} 3x + y + 2z = 2 \\ 4x - 2y + z = -4 \\ 11x - 3y + 4z = -6 \end{cases}$

15. $\begin{cases} 3x - 2y + 3z = 11 \\ 2x + 3y + z = 3 \\ 5x + 14y - z = 1 \end{cases}$

▶ **16.** $\begin{cases} 2x + 3y + 2z = 14 \\ x - 3y + 4z = 4 \\ -x + 12y - 6z = 2 \end{cases}$

17. $\begin{cases} 2x - 3y + 6z = 3 \\ x + 2y - 4z = 5 \\ 3x + 4y - 8z = 7 \end{cases}$

▶ **18.** $\begin{cases} 2x + 3y - 6z = 4 \\ 3x - 2y - 9z = -7 \\ 2x + 5y - 6z = 8 \end{cases}$

19. $\begin{cases} 2x - 3y + 5z = 14 \\ x + 4y - 3z = -2 \end{cases}$

▶ **20.** $\begin{cases} x - 3y + 4z = 9 \\ 3x - 8y - 2z = 4 \end{cases}$

21. $\begin{cases} 6x - 9y + 6z = 7 \\ 4x - 6y + 4z = 9 \end{cases}$

22. $\begin{cases} 4x - 2y + 6z = 5 \\ 2x - y + 3z = 2 \end{cases}$

23. $\begin{cases} 5x + 3y + 2z = 10 \\ 3x - 4y - 4z = -5 \end{cases}$

24. $\begin{cases} 3x - 4y - 7z = -5 \\ 2x + 3y - 5z = 2 \end{cases}$

In Exercises 25 to 32, solve each homogeneous system of equations.

25. $\begin{cases} x + 3y - 4z = 0 \\ 2x + 7y + z = 0 \\ 3x - 5y - 2z = 0 \end{cases}$

26. $\begin{cases} x - 2y + 3z = 0 \\ 3x - 7y - 4z = 0 \\ 4x - 4y + z = 0 \end{cases}$

27. $\begin{cases} 2x - 3y + z = 0 \\ 2x + 4y - 3z = 0 \\ 6x - 2y - z = 0 \end{cases}$

28. $\begin{cases} 5x - 4y - 3z = 0 \\ 2x + y + 2z = 0 \\ x - 6y - 7z = 0 \end{cases}$

29. $\begin{cases} 3x - 5y + 3z = 0 \\ 2x - 3y + 4z = 0 \\ 7x - 11y + 11z = 0 \end{cases}$

30. $\begin{cases} 5x - 2y - 3z = 0 \\ 3x - y - 4z = 0 \\ 4x - y - 9z = 0 \end{cases}$

31. $\begin{cases} 4x - 7y - 2z = 0 \\ 2x + 4y + 3z = 0 \\ 3x - 2y - 5z = 0 \end{cases}$

▶ **32.** $\begin{cases} 5x + 2y + 3z = 0 \\ 3x + y - 2z = 0 \\ 4x - 7y + 5z = 0 \end{cases}$

In Exercises 33 to 44, solve each exercise by solving a system of equations.

33. **CURVE FITTING** Find an equation of the form $y = ax^2 + bx + c$ whose graph passes through the points $(2, 3)$, $(-2, 7)$, and $(1, -2)$.

34. **CURVE FITTING** Find an equation of the form $y = ax^2 + bx + c$ whose graph passes through the points $(1, -2)$, $(3, -4)$, and $(2, -2)$.

35. **CURVE FITTING** Find the equation of the circle whose graph passes through the points $(5, 3)$, $(-1, -5)$, and $(-2, 2)$. (*Hint:* Use the equation $x^2 + y^2 + ax + by + c = 0$.)

▶ **36.** **CURVE FITTING** Find the equation of the circle whose graph passes through the points $(0, 6)$, $(1, 5)$, and $(-7, -1)$. (*Hint:* See Exercise 35.)

37. **CURVE FITTING** Find the center and radius of the circle whose graph passes through the points $(-2, 10)$, $(-12, -14)$, and $(5, 3)$. (*Hint:* See Exercise 35.)

38. **CURVE FITTING** Find the center and radius of the circle whose graph passes through the points $(2, 5)$, $(-4, -3)$, and $(3, 4)$. (*Hint:* See Exercise 35.)

39. **TRAFFIC FLOW** Suppose that the traffic flow for some one-way streets can be modeled by the diagram below, where each number or variable represents the number of cars entering or leaving an intersection per hour.

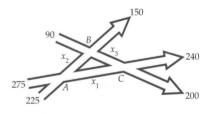

What is the minimum number of cars that can travel between A and C?

40. **TRAFFIC FLOW** A *roundabout* is a type of intersection that accommodates traffic flow in one direction, around a circular island. The graphic model on the following page shows the numbers of cars per hour that are entering or leaving a roundabout.

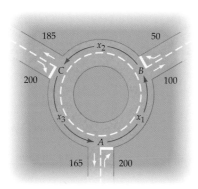

If the portion of the roundabout between A and B has an estimated traffic flow of from 60 to 80 cars per hour, what is the estimated traffic flow between C and A and between B and C?

41. TRAFFIC FLOW Suppose that the traffic flow for some one-way streets can be modeled by the diagram below, where each number or variable represents the number of cars entering or leaving an intersection per hour.

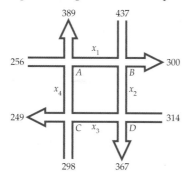

If the street connecting intersections A and B has an estimated traffic flow of from 125 to 175 cars per hour, what is the estimated traffic flow between C and A, D and C, and B and D?

▶ 42. TRAFFIC FLOW A *roundabout* is a type of intersection that accommodates traffic flow in one direction, around a circular island. The graphic model below shows the numbers of cars per hour that are entering and leaving a roundabout.

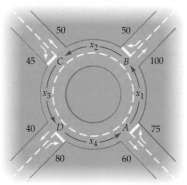

What is the minimum number of cars per hour that can travel between B and C?

43. ART A sculptor is creating a windchime consisting of three chimes that will be suspended from a rod 13 inches long. The weights, in ounces, of the chimes are shown in the diagram. For the rod to remain horizontal, the chimes must be positioned so that $w_1 d_1 + w_2 d_2 = w_3 d_3$. If the sculptor wants d_2 to be one-third of d_1, find the position of each chime so that the windchime will balance.

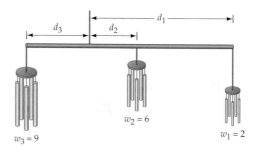

44. ART A designer wants to create a mobile of colored blocks as shown in the diagram below. The weight, in ounces, of each of the blocks is shown next to the block.

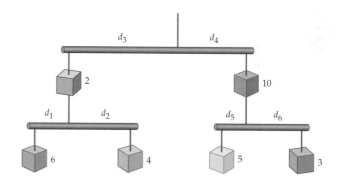

Given that $d_3 + d_4 = 20$ inches, $d_1 + d_2 = 10$ inches, and $d_5 + d_6 = 8$ inches, find the values of d_1 through d_6 so that each bar is horizontal. (A bar is horizontal when the value of weight times distance on each side of a vertical support is equal. For instance, for the above diagram, $6d_1$ must equal $4d_2$. Because there are six variables, the resulting system of equations must contain six equations.)

CONNECTING CONCEPTS

In Exercises 45 to 50, solve each system of equations.

45. $\begin{cases} 2x + y - 3z + 2w = -1 \\ 2y - 5z - 3w = 9 \\ 3y - 8z + w = -4 \\ 2y - 2z + 3w = -3 \end{cases}$

46. $\begin{cases} 3x - y + 2z - 3w = 5 \\ 2y - 5z + 2w = -7 \\ 4y - 9z + w = -19 \\ 3y + z - 2w = -12 \end{cases}$

47. $\begin{cases} x - 3y + 2z - w = 2 \\ 2x - 5y - 3z + 2w = 21 \\ 3x - 8y - 2z - 3w = 12 \\ -2x + 8y + z + 2w = -13 \end{cases}$

48. $\begin{cases} x - 2y + 3z + 2w = 8 \\ 3x - 7y - 2z + 3w = 18 \\ 2x - 5y + 2z - w = 19 \\ 4x - 8y + 3z + 2w = 29 \end{cases}$

49. $\begin{cases} x + 2y - 2z + 3w = 2 \\ 2x + 5y + 2z + 4w = 9 \\ 4x + 9y - 2z + 10w = 13 \\ -x - y + 8z - 5w = 3 \end{cases}$

50. $\begin{cases} x - 2y + 3z - 2w = -1 \\ 3x - 7y - 2z - 3w = -19 \\ 2x - 5y + 2z - w = -11 \\ -x + 3y - 2z - w = 3 \end{cases}$

In Exercises 51 and 52, use the system of equations

$$\begin{cases} x - 3y - 2z = A^2 \\ 2x - 5y + Az = 9 \\ 2x - 8y + z = 18 \end{cases}$$

51. Find all values of A for which the system has no solution.

52. Find all values of A for which the system has a unique solution.

In Exercises 53 to 55, use the system of equations

$$\begin{cases} x + 2y + z = A^2 \\ -2x - 3y + Az = 1 \\ 7x + 12y + A^2z = 4A^2 - 3 \end{cases}$$

53. Find all values of A for which the system has a unique solution.

54. Find all values of A for which the system has an infinite number of solutions.

55. Find all values of A for which the system has no solution.

56. Find an equation of the plane that contains the points $(2, 1, 1)$, $(-1, 2, 12)$, and $(3, 2, 0)$. (*Hint:* The equation of a plane can be written as $z = ax + by + c$.)

57. Find an equation of the plane that contains the points $(1, -1, 5)$, $(2, -2, 9)$, and $(-3, -1, -1)$. (*Hint:* The equation of a plane can be written as $z = ax + by + c$.)

PREPARE FOR SECTION 5.3

58. Solve $3x - 2 \le 5x + 4$ for x. [1.5]

59. Solve $x^2 + 2x - 2 = 0$ for x. [1.3]

60. Solve: $\begin{cases} x + 4y = -11 \\ 3x - 2y = 9 \end{cases}$ [5.1]

61. Graph: $y = -2x + 3$ [2.3]

62. Graph: $y = -x^2 + 3x + 4$ [2.4]

63. Graph: $y = |x| + 1$ [2.2]

PROJECTS

1. CONCEPT OF DIMENSION In this chapter we graphed first-degree equations in three variables. If we were to attempt to graph an equation in four variables, we would need a fourth axis perpendicular to the three axes of an *xyz*-coordinate system. It seems impossible to imagine a fourth dimension, but incorporating it is really a quite practical matter in mathematics. In fact, there are some systems that require an infinite-dimensional coordinate system. To gain some insight into the concept of dimension, read the book *Flatland* by Edwin A. Abbott, and then write an essay explaining what this book has to do with dimension.

2. ABILITIES OF A FOUR-DIMENSIONAL HUMAN There have been a number of attempts to describe the abilities of a four-dimensional human in a three-dimensional world. Read some of these accounts, and then write an essay on some of the actions a four-dimensional person could perform. Answer the following question in your essay. Can a four-dimensional person remove the money from a locked safe without first opening the safe?

SECTION 5.3

INEQUALITIES IN TWO VARIABLES AND SYSTEMS OF INEQUALITIES

- GRAPH AN INEQUALITY
- SYSTEMS OF INEQUALITIES IN TWO VARIABLES
- LINEAR PROGRAMMING

● GRAPH AN INEQUALITY

Two examples of inequalities in two variables are

$$2x + 3y > 6 \quad \text{and} \quad xy \leq 1$$

A solution of an inequality in two variables is an ordered pair (x, y) that satisfies the inequality. For example, $(-2, 4)$ is a solution of the first inequality because $2(-2) + 3(4) > 6$. The ordered pair $(2, 1)$ is not a solution of the second inequality because $(2)(1) \not\leq 1$.

The **solution set of an inequality** in two variables is the set of all ordered pairs that satisfy the inequality. The **graph** of an inequality is the graph of the solution set.

To sketch the graph of an inequality, first replace the inequality symbol by an equality sign and sketch the graph of the equation. Use a dashed graph for $<$ or $>$ to indicate that the curve is not part of the solution set. Use a solid graph for $\leq$ or $\geq$ to show that the graph *is* part of the solution set.

It is important to test an ordered pair in each region of the plane defined by the graph. If the ordered pair satisfies the inequality, shade that entire region. Do this for each region into which the graph divides the plane. For example, consider the inequality $xy \geq 1$. **Figure 5.20** shows the three regions of the plane defined by this inequality. Because the inequality is $\geq$, a solid graph is used.

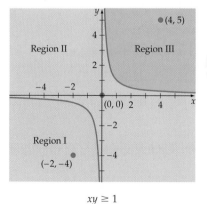

$xy \geq 1$

FIGURE 5.20

Choose an ordered pair in each of the three regions and determine whether that ordered pair satisfies the inequality. In Region I, choose a point, say, $(-2, -4)$. Because $(-2)(-4) \geq 1$, Region I is part of the solution set. In Region II, choose a point, say, $(0, 0)$. Because $0 \cdot 0 \ngeq 1$, Region II is not part of the solution set. In Region III, choose $(4, 5)$. Because $4 \cdot 5 \geq 1$, Region III is part of the solution set.

You may choose the coordinates of any point not on the graph of the equation as a test ordered pair; $(0, 0)$ is usually a good choice.

❓ QUESTION Is $(0, 0)$ a solution of $y \geq x^2 + 2x - 3$?

EXAMPLE 1   **Graph a Linear Inequality**

Graph: $3x + 4y > 12$

Solution

Graph the line $3x + 4y = 12$ using a dashed line.

$$\text{Test the ordered pair } (0, 0): \quad 3(0) + 4(0) = 0 \ngtr 12$$

Because $(0, 0)$ does not satisfy the inequality, do not shade this region.

$$\text{Test the ordered pair } (2, 3): \quad 3(2) + 4(3) = 18 > 12$$

Because $(2, 3)$ satisfies the inequality, the half-plane that includes $(2, 3)$ is the solution set. See **Figure 5.21.**

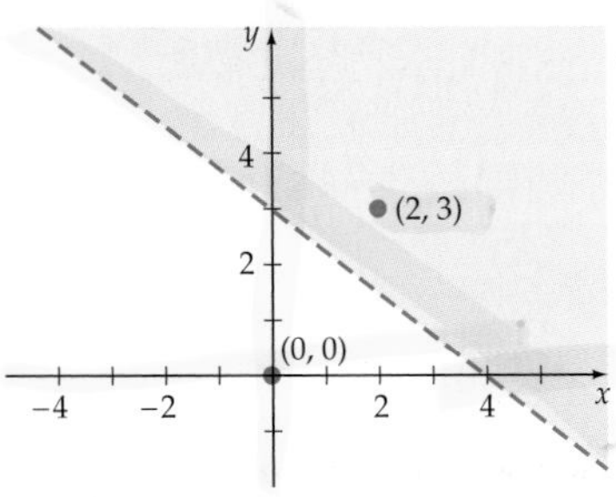

▶ **TRY EXERCISE 6, PAGE 451**

$3x + 4y > 12$

FIGURE 5.21

In general, the solution set of a *linear inequality in two variables* will be one of the regions of the plane separated by a line. Each region is called a **half-plane.**

EXAMPLE 2   **Graph a Nonlinear Inequality**

Graph: $y \leq x^2 + 2x - 3$

Solution

Graph the parabola $y = x^2 + 2x - 3$ using a solid curve.

$$\text{Test the ordered pair } (0, 0): \quad 0 \nleq 0^2 + 2(0) - 3$$

Because $(0, 0)$ does not satisfy the inequality, do not shade this region.

$$\text{Test the ordered pair } (3, 2): \quad 2 \leq (3)^2 + 2(3) - 3$$

Because $(3, 2)$ satisfies the inequality, shade this region of the plane. See **Figure 5.22.**

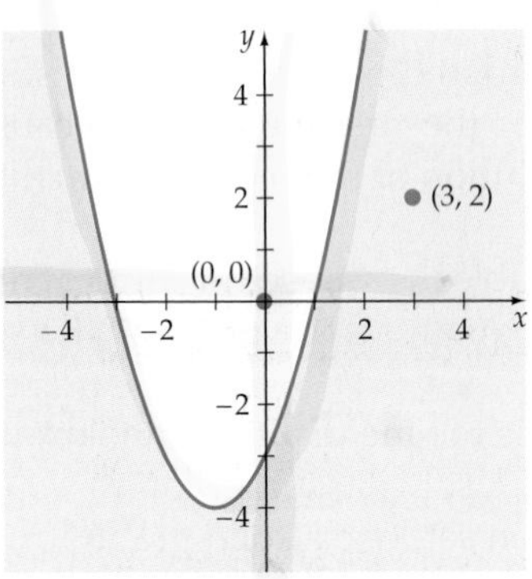

▶ **TRY EXERCISE 10, PAGE 451**

$y \leq x^2 + 2x - 3$

FIGURE 5.22

❓ ANSWER Yes.

EXAMPLE 3   **Graph an Absolute Value Inequality**

Graph: $y \geq |x| + 1$

Solution

Graph the equation $y = |x| + 1$ using a solid graph.

Test the ordered pair $(0, 0)$: $0 \not\geq |0| + 1$

Because $0 \not\geq 1$, $(0, 0)$ does not belong to the solution set. Do not shade the portion of the plane that contains $(0, 0)$.

Test the ordered pair $(0, 4)$: $4 \geq |0| + 1$

Because $(0, 4)$ satisfies the inequality, shade this region. See **Figure 5.23**.

▶ **TRY EXERCISE 12, PAGE 451**

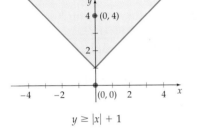

$y \geq |x| + 1$

FIGURE 5.23

● SYSTEMS OF INEQUALITIES IN TWO VARIABLES

The **solution set of a system of inequalities** is the intersection of the solution sets of the individual inequalities. To graph the solution set of a system of inequalities, first graph the solution set of each inequality. The solution set of the system of inequalities is the region of the plane represented by the intersection of the shaded regions.

EXAMPLE 4   **Graph a System of Linear Inequalities**

Graph the solution set of the system of inequalities.

$$\begin{cases} 3x - 2y > 6 \\ 2x - 5y \leq 10 \end{cases}$$

Solution

Graph the line $3x - 2y = 6$ using a dashed line. Test the ordered pair $(0, 0)$. Because $3(0) - 2(0) \not> 6$, $(0, 0)$ does not belong to the solution set. Do not shade the region that contains $(0, 0)$. Instead, shade the region below and to the right of the graph of $3x - 2y = 6$, because any ordered pair from this region satisfies $3x - 2y > 6$.

Graph the line $2x - 5y = 10$ using a solid line. Test the ordered pair $(0, 0)$. Because $2(0) - 5(0) \leq 10$, shade the region that contains $(0, 0)$.

The solution set is the region of the plane represented by the intersection of the solution sets of the individual inequalities. See **Figure 5.24**.

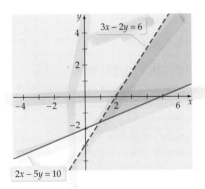

FIGURE 5.24

▶ **TRY EXERCISE 20, PAGE 451**

EXAMPLE 5 **Graph a System of Four Inequalities**

Graph the solution set of the system of inequalities.

$$\begin{cases} 2x - 3y \le 2 \\ 3x + 4y \ge 12 \\ x \ge -1, y \ge 2 \end{cases}$$

Solution

First graph the inequalities $x \ge -1$ and $y \ge 2$. Because $x \ge -1$ and $y \ge 2$, the solution set for this system will be on or above the line $y = 2$ and on or to the right of the line $x = -1$. See **Figure 5.25.**

Graph the solution set of $2x - 3y = 2$ by using a solid graph. Because $2(0) - 3(0) \le 2$, shade the region above the line.

Graph the solution set of $3x + 4y = 12$ by using a solid graph. Test an ordered pair, say $(3, 3)$, to determine that we need to shade above the line $3x + 4y = 12$.

The solution set of the system of inequalities is the region where the graphs of the solution sets of all four inequalities intersect. This intersection is indicated by the dark color in **Figure 5.26.**

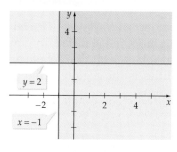

FIGURE 5.25   **FIGURE 5.26**

▶ **TRY EXERCISE 28, PAGE 451**

● LINEAR PROGRAMMING

Consider a business analyst who is trying to maximize the profit from the production of a product or an engineer who is trying to minimize the amount of energy an electrical circuit needs to operate. Generally, problems that seek to maximize or minimize a situation are called **optimization problems.** One strategy for solving certain of these problems was developed in the 1940s and is called **linear programming.**

A linear programming problem involves a **linear objective function**, which is the function that must be maximized or minimized. This objective function is subject to some **constraints,** which are inequalities or equations that restrict the values of the variables. To illustrate these concepts, suppose a manufacturer produces two types of computer monitors: LCD and plasma. Past sales experience shows that at least twice as many LCD monitors are sold as plasma monitors. Suppose further that the manufacturing plant is capable of producing 12 monitors per day. Let x represent the number of LCD monitors produced, and let y represent the number of plasma monitors produced. Then

$$\begin{cases} x \geq 2y \\ x + y \leq 12 \end{cases}$$ • These are the constraints.

These two inequalities place a constraint, or restriction, on the manufacturer. For example, the manufacturer cannot produce five plasma monitors, because that would require producing at least ten LCD monitors, and $5 + 10 \not\leq 12$.

Suppose a profit of $50 is earned on each LCD monitor sold and $75 is earned on each plasma monitor sold. Then the manufacturer's profit P, in dollars, is given by the equation

$$P = 50x + 75y$$ • Objective function

The equation $P = 50x + 75y$ defines the objective function. The goal of this linear programming problem is to determine how many of each monitor should be produced to maximize the manufacturer's profit and at the same time satisfy the constraints.

Because the manufacturer cannot produce fewer than zero units of either monitor, there are two other implied constraints, $x \geq 0$ and $y \geq 0$. Our linear programming problem now looks like

Objective function: $P = 50x + 75y$

Constraints: $$\begin{cases} x - 2y \geq 0 \\ x + y \leq 12 \\ x \geq 0, y \geq 0 \end{cases}$$

To solve this problem, graph the solution set of the constraints. The solution set of the constraints is called the **set of feasible solutions**. Ordered pairs in this set are used to evaluate the objective function to determine which ordered pair maximizes the profit. For example, (5, 2), (8, 3), and (10, 1) are three ordered pairs in the set. See **Figure 5.27**. For these ordered pairs, the profit would be

$$P = 50(5) + 75(2) \ = 400$$ • x = 5, y = 2
$$P = 50(8) + 75(3) \ = 625$$ • x = 8, y = 3
$$P = 50(10) + 75(1) = 575$$ • x = 10, y = 1

It would be impossible to check every ordered pair in the set of feasible solutions to find which maximizes profit. Fortunately, we can find that ordered pair by solving the objective function $P = 50x + 75y$ for y.

$$y = -\frac{2}{3}x + \frac{P}{75}$$

In this form, the objective function is a linear equation whose graph has slope $-\dfrac{2}{3}$ and y-intercept $\dfrac{P}{75}$. If P is as large as possible (P a maximum), then the y-intercept will be as large as possible. Thus the maximum profit will occur on the line that has a slope of $-\dfrac{2}{3}$, has the largest possible y-intercept, and intersects the set of feasible solutions.

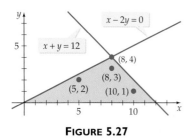

FIGURE 5.27

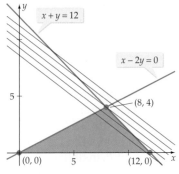

FIGURE 5.28

From **Figure 5.28,** the largest possible y-intercept occurs when the line passes through the point with coordinates $(8, 4)$. Observe that this point is a vertex of the set of feasible solutions. At this point, the profit is

$$P = 50(8) + 75(4) = 700$$

The manufacturer will maximize profit by producing eight LCD monitors and four plasma monitors each day. The profit will be $700 per day.

The following theorem states that the solution of a linear programming problem occurs at a vertex of the set of feasible solutions.

Fundamental Linear Programming Theorem

If an objective function has an optimal solution, then that solution will be at a vertex of the set of feasible solutions.

Linear programming can be used to determine the best allocation of the resources available to a company. In fact, the word *programming* refers to a "program to allocate resources."

EXAMPLE 6   Solve an Applied Minimization Problem

A manufacturer of animal food makes two grain mixtures, G_1 and G_2. Each kilogram of G_1 contains 300 grams of vitamins, 400 grams of protein, and 100 grams of carbohydrate. Each kilogram of G_2 contains 100 grams of vitamins, 300 grams of protein, and 200 grams of carbohydrate. Minimum nutritional guidelines require that a feed mixture made from these grains contain at least 900 grams of vitamins, 2200 grams of protein, and 800 grams of carbohydrate. G_1 costs $2.00 per kilogram to produce, and G_2 costs $1.25 per kilogram to produce. Find the number of kilograms of each grain mixture that should be produced to minimize cost.

Solution

Let

$$x = \text{the number of kilograms of } G_1$$
$$y = \text{the number of kilograms of } G_2$$

The objective function is the cost function $C = 2x + 1.25y$.

Because x kilograms of G_1 contain $300x$ grams of vitamins and y kilograms of G_2 contain $100y$ grams of vitamins, the total amount of vitamins contained in x kilograms of G_1 and y kilograms of G_2 is $300x + 100y$. At least 900 grams of vitamins are necessary, so $300x + 100y \geq 900$. Following similar reasoning, we have the constraints

$$\begin{cases} 300x + 100y \geq 900 \\ 400x + 300y \geq 2200 \\ 100x + 200y \geq 800 \\ x \geq 0, y \geq 0 \end{cases}$$

Two of the vertices of the set of feasible solutions (see **Figure 5.29**) can be found by solving two systems of equations. These systems are formed by the equations of the lines that intersect to form a vertex of the set of feasible solutions.

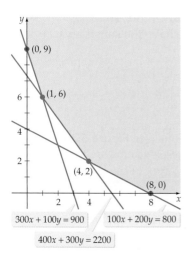

$$\begin{cases} 300x + 100y = 900 \\ 400x + 300y = 2200 \end{cases}$$

• The vertex is **(1, 6)**.

$$\begin{cases} 100x + 200y = 800 \\ 400x + 300y = 2200 \end{cases}$$

• The vertex is **(4, 2)**.

FIGURE 5.29

The vertices on the x- and y-axes are the x- and y-intercepts $(8, 0)$ and $(0, 9)$. Substitute the coordinates of the vertices into the objective function.

(x, y)	$C = 2x + 1.25y$	
$(0, 9)$	$C = 2(0) + 1.25(9) = 11.25$	
$(1, 6)$	$C = 2(1) + 1.25(6) = 9.50$	• Minimum
$(4, 2)$	$C = 2(4) + 1.25(2) = 10.50$	
$(8, 0)$	$C = 2(8) + 1.25(0) = 16.00$	

The minimum value of the objective function is $9.50. It occurs when the company produces a feed mixture that contains 1 kilogram of G_1 and 6 kilograms of G_2.

▶ **TRY EXERCISE 46, PAGE 452**

EXAMPLE 7   **Solve an Applied Maximization Problem**

A chemical firm produces two types of industrial solvents, S_1 and S_2. Each solvent is a mixture of three chemicals. Each kiloliter of S_1 requires 12 liters of chemical 1, 9 liters of chemical 2, and 30 liters of chemical 3. Each kiloliter of S_2 requires 24 liters of chemical 1, 5 liters of chemical 2, and 30 liters of chemical 3. The profit per kiloliter of S_1 is $100, and the profit per kiloliter of S_2 is $85. The inventory of the company shows 480 liters of chemical 1, 180 liters of chemical 2, and 720 liters of chemical 3. Assuming the company can sell all the solvent it makes, find the number of kiloliters of each solvent the company should make to maximize profit.

Solution

Let

$$x = \text{the number of kiloliters of } S_1$$

$$y = \text{the number of kiloliters of } S_2$$

The objective function is the profit function $P = 100x + 85y$.

Because x kiloliters of S_1 require $12x$ liters of chemical 1, and y kiloliters of S_2 require $24y$ liters of chemical 1, the total amount of chemical 1 needed

Continued ▶

is $12x + 24y$. There are 480 liters of chemical 1 in inventory, so $12x + 24y \leq 480$. Following similar reasoning, we have the constraints

$$\begin{cases} 12x + 24y \leq 480 \\ 9x + 5y \leq 180 \\ 30x + 30y \leq 720 \\ x \geq 0, y \geq 0 \end{cases}$$

Two of the vertices of the set of feasible solutions (see **Figure 5.30**) can be found by solving two systems of equations. These systems are formed by the equations of the lines that intersect to form a vertex of the set of feasible solutions.

$$\begin{cases} 12x + 24y = 480 \\ 30x + 30y = 720 \end{cases}$$ • **The vertex is (8, 16).**

$$\begin{cases} 9x + 5y = 180 \\ 30x + 30y = 720 \end{cases}$$ • **The vertex is (15, 9).**

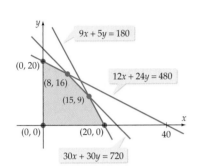

FIGURE 5.30

The vertices on the x- and y-axes are the x- and y-intercepts $(20, 0)$ and $(0, 20)$.

Substitute the coordinates of the vertices into the objective function.

$$\begin{array}{ll} (x, y) & P = 100x + 85y \\ (0, 20) & P = 100(0) + 85(20) = 1700 \\ (8, 16) & P = 100(8) + 85(16) = 2160 \\ (15, 9) & P = 100(15) + 85(9) = 2265 \qquad \bullet \text{ **Maximum**} \\ (20, 0) & P = 100(20) + 85(0) = 2000 \end{array}$$

The maximum value of the objective function is \$2265 when the company produces 15 kiloliters of S_1 and 9 kiloliters of S_2.

▶ **TRY EXERCISE 48, PAGE 452**

 ## TOPICS FOR DISCUSSION

1. Does the graph of a linear inequality in two variables represent the graph of a function? Why or why not?

2. What is a half-plane?

3. What is a constraint for a linear programming problem? Explain what type of condition might be a constraint.

4. What is the objective function for a linear programming problem?

5. What is the set of feasible solutions for a linear programming problem?

EXERCISE SET 5.3

In Exercises 1 to 16, sketch the graph of each inequality.

1. $y \le -2$

2. $x + y > -2$

3. $y \ge 2x + 3$

4. $y < -2x + 1$

5. $2x - 3y < 6$

▶ 6. $3x + 4y \le 4$

7. $4x + 3y \le 12$

8. $5x - 2y < 8$

9. $y \ge x^2 - 2x - 3$

▶ 10. $y < 2x^2 - x - 3$

11. $y \ge |2x - 4|$

▶ 12. $y < |x|$

13. $y < 2^{x-1}$

14. $y > \log_3 x$

15. $y \le \log_2 (x - 1)$

16. $y > 3^x + 1$

In Exercises 17 to 32, sketch the graph of the solution set of each system of inequalities.

17. $\begin{cases} 1 \le x < 3 \\ -2 < y \le 4 \end{cases}$

18. $\begin{cases} -2 < x < 4 \\ y \ge -1 \end{cases}$

19. $\begin{cases} 3x + 2y \ge 1 \\ x + 2y < -1 \end{cases}$

▶ 20. $\begin{cases} 2x - 5y < -6 \\ 3x + y < 8 \end{cases}$

21. $\begin{cases} 2x - y \ge -4 \\ 4x - 2y \le -17 \end{cases}$

22. $\begin{cases} 4x + 2y > 5 \\ 6x + 3y > 10 \end{cases}$

23. $\begin{cases} 4x - 3y < 14 \\ 2x + 5y \le -6 \end{cases}$

24. $\begin{cases} 3x + 5y \ge -8 \\ 2x - 3y \ge 1 \end{cases}$

25. $\begin{cases} y < 2x + 3 \\ y > 2x - 2 \end{cases}$

26. $\begin{cases} y > 3x + 1 \\ y < 3x - 2 \end{cases}$

27. $\begin{cases} 2x - 3y \ge -5 \\ x + 2y \le 7 \\ x \ge -1, y \ge 0 \end{cases}$

▶ 28. $\begin{cases} 5x + y \le 9 \\ 2x + 3y \le 14 \\ x \ge -2, y \ge 2 \end{cases}$

29. $\begin{cases} 3x + 2y \ge 14 \\ x + 3y \ge 14 \\ x \le 10, y \le 8 \end{cases}$

30. $\begin{cases} 4x + y \ge 13 \\ 3x + 2y \ge 16 \\ x \le 15, y \le 12 \end{cases}$

31. $\begin{cases} 3x + 4y \le 12 \\ 2x + 5y \le 10 \\ x \ge 0, y \ge 0 \end{cases}$

32. $\begin{cases} 5x + 3y \le 15 \\ x + 4y \le 8 \\ x \ge 0, y \ge 0 \end{cases}$

In Exercises 33 to 44, solve the linear programming problem. Assume $x \ge 0$ and $y \ge 0$.

33. Minimize $C = 4x + 2y$ with the constraints
$$\begin{cases} x + y \ge 7 \\ 4x + 3y \ge 24 \\ x \le 10, y \le 10 \end{cases}$$

34. Minimize $C = 5x + 4y$ with the constraints
$$\begin{cases} 3x + 4y \ge 32 \\ x + 4y \ge 24 \\ x \le 12, y \le 15 \end{cases}$$

35. Maximize $C = 6x + 7y$ with the constraints
$$\begin{cases} x + 2y \le 16 \\ 5x + 3y \le 45 \end{cases}$$

36. Maximize $C = 6x + 5y$ with the constraints
$$\begin{cases} 2x + 3y \le 27 \\ 7x + 3y \le 42 \end{cases}$$

37. Minimize $C = 5x + 6y$ with the constraints
$$\begin{cases} 4x - 3y \le 2 \\ 2x + 3y \ge 10 \end{cases}$$

38. Maximize $C = 4x + 5y$ with the constraints
$$\begin{cases} 2x - y \le 0 \\ 0 \le y \le 10 \\ 0 \le x \le 10 \end{cases}$$

39. Maximize $C = 2x + 7y$ with the constraints
$$\begin{cases} x + y \le 10 \\ x + 2y \le 16 \\ 2x + y \le 16 \end{cases}$$

40. Minimize $C = 4x + 3y$ with the constraints
$$\begin{cases} 2x + y \ge 8 \\ 2x + 3y \ge 16 \\ x + 3y \ge 11 \\ x \le 20, y \le 20 \end{cases}$$

41. Minimize $C = 3x + 2y$ with the constraints
$$\begin{cases} 3x + y \ge 12 \\ 2x + 7y \ge 21 \\ x + y \ge 8 \end{cases}$$

42. Maximize $C = 2x + 6y$ with the constraints

$$\begin{cases} x + y \le 12 \\ 3x + 4y \le 40 \\ x + 2y \le 18 \end{cases}$$

43. Maximize $C = 3x + 4y$ with the constraints

$$\begin{cases} 2x + y \le 10 \\ 2x + 3y \le 18 \\ x - y \le 2 \end{cases}$$

44. Minimize $C = 3x + 7y$ with the constraints

$$\begin{cases} x + y \ge 9 \\ 3x + 4y \ge 32 \\ x + 2y \ge 12 \end{cases}$$

45. MAXIMIZE PROFIT A farmer is planning to raise wheat and barley. Each acre of wheat yields a profit of $50, and each acre of barley yields a profit of $70. To sow the crop, two machines, a tractor and a tiller, are rented. The tractor is available for 200 hours, and the tiller is available for 100 hours. Sowing an acre of barley requires 3 hours of tractor time and 2 hours of tilling. Sowing an acre of wheat requires 4 hours of tractor time and 1 hour of tilling. How many acres of each crop should be planted to maximize the farmer's profit?

▶ **46. MINIMIZE COST** An ice cream supplier has two machines that produce vanilla and chocolate ice cream. To meet one of its contractual obligations, the company must produce at least 60 gallons of vanilla ice cream and 100 gallons of

chocolate ice cream per day. One machine makes 4 gallons of vanilla and 5 gallons of chocolate ice cream per hour. The second machine makes 3 gallons of vanilla and 10 gallons of chocolate ice cream per hour. It costs $28 per hour to run machine 1 and $25 per hour to run machine 2. How many hours should each machine be operated to fulfill the contract at the least expense?

47. MAXIMIZE PROFIT A manufacturer makes two types of golf clubs: a starter model and a professional model. The starter model requires 4 hours in the assembly room and 1 hour in the finishing room. The professional model requires 6 hours in the assembly room and 1 hour in the finishing room. The total number of hours available in the assembly room is 108. There are 24 hours available in the finishing room. The profit for each starter model is $35, and the profit for each professional model is $55. Assuming all the sets produced can be sold, find how many of each set should be manufactured to maximize profit.

▶ **48. MAXIMIZE PROFIT** A company makes two types of telephone answering machines: the standard model and the deluxe model. Each machine passes through three processes: P_1, P_2, and P_3. One standard answering machine requires 1 hour in P_1, 1 hour in P_2, and 2 hours in P_3. One deluxe answering machine requires 3 hours in P_1, 1 hour in P_2, and 1 hour in P_3. Because of employee work schedules, P_1 is available for 24 hours, P_2 is available for 10 hours, and P_3 is available for 16 hours. If the profit is $25 for each standard model and $35 for each deluxe model, how many units of each type should the company produce to maximize profit?

─── *CONNECTING CONCEPTS* ───

In Exercises 49 to 54, sketch the graph of the inequality.

49. $|y| \ge |x|$

50. $|y| \le |x - 1|$

51. $|x + y| \le 1$

52. $|x - y| > 1$

53. $|x| + |y| \le 1$

54. $|x| - |y| > 1$

55. ✎ Sketch the graphs of $xy > 1$ and $y > \dfrac{1}{x}$. Note that the two graphs are not the same, yet the second inequality can be derived from the first by dividing each side by x. Explain.

56. ✎ Sketch the graph of $\dfrac{x}{y} < 1$ and the graph of $x < y$. Note that the two graphs are not the same, yet the second inequality can be derived from the first by multiplying each side by y. Explain.

57. MINIMIZE COST A dietitian formulates a special diet from two food groups: A and B. Each ounce of food group A contains 3 units of vitamin A, 1 unit of vitamin C, and 1 unit of vitamin D. Each ounce of food group B contains 1 unit of vitamin A, 1 unit of vitamin C, and 3 units of vitamin D. Each ounce of food group A costs 40 cents, and each ounce of food group B costs 10 cents. The dietary constraints are such that at least 24 units of vitamin A, 16 units of vitamin C, and 30 units of vitamin D are required. Find the amount of each food group that should be used to minimize the cost. What is the minimum cost?

58. MAXIMIZE PROFIT Among the many products it produces, an oil refinery makes two specialized petroleum distillates: Pymex A and Pymex B. Each distillate passes through three stages: S_1, S_2, and S_3. Each liter of Pymex A requires 1 hour in S_1, 3 hours in S_2, and 3 hours in S_3. Each liter of Pymex B requires 1 hour in S_1, 4 hours in S_2, and 2 hours in S_3. There are 10 hours available for S_1, 36 hours available for S_2, and 27 hours available for S_3. The profit per liter of Pymex A is \$12, and the profit per liter of Pymex B is \$9. How many liters of each distillate should be produced to maximize profit? What is the maximum profit?

59. MAXIMIZE PROFIT An engine reconditioning company works on 4- and 6-cylinder engines. Each 4-cylinder engine requires 1 hour for cleaning, 5 hours for overhauling, and 3 hours for testing. Each 6-cylinder engine requires 1 hour for cleaning, 10 hours for overhauling, and 2 hours for testing. The cleaning station is available for at most 9 hours. The overhauling equipment is available for at most 80 hours, and the testing equipment is available for at most 24 hours. For each reconditioned 4-cylinder engine, the company makes a profit of \$150. A reconditioned 6-cylinder engine yields a profit of \$250. The company can sell all the reconditioned engines it produces. How many of each type should be produced to maximize profit? What is the maximum profit?

60. MINIMIZE COST A producer of animal feed makes two food products: F_1 and F_2. The products contain three major ingredients: M_1, M_2, and M_3. Each ton of F_1 requires 200 pounds of M_1, 100 pounds of M_2, and 100 pounds of M_3. Each ton of F_2 requires 100 pounds of M_1, 200 pounds of M_2, and 400 pounds of M_3. There are at least 5000 pounds of M_1 available, at least 7000 pounds of M_2 available, and at least 10,000 pounds of M_3 available. Each ton of F_1 costs \$450 to make, and each ton of F_2 costs \$300 to make. How many tons of each food product should the feed producer make to minimize cost? What is the minimum cost?

PROJECTS

1.  **A PARALLELOGRAM COORDINATE SYSTEM** The xy-coordinate system described in this chapter consisted of two coordinate lines that intersected at right angles. It is not necessary that coordinate lines intersect at right angles for a coordinate system to exist. Draw two coordinate axes that intersect at 0 but for which the angle between the two axes is 45°. You now have a *parallelogram* coordinate system rather than a *rectangular* coordinate system. Explain the last sentence. Now experiment in this system. For example, is the graph of $3x + 4y = 12$ a straight line in the *parallelogram* coordinate system? In a parallelogram coordinate system, is the graph of $y = x^2$ a parabola?

EXPLORING CONCEPTS WITH TECHNOLOGY

Ill-Conditioned Systems of Equations

Solving systems of equations algebraically as we did in this chapter is not practical for systems of equations that contain a large number of variables. In such cases, a computer solution is the only hope. Computer solutions are not without some problems, however.

Consider the system of equations

$$\begin{cases} 0.24567x + 0.49133y = 0.73700 \\ 0.84312x + 1.68623y = 2.52935 \end{cases}$$

It is easy to verify that the solution of this system of equations is $(1, 1)$. However, change the constant 0.73700 to 0.73701 (add 0.00001) and the constant 2.52935 to 2.52936 (add 0.00001), and the solution is now $(3, 0)$. Thus a very small change in the constant terms produces a dramatic change in the solution. A system of equations of this sort is said to be *ill-conditioned*.

These types of systems are important because computers generally cannot store numbers beyond a certain number of significant digits. Your calculator, for example, probably allows you to enter no more than 10 significant digits. If an exact number cannot be entered, then an approximation to that number is necessary. When a computer is solving an equation or a system of equations, the hope is that approximations of the coefficients it uses will give reasonable approximations to the solutions. For ill-conditioned systems of equations, this is not always true.

In the preceding system of equations, small changes in the constant terms caused a large change in the solution. It is possible that small changes in the coefficients of the variables will also cause large changes in the solution.

In the two systems of equations that follow, examine the effects of approximating the fractional coefficients on the solutions. Try approximating each fraction to the nearest hundredth, to the nearest thousandth, to the nearest ten-thousandth, and then to the limits of your calculator. The exact solution of the first system of equations is $(27, -192, 210)$. The exact solution of the second system of equations is $(-64, 900, -2520, 1820)$.

$$\begin{cases} x + \dfrac{1}{2}y + \dfrac{1}{3}z = 1 \\[2mm] \dfrac{1}{2}x + \dfrac{1}{3}y + \dfrac{1}{4}z = 2 \\[2mm] \dfrac{1}{3}x + \dfrac{1}{4}y + \dfrac{1}{5}z = 3 \end{cases} \qquad \begin{cases} x + \dfrac{1}{2}y + \dfrac{1}{3}z + \dfrac{1}{4}w = 1 \\[2mm] \dfrac{1}{2}x + \dfrac{1}{3}y + \dfrac{1}{4}z + \dfrac{1}{5}w = 2 \\[2mm] \dfrac{1}{3}x + \dfrac{1}{4}y + \dfrac{1}{5}z + \dfrac{1}{6}w = 3 \\[2mm] \dfrac{1}{4}x + \dfrac{1}{5}y + \dfrac{1}{6}z + \dfrac{1}{7}w = 4 \end{cases}$$

Note how the solutions change as the approximations change and thus how important it is to know whether a system of equations is ill-conditioned. For systems that are not ill-conditioned, approximations of the coefficients yield reasonable approximations of the solution. For ill-conditioned systems of equations, this is not always true.

CHAPTER 5 SUMMARY

5.1 Systems of Linear Equations in Two Variables

- A system of equations is two or more equations considered together. A solution of a system of equations in two variables is an ordered pair that satisfies each equation of the system. Equivalent systems of equations have the same solution set.

- A system of equations is consistent if it has one or more solutions. A system of linear equations is independent if it has exactly one solution. A system is dependent if it has infinitely many solutions. An inconsistent system of equations has no solution.

- **Operations That Produce Equivalent Systems of Equations**

  1. Interchange any two equations.

  2. Replace an equation with a nonzero multiple of that equation.

  3. Replace an equation with the sum of that equation and a nonzero constant multiple of another equation in the system.

5.2 Systems of Linear Equations in More Than Two Variables

- An equation of the form $ax + by + cz = d$, with a, b, and c not all zero, is a linear equation in three variables. A solution of a system of equations in three variables is an ordered triple that satisfies each equation of the system.

- The graph of a linear equation in three variables is a plane.

- A linear system of equations for which the constant term is zero for all equations of the system is called a homogeneous system of equations.

5.3 Inequalities in Two Variables and Systems of Inequalities

- The graph of an inequality in two variables frequently separates the plane into two or more regions.

- The solution set of a system of inequalities is the intersection of the solution sets of the individual inequalities.

- A linear programming problem consists of a linear objective function and a number of constraints, which are inequalities or equations that restrict the values of the variables.

- The Fundamental Linear Programming Theorem states that if an objective function has an optimal solution, then that solution will be at a vertex of the set of feasible solutions.

CHAPTER 5 TRUE/FALSE EXERCISES

In Exercises 1 to 9, answer true or false. If the statement is false, give an example or state a reason to show that the statement is false.

1. A system of equations will always have a solution as long as the number of equations is equal to the number of variables.

2. A system of two different quadratic equations can have at most four solutions.

3. A homogeneous system of equations is one in which all the variables have the same exponent.

4. In an xyz-coordinate system, the graph of the set of points formed by the intersection of two different planes is a straight line.

5. Two systems of equations with the same solution set have the same equations in their respective systems.

6. The systems of equations
$$\begin{cases} x = 0 \\ y = 0 \end{cases} \text{ and } \begin{cases} y = x \\ y = -x \end{cases}$$
are equivalent systems of equations.

7. For a linear programming problem, one or more constraints are used to define the set of feasible solutions.

8. A system of three linear equations in three variables for which two of the planes are parallel and the third plane intersects the first two is a dependent system of equations.

9. The inequality $xy < 1$ and the inequality $y < \dfrac{1}{x}$ are equivalent inequalities.

CHAPTER 5 REVIEW EXERCISES

In Exercises 1 to 18, solve each system of equations.

1. $\begin{cases} 2x - 4y = -3 \\ 3x + 8y = -12 \end{cases}$

2. $\begin{cases} 4x - 3y = 15 \\ 2x + 5y = -12 \end{cases}$

3. $\begin{cases} 3x - 4y = -5 \\ y = \dfrac{2}{3}x + 1 \end{cases}$

4. $\begin{cases} 7x + 2y = -14 \\ y = -\dfrac{5}{2}x - 3 \end{cases}$

5. $\begin{cases} y = 2x - 5 \\ x = 4y - 1 \end{cases}$

6. $\begin{cases} y = 3x + 4 \\ x = 4y - 5 \end{cases}$

7. $\begin{cases} 6x + 9y = 15 \\ 10x + 15y = 25 \end{cases}$

8. $\begin{cases} 4x - 8y = 9 \\ 2x - 4y = 5 \end{cases}$

9. $\begin{cases} 2x - 3y + z = -9 \\ 2x + 5y - 2z = 18 \\ 4x - y + 3z = -4 \end{cases}$

10. $\begin{cases} x - 3y + 5z = 1 \\ 2x + 3y - 5z = 15 \\ 3x + 6y + 5z = 15 \end{cases}$

11. $\begin{cases} x + 3y - 5z = -12 \\ 3x - 2y + z = 7 \\ 5x + 4y - 9z = -17 \end{cases}$ **12.** $\begin{cases} 2x - y + 2z = 5 \\ x + 3y - 3z = 2 \\ 5x - 9y + 8z = 13 \end{cases}$

13. $\begin{cases} 3x + 4y - 6z = 10 \\ 2x + 2y - 3z = 6 \\ x - 6y + 9z = -4 \end{cases}$ **14.** $\begin{cases} x - 6y + 4z = 6 \\ 4x + 3y - 4z = 1 \\ 5x - 9y + 8z = 13 \end{cases}$

15. $\begin{cases} 2x + 3y - 2z = 0 \\ 3x - y - 4z = 0 \\ 5x + 13y - 4z = 0 \end{cases}$ **16.** $\begin{cases} 3x - 5y + z = 0 \\ x + 4y - 3z = 0 \\ 2x + y - 2z = 0 \end{cases}$

17. $\begin{cases} x - 2y + z = 1 \\ 3x + 2y - 3z = 1 \end{cases}$ **18.** $\begin{cases} 2x - 3y + z = 1 \\ 4x + 2y + 3z = 21 \end{cases}$

In Exercises 19 to 22, graph the solution set of each inequality.

19. $4x - 5y < 20$ **20.** $2x + 7y \geq -14$

21. $y \geq 2x^2 - x - 1$ **22.** $y < x^2 - 5x - 6$

In Exercises 23 to 30, graph the solution set of each system of inequalities.

23. $\begin{cases} 2x - 5y < 9 \\ 3x + 4y \geq 2 \end{cases}$ **24.** $\begin{cases} 3x + y > 7 \\ 2x + 5y < 9 \end{cases}$

25. $\begin{cases} 2x + 3y > 6 \\ 2x - y > -2 \\ x \leq 4 \end{cases}$ **26.** $\begin{cases} 2x + 5y > 10 \\ x - y > -2 \\ x \leq 4 \end{cases}$

27. $\begin{cases} 2x + 3y \leq 18 \\ x + y \leq 7 \\ x \geq 0, y \geq 0 \end{cases}$ **28.** $\begin{cases} 3x + 5y \geq 25 \\ 2x + 3y \geq 16 \\ x \geq 0, y \geq 0 \end{cases}$

29. $\begin{cases} 3x + y \geq 6 \\ x + 4y \geq 14 \\ 2x + 3y \geq 16 \\ x \geq 0, y \geq 0 \end{cases}$ **30.** $\begin{cases} 3x + 2y \geq 14 \\ x + y \geq 6 \\ 11x + 4y \leq 48 \\ x \geq 0, y \geq 0 \end{cases}$

In Exercises 31 to 36, solve the linear programming problem. In each problem, assume $x \geq 0$ and $y \geq 0$.

31. Objective function: $P = 2x + 2y$
Constraints: $\begin{cases} x + 2y \leq 14 \\ 5x + 2y \leq 30 \end{cases}$
Maximize the objective function.

32. Objective function: $P = 4x + 5y$
Constraints: $\begin{cases} 2x + 3y \leq 24 \\ 4x + 3y \leq 36 \end{cases}$
Maximize the objective function.

33. Objective function: $P = 4x + y$
Constraints: $\begin{cases} 5x + 2y \geq 16 \\ x + 2y \geq 8 \\ x \leq 20, y \leq 20 \end{cases}$
Minimize the objective function.

34. Objective function: $P = 2x + 7y$
Constraints: $\begin{cases} 4x + 3y \geq 24 \\ 4x + 7y \geq 40 \\ x \leq 10, y \leq 10 \end{cases}$
Minimize the objective function.

35. Objective function: $P = 6x + 3y$
Constraints: $\begin{cases} 5x + 2y \geq 20 \\ x + y \geq 7 \\ x + 2y \geq 10 \\ x \leq 15, y \leq 15 \end{cases}$
Minimize the objective function.

36. Objective function: $P = 5x + 4y$
Constraints: $\begin{cases} x + y \leq 10 \\ 2x + y \leq 13 \\ 3x + y \leq 18 \end{cases}$
Maximize the objective function.

In Exercises 37 to 43, solve each exercise by solving a system of equations.

37. Find an equation of the form $y = ax^2 + bx + c$ whose graph passes through the points $(1, 0)$, $(-1, 5)$, and $(2, 3)$.

38. Find an equation of the circle that passes through the points $(4, 2)$, $(0, 1)$, and $(3, -1)$.

39. Find an equation of the plane that passes through the points $(2, 1, 2)$, $(3, 1, 0)$, and $(-2, -3, -2)$. Use the equation $z = ax + by + c$.

40. How many liters of a 20% acid solution should be mixed with 10 liters of a 10% acid solution so that the resulting solution is a 16% acid solution?

41. Flying with the wind, a small plane traveled 855 miles in 5 hours. Flying against the wind, the same plane traveled 575 miles in the same time. Find the rate of the wind and the rate of the plane in calm air.

42. A collection of 10 coins has a value of $1.25. The collection consists of only nickels, dimes, and quarters. How many of each coin are in the collection? (*Hint:* There is more than one solution.)

43. Consider the ordered triple (a, b, c). Find all real number values for a, b, and c so that the product of any two numbers equals the remaining number.

CHAPTER 5 TEST

In Exercises 1 to 6, solve each system of equations. If a system of equations is inconsistent, so state.

1. $\begin{cases} 3x + 2y = -5 \\ 2x - 5y = -16 \end{cases}$

2. $\begin{cases} x - \dfrac{1}{2}y = 3 \\ 2x - y = 6 \end{cases}$

3. $\begin{cases} x + 3y - z = 8 \\ 2x - 7y + 2z = 1 \\ 4x - y + 3z = 13 \end{cases}$

4. $\begin{cases} 3x - 2y + z = 2 \\ x + 2y - 2z = 1 \\ 4x - z = 3 \end{cases}$

5. $\begin{cases} 2x - 3y + z = -1 \\ x + 5y - 2z = 5 \end{cases}$

6. $\begin{cases} 4x + 2y + z = 0 \\ x - 3y - 2z = 0 \\ 3x + 5y + 3z = 0 \end{cases}$

In Exercises 7 to 10, graph each inequality.

7. $y \le \dfrac{1}{2}x - 3$

8. $3x - 4y > 8$

9. $y \le |x - 1| - 1$

10. $y \le x^2 - 2x - 3$

In Exercises 11 to 13, graph each system of inequalities. If the solution set is empty, so state.

11. $\begin{cases} 2x - 5y \le 16 \\ x + 3y \ge -3 \end{cases}$

12. $\begin{cases} x + y \ge 8 \\ 2x + y \ge 11 \\ x \ge 0, y \ge 0 \end{cases}$

13. $\begin{cases} 2x + 3y \le 12 \\ x + y \le 5 \\ 3x + 2y \le 11 \\ x \ge 0, y \ge 0 \end{cases}$

14. A farmer has 160 acres available on which to plant oats and barley. It costs $15 per acre for oat seed and $13 per acre for barley seed. The labor cost is $15 per acre for oats and $20 per acre for barley. The farmer has $2200 available to purchase seed and has set aside $2600 for labor. The profit per acre for oats is $120, and the profit per acre for barley is $150. How many acres of oats should the farmer plant to maximize profit?

15. Find an equation of the circle that passes through the points $(3, 5)$, $(-3, -3)$, and $(4, 4)$. (*Hint:* Use $x^2 + y^2 + ax + by + c = 0$.)

CUMULATIVE REVIEW EXERCISES

1. Find the slope of the line that passes through the points $\left(-\dfrac{1}{2}, 2\right)$ and $\left(4, -\dfrac{1}{3}\right)$.

2. Find the range of $f(x) = -x^2 + 2x - 4$.

3. Evaluate $3x^4 - 4x^3 + 2x^2 - x + 1$ for $x = -2$.

4. Write $\log_6(x - 5) + 3\log_6(2x)$ as a single logarithm with a coefficient of 1.

5. Given that -3 is a zero of $f(x) = x^3 + 2x^2 - 5x - 6$, find the remaining zeros of f.

6. Solve $\dfrac{1}{F} = \dfrac{1}{d_0} + \dfrac{1}{d_1}$ for d_0.

7. Find the equation of the line that passes through $P_1(-4, 2)$ and $P_2(2, -1)$.

8. Let $f(x) = \dfrac{x^2 - 1}{x^4}$. Is f an even function, an odd function, or neither?

9. Solve: $\log x - \log(2x - 3) = 2$

10. Let $f(x) = x^2 - x + 1$. Write the difference quotient $\dfrac{f(2 + h) - f(2)}{h}$ in simplest form.

11. Given $g(x) = \dfrac{x - 2}{x}$, find $g\left(-\dfrac{1}{2}\right)$.

12. Given $f(x) = x^2 - 1$ and $g(x) = x^2 - 4x - 2$, find $(f \cdot g)(-2)$.

13. Evaluate: $\log_{0.25} 0.015625$

14. Given $f(x) = 1 - 2x$ and $g(x) = x^2 - 1$, write $(f \circ g)(x)$ in simplest form.

15. Find the polynomial of lowest degree that has zeros of -2, $3i$, and $-3i$.

16. Find the inverse function of $Q(r) = \dfrac{2}{1-r}$.

17. Find the slant asymptote of the graph of $H(x) = \dfrac{2x^3 - x^2 - 2}{x^2 - x - 1}$.

18. Given that $f(x) = 2^x$ and $g(x) = 3^{2x}$, find $g[f(1)]$.

19. Sketch the graph of $F(x) = \dfrac{2^x - 2^{-x}}{3}$.

20. How long will it take $2000 to double if it is invested at an annual interest rate of 6.5% compounded continuously? Round to the nearest year.

SOLUTIONS
TO THE TRY EXERCISES

Exercise Set P.1, page 15

2. a. Integers: 31, 51

    b. Rational numbers: $\dfrac{5}{7}$, 31, $-2\dfrac{1}{2}$, 4.235653907493, 51,

       0.888...

    c. Irrational number: $\dfrac{5}{\sqrt{7}}$

    d. Prime number: 31

    e. Real numbers: All the numbers are real numbers.

6. In absolute value, the four smallest integers are 0, 1, 2, and 3. Replacing x in $x^2 - 1$ by these values, we obtain $\{-1, 0, 3, 8\}$.

16. $A \cap B = \{-2, 0, 2\}$ and $A \cap C = \{0, 1, 2, 3\}$. Therefore, $(A \cap B) \cup (A \cap C) = \{-2, 0, 1, 2, 3\}$.

48. $d(z, 5) = |z - 5|$; therefore $|z - 5| > 7$.

54. The interval $(-\infty, 3]$ includes all real numbers from $-\infty$ to 3, including 3. The interval $(2, 6)$ includes all real numbers between 2 and 6, not including 2 and not including 6. Therefore, $(-\infty, 3] \cap (2, 6) = (2, 3]$. The graph is

    $\underset{-5\,-4\,-3\,-2\,-1\ \ 0\ \ 1\ \ 2\ \ 3\ \ 4\ \ 5}{\longleftrightarrow}$

64. $\{x \mid -3 \le x < 0\} \cup \{x \mid x \ge 2\}$ is the set of all real numbers between -3 and 0, including -3 but excluding 0, together with (union) all real numbers greater than or equal to 2. The graph is

    $\underset{-5\,-4\,-3\,-2\,-1\ \ 0\ \ 1\ \ 2\ \ 3\ \ 4\ \ 5}{\longleftrightarrow}$

74. $(z - 2y)^2 - 3z^3$

    $[(-1) - 2(-2)]^2 - 3(-1)^3 = [-1 + 4]^2 - 3(-1)$

                            $= 3^2 + 3$

                            $= 9 + 3 = 12$

86. Commutative property of addition

90. Substitution

106. $6 + 3[2x - 4(3x - 2)] = 6 + 3[2x - 12x + 8]$

                           $= 6 + 3[-10x + 8]$

                           $= 6 - 30x + 24 = -30x + 30$

Exercise Set P.2, page 31

10. $\dfrac{4^{-2}}{2^{-3}} = \dfrac{2^3}{4^2} = \dfrac{8}{16} = \dfrac{1}{2}$

30. $\dfrac{(-3a^2b^3)^2}{(-2ab^4)^3} = \dfrac{(-3)^{1\cdot2}a^{2\cdot2}b^{3\cdot2}}{(-2)^{1\cdot3}a^{1\cdot3}b^{4\cdot3}}$

                  $= \dfrac{9a^4b^6}{-8a^3b^{12}} = -\dfrac{9a}{8b^6}$

46. $\dfrac{(6.9 \times 10^{27})(8.2 \times 10^{-13})}{4.1 \times 10^{15}} = \dfrac{(6.9)(8.2) \times 10^{27-13}}{4.1 \times 10^{15}}$

                                $= \dfrac{56.58 \times 10^{14}}{4.1 \times 10^{15}}$

                                $= 13.8 \times 10^{-1} = 1.38$

62. $(-5x^{1/3})(-4x^{1/2}) = (-5)(-4)x^{1/3+1/2}$

                         $= 20x^{2/6+3/6} = 20x^{5/6}$

78. $\sqrt{18x^2y^5} = \sqrt{9x^2y^4}\,\sqrt{2y} = 3|x|y^2\sqrt{2y}$

86. $-3x\sqrt[3]{54x^4} + 2\sqrt[3]{16x^7} = -3x\sqrt[3]{3^3 \cdot 2x^4} + 2\sqrt[3]{2^4x^7}$

                            $= -3x\sqrt[3]{3^3x^3}\,\sqrt[3]{2x} + 2\sqrt[3]{2^3x^6}\,\sqrt[3]{2x}$

                            $= -3x(3x\sqrt[3]{2x}) + 2(2x^2\sqrt[3]{2x})$

                            $= -9x^2\sqrt[3]{2x} + 4x^2\sqrt[3]{2x}$

                            $= -5x^2\sqrt[3]{2x}$

96. $(3\sqrt{5y} - 4)^2 = (3\sqrt{5y} - 4)(3\sqrt{5y} - 4)$

                      $= 9 \cdot 5y - 12\sqrt{5y} - 12\sqrt{5y} + 16$

                      $= 45y - 24\sqrt{5y} + 16$

106. $\dfrac{2}{\sqrt[4]{4y}} = \dfrac{2}{\sqrt[4]{4y}} \cdot \dfrac{\sqrt[4]{4y^3}}{\sqrt[4]{4y^3}} = \dfrac{2\sqrt[4]{4y^3}}{2y} = \dfrac{\sqrt[4]{4y^3}}{y}$

110. $-\dfrac{7}{3\sqrt{2} - 5} = -\dfrac{7}{3\sqrt{2} - 5} \cdot \dfrac{3\sqrt{2} + 5}{3\sqrt{2} + 5}$

                  $= \dfrac{-21\sqrt{2} - 35}{18 - 25}$

                  $= \dfrac{-21\sqrt{2} - 35}{-7} = 3\sqrt{2} + 5$

Exercise Set P.3, page 46

22. $(5y^2 - 7y + 3) + (2y^2 + 8y + 1) = 7y^2 + y + 4$

26.

$$
\begin{array}{r}
3x^2 - 8x - 5 \\
5x - 7 \\
\hline
-21x^2 + 56x + 35 \\
15x^3 - 40x^2 - 25x \quad\quad \\
\hline
15x^3 - 61x^2 + 31x + 35
\end{array}
$$

46. $(4x^2 - 3y)(4x^2 + 3y) = (4x^2)^2 - (3y)^2 = 16x^4 - 9y^2$

56. $6a^3b^2 - 12a^2b + 72ab^3 = 6ab(a^2b - 2a + 12b^2)$

62. $b^2 + 12b - 28 = (b + 14)(b - 2)$

66. $57y^2 + y - 6 = (19y - 6)(3y + 1)$

78. $8x^6 - 10x^3 - 3 = (4x^3 + 1)(2x^3 - 3)$

82. $81b^2 - 16c^2 = (9b - 4c)(9b + 4c)$

88. $b^2 - 24b + 144 = (b - 12)^2$

94. $b^3 + 64 = (b + 4)(b^2 - 4b + 16)$

102. $a^2y^2 - ay^3 + ac - cy = ay^2(a - y) + c(a - y)$
$$= (a - y)(ay^2 + c)$$

Exercise Set P.4, page 56

2. $\dfrac{2x^2 - 5x - 12}{2x^2 + 5x + 3} = \dfrac{(2x + 3)(x - 4)}{(2x + 3)(x + 1)} = \dfrac{x - 4}{x + 1}$

16. $\dfrac{x^2 - 16}{x^2 + 7x + 12} \cdot \dfrac{x^2 - 4x - 21}{x^2 - 4x}$

$$= \dfrac{(x - 4)(x + 4)(x + 3)(x - 7)}{(x + 3)(x + 4)x(x - 4)} = \dfrac{x - 7}{x}$$

30. $\dfrac{3y - 1}{3y + 1} - \dfrac{2y - 5}{y - 3} = \dfrac{(3y - 1)(y - 3)}{(3y + 1)(y - 3)} - \dfrac{(2y - 5)(3y + 1)}{(y - 3)(3y + 1)}$

$$= \dfrac{(3y^2 - 10y + 3) - (6y^2 - 13y - 5)}{(3y + 1)(y - 3)}$$

$$= \dfrac{-3y^2 + 3y + 8}{(3y + 1)(y - 3)}$$

42. $\dfrac{3 - \dfrac{2}{a}}{5 + \dfrac{3}{a}} = \dfrac{\left(3 - \dfrac{2}{a}\right)a}{\left(5 + \dfrac{3}{a}\right)a} = \dfrac{3a - 2}{5a + 3}$

60. $\dfrac{e^{-2} - f^{-1}}{ef} = \dfrac{\dfrac{1}{e^2} - \dfrac{1}{f}}{ef} = \dfrac{f - e^2}{e^2f} \div \dfrac{ef}{1}$

$$= \dfrac{f - e^2}{e^2f} \cdot \dfrac{1}{ef} = \dfrac{f - e^2}{e^3f^2}$$

64. a. $\dfrac{v_1 + v_2}{1 + \dfrac{v_1v_2}{c^2}} = \dfrac{1.2 \times 10^8 + 2.4 \times 10^8}{1 + \dfrac{(1.2 \times 10^8)(2.4 \times 10^8)}{(6.7 \times 10^8)^2}} \approx 3.4 \times 10^8 \text{ mph}$

b. $\dfrac{v_1 + v_2}{1 + \dfrac{v_1 \cdot v_2}{c^2}} = \dfrac{c^2(v_1 + v_2)}{c^2\left(1 + \dfrac{v_1 \cdot v_2}{c^2}\right)} = \dfrac{c^2(v_1 + v_2)}{c^2 + v_1 \cdot v_2}$

Exercise Set P.5, page 65

8. $6 - \sqrt{-1} = 6 - i$

18. $(5 - 3i) - (2 + 9i) = 5 - 3i - 2 - 9i = 3 - 12i$

34. $\left(5 + 2\sqrt{-16}\right)\left(1 - \sqrt{-25}\right) = [5 + 2(4i)](1 - 5i)$
$$= (5 + 8i)(1 - 5i)$$
$$= 5 - 25i + 8i - 40i^2$$
$$= 5 - 25i + 8i - 40(-1)$$
$$= 5 - 25i + 8i + 40$$
$$= 45 - 17i$$

48. $\dfrac{8 - i}{2 + 3i} = \dfrac{8 - i}{2 + 3i} \cdot \dfrac{2 - 3i}{2 - 3i}$

$$= \dfrac{16 - 24i - 2i + 3i^2}{2^2 + 3^2}$$

$$= \dfrac{16 - 24i - 2i + 3(-1)}{4 + 9}$$

$$= \dfrac{16 - 26i - 3}{13}$$

$$= \dfrac{13 - 26i}{13} = \dfrac{13(1 - 2i)}{13}$$

$$= 1 - 2i$$

60. $\dfrac{1}{i^{83}} = \dfrac{1}{i^{80} \cdot i^3} = \dfrac{1}{i^3} = \dfrac{1}{-i}$

$$= \dfrac{1}{-i} \cdot \dfrac{i}{i} = \dfrac{i}{-i^2} = \dfrac{i}{-(-1)} = \dfrac{i}{1} = i$$

Exercise Set 1.1, page 82

2. $-3y + 20 = 2$
$$-3y = -18 \quad \text{• Subtract 20 from each side.}$$
$$y = 6 \quad \text{• Divide each side by } -3.$$

12. $\dfrac{1}{2}x + 7 - \dfrac{1}{4}x = \dfrac{19}{2}$

$$4\left(\dfrac{1}{2}x + 7 - \dfrac{1}{4}x\right) = 4\left(\dfrac{19}{2}\right) \quad \text{• Multiply each side by 4.}$$

$$2x + 28 - x = 38$$
$$x = 38 - 28 \quad \text{• Collect like terms.}$$
$$x = 10$$

18. $5(x + 4)(x - 4) = (x - 3)(5x + 4)$
$$5(x^2 - 16) = 5x^2 - 11x - 12$$
$$5x^2 - 80 = 5x^2 - 11x - 12$$
$$-80 + 12 = -11x$$
$$-68 = -11x$$
$$\dfrac{68}{11} = x$$

24. $2x + \dfrac{1}{3} = \dfrac{6x + 1}{3} \quad \text{• Rewrite the left side.}$

$$\dfrac{6x + 1}{3} = \dfrac{6x + 1}{3}$$

The left side of this equation is now identical to the right side. Thus the original equation is an identity. The solution set of the original equation consists of all real numbers.

38.
$$|2x - 3| = 21$$

$2x - 3 = 21$ or $2x - 3 = -21$

$\qquad 2x = 24 \qquad\qquad\qquad 2x = -18$

$\qquad\quad x = 12 \qquad\qquad\qquad\quad x = -9$

The solutions of $|2x - 3| = 21$ are -9 and 12.

50. Substitute 175 for P, the number of patents measured in thousands, and solve for x.

$$P = 5.4x + 110$$

$$175 = 5.4x + 110$$

$\qquad 65 = 5.4x$ • Subtract 110 from each side.

$\qquad x = \dfrac{65}{5.4}$ • Solve for x.

$\qquad x \approx 12.04$

Adding 12.04 to 1993 yields 2005.04. Thus, according to the model, the number of patents will first exceed 175,000 in the year 2005.

52. Substitute 22 for m in the given equation and solve for s.

$22 = -\dfrac{1}{2}|s - 55| + 25$ • Substitute 22 for m.

$-44 = |s - 55| - 50$ • Multiply each side by -2 to clear the equation of fractions.

$\quad 6 = |s - 55|$ • Add 50 to each side.

$s - 55 = 6$ or $s - 55 = -6$

$\qquad s = 61 \qquad\qquad\qquad s = 49$

Kate should drive her car at either 61 miles per hour or 49 miles per hour to obtain a gas mileage of 22 miles per gallon.

Exercise Set 1.2, page 92

4. $A = P + Prt$

$A = P(1 + rt)$ • Factor.

$P = \dfrac{A}{(1 + rt)}$ • Solve for P.

14. Substitute 105 for w.

SMOG reading grade level $= \sqrt{105} + 3$

$\approx 10.2 + 3$

$= 13.2$

According to the SMOG formula, the estimated reading grade level required to fully understand *A Tale of Two Cities* is 13.2. (*Note:* A different sample of 30 sentences likely would produce a different result. It is for this reason that reading grade levels are often estimated by using several different samples and then computing an average of the results.)

20. $P = 2l + 2w, \qquad w = \dfrac{1}{2}l + 1$

$110 = 2l + 2\left(\dfrac{1}{2}l + 1\right)$ • Substitute for w.

$110 = 2l + l + 2$ • Simplify.

$108 = 3l$

$\ 36 = l$

$\quad l = 36$ meters

$w = \dfrac{1}{2}l + 1 = \dfrac{1}{2}(36) + 1 = 19$ meters

24. Let $t_1 =$ the time it takes to travel to the island. Let $t_2 =$ the time it takes to make the return trip.

$t_1 + t_2 = 7.5$

$\qquad t_2 = 7.5 - t_1$

$\quad 15t_1 = 10t_2$

$\quad 15t_1 = 10(7.5 - t_1)$ • Substitute for t_2.

$\quad 15t_1 = 75 - 10t_1$

$\quad 25t_1 = 75$

$\qquad t_1 = 3$ hours

$\qquad D = 15t_1 = 15(3) = 45$ nautical miles

32. Let $x =$ the number of glasses of orange juice.

Profit $=$ revenue $-$ cost

$\$2337 = 0.75x - 0.18x$

$2337 = 0.57x$

$\quad x = \dfrac{2337}{0.57}$

$\quad x = 4100$

36. Let $x =$ the amount of money invested at 5%. Then $7500 - x$ is the amount of money invested at 7%.

5%	x
7%	$7500 - x$

$0.05x + 0.07(7500 - x) = 405$

$0.05x + 525 - 0.07x = 405$

$-0.02x = -120$

$x = 6000$

$7500 - x = 1500$

$\$6000$ was invested at 5%. $\$1500$ was invested at 7%.

40. Let $x =$ the number of liters of the 40% solution to be mixed with the 24% solution.

0.40	x
0.24	4
0.30	$4 + x$

$0.40x + 0.24(4) = 0.30(4 + x)$

$0.40x + 0.96 = 1.2 + 0.30x$

$0.10x = 0.24$

$x = 2.4$

Thus 2.4 liters of 40% sulfuric acid should be mixed with 4 liters of a 24% sulfuric acid solution to produce the 30% solution.

50. Let $x =$ the number of hours needed to print the report if both the printers are used.

Printer A prints $\dfrac{1}{3}$ of the report every hour.

Printer B prints $\dfrac{1}{4}$ of the report every hour.

Thus

$$\frac{1}{3}x + \frac{1}{4}x = 1$$
$$4x + 3x = 12 \cdot 1$$
$$7x = 12$$
$$x = \frac{12}{7} \approx 1.71$$

It would take approximately 1.71 hours to print the report.

Exercise Set 1.3, page 107

6. $12x^2 - 41x + 24 = 0$

$(4x - 3)(3x - 8) = 0$ • Factor.

$4x - 3 = 0$ or $3x - 8 = 0$ • Apply the zero product property.

$$x = \frac{3}{4} \qquad\qquad x = \frac{8}{3}$$

A check shows that $\dfrac{3}{4}$ and $\dfrac{8}{3}$ are both solutions of $12x^2 - 41x + 24 = 0$.

20. $(x + 2)^2 + 28 = 0$

$$(x + 2)^2 = -28$$
$$\sqrt{(x + 2)^2} = \sqrt{-28}$$
$$x + 2 = \pm i\sqrt{28} = \pm 2i\sqrt{7}$$
$$x = -2 \pm 2i\sqrt{7}$$

The solutions are $-2 - 2i\sqrt{7}$ and $-2 + 2i\sqrt{7}$.

26. $x^2 - 6x + 10 = 0$

$$x^2 - 6x = -10$$

$x^2 - 6x + 9 = -10 + 9$ • Add $\left[\dfrac{1}{2}(-6)\right]^2$ to each side.

$(x - 3)^2 = -1$ • Factor the left side.

$\sqrt{(x - 3)^2} = \pm\sqrt{-1}$ • The square root procedure

$$x - 3 = \pm i$$
$$x = 3 \pm i$$

The solutions are $3 - i$ and $3 + i$.

30. $2x^2 + 10x - 3 = 0$

$$2x^2 + 10x = 3$$
$$2(x^2 + 5x) = 3$$

$$x^2 + 5x = \frac{3}{2}$$ • Divide each side by 2.

$x^2 + 5x + \dfrac{25}{4} = \dfrac{3}{2} + \dfrac{25}{4}$ • Complete the square.

$$\left(x + \frac{5}{2}\right)^2 = \frac{31}{4}$$
$$x + \frac{5}{2} = \pm\sqrt{\frac{31}{4}}$$
$$x = -\frac{5}{2} \pm \frac{\sqrt{31}}{2}$$
$$x = \frac{-5 + \sqrt{31}}{2} \quad\text{or}\quad x = \frac{-5 - \sqrt{31}}{2}$$

38. $2x^2 + 4x - 1 = 0$

$$x = \frac{-4 \pm \sqrt{4^2 - 4(2)(-1)}}{4}$$
$$x = \frac{-4 \pm \sqrt{16 + 8}}{4} = \frac{-4 \pm \sqrt{24}}{4}$$
$$x = \frac{-4 \pm 2\sqrt{6}}{4} = \frac{-2 \pm \sqrt{6}}{2}$$
$$x = \frac{-2 + \sqrt{6}}{2} \quad\text{or}\quad x = \frac{-2 - \sqrt{6}}{2}$$

48. $x^2 + 3x - 11 = 0$

$$b^2 - 4ac = 3^2 - 4(1)(-11) = 9 + 44 = 53 > 0$$

Thus the equation has two distinct real roots.

58. Home plate, first base, and second base form a right triangle. The legs of this right triangle each measure 90 ft. The distance from home plate to second base is the length of the hypotenuse c of this right triangle.

$$c^2 = 90^2 + 90^2$$
$$c^2 = 16{,}200$$
$$c = \sqrt{16{,}200}$$
$$c \approx 127.3$$

To the nearest tenth of a foot, the distance from home plate to second base is 127.3 feet. (*Note:* We have not considered $-\sqrt{16{,}200}$ as a solution because we know that the distance must be positive.)

70. Let w be the width, in inches, of the new candy bar. Then the length, in inches, of the new candy bar is $2.5w$, and the height is 0.5 inch. The volume of the original candy bar is $5 \cdot 2 \cdot 0.5 = 5$ cubic inches. Thus the volume of the new candy bar is $0.80(5) = 4$ cubic inches. Substitute in the formula for the volume of a rectangular solid to produce

$$lwh = V$$
$$(2.5w)(w)(0.5) = 4$$
$$1.25w^2 = 4$$

$$w^2 = 3.2$$
$$w = \sqrt{3.2}$$
$$\approx 1.8$$

The width of the new candy bar should be about 1.8 inches and the length $2.5(1.8) \approx 4.5$ inches.

72. When the ball hits the ground, $h = 0$. Thus we need to solve $0 = -16t^2 + 52t + 4.5$ for t.

$$0 = -16t^2 + 52t + 4.5$$
$$t = \frac{-(52) \pm \sqrt{(52)^2 - 4(-16)(4.5)}}{2(-16)}$$
$$= \frac{-52 \pm \sqrt{2992}}{-32}$$
$$\approx 3.3$$

The ball will hit the ground in about 3.3 seconds. Disregard the negative solution because the time must be positive.

Exercise Set 1.4, page 119

6.
$$x^4 - 36x^2 = 0$$
$$x^2(x^2 - 36) = 0$$
$$x^2(x - 6)(x + 6) = 0$$
$$x = 0, x = 6, x = -6$$

14. Multiply each side of the equation by $(y + 2)(y - 4)$ to clear the equation of fractions.

$$\frac{4}{y + 2} = \frac{7}{y - 4} \qquad y \neq -2, y \neq 4$$
$$(y + 2)(y - 4)\left(\frac{4}{y + 2}\right) = (y + 2)(y - 4)\left(\frac{7}{y - 4}\right)$$
$$(y - 4)4 = (y + 2)7$$
$$4y - 16 = 7y + 14$$
$$4y - 7y = 14 + 16$$
$$-3y = 30$$
$$y = -10$$

Check to verify that -10 is the solution.

28. $\sqrt{10 - x} = 4$ *Check:* $\sqrt{10 - (-6)} = 4$
      $10 - x = 16$ $\sqrt{16} = 4$
        $-x = 6$ $4 = 4$
          $x = -6$

The solution is -6.

30.
$$x = \sqrt{5 - x} + 5$$
$$(x - 5)^2 = \left(\sqrt{5 - x}\right)^2$$
$$x^2 - 10x + 25 = 5 - x$$
$$x^2 - 9x + 20 = 0$$

$$(x - 5)(x - 4) = 0$$
$$x = 5 \quad \text{or} \quad x = 4$$

Check: $5 = \sqrt{5 - 5} + 5$ $4 = \sqrt{5 - 4} + 5$
            $5 = 0 + 5$ $4 = 1 + 5$
            $5 = 5$ $4 = 6$ False

The solution is 5.

34.
$$\sqrt{x + 7} - 2 = \sqrt{x - 9}$$
$$\left(\sqrt{x + 7} - 2\right)^2 = \left(\sqrt{x - 9}\right)^2$$
$$x + 7 - 4\sqrt{x + 7} + 4 = x - 9$$
$$-4\sqrt{x + 7} = -20$$
$$\left(\sqrt{x + 7}\right)^2 = (5)^2$$
$$x + 7 = 25$$
$$x = 18$$

Check: $\sqrt{18 + 7} - 2 = \sqrt{18 - 9}$
          $\sqrt{25} - 2 = \sqrt{9}$
              $5 - 2 = 3$
                $3 = 3$

The solution is 18.

42. $x^4 - 10x^2 + 9 = 0$ • Let $u = x^2$.
$$u^2 - 10u + 9 = 0$$
$$(u - 9)(u - 1) = 0$$
$$u = 9 \qquad \text{or} \qquad u = 1$$
$$x^2 = 9 \qquad\qquad x^2 = 1$$
$$x = \pm 3 \qquad\qquad x = \pm 1$$

The solutions are 3, -3, 1, and -1.

50. $6x^{2/3} - 7x^{1/3} - 20 = 0$ • Let $u = x^{1/3}$.
$$6u^2 - 7u - 20 = 0$$
$$(3u + 4)(2u - 5) = 0$$
$$u = -\frac{4}{3} \qquad \text{or} \qquad u = \frac{5}{2}$$
$$x^{1/3} = -\frac{4}{3} \qquad\qquad x^{1/3} = \frac{5}{2}$$
$$(x^{1/3})^3 = \left(-\frac{4}{3}\right)^3 \qquad (x^{1/3})^3 = \left(\frac{5}{2}\right)^3$$
$$x = -\frac{64}{27} \qquad\qquad x = \frac{125}{8}$$

The solutions are $-\dfrac{64}{27}$ and $\dfrac{125}{8}$.

60. Sandy is to receive $\dfrac{1}{2}$ of an adult dosage of a particular medication. Substituting into Young's rule yields:

$$\frac{1}{2} = \frac{x}{x + 12}$$

(continued)

$$2(x + 12)\frac{1}{2} = 2(x + 12)\left(\frac{x}{x + 12}\right)$$

$$x + 12 = 2x$$

$$12 = x$$

Sandy is 12 years old.

62. Substitute 4 for the reading grade level in the SMOG formula to produce

$$4 = \sqrt{w} + 3$$

$1 = \sqrt{w}$ • Subtract 3 from each side.

$1 = w$ • Square each side.

The writer should strive for a maximum of one word with three or more syllables in any sample of 30 sentences.

Exercise Set 1.5, page 134

6. $-4(x - 5) \geq 2x + 15$

$-4x + 20 \geq 2x + 15$

$-6x \geq -5$

$x \leq \dfrac{5}{6}$

The solution set is $\left\{x \,\middle|\, x \leq \dfrac{5}{6}\right\}$.

10. $2x + 5 > -16 \quad$ and $\quad 2x + 5 < 9$

$\quad 2x > -21 \quad$ and $\quad 2x < 4$

$\quad x > -\dfrac{21}{2} \quad$ and $\quad x < 2$

$$\left\{x \,\middle|\, x > -\frac{21}{2}\right\} \cap \{x \mid x < 2\} = \left\{x \,\middle|\, -\frac{21}{2} < x < 2\right\}$$

The solution set is $\left\{x \,\middle|\, -\dfrac{21}{2} < x < 2\right\}$.

18. $|2x - 9| < 7$

$-7 < 2x - 9 < 7$

$2 < \quad 2x \quad < 16$

$1 < \quad x \quad < 8$

In interval notation, the solution set is $(1, 8)$.

34. $x^2 + 5x + 6 < 0$

$(x + 2)(x + 3) = 0$

$x = -2 \quad$ and $\quad x = -3$ • Critical values

Use a test number from each of the intervals $(-\infty, -3)$, $(-3, -2)$, and $(-2, \infty)$ to determine where $x^2 + 5x + 6$ is negative.

$$+++++++\,|\,-\,|\,+++++$$

In interval notation, the solution set is $(-3, -2)$.

46.
$$\frac{3x + 1}{x - 2} \geq 4$$

$$\frac{3x + 1}{x - 2} - 4 \geq 0$$

$$\frac{3x + 1 - 4(x - 2)}{x - 2} \geq 0$$

$$\frac{-x + 9}{x - 2} \geq 0$$

$x = 2 \quad$ and $\quad x = 9$ • Critical values

Use a test number from each of the intervals $(-\infty, 2)$, $(2, 9)$, and $(9, \infty)$ to determine where $\dfrac{-x + 9}{x - 2}$ is positive.

The solution set is $(2, 9]$.

52. Let $m =$ the number of miles driven.

Company A: $29 + 0.12m$
Company B: $22 + 0.21m$

$29 + 0.12m < 22 + 0.21m$

$77.\overline{7} < m$

Company A is less expensive if you drive at least 78 miles.

54. Substitute 6.50 for P. Then solve the following inequality.

$0.218t + 4.02 > 6.50$

$0.218t > 2.48$

$t > \dfrac{2.48}{0.218} \approx 11.38$

Adding 11.38 to 1994 yields 2005.38. According to the given mathematical model, we should first expect to see the average price of a movie ticket exceed $6.50 in the year 2005.

58. $41 \leq \quad F \quad \leq 68$

$41 \leq \dfrac{9}{5}C + 32 \leq 68$

$9 \leq \dfrac{9}{5}C \leq 36$

$\dfrac{5}{9}(9) \leq \left(\dfrac{5}{9}\right)\left(\dfrac{9}{5}\right)C \leq \dfrac{5}{9}(36)$

$5 \leq \quad C \quad \leq 20$

The Celsius temperature is between 5°C and 20°C, inclusive.

66. We need to solve:

$$\frac{0.00014x^2 + 12x + 400,000}{x} < 30 \qquad \text{(I)}$$

Because $x > 0$, we can multiply both sides by x to produce

$$0.00014x^2 + 12x + 400,000 < 30x$$
$$0.00014x^2 - 18x + 400,000 < 0$$

Use the quadratic formula to find the critical values.

$$x = \frac{-(-18) \pm \sqrt{(-18)^2 - 4(0.00014)(400,000)}}{2(0.00014)}$$

$$= \frac{18 \pm \sqrt{100}}{0.00028}$$

Use a calculator to find that the critical values are approximately 28,571.4 and 100,000. A test value can be used to show that our original inequality (I) is true on the interval (28,571.4, 100,000). Thus the company should manufacture from 28,572 to 99,999 pairs of running shoes if it wishes to bring the average cost below $30 per pair.

Exercise Set 2.1, page 158

6. $d = \sqrt{(x_2 - x_1)^2 + (y_2 - y_1)^2}$
$d = \sqrt{[-10 - (-5)]^2 + (14 - 8)^2}$
$= \sqrt{(-5)^2 + 6^2} = \sqrt{25 + 36}$
$= \sqrt{61}$

26.

30.

32.

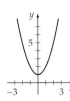

40. y-intercept: $\left(0, -\dfrac{15}{4}\right)$

x-intercept: $(5, 0)$

64. $r = \sqrt{(1 - (-2))^2 + (7 - 5)^2}$
$= \sqrt{9 + 4} = \sqrt{13}$

Using the standard form

$$(x - h)^2 + (y - k)^2 = r^2$$

with $h = -2$, $k = 5$, and $r = \sqrt{13}$ yields

$$(x + 2)^2 + (y - 5)^2 = \left(\sqrt{13}\right)^2$$

66. $\quad x^2 + y^2 - 6x - 4y + 12 = 0$
$$x^2 - 6x + y^2 - 4y = -12$$
$$x^2 - 6x + 9 + y^2 - 4y + 4 = -12 + 9 + 4$$
$$(x - 3)^2 + (y - 2)^2 = 1^2$$

center (3, 2), radius 1

Exercise Set 2.2, page 174

2. Given $g(x) = 2x^2 + 3$

   a. $g(3) = 2(3)^2 + 3 = 18 + 3 = 21$

   b. $g(-1) = 2(-1)^2 + 3 = 2 + 3 = 5$

   c. $g(0) = 2(0)^2 + 3 = 0 + 3 = 3$

   d. $g\left(\dfrac{1}{2}\right) = 2\left(\dfrac{1}{2}\right)^2 + 3 = \dfrac{1}{2} + 3 = \dfrac{7}{2}$

   e. $g(c) = 2(c)^2 + 3 = 2c^2 + 3$

   f. $g(c + 5) = 2(c + 5)^2 + 3 = 2c^2 + 20c + 50 + 3$
   $= 2c^2 + 20c + 53$

10. a. Because $0 \le 0 \le 5$, $Q(0) = 4$.

   b. Because $6 < e < 7$, $Q(e) = -e + 9$.

   c. Because $1 < n < 2$, $Q(n) = 4$.

   d. Because $1 < m \le 2$, $8 < m^2 + 7 \le 11$. Thus
   $Q(m^2 + 7) = \sqrt{(m^2 + 7) - 7} = \sqrt{m^2} = m$

14. $x^2 - 2y = 2$ • **Solve for y.**
   $-2y = -x^2 + 2$

   $$y = \frac{1}{2}x^2 - 1$$

y is a function of x because each x value will yield one and only one y value.

28. Domain is the set of all real numbers.

40. Domain is the set of all real numbers.

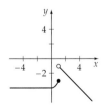

48. a. $[0, \infty)$

   b. Since $31,250 is between $27,950 and $67,700, use $T(x) = 0.27(x - 27,950) + 3892.50$. Then,
   $T(31,250) = 0.27(31,250 - 27,950) + 3892.50 = \$4783.50$.

   c. Since $72,000 is between $67,700 and $141,250, use $T(x) = 0.30(x - 67,700) + 14,625$. Then,
   $T(72,000) = 0.30(72,000 - 67,700) + 14,625 = \$15,915$.

50. a. This is the graph of a function. Every vertical line intersects the graph in at most one point.

(continued)

b. This is not the graph of a function. Some vertical lines intersect the graph at two points.

c. This is not the graph of a function. The vertical line at $x = -2$ intersects the graph at more than one point.

d. This is the graph of a function. Every vertical line intersects the graph at exactly one point.

66. $v(t) = 44{,}000 - 4200t, 0 \le t \le 8$

68. a. $V(x) = (30 - 2x)^2 x$

$\qquad = (900 - 120x + 4x^2)x$

$\qquad = 900x - 120x^2 + 4x^3$

b. Domain: $\{x \mid 0 < x < 15\}$

72. $d(A, B) = \sqrt{1 + x^2}$. The time required to swim from A to B at 2 mph is $\dfrac{\sqrt{1 + x^2}}{2}$ hours.

$d(B, C) = 3 - x$. The time required to run from B to C at 8 mph is $\dfrac{3 - x}{8}$ hours.

Thus the total time to reach point C is

$$t = \frac{\sqrt{1 + x^2}}{2} + \frac{3 - x}{8} \text{ hours}$$

Exercise Set 2.3, page 191

2. $m = \dfrac{1 - 4}{5 - (-2)} = -\dfrac{3}{7}$

16. $m = -1$
$b = 1$

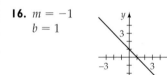

28. $y - 5 = -2(x - 0)$

$\qquad y = -2x + 5$

42. $f(x) = \dfrac{2x}{3} + 2$

$4 = \dfrac{2x}{3} + 2$ • **Replace** $f(x)$ **by 4 and solve for** x.

$2 = \dfrac{2x}{3}$

$3 = x$

When $x = 3, f(x) = 4$.

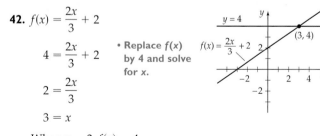

46. $f(x) = 0$

$\quad -2x - 4 = 0$

$\qquad -2x = 4$

$\qquad\quad x = -2$

50. $f_1(x) = f_2(x)$

$-2x - 11 = 3x + 7$

$-5x - 11 = 7$

$\qquad -5x = 18$

$\qquad\quad x = -\dfrac{18}{5} = -3.6$

$(-3.6, -3.8)$

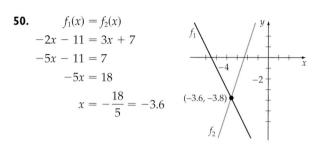

56. a. Using the data for 1997 and 2003, two ordered pairs on the line are (1997, 531.0) and (2003, 725.0). Find the slope of the line.

$$m = \frac{725.0 - 531.0}{2003 - 1997} \approx 32.3$$

Use the point-slope formula to find the equation of the line between the given points.

$y - y_1 = m(x - x_1)$

$y - 531.0 = 32.3(x - 1997)$

$y - 531.0 = 32.3x - 64{,}503.1$

$\qquad\quad y = 32.3x - 63{,}972.1$

Using functional notation, the linear function is $C(t) = 32.3t - 63{,}972.1$.

b. To find the year when consumer debt first exceeds $850 billion, let $C(t) = 850$ and solve for t.

$C(t) = 32.3t - 63{,}972.1$

$\quad 850 = 32.3t - 63{,}972.1$

$64{,}822.1 = 32.3t$

$\quad 2006 \approx t$

According to the model, revolving consumer debt will first exceed $850 billion in 2006.

66. $\qquad P(x) = R(x) - C(x)$

$\qquad P(x) = 124x - (78.5x + 5005)$

$\qquad P(x) = 45.5x - 5005$

$45.5x - 5005 = 0$

$\qquad 45.5x = 5005$

$\qquad\quad x = 110$ • **The break-even point**

78. a. The slope of the radius from $(0, 0)$ to $\left(\sqrt{15}, 1\right)$ is $\dfrac{1}{\sqrt{15}}$. The slope of the linear path of the rock is $-\sqrt{15}$. The path of the rock is given by

$y - 1 = -\sqrt{15}\left(x - \sqrt{15}\right)$

$y - 1 = -\sqrt{15}x + 15$

$\qquad y = -\sqrt{15}x + 16$

Every point on the wall has a y value of 14. Thus

$$14 = -\sqrt{15}x + 16$$
$$-2 = -\sqrt{15}x$$
$$x = \frac{2}{\sqrt{15}} \approx 0.52$$

The rock hits the wall at $(0.52, 14)$.

Exercise Set 2.4, page 206

10. $f(x) = x^2 + 6x - 1$

$\quad\quad = x^2 + 6x + 9 + (-1 - 9)$

$\quad\quad = (x + 3)^2 - 10$

vertex $(-3, -10)$

axis of symmetry $x = -3$

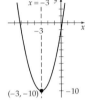

20. $h = -\dfrac{b}{2a} = -\dfrac{-6}{2(1)} = 3$

$k = f(3) = 3^2 - 6(3) = -9$

vertex $(3, -9)$

$f(x) = (x - 3)^2 - 9$

32. Determine the y-coordinate of the vertex of the graph of $f(x) = 2x^2 + 6x - 5$.

$f(x) = 2x^2 + 6x - 5$ • $a = 2, b = 6, c = -5$

$h = -\dfrac{b}{2a} = -\dfrac{6}{2(2)} = -\dfrac{3}{2}$ • **Find the x-coordinate of the vertex.**

$k = f\left(-\dfrac{3}{2}\right) = 2\left(-\dfrac{3}{2}\right)^2 + 6\left(-\dfrac{3}{2}\right) - 5 = -\dfrac{19}{2}$

                           • **Find the y-coordinate of the vertex.**

The vertex is $\left(-\dfrac{3}{2}, -\dfrac{19}{2}\right)$. Because the parabola opens up, $-\dfrac{19}{2}$ is the minimum value of f. Therefore, the range of f is $\left\{y \mid y \geq -\dfrac{19}{2}\right\}$. To determine the values of x for which $f(x) = 15$, replace $f(x)$ by $2x^2 + 6x - 5$ and solve for x.

$\quad\quad\quad\quad f(x) = 15$

$\quad\quad 2x^2 + 6x - 5 = 15$ • **Replace $f(x)$ by $2x^2 + 6x - 5$.**

$\quad\quad 2x^2 + 6x - 20 = 0$ • **Solve for x.**

$\quad\quad 2(x - 2)(x + 5) = 0$ • **Factor.**

$\quad x - 2 = 0 \quad\quad x + 5 = 0$ • **Use the Principle of Zero Products to solve for x.**

$\quad\quad x = 2 \quad\quad\quad x = -5$

The values of x for which $f(x) = 15$ are 2 and -5.

36. $f(x) = -x^2 - 6x$

$\quad\quad = -(x^2 + 6x)$

$\quad\quad = -(x^2 + 6x + 9) + 9$

$\quad\quad = -(x + 3)^2 + 9$

Maximum value of f is 9 when $x = -3$.

46. a. $l + w = 240$, so $w = 240 - l$.

    b. $A = lw = l(240 - l) = 240l - l^2$

    c. The l value of the vertex point of the graph of $A = 240l - l^2$ is

$$-\frac{b}{2a} = -\frac{240}{2(-1)} = 120$$

Thus $l = 120$ meters and $w = 240 - 120 = 120$ meters are the dimensions that produce the greatest area.

68. Let $x =$ the number of parcels.

    a. $R(x) = xp = x(22 - 0.01x) = -0.01x^2 + 22x$

    b. $P(x) = R(x) - C(x)$

$\quad\quad\quad = (-0.01x^2 + 22x) - (2025 + 7x)$

$\quad\quad\quad = -0.01x^2 + 15x - 2025$

    c. $-\dfrac{b}{2a} = -\dfrac{15}{2(-0.01)} = 750$

The maximum profit is

$P(750) = -0.01(750)^2 + 15(750) - 2025 = \$3600$

    d. The price per parcel that yields the maximum profit is

$p(750) = 22 - 0.01(750) = \$14.50$

    e. The break-even point(s) occur when $R(x) = C(x)$.

$\quad\quad -0.01x^2 + 22x = 2025 + 7x$

$\quad\quad\quad 0 = 0.01x^2 - 15x + 2025$

$$x = \frac{-(-15) \pm \sqrt{(-15)^2 - 4(0.01)(2025)}}{2(0.01)}$$

$x = 150$ and $x = 1350$ are the break-even points.

Thus the minimum number of parcels the air freight company must ship to break even is 150.

70. $h(t) = -16t^2 + 64t + 80$

$$t = -\frac{b}{2a} = -\frac{64}{2(-16)} = 2$$

$h(2) = -16(2)^2 + 64(2) + 80$

$\quad\quad = -64 + 128 + 80 = 144$

    a. The vertex $(2, 144)$ gives us the maximum height of 144 feet.

    b. The vertex of the graph of h is $(2, 144)$, so the time when the projectile achieves this maximum height is at time $t = 2$ seconds.

(continued)

c. $-16t^2 + 64t + 80 = 0$ • Solve for t with $h = 0$.

$-16(t^2 - 4t - 5) = 0$

$-16(t + 1)(t - 5) = 0$

$t = -1$ $t - 5 = 0$

no $t = 5$

The projectile will have a height of 0 feet at time $t = 5$ seconds.

Exercise Set 2.5, page 222

14. The graph is symmetric with respect to the x-axis, because replacing y with $-y$ leaves the equation unaltered. The graph is not symmetric with respect to the y-axis, because replacing x with $-x$ alters the equation.

24. The graph is symmetric with respect to the origin because $(-y) = (-x)^3 - (-x)$ simplifies to $-y = -x^3 + x$, which is equivalent to the original equation $y = x^3 - x$.

44. Even, because $h(-x) = (-x)^2 + 1 = x^2 + 1 = h(x)$.

58.

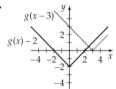

68.

70.

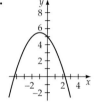

72. a.

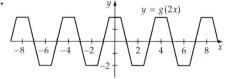

b.

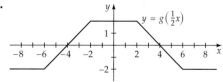

Exercise Set 2.6, page 235

10. $f(x) + g(x) = \sqrt{x - 4} - x$ Domain: $\{x \mid x \geq 4\}$

$f(x) - g(x) = \sqrt{x - 4} + x$ Domain: $\{x \mid x \geq 4\}$

$f(x)g(x) = -x\sqrt{x - 4}$ Domain: $\{x \mid x \geq 4\}$

$\dfrac{f(x)}{g(x)} = -\dfrac{\sqrt{x - 4}}{x}$ Domain: $\{x \mid x \geq 4\}$

14. $(f + g)(x) = (x^2 - 3x + 2) + (2x - 4) = x^2 - x - 2$

$(f + g)(-7) = (-7)^2 - (-7) - 2 = 49 + 7 - 2 = 54$

30. $\dfrac{f(x + h) - f(x)}{h} = \dfrac{[4(x + h) - 5] - (4x - 5)}{h}$

$= \dfrac{4x + 4(h) - 5 - 4x + 5}{h}$

$= \dfrac{4(h)}{h} = 4$

38. $(g \circ f)(x) = g[f(x)] = g[2x - 7]$

$= 3[2x - 7] + 2 = 6x - 19$

$(f \circ g)(x) = f[g(x)] = f[3x + 2]$

$= 2[3x + 2] - 7 = 6x - 3$

50. $(f \circ g)(4) = f[g(4)]$

$= f[4^2 - 5(4)]$

$= f[-4] = 2(-4) + 3 = -5$

66. a. $l = 3 - 0.5t$ for $0 \leq t \leq 6$. $l = -3 + 0.5t$ for $t > 6$. In either case, $l = |3 - 0.5t|$. $w = |2 - 0.2t|$ as in Example 7.

b. $A(t) = |3 - 0.5t||2 - 0.2t|$

c. A is decreasing on $[0, 6]$ and on $[8, 10]$. A is increasing on $[6, 8]$ and on $[10, 14]$.

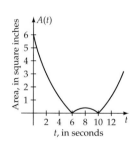

d. The highest point on the graph of A for $0 \leq t \leq 14$ occurs when $t = 0$ seconds.

72. a. On $[2, 3]$,

$a = 2$

$\Delta t = 3 - 2 = 1$

$s(a + \Delta t) = s(3) = 6 \cdot 3^2 = 54$

$s(a) = s(2) = 6 \cdot 2^2 = 24$

Average velocity $= \dfrac{s(a + \Delta t) - s(a)}{\Delta t}$

$= \dfrac{s(3) - s(2)}{1}$

$= 54 - 24 = 30$ feet per second

This is identical to the slope of the line through $(2, f(2))$ and $(3, f(3))$ because

$$m = \frac{s(3) - s(2)}{3 - 2} = s(3) - s(2) = 54 - 24 = 30$$

b. On $[2, 2.5]$,

$a = 2$

$\Delta t = 2.5 - 2 = 0.5$

$s(a + \Delta t) = s(2.5) = 6(2.5)^2 = 37.5$

$$\text{Average velocity} = \frac{s(2.5) - s(2)}{0.5}$$

$$= \frac{37.5 - 24}{0.5}$$

$$= \frac{13.5}{0.5} = 27 \text{ feet per second}$$

c. On $[2, 2.1]$,

$a = 2$

$\Delta t = 2.1 - 2 = 0.1$

$s(a + \Delta t) = s(2.1) = 6(2.1)^2 = 26.46$

$$\text{Average velocity} = \frac{s(2.1) - s(2)}{0.1}$$

$$= \frac{26.46 - 24}{0.1}$$

$$= \frac{2.46}{0.1} = 24.6 \text{ feet per second}$$

d. On $[2, 2.01]$,

$a = 2$

$\Delta t = 2.01 - 2 = 0.01$

$s(a + \Delta t) = s(2.01) = 6(2.01)^2 = 24.2406$

$$\text{Average velocity} = \frac{s(2.01) - s(2)}{0.01}$$

$$= \frac{24.2406 - 24}{0.01}$$

$$= \frac{0.2406}{0.01} = 24.06 \text{ feet per second}$$

e. On $[2, 2.001]$,

$a = 2$

$\Delta t = 2.001 - 2 = 0.001$

$s(a + \Delta t) = s(2.001) = 6(2.001)^2 = 24.024006$

$$\text{Average velocity} = \frac{s(2.001) - s(2)}{0.001}$$

$$= \frac{24.024006 - 24}{0.001}$$

$$= \frac{0.024006}{0.001} = 24.006 \text{ feet per second}$$

f. On $[2, 2 + \Delta t]$,

$$\frac{s(2 + \Delta t) - s(2)}{\Delta t} = \frac{6(2 + \Delta t)^2 - 24}{\Delta t}$$

$$= \frac{6(4 + 4(\Delta t) + (\Delta t)^2) - 24}{\Delta t}$$

$$= \frac{24 + 24(\Delta t) + 6(\Delta t)^2 - 24}{\Delta t}$$

$$= \frac{24\Delta t + 6(\Delta t)^2}{\Delta t} = 24 + 6(\Delta t)$$

As Δt approaches zero, the average velocity approaches 24 feet per second.

Exercise Set 3.1, page 257

2.

$$
\begin{array}{r}
6x^2 - 9x + 28 \\
x + 4 \overline{\smash{\big)}\ 6x^3 + 15x^2 - 8x + 2} \\
\underline{6x^3 + 24x^2} \\
-9x^2 - 8x \\
\underline{-9x^2 - 36x} \\
28x + 2 \\
\underline{28x + 112} \\
-110
\end{array}
$$

$$\frac{6x^3 + 15x^2 - 8x + 2}{x + 4} = 6x^2 - 9x + 28 - \frac{110}{x + 4}$$

12.

$$
\begin{array}{r|rrrr}
5 & 5 & 6 & -8 & 1 \\
 & & 25 & 155 & 735 \\
\hline
 & 5 & 31 & 147 & 736
\end{array}
$$

$$\frac{5x^3 + 6x^2 - 8x + 1}{x - 5} = 5x^2 + 31x + 147 + \frac{736}{x - 5}$$

26.

$$
\begin{array}{r|rrrr}
3 & 2 & -1 & 3 & -1 \\
 & & 6 & 15 & 54 \\
\hline
 & 2 & 5 & 18 & 53
\end{array}
$$

$P(c) = P(3) = 53$

36.

$$
\begin{array}{r|rrrr}
-6 & 1 & 4 & -27 & -90 \\
 & & -6 & 12 & 90 \\
\hline
 & 1 & -2 & -15 & 0
\end{array}
$$

A remainder of 0 indicates that $x + 6$ is a factor of $P(x)$.

56.

$$
\begin{array}{r|rrrrr}
-1 & 1 & 5 & 3 & -5 & -4 \\
 & & -1 & -4 & 1 & 4 \\
\hline
 & 1 & 4 & -1 & -4 & 0
\end{array}
$$

The reduced polynomial is $x^3 + 4x^2 - x - 4$.

$x^4 + 5x^3 + 3x^2 - 5x - 4 = (x + 1)(x^3 + 4x^2 - x - 4)$

Section 3.2, page 271

2. Because $a_n = -2$ is negative and $n = 3$ is odd, the graph of P goes up to the far left and down to the far right.

22. $P(x) = x^3 - 6x^2 + 8x$

$= x(x^2 - 6x + 8)$

$= x(x - 2)(x - 4)$

The factor x can be written as $(x - 0)$. Apply the Factor Theorem to determine that the real zeros of P are 0, 2, and 4.

28.
$$\begin{array}{r|rrrr} 0 & 4 & -1 & -6 & 1 \\ & & 0 & 0 & 0 \\ \hline & 4 & -1 & -6 & 1 \end{array}$$ • $P(0) = 1$

$$\begin{array}{r|rrrr} 1 & 4 & -1 & -6 & 1 \\ & & 4 & 3 & -3 \\ \hline & 4 & 3 & -3 & -2 \end{array}$$ • $P(1) = -2$

Because P is a polynomial function, the graph of P is continuous. Also, $P(0)$ and $P(1)$ have opposite signs. Thus by the Zero Location Theorem we know that P must have a real zero between 0 and 1.

34. The exponent of $(x + 2)$ is 1, which is odd. Thus the graph of P crosses the x-axis at the x-intercept $(-2, 0)$. The exponent of $(x - 6)^2$ is even. Thus the graph of P intersects but does not cross the x-axis at $(6, 0)$.

42. *Far-left and far-right behavior.* The leading term of $P(x) = x^3 + 2x^2 - 3x$ is $1x^3$. The leading coefficient 1 is positive and the degree of the polynomial 3 is odd. Thus the graph of P goes down to its far left and up to its far right.

The y-intercept. $P(0) = 0^3 + 2(0)^2 - 3(0) = 0$. The y-intercept is $(0, 0)$.

The x-intercept(s). Try to factor $x^3 + 2x^2 - 3x$.

$x^3 + 2x^2 - 3x = x(x^2 + 2x - 3)$

$= x(x + 3)(x - 1)$

Use the Factor Theorem to determine that $(0, 0)$, $(-3, 0)$, and $(1, 0)$ are the x-intercepts. Apply the Even and Odd Powers of $(x - c)$ Theorem to determine that the graph of P will cross the x-axis at each of its x-intercepts.

Additional points: $(-2, 6)$, $(-1, 4)$, $(0.5, -0.875)$, $(1.5, 3.375)$

Symmetry: The function P is not an even or an odd function. Thus the graph of P is *not* symmetric with respect to either the y-axis or the origin.

Sketch the graph.

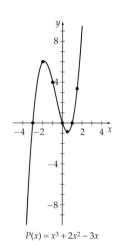

$P(x) = x^3 + 2x^2 - 3x$

48. The volume of the box is $V = lwh$, with $h = x$, $l = 18 - 2x$, and $w = \dfrac{42 - 3x}{2}$. Therefore, the volume is

$$V(x) = (18 - 2x)\left(\frac{42 - 3x}{2}\right)x$$

$$= 3x^3 - 69x^2 + 378x$$

Use a graphing utility to graph $V(x)$. The graph is shown below. The value of x that produces the maximum volume is 3.571 inches (to the nearest 0.001 inch). *Note:* Your x-value may differ slightly from 3.5705971 depending on the values you use for Xmin and Xmax. The maximum volume is approximately 606.6 cubic inches.

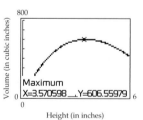

Section 3.3, page 286

10. $p = \pm 1, \pm 2, \pm 4, \pm 8$

$q = \pm 1, \pm 3$

$\dfrac{p}{q} = \pm 1, \pm 2, \pm 4, \pm 8, \pm \dfrac{1}{3}, \pm \dfrac{2}{3}, \pm \dfrac{4}{3}, \pm \dfrac{8}{3}$

18.
$$\begin{array}{r|rrrr} 1 & 1 & 0 & -19 & -28 \\ & & 1 & 1 & -18 \\ \hline & 1 & 1 & -18 & -46 \end{array}$$

$$\begin{array}{r|rrrr} 2 & 1 & 0 & -19 & -28 \\ & & 2 & 4 & -30 \\ \hline & 1 & 2 & -15 & -58 \end{array}$$

$$\begin{array}{r|rrrr} -1 & 1 & 0 & -19 & -28 \\ & & -1 & 1 & 18 \\ \hline & 1 & -1 & -18 & -10 \end{array}$$

$$\begin{array}{r|rrrr} -2 & 1 & 0 & -19 & -28 \\ & & -2 & 4 & 30 \\ \hline & 1 & -2 & -15 & 2 \end{array}$$

$$\begin{array}{r|rrrr} 3 & 1 & 0 & -19 & -28 \\ & & 3 & 9 & -30 \\ \hline & 1 & 3 & -10 & -58 \end{array}$$

$$\begin{array}{r|rrrr} 4 & 1 & 0 & -19 & -28 \\ & & 4 & 16 & -12 \\ \hline & 1 & 4 & -3 & -40 \end{array}$$

$$\begin{array}{r|rrrr} 5 & 1 & 0 & -19 & -28 \\ & & 5 & 25 & 30 \\ \hline & 1 & 5 & 6 & 2 \end{array}$$

None of these numbers are negative, so 5 is an upper bound.

$$\begin{array}{r|rrrr} -3 & 1 & 0 & -19 & -28 \\ & & -3 & 9 & 30 \\ \hline & 1 & -3 & -10 & 2 \end{array}$$

$$\begin{array}{r|rrrr} -4 & 1 & 0 & -19 & -28 \\ & & -4 & 16 & 12 \\ \hline & 1 & -4 & -3 & -16 \end{array}$$

$$\begin{array}{r|rrrr} -5 & 1 & 0 & -19 & -28 \\ & & -5 & 25 & -30 \\ \hline & 1 & -5 & 6 & -58 \end{array}$$

These numbers alternate in sign, so -5 is a lower bound.

28. $P(x)$ has one positive real zero because P has one variation in sign.

$$P(-x) = (-x)^3 - 19(-x) - 30 = -x^3 + 19x - 30$$

$P(x)$ has two or no negative real zeros because $P(-x) = -x^3 + 19x - 30$ has two variations in sign.

38. $P(x)$ has one positive and two or no negative real zeros (see Exercise 28 above).

$$\begin{array}{r|rrrr} 5 & 1 & 0 & -19 & -30 \\ & & 5 & 25 & 30 \\ \hline & 1 & 5 & 6 & 0 \end{array}$$

The reduced polynomial is $x^2 + 5x + 6 = (x + 3)(x + 2)$, which has -3 and -2 as zeros. Thus the zeros of $P(x) = x^3 - 19x - 30$ are -3, -2, and 5.

68. The volume of the tank is equal to the volume of the two hemispheres plus the volume of the cylinder. Thus

$$\frac{4}{3}\pi x^3 + 6\pi x^2 = 9\pi$$

Dividing each term by π and multiplying by 3 produces

$$4x^3 + 18x^2 = 27$$

Intersection Method   Use a graphing utility to graph $y = 4x^3 + 18x^2$ and $y = 27$ on the same screen, with $x > 0$. The x-coordinate of the point of intersection of the two graphs is the desired solution. The graphs intersect at $x \approx 1.098$ (rounded to the nearest thousandth of a foot). The length of the radius is approximately 1.098 feet.

72. We need to find the natural number solution of $n^3 - 3n^2 + 2n = 504$, which can be written as

$$n^3 - 3n^2 + 2n - 504 = 0$$

The constant term has many natural number divisors, but the following synthetic division shows that 10 is an upper bound for the zeros of $P(n) = n^3 - 3n^2 + 2n - 504$.

$$\begin{array}{r|rrrr} 10 & 1 & -3 & 2 & -504 \\ & & 10 & 70 & 720 \\ \hline & 1 & 7 & 72 & 216 \end{array}$$

The following synthetic division shows that 9 is a zero of $P(n)$.

$$\begin{array}{r|rrrr} 9 & 1 & -3 & 2 & -504 \\ & & 9 & 54 & 504 \\ \hline & 1 & 6 & 56 & 0 \end{array}$$

Thus the given group of cards consists of exactly nine cards. There is no need to seek additional solutions, because any increase (decrease) in the number of cards will increase (decrease) the number of ways one can select three cards from the group of cards.

Section 3.4, page 297

2. Use the Rational Zero Theorem to determine the possible rational zeros.

$$\frac{p}{q} = \pm 1, \pm 5$$

The following synthetic division shows that 1 is a zero of $P(x)$.

$$\begin{array}{r|rrrr} 1 & 1 & -3 & 7 & -5 \\ & & 1 & -2 & 5 \\ \hline & 1 & -2 & 5 & 0 \end{array}$$

Use the quadratic formula to find the zeros of the reduced polynomial $x^2 - 2x + 5$.

$$x = \frac{-(-2) \pm \sqrt{(-2)^2 - 4(1)(5)}}{2(1)} = 1 \pm 2i$$

The zeros of $P(x) = x^3 - 3x^2 + 7x - 5$ are 1, $1 - 2i$, and $1 + 2i$.

The linear factored form of $P(x)$ is

$$P(x) = (x - 1)(x - [1 - 2i])(x - [1 + 2i])$$

or

$$P(x) = (x - 1)(x - 1 + 2i)(x - 1 - 2i)$$

12.

$$\begin{array}{r|rrrr} 5 + 3i & 3 & -29 & 92 & 34 \\ & & 15 + 9i & -97 + 3i & -34 \\ \hline & 3 & -14 + 9i & -5 + 3i & 0 \end{array}$$

$$\begin{array}{r|rrr} 5 - 3i & 3 & -14 + 9i & -5 + 3i \\ & & 15 - 9i & 5 - 3i \\ \hline & 3 & 1 & 0 \end{array}$$

The reduced polynomial $3x + 1$ has $-\dfrac{1}{3}$ as a zero. The zeros of $3x^3 - 29x^2 + 92x + 34$ are $5 + 3i$, $5 - 3i$, and $-\dfrac{1}{3}$.

16.

$$
3i\,|\;1 \quad -6 + 0i \quad\; 22 + \;\;0i \quad -64 + \;\;0i \quad\;\; 117 + \;\;0i \quad -90
$$
$$
\underline{\qquad\quad 0 + 3i \quad -9 - 18i \quad\;\; 54 + 39i \quad -117 - 30i \quad\;\; 90}
$$
$$
-3i\,|\;1 \quad -6 + 3i \quad 13 - 18i \quad -10 + 39i \quad\quad\; 0 - 30i \quad\;\; 0
$$
$$
\underline{\qquad\quad 0 - 3i \quad\;\; 0 + 18i \quad\;\;\; 0 - 39i \quad\qquad\quad 30i}
$$
$$
\;\;\;\;\;\; 1 \quad -6 \quad\quad 13 \quad\quad -10 \quad\qquad\quad 0
$$

$$\frac{p}{q} = \pm 1, \pm 2, \pm 5, \pm 10$$

$$
2\,|\;1 \quad -6 \quad 13 \quad -10
$$
$$
\underline{\qquad\quad 2 \quad -8 \quad\;\; 10}
$$
$$
\;\;\;\; 1 \quad -4 \quad\;\; 5 \quad\quad 0
$$

Use the quadratic formula to solve $x^2 - 4x + 5 = 0$.

$$x = \frac{-(-4) \pm \sqrt{(-4)^2 - 4(1)(5)}}{2(1)} = \frac{4 \pm \sqrt{-4}}{2}$$

$$= \frac{4 \pm 2i}{2} = 2 \pm i$$

The zeros of $x^5 - 6x^4 + 22x^3 - 64x^2 + 117x - 90$ are $3i$, $-3i, 2, 2 + i$, and $2 - i$.

24. The graph of $P(x) = 4x^3 + 3x^2 + 16x + 12$ is shown below. Applying Descartes' Rule of Signs, we find that the real zeros are all negative numbers. From the Upper- and Lower-Bound Theorem there is no real zero less than -1, and from the Rational Zero Theorem the possible rational zeros (that are negative and greater than -1) are $\dfrac{p}{q} = -\dfrac{1}{2}, -\dfrac{1}{4}$, and $-\dfrac{3}{4}$. From the graph, it appears that $-\dfrac{3}{4}$ is a zero.

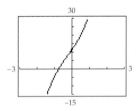

Use synthetic division with $c = -\dfrac{3}{4}$.

$$
-\frac{3}{4}\,\Big|\;4 \quad\; 3 \quad 16 \quad\;\; 12
$$
$$
\underline{\qquad\qquad -3 \quad\;\; 0 \quad -12}
$$
$$
\;\;\;\;\;\;\;\; 4 \quad\; 0 \quad 16 \quad\quad 0
$$

Thus $-\dfrac{3}{4}$ is a zero, and by the Factor Theorem,

$$4x^3 + 3x^2 + 16x + 12 = \left(x + \frac{3}{4}\right)(4x^2 + 16) = 0$$

Solve $4x^2 + 16 = 0$ to find that $x = -2i$ and $x = 2i$. The solutions of the original equation are $-\dfrac{3}{4}, -2i$, and $2i$.

42. Because P has real coefficients, use the Conjugate Pair Theorem.

$$
\begin{aligned}
P &= (x - [3 + 2i])(x - [3 - 2i])(x - 7) \\
&= (x - 3 - 2i)(x - 3 + 2i)(x - 7) \\
&= (x^2 - 6x + 13)(x - 7) \\
&= x^3 - 13x^2 + 55x - 91
\end{aligned}
$$

Section 3.5, page 311

2. Set the denominator equal to zero.

$$x^2 - 4 = 0$$
$$(x - 2)(x + 2) = 0$$
$$x = 2 \quad \text{or} \quad x = -2$$

The vertical asymptotes are $x = 2$ and $x = -2$.

6. The horizontal asymptote is $y = 0$ (x-axis) because the degree of the denominator is larger than the degree of the numerator.

10. Vertical asymptote: $x - 2 = 0$

$$x = 2$$

Horizontal asymptote: $y = 0$

No x-intercepts.

y-intercept: $\left(0, -\dfrac{1}{2}\right)$

26. Vertical asymptote: $\quad x^2 - 6x + 9 = 0$

$$(x - 3)(x - 3) = 0$$
$$x = 3$$

The horizontal asymptote is $y = \dfrac{1}{1} = 1$ (the Theorem on Horizontal Asymptotes) because the numerator and denominator both have degree 2. The graph crosses the horizontal asymptote at $\left(\dfrac{3}{2}, 1\right)$. The graph intersects, but does not cross, the x-axis at $(0, 0)$. See the graph below.

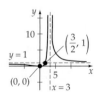

32.
$$
\require{enclose}
\begin{array}{r}
x + 1 \\
x^2 - 3x + 5 \enclose{longdiv}{x^3 - 2x^2 + 3x + 4} \\
\underline{x^3 - 3x^2 + 5x} \\
x^2 - 2x + 4 \\
\underline{x^2 - 3x + 5} \\
x - 1
\end{array}
$$

$$F(x) = x + 1 + \frac{x - 1}{x^2 - 3x + 5}$$

Slant asymptote: $y = x + 1$

48. $F(x) = \dfrac{x^2 - x - 12}{x^2 - 2x - 8} = \dfrac{(x - 4)(x + 3)}{(x - 4)(x + 2)} = \dfrac{x + 3}{x + 2}, x \neq 4$

The function F is undefined at $x = 4$. Thus the graph of F

is the graph of $y = \dfrac{x + 3}{x + 2}$ with an open circle at $\left(4, \dfrac{7}{6}\right)$.

The height of the open circle was found by evaluating

$y = \dfrac{x + 3}{x + 2}$ at $x = 4$.

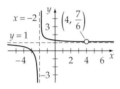

56. a.

$$\overline{C}(1000) = \frac{0.0006(1000)^2 + 9(1000) + 401{,}000}{1000}$$

$$= \$410.60$$

$$\overline{C}(10{,}000) = \frac{0.0006(10{,}000)^2 + 9(10{,}000) + 401{,}000}{10{,}000}$$

$$= \$55.10$$

$$\overline{C}(100{,}000) = \frac{0.0006(100{,}000)^2 + 9(100{,}000) + 401{,}000}{100{,}000}$$

$$= \$73.01$$

b. Graph $\overline{C}$ and use the minimum feature of a graphing utility.

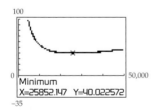

The minimum average cost per telephone is $40.02. The minimum is achieved by producing approximately 25,852 telephones.

Exercise Set 4.1, page 334

10. Because the graph of the given function is a line that passes through $(0, 6)$, $(2, 3)$, and $(6, -3)$, the graph of the inverse will be a line that passes through $(6, 0)$, $(3, 2)$, and $(-3, 6)$. See the following figure. Notice that the line shown below is a reflection of the line given in Exercise 10 across the line given by $y = x$.

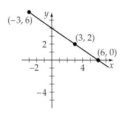

20. Check to see if $f[g(x)] = x$ for all x in the domain of g and $g[f(x)] = x$ for all x in the domain of f. The following shows that $f[g(x)] = x$ for all real numbers x.

$$f[g(x)] = f[2x + 3]$$

$$= \frac{1}{2}(2x + 3) - \frac{3}{2}$$

$$= x + \frac{3}{2} - \frac{3}{2}$$

$$= x$$

The following shows that $g[f(x)] = x$ for all real numbers x.

$$g[f(x)] = g\left[\frac{1}{2}x - \frac{3}{2}\right]$$

$$= 2\left(\frac{1}{2}x - \frac{3}{2}\right) + 3$$

$$= x - 3 + 3$$

$$= x$$

Thus f and g are inverses.

28.
$$f(x) = 4x - 8$$
$$y = 4x - 8 \qquad \text{• Replace } f(x) \text{ by } y.$$
$$x = 4y - 8 \qquad \text{• Interchange } x \text{ and } y.$$
$$x + 8 = 4y \qquad \text{• Solve for } y.$$
$$\frac{1}{4}(x + 8) = y$$
$$y = \frac{1}{4}x + 2$$
$$f^{-1}(x) = \frac{1}{4}x + 2 \qquad \text{• Replace } y \text{ by } f^{-1}(x).$$

34. $f(x) = \dfrac{x}{x-2}, x \neq 2$

$y = \dfrac{x}{x-2}$ • Replace $f(x)$ by y.

$x = \dfrac{y}{y-2}$ • Interchange x and y.

$x(y-2) = y$

$xy - 2x = y$ • Solve for y.

$xy - y = 2x$

$y(x-1) = 2x$

$y = \dfrac{2x}{x-1}$

$f^{-1}(x) = \dfrac{2x}{x-1}, x \neq 1$ • Replace y by $f^{-1}(x)$ and
                                               indicate any restrictions.

40. $f(x) = \sqrt{4-x}, x \leq 4$

$y = \sqrt{4-x}$ • Replace $f(x)$ by y.

$x = \sqrt{4-y}$ • Interchange x and y.

$x^2 = 4 - y$ • Solve for y.

$x^2 - 4 = -y$

$-x^2 + 4 = y$

$f^{-1}(x) = -x^2 + 4, x \geq 0$ • Replace y by $f^{-1}(x)$ and
                                        indicate any restrictions.

The range of f is $\{y \mid y \geq 0\}$. Therefore, the domain of f^{-1}
is $\{x \mid x \geq 0\}$, as indicated above.

50. $K(x) = 1.3x - 4.7$

$y = 1.3x - 4.7$ • Replace $K(x)$ by y.

$x = 1.3y - 4.7$ • Interchange x and y.

$x + 4.7 = 1.3y$ • Solve for y.

$\dfrac{x + 4.7}{1.3} = y$

$K^{-1}(x) = \dfrac{x + 4.7}{1.3}$ • Replace y by $K^{-1}(x)$.

The function $K^{-1}(x) = \dfrac{x + 4.7}{1.3}$ can be used to convert a
United Kingdom men's shoe size to its equivalent U.S.
shoe size.

Exercise Set 4.2, page 346

2. $f(3) = 5^3 = 5 \cdot 5 \cdot 5 = 125$

$f(-2) = 5^{-2} = \dfrac{1}{5^2} = \dfrac{1}{5 \cdot 5} = \dfrac{1}{25}$

22. The graph of $f(x) = \left(\dfrac{5}{2}\right)^x$ has a y-intercept of $(0, 1)$
and passes through $\left(1, \dfrac{5}{2}\right)$. Plot a few additional
points, such as $\left(-1, \dfrac{2}{5}\right)$ and $\left(2, \dfrac{25}{4}\right)$. Because the base
$\dfrac{5}{2}$ is greater than 1, we know that the graph must have
all the properties of an increasing exponential function.
Draw a smooth increasing curve through the points. The
graph should be asymptotic to the negative portion of
the x-axis, as shown in the following figure.

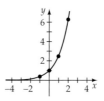

28. Because $F(x) = 6^{x+5} = f(x + 5)$, the graph of $F(x)$ can be
produced by shifting the graph of f horizontally to the
left 5 units.

30. Because $F(x) = -\left[\left(\dfrac{5}{2}\right)^x\right] = -f(x)$, the graph of $F(x)$
can be produced by reflecting the graph of f across the
x-axis.

44. a. $A(45) = 200e^{-0.014(45)}$

≈ 106.52

After 45 minutes the patient will have about
107 milligrams of medication in his or her
bloodstream.

b. Use a graphing calculator to graph $y = 200e^{-0.014x}$
and $y = 50$ in the same viewing window as shown
below.

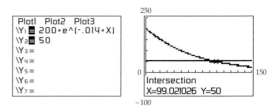

The x-coordinate (which represents time in minutes) of
the point of intersection is about 99.02. Thus it will take
about 99 minutes before the patient's medication level is
reduced to 50 milligrams.

54. a. $P(0) = \dfrac{3600}{1 + 7e^{-0.05(0)}}$

$= \dfrac{3600}{1 + 7}$

$= \dfrac{3600}{8}$

$= 450$

Immediately after the lake was stocked, the lake contained 450 bass.

b. $P(12) = \dfrac{3600}{1 + 7e^{-0.05(12)}}$

≈ 743.54

After 1 year (12 months) there were about 744 bass in the lake.

c. As $t \to \infty$, $7e^{-0.05t} = \dfrac{7}{e^{0.05t}}$ approaches 0. Thus as $t \to \infty$,

$P(t) = \dfrac{3600}{1 + 7e^{-0.05t}}$ will approach $\dfrac{3600}{1 + 0} = 3600$. As time goes by the bass population will increase, approaching 3600.

Exercise Set 4.3, page 361

4. The exponential form of $\log_b x = y$ is $b^y = x$. Thus the exponential form of $\log_4 64 = 3$ is $4^3 = 64$.

12. The logarithmic form of $b^y = x$ is $y = \log_b x$. Thus the logarithmic form of $5^3 = 125$ is $3 = \log_5 125$.

28. $\log 1{,}000{,}000 = \log_{10} 10^6 = 6$

32. To graph $y = \log_6 x$, use the equivalent exponential equation $x = 6^y$. Choose some y-values, such as $-1, 0, 1$, and calculate the corresponding x-values. This yields the ordered pairs $\left(\dfrac{1}{6}, -1 \right)$, $(1, 0)$, and $(6, 1)$. Plot these ordered pairs and draw a smooth curve through the points to produce the following graph.

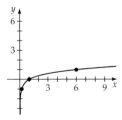

40. $\log_4(5 - x)$ is defined only for $5 - x > 0$, which is equivalent to $x < 5$. Using interval notation, the domain of $k(x) = \log_4(5 - x)$ is $(-\infty, 5)$.

50. The graph of $f(x) = \log_6(x + 3)$ can be produced by shifting the graph of $f(x) = \log_6 x$ (from Exercise 32) 3 units to the left. See the figure at the top of the next column.

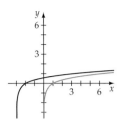

70. a. $S(0) = 5 + 29 \ln(0 + 1) = 5 + 0 = 5$. When starting, the student had an average typing speed of 5 words per minute. $S(3) = 5 + 29 \ln(3 + 1) \approx 45.2$. After 3 months the student's average typing speed was about 45 words per minute.

b. Use the intersection feature of a graphing utility to find the x-coordinate of the point of intersection of the graphs of $y = 5 + 29 \ln(x + 1)$ and $y = 65$.

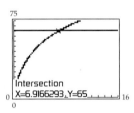

The graphs intersect at about $(6.9, 65)$. The student will achieve a typing speed of 65 words per minute in about 6.9 months.

Exercise Set 4.4, page 373

2. $\ln \dfrac{z^3}{\sqrt{xy}} = \ln z^3 - \ln \sqrt{xy}$

$= \ln z^3 - \ln(xy)^{1/2}$

$= 3 \ln z - \dfrac{1}{2} \ln(xy)$

$= 3 \ln z - \dfrac{1}{2}(\ln x + \ln y)$

$= 3 \ln z - \dfrac{1}{2} \ln x - \dfrac{1}{2} \ln y$

10. $3 \log_2 t - \dfrac{1}{3} \log_2 u + 4 \log_2 v = \log_2 t^3 - \log_2 u^{1/3} + \log_2 v^4$

$= \log_2 \dfrac{t^3}{u^{1/3}} + \log_2 v^4$

$= \log_2 \dfrac{t^3 v^4}{u^{1/3}}$

$= \log_2 \dfrac{t^3 v^4}{\sqrt[3]{u}}$

16. $\log_5 37 = \dfrac{\log 37}{\log 5} \approx 2.2436$

24. $\log_8(5 - x) = \dfrac{\ln(5 - x)}{\ln 8}$, so enter $\dfrac{\ln(5 - x)}{\ln 8}$ into Y1 on a graphing calculator.

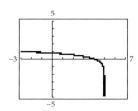

48. $\text{pH} = -\log[H^+] = -\log(1.26 \times 10^{-3}) \approx 2.9$

50.
$$\text{pH} = -\log[H^+]$$
$$5.6 = -\log[H^+]$$
$$-5.6 = \log[H^+]$$
$$10^{-5.6} = H^+$$

The hydronium-ion concentration is $10^{-5.6} \approx 2.51 \times 10^{-6}$ mole per liter.

56. $M = \log\left(\dfrac{I}{I_0}\right) = \log\left(\dfrac{398{,}107{,}000I_0}{I_0}\right) = \log(398{,}107{,}000)$

≈ 8.6

58. $\log\left(\dfrac{I}{I_0}\right) = 9.5$

$\dfrac{I}{I_0} = 10^{9.5}$

$I = 10^{9.5}I_0$

$I \approx 3{,}162{,}277{,}660 I_0$

60. In Example 7 we noticed that if an earthquake has a Richter scale magnitude of M_1 and a smaller earthquake has a Richter scale magnitude of M_2, then the first earthquake is $10^{M_1-M_2}$ times as intense as the smaller earthquake. In this exercise, $M_1 = 9.5$ and $M_2 = 8.3$. Thus $10^{M_1-M_2} = 10^{9.5-8.3} = 10^{1.2} \approx 15.8$. The 1960 earthquake in Chile was about 15.8 times as intense as the San Francisco earthquake of 1906.

64. $M = \log A + 3\log 8t - 2.92$

$= \log 26 + 3\log[8 \cdot 17] - 2.92$ • **Substitute 26 for**

$\approx 1.4150 + 6.4006 - 2.92$ **A and 17 for t.**

≈ 4.9

Exercise Set 4.5, page 385

2. $3^x = 243$

$3^x = 3^5$

$x = 5$

10.
$$6^x = 50$$
$$\log(6^x) = \log 50$$
$$x \log 6 = \log 50$$
$$x = \dfrac{\log 50}{\log 6} \approx 2.18$$

18.
$$3^{x-2} = 4^{2x+1}$$
$$\log 3^{x-2} = \log 4^{2x+1}$$
$$(x - 2)\log 3 = (2x + 1)\log 4$$
$$x \log 3 - 2\log 3 = 2x \log 4 + \log 4$$
$$x \log 3 - 2\log 3 - 2x \log 4 = \log 4$$
$$x \log 3 - 2x \log 4 = \log 4 + 2\log 3$$
$$x(\log 3 - 2\log 4) = \log 4 + 2\log 3$$
$$x = \dfrac{\log 4 + 2\log 3}{\log 3 - 2\log 4}$$
$$x \approx -2.141$$

22. $\log(x^2 + 19) = 2$

$x^2 + 19 = 10^2$

$x^2 + 19 = 100$

$x^2 = 81$

$x = \pm 9$

A check shows that 9 and -9 are both solutions of the original equation.

26. $\log_3 x + \log_3(x + 6) = 3$

$\log_3[x(x + 6)] = 3$

$3^3 = x(x + 6)$

$27 = x^2 + 6x$

$x^2 + 6x - 27 = 0$

$(x + 9)(x - 3) = 0$

$x = -9$ or $x = 3$

Because $\log_3 x$ is defined only for $x > 0$, the only solution is $x = 3$.

36. $\ln x = \dfrac{1}{2}\ln\left(2x + \dfrac{5}{2}\right) + \dfrac{1}{2}\ln 2$

$= \dfrac{1}{2}\left[\ln\left(2x + \dfrac{5}{2}\right) + \ln 2\right]$

$\ln x = \dfrac{1}{2}\ln\left[2\left(2x + \dfrac{5}{2}\right)\right]$

$\ln x = \dfrac{1}{2}\ln(4x + 5)$

$\ln x = \ln(4x + 5)^{1/2}$

$x = \sqrt{4x + 5}$

$x^2 = 4x + 5$

$0 = x^2 - 4x - 5$

$0 = (x - 5)(x + 1)$

$x = 5$ or $x = -1$

Check: $\ln 5 = \dfrac{1}{2}\ln\left(10 + \dfrac{5}{2}\right) + \dfrac{1}{2}\ln 2$

$1.6094 \approx 1.2629 + 0.3466$

Because $\ln(-1)$ is not defined, -1 is not a solution. Thus the only solution is $x = 5$.

40. $\dfrac{10^x + 10^{-x}}{2} = 8$

$10^x + 10^{-x} = 16$

$10^x(10^x + 10^{-x}) = (16)10^x$ • **Multiply each side**

$10^{2x} + 1 = 16(10^x)$ **by 10^x.**

$10^{2x} - 16(10^x) + 1 = 0$

$u^2 - 16u + 1 = 0$ • **Let $u = 10^x$.**

$u = \dfrac{16 \pm \sqrt{16^2 - 4(1)(1)}}{2} = 8 \pm 3\sqrt{7}$

$10^x = 8 \pm 3\sqrt{7}$ • **Replace u with 10^x.**

$\log 10^x = \log(8 \pm 3\sqrt{7})$

$x = \log(8 \pm 3\sqrt{7}) \approx \pm 1.20241$

68. a. $t = \dfrac{9}{24}\ln\dfrac{24 + v}{24 - v}$

$1.5 = \dfrac{9}{24}\ln\dfrac{24 + v}{24 - v}$

$4 = \ln\dfrac{24 + v}{24 - v}$

$e^4 = \dfrac{24 + v}{24 - v}$ • **$N = \ln M$ means $e^N = M$.**

$(24 - v)e^4 = 24 + v$

$-v - ve^4 = 24 - 24e^4$

$v(-1 - e^4) = 24 - 24e^4$

$v = \dfrac{24 - 24e^4}{-1 - e^4} \approx 23.14$

The velocity is about 23.14 feet per second.

b. The vertical asymptote is $v = 24$.

c. Due to the air resistance, the object cannot reach or exceed a velocity of 24 feet per second.

Exercise Set 4.6, page 400

4. a. $P = 12{,}500, r = 0.08, t = 10, n = 1.$

$A = 12{,}500\left(1 + \dfrac{0.08}{1}\right)^{10} \approx \$26{,}986.56$

b. $n = 365$

$A = 12{,}500\left(1 + \dfrac{0.08}{365}\right)^{3650} \approx \$27{,}816.82$

c. $n = 8760$

$A = 12{,}500\left(1 + \dfrac{0.08}{8760}\right)^{87600} \approx \$27{,}819.16$

6. $P = 32{,}000, r = 0.08, t = 3.$

$A = Pe^{rt} = 32{,}000e^{3(0.08)} \approx \$40{,}679.97$

10. $t = \dfrac{\ln 3}{r}$ $r = 0.055$

$t = \dfrac{\ln 3}{0.055}$

$t \approx 20$ years (to the nearest year)

18. a. $P(12) = 20{,}899(1.027)^{12} \approx 28{,}722$ thousands, or 28,772,000.

b. P is in thousands, so

$35{,}000 = 20{,}899(1.027)^t$

$\dfrac{35{,}000}{20{,}899} = 1.027^t$

$\ln\left(\dfrac{35{,}000}{20{,}899}\right) = t\ln 1.027$

$\dfrac{\ln\left(\dfrac{35{,}000}{20{,}899}\right)}{\ln 1.027} = t$

$19.35 \approx t$

According to the growth function, the population will first exceed 35 million in 19.35 years—that is, in the year $1991 + 19 = 2010$.

20. $N(t) = N_0 e^{kt}$

$N(138) = N_0 e^{138k}$

$0.5N_0 = N_0 e^{138k}$

$0.5 = e^{138k}$

$\ln 0.5 = 138k$

$k = \dfrac{\ln 0.5}{138} \approx -0.005023$

$N(t) = N_0(0.5)^{t/138} \approx N_0 e^{-0.005023t}$

24. $N(t) = N_0(0.5)^{t/5730}$

$0.65N_0 = N_0(0.5)^{t/5730}$

$0.65 = (0.5)^{t/5730}$

$\ln 0.65 = \ln(0.5)^{t/5730}$

$t = 5730\dfrac{\ln 0.65}{\ln 0.5} \approx 3600$

The bone is approximately 3600 years old.

32. a.

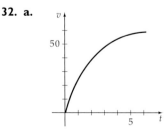

b. Here is an algebraic solution. An approximate solution can be obtained from the graph.

$$v = 64(1 - e^{-t/2})$$

$$50 = 64(1 - e^{-t/2})$$

$$\frac{50}{64} = (1 - e^{-t/2})$$

$$1 - \frac{50}{64} = e^{-t/2}$$

$$\ln\left(1 - \frac{50}{64}\right) = -\frac{t}{2}$$

$$t = -2\ln\left(1 - \frac{50}{64}\right) \approx 3.0$$

The velocity is 50 feet per second in approximately 3.0 seconds.

c. As $t \to \infty$, $e^{-t/2} \to 0$. Therefore, $64(1 - e^{-t/2}) \to 64$. The horizontal asymptote is $v = 64$.

d. Because of the air resistance, the velocity of the object will never reach or exceed 64 feet per second.

48. a. Represent the year 2001 by $t = 0$; then the year 2002 will be represented by $t = 1$. Use the following substitutions: $P_0 = 240$, $P(1) = 310$, $c = 3400$, and

$$a = \frac{c - P_0}{P_0} = \frac{3400 - 240}{240} \approx 13.16667.$$

$$P(t) = \frac{c}{1 + ae^{-bt}}$$

$$P(1) = \frac{3400}{1 + 13.16667e^{-b(1)}}$$

$$310 = \frac{3400}{1 + 13.16667e^{-b}}$$

$$310(1 + 13.16667e^{-b}) = 3400$$

$$1 + 13.16667e^{-b} = \frac{3400}{310}$$

$$13.16667e^{-b} = \frac{3400}{310} - 1$$

$$13.16667e^{-b} \approx 9.96774$$

$$e^{-b} \approx \frac{9.96774}{13.16667}$$

$$-b \approx \ln\frac{9.96774}{13.16667}$$

$$b \approx 0.27833$$

Using $a = 13.16667$, $b = 0.27833$, and $c = 3400$ gives the following logistic model.

$$P(t) \approx \frac{3400}{1 + 13.16667e^{-0.27833t}}$$

b. Because 2008 is 7 years past 2001, the year 2008 is represented by $t = 7$.

$$P(7) \approx \frac{3400}{1 + 13.16667e^{-0.27833(7)}} \approx 1182$$

According to the model there will be about 1182 groundhogs in 2008.

Exercise Set 5.1, page 425

6. $\begin{cases} 8x + 3y = -7 & (1) \\ x = 3y + 15 & (2) \end{cases}$

$\quad 8(3y + 15) + 3y = -7$ • Replace x in Eq. (1).

$\quad 24y + 120 + 3y = -7$ • Simplify.

$\quad\quad\quad\quad 27y = -7$

$$y = -\frac{127}{27}$$

$x = 3\left(-\dfrac{127}{27}\right) + 15 = \dfrac{8}{9}$ • Substitute $-\dfrac{127}{27}$ for y in Eq. (2).

The solution is $\left(\dfrac{8}{9}, -\dfrac{127}{27}\right)$.

18. $\begin{cases} 3x - 4y = 8 & (1) \\ 6x - 8y = 9 & (2) \end{cases}$

$\quad\quad\quad 8y = 6x - 9$ • Solve Eq. (2) for y.

$$y = \frac{3}{4}x - \frac{9}{8}$$

$3x - 4\left(\dfrac{3}{4}x - \dfrac{9}{8}\right) = 8$ • Replace y in Eq. (1).

$\quad\quad 3x - 3x + \dfrac{9}{2} = 8$ • Simplify.

$$\frac{9}{2} = 8$$

This is a false equation. Therefore, the system of equations is inconsistent and has no solution.

20. $\begin{cases} 5x + 2y = 2 & (1) \\ y = -\dfrac{5}{2}x + 1 & (2) \end{cases}$

$5x + 2\left(-\dfrac{5}{2}x + 1\right) = 2$ • Replace y in Eq. (1).

$\quad\quad 5x - 5x + 2 = 2$ • Simplify.

$$2 = 2$$

This is a true statement; therefore, the system of equations is dependent. Let $x = c$. Then $y = -\dfrac{5}{2}c + 1$. Thus the solutions are $\left(c, -\dfrac{5}{2}c + 1\right)$.

24. $\begin{cases} 3x - 8y = -6 & (1) \\ -5x + 4y = 10 & (2) \end{cases}$

$3x - 8y = -6$

$\underline{-10x + 8y = 20} \qquad \bullet \text{ 2 times Eq. (2)}$

$-7x \qquad\quad = 14$

$x = -2$

$3(-2) - 8y = -6 \qquad \bullet \text{ Substitute } -2 \text{ for } x \text{ in Eq. (1).}$
$\qquad\qquad\qquad\qquad\quad \text{Solve for } y.$

$-8y = 0$

$y = 0$

The solution is $(-2, 0)$.

28. $\begin{cases} 4x + 5y = 2 & (1) \\ 8x - 15y = 9 & (2) \end{cases}$

$12x + 15y = 6 \qquad \bullet \text{ 3 times Eq. (1)}$

$\underline{8x - 15y = 9}$

$20x \qquad\quad = 15$

$x = \dfrac{3}{4}$

$4\left(\dfrac{3}{4}\right) + 5y = 2 \qquad \bullet \text{ Substitute } \dfrac{3}{4} \text{ for } x \text{ in Eq. (1).}$

$3 + 5y = 2 \qquad \bullet \text{ Solve for } y.$

$y = -\dfrac{1}{5}$

The solution is $\left(\dfrac{3}{4}, -\dfrac{1}{5}\right)$.

42. Solve the system of equations $\begin{cases} x = 25p - 500 \\ x = -7p + 1100 \end{cases}$ by the substitution method.

$25p - 500 = -7p + 1100$

$32p = 1600$

$p = 50$

The equilibrium price is $50.

46. Let r = the rate of the canoeist.
Let w = the rate of the current.
Rate of canoeist with the current: $r + w$
Rate of canoeist against the current: $r - w$

$r \cdot t = d$

$(r + w) \cdot 2 = 12 \quad (1)$

$(r - w) \cdot 4 = 12 \quad (2)$

$r + w = 6 \qquad \bullet \text{ Divide Eq. (1) by 2.}$

$\underline{r - w = 3} \qquad \bullet \text{ Divide Eq. (2) by 4.}$

$2r \qquad = 9$

$r = 4.5$

$4.5 + w = 6$

$w = 1.5$

Rate of canoeist = 4.5 miles per hour
Rate of current = 1.5 miles per hour

Exercise Set 5.2, page 439

12. $\begin{cases} 3x + 2y - 5z = 6 & (1) \\ 5x - 4y + 3z = -12 & (2) \\ 4x + 5y - 2z = 15 & (3) \end{cases}$

$15x + 10y - 25z = 30 \qquad \bullet \text{ 5 times Eq. (1)}$

$\underline{-15x + 12y - 9z = 36} \qquad \bullet \text{ −3 times Eq. (2)}$

$22y - 34z = 66 \qquad \bullet \text{ Divide by 2.}$

$11y - 17z = 33 \quad (4)$

$12x + 8y - 20z = 24 \qquad \bullet \text{ 4 times Eq. (1)}$

$\underline{-12x - 15y + 6z = -45} \qquad \bullet \text{ −3 times Eq. (3)}$

$-7y - 14z = -21 \qquad \bullet \text{ Divide by −7.}$

$y + 2z = 3 \quad (5)$

$11y - 17z = 33 \quad (4)$

$\underline{-11y - 22z = -33} \qquad \bullet \text{ −11 times Eq. (5)}$

$-39z = 0$

$z = 0 \quad (6)$

$11y - 17(0) = 33$

$y = 3$

$3x + 2(3) - 5(0) = 6$

$x = 0$

The solution is $(0, 3, 0)$.

16. $\begin{cases} 2x + 3y + 2z = 14 & (1) \\ x - 3y + 4z = 4 & (2) \\ -x + 12y - 6z = 2 & (3) \end{cases}$

$2x + 3y + 2z = 14 \quad (1)$

$\underline{-2x + 6y - 8z = -8} \qquad \bullet \text{ −2 times Eq. (2)}$

$9y - 6z = 6 \qquad \bullet \text{ Divide by 3.}$

$3y - 2z = 2 \quad (4)$

$2x + 3y + 2z = 14 \quad (1)$

$\underline{-2x + 24y - 12z = 4} \qquad \bullet \text{ 2 times Eq. (3)}$

$27y - 10z = 18 \quad (5)$

$-27y + 18z = -18 \qquad \bullet \text{ −9 times Eq. (4)}$

$\underline{27y - 10z = 18} \quad (5)$

$8z = 0$

$z = 0 \quad (6)$

$3y - 2(0) = 2 \qquad \bullet \text{ Substitute } z = 0 \text{ in Eq. (4).}$

$y = \dfrac{2}{3}$

(continued)

$2x + 3\left(\dfrac{2}{3}\right) + 2(0) = 14$ • Substitute $y = \dfrac{2}{3}$ and

$z = 0$ in Eq. (1).

$$x = 6$$

The solution is $\left(6, \dfrac{2}{3}, 0\right)$.

18. $\begin{cases} 2x + 3y - 6z = 4 \quad (1) \\ 3x - 2y - 9z = -7 \quad (2) \\ 2x + 5y - 6z = 8 \quad (3) \end{cases}$

$\begin{aligned} 6x + 9y - 18z &= 12 \\ -6x + 4y + 18z &= 14 \\ \hline 13y &= 26 \end{aligned}$ • 3 times Eq. (1)
    • −2 times Eq. (2)

$y = 2 \quad (4)$

$\begin{aligned} 2x + 3y - 6z &= 4 \quad (1) \\ -2x - 5y + 6z &= -8 \\ \hline -2y &= -4 \end{aligned}$ • −1 times Eq. (3)

$y = 2 \quad (5)$

$\begin{aligned} y &= 2 \quad (4) \\ -y &= -2 \\ \hline 0 &= 0 \quad (6) \end{aligned}$ • −1 times Eq. (5)

The equations are dependent. Let $z = c$.

$2x + 3(2) - 6c = 4$ • Substitute $y = 2$ and $z = c$ in Eq. (1).

$$x = 3c - 1$$

The solutions are $(3c - 1, 2, c)$.

20. $\begin{cases} x - 3y + 4z = 9 \quad (1) \\ 3x - 8y - 2z = 4 \quad (2) \end{cases}$

$\begin{aligned} -3x + 9y - 12z &= -27 \\ 3x - 8y - 2z &= 4 \quad (2) \\ \hline y - 14z &= -23 \quad (3) \end{aligned}$ • −3 times Eq. (1)

$y = 14z - 23$ • Solve Eq. (3) for y.

$x - 3(14z - 23) + 4z = 9$ • Substitute $14z - 23$ for y in Eq. (1).

$x = 38z - 60$ • Solve for x.

Let $z = c$. The solutions are $(38c - 60, 14c - 23, c)$.

32. $\begin{cases} 5x + 2y + 3z = 0 \quad (1) \\ 3x + y - 2z = 0 \quad (2) \\ 4x - 7y + 5z = 0 \quad (3) \end{cases}$

$\begin{aligned} 15x + 6y + 9z &= 0 \\ -15x - 5y + 10z &= 0 \\ \hline y + 19z &= 0 \quad (4) \end{aligned}$ • 3 times Eq. (1)
    • −5 times Eq. (2)

$\begin{aligned} 20x + 8y + 12z &= 0 \\ -20x + 35y - 25z &= 0 \\ \hline 43y - 13z &= 0 \quad (5) \end{aligned}$ • 4 times Eq. (1)
    • −5 times Eq. (3)

$\begin{aligned} -43y - 817z &= 0 \\ 43y - 13z &= 0 \quad (5) \\ \hline -830z &= 0 \end{aligned}$ • −43 times Eq. (4)

$z = 0 \quad (6)$

Solving by back substitution, the only solution is $(0, 0, 0)$.

36. $x^2 + y^2 + ax + by + c = 0$

$\begin{cases} 0 + 36 + a(0) + b(6) + c = 0 & \text{• Let } x = 0, y = 6. \\ 1 + 25 + a(1) + b(5) + c = 0 & \text{• Let } x = 1, y = 5. \\ 49 + 1 + a(-7) + b(-1) + c = 0 & \text{• Let } x = -7, y = -1. \end{cases}$

$\begin{cases} 6b + c = -36 \quad (1) \\ a + 5b + c = -26 \quad (2) \\ -7a - b + c = -50 \quad (3) \end{cases}$

$\begin{aligned} 7a + 35b + 7c &= -182 \\ -7a - b + c &= -50 \quad (3) \\ \hline 34b + 8c &= -232 \end{aligned}$ • 7 times Eq. (2)

$17b + 4c = -116 \quad (4)$

$\begin{aligned} -24b - 4c &= 144 \\ 17b + 4c &= -116 \quad (4) \\ \hline -7b &= 28 \end{aligned}$ • −4 times Eq. (1)

$b = -4$

$17(-4) + 4c = -116$ • Substitute −4 for b in Eq. (4).

$c = -12$

$-7a - (-4) - 12 = -50$ • Substitute −4 for b and −12 for c in Eq. (3).

$a = 6$

An equation of the circle whose graph passes through the three given points is $x^2 + y^2 + 6x - 4y - 12 = 0$.

42. Let x_1, x_2, x_3, and x_4 represent the numbers of cars per hour that travel AB, BC, CD, and DA, respectively. Using the principle that the number of cars entering an intersection must equal the number of cars leaving the intersection, we can write the following equations.

$A: 75 + x_4 = x_1 + 60$

$B: x_1 + 50 = x_2 + 100$

$C: x_2 + 45 = x_3 + 50$

$D: x_3 + 80 = x_4 + 40$

The equations for the traffic intersections result in the following system of equations.

$\begin{cases} x_1 - x_4 = 15 \quad (1) \\ x_1 - x_2 = 50 \quad (2) \\ x_2 - x_3 = 5 \quad (3) \\ x_3 - x_4 = -40 \quad (4) \end{cases}$

Subtracting Equation (2) from Equation (1) gives

$$x_1 - x_4 = 15$$
$$\underline{x_1 - x_2 = 50}$$
$$x_2 - x_4 = -35 \quad (5)$$

Adding Equation (3) and Equation (4) gives

$$x_2 - x_3 = 5$$
$$\underline{x_3 - x_4 = -40}$$
$$x_2 - x_4 = -35 \quad (6)$$

Because Equation (5) and Equation (6) are the same, the system of equations is dependent. Because we want to know the cars per hour between B and C, solve the system in terms of x_2.

$$x_1 = x_2 + 50$$
$$x_3 = x_2 - 5$$
$$x_4 = x_2 + 35$$

Because there cannot be a negative number of cars per hour between two intersections, to ensure that $x_3 \geq 0$, we must have $x_2 \geq 5$. The minimum number of cars traveling between B and C is 5 cars per hour.

Exercise Set 5.3, page 451

6.

10.

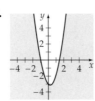

12.

20.

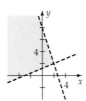

28.

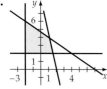

46. $x =$ hours of machine 1 use

$y =$ hours of machine 2 use

Cost $= 28x + 25y$

Constraints: $\begin{cases} 4x + 3y \geq 60 \\ 5x + 10y \geq 100 \\ x \geq 0, y \geq 0 \end{cases}$

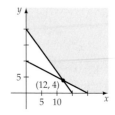

(x, y)	Cost	
(0, 20)	500	
(12, 4)	436	• Minimum
(20, 0)	560	

To achieve the minimum cost, use machine 1 for 12 hours and machine 2 for 4 hours.

48. Let $x =$ number of standard models.

Let $y =$ number of deluxe models.

Profit $= 25x + 35y$

Constraints: $\begin{cases} x + 3y \leq 24 \\ x + y \leq 10 \\ 2x + y \leq 16 \\ x \geq 0, y \geq 0 \end{cases}$

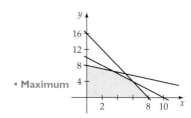

(x, y)	Profit	
(0, 0)	0	
(0, 8)	280	
(6, 4)	290	
(3, 7)	320	• Maximum
(8, 0)	200	

To maximize profits, produce three standard models and seven deluxe models.

ANSWERS TO SELECTED EXERCISES

Exercise Set P.1, page 15

1. Integers: $0, -44, \sqrt{81}, 53$; Rational numbers: $-\dfrac{1}{5}, 0, -44, 3.14, \sqrt{81}, 53$; Irrational numbers: $\pi, 5.05005000500005\ldots$; Prime number: 5
Real numbers: All the numbers are real numbers. **3.** $2, 4, 6, 8$ **5.** $3, 5, 7, 9$ **7.** $0, 1, 2, 3$ **9.** $\{-3, -2, -1, 0, 1, 2, 3, 4, 6\}$
11. $\{0, 1, 2, 3\}$ **13.** $\varnothing$ **15.** $\{1, 3\}$ **17.** $\{-2, 0, 1, 2, 3, 4, 6\}$ **19.** $\{x\,|\,-2 < x < 3\}$,

21. $\{x\,|\,-5 \le x \le -1\}$, **23.** $\{x\,|\,x \ge 2\}$,

25. $(3, 5)$, **27.** $[-2, \infty)$, **29.** $[0, 1]$,

31. -5 **33.** 12 **35.** $\pi^2 + 10$ **37.** 9 **39.** $3x - 1$ **41.** $|x - 3|$ **43.** $|x + 2| = 4$ **45.** $|m - n|$ **47.** $|a - 4| < 5$

49. $|x + 2| > 4$ **51.** **53.** **55.**

57. **59.** **61.**

63. **65.** **67.** 8 **69.** 12 **71.** -72 **73.** 19 **75.** 13 **77.** -3

79. Associative property of multiplication **81.** Distributive property **83.** Commutative property of multiplication **85.** Identity property of multiplication **87.** Reflexive property of equality **89.** Transitive property of equality **91.** Inverse property of multiplication

93. $6x$ **95.** $3x + 6$ **97.** $\dfrac{3}{2}a$ **99.** $6x - 13$ **101.** $-12x + 6y + 5$ **103.** $2a$ **105.** $21a + 6$ **107.** 6 square inches

109. \$5150 **111.** 66 beats per minute **113.** 100 feet **115.** A **117.** $\varnothing$ **119.** B is a subset of A.
121. No. $(8 \div 4) \div 2 = 2 \div 2 = 1, 8 \div (4 \div 2) = 8 \div 2 = 4$ **123.** All but the multiplicative inverse property **125.** $x + 7$
127. $|x - 2| < |x - 6|$ **129.** $|x - 3| > |x + 7|$ **131.** $2 < |x - 4| < 7$

Prepare for Section P.2, page 18

133. 32 **134.** $\dfrac{1}{16}$ **135.** 64 **136.** $314{,}000$ **137.** False **138.** False

Exercise Set P.2, page 31

1. -125 **3.** 1 **5.** $\dfrac{1}{16}$ **7.** 32 **9.** 27 **11.** -2 **13.** $\dfrac{2}{x^4}$ **15.** $6a^3b^8$ **17.** $\dfrac{3}{4a^4}$ **19.** $\dfrac{2y^2}{3x^2}$ **21.** $\dfrac{12}{a^3b}$ **23.** $-18m^5n^6$ **25.** $\dfrac{1}{x^6}$

27. $\dfrac{1}{4a^4b^2}$ **29.** $2x$ **31.** $\dfrac{b^{10}}{a^{10}}$ **33.** 2.011×10^{12} **35.** 5.62×10^{-10} **37.** $31{,}400{,}000$ **39.** -0.0000023 **41.** 2.7×10^8 **43.** 1.5×10^{-11}

45. 7.2×10^{12} **47.** 8×10^{-16} **49.** 8 **51.** -16 **53.** $\dfrac{1}{27}$ **55.** $\dfrac{2}{3}$ **57.** 16 **59.** $8ab^2$ **61.** $-12x^{11/12}$ **63.** $3x^2y^3$ **65.** $\dfrac{4z^{2/5}}{3}$

67. $6x^{5/6}y^{5/6}$ **69.** $\dfrac{3a^{1/12}}{b}$ **71.** $3\sqrt{5}$ **73.** $2\sqrt[3]{3}$ **75.** $-3\sqrt[3]{5}$ **77.** $2|x|y\sqrt{6y}$ **79.** $2ay^2\sqrt[3]{2y}$ **81.** $-13\sqrt{2}$ **83.** $-10\sqrt[4]{3}$

85. $17y\sqrt[3]{4y}$ **87.** $-14x^2y\sqrt[3]{y}$ **89.** $17 + 7\sqrt{5}$ **91.** -7 **93.** $12z + \sqrt{z} - 6$ **95.** $x + 4\sqrt{x} + 4$ **97.** $x + 4\sqrt{x - 3} + 1$ **99.** $\sqrt{2}$

101. $\dfrac{\sqrt{10}}{6}$ **103.** $\dfrac{3\sqrt[3]{4}}{2}$ **105.** $\dfrac{2\sqrt[3]{x}}{x}$ **107.** $-\dfrac{3\sqrt{3} - 12}{13}$ **109.** $\dfrac{3\sqrt{5} - 3}{4}$ **111.** $\dfrac{3\sqrt{5} - 3\sqrt{x}}{5 - x}$ **113.** $\approx 2.21 \times 10^4$ dollars

115. $\approx 3.13 \times 10^7$ **117.** $\approx 1.38 \times 10^{-2}$ **119.** 8 minutes **121.** \$22,688 **123.** ≈ 8.91 billion **125. a.** 56% **b.** 24%

127. No. $2 < 3$, but $\dfrac{1}{2} > \dfrac{1}{3}$. **129.** $\dfrac{8}{5}$ **131.** $-\dfrac{19}{12}$ **133.** $\dfrac{1}{\sqrt{4 + h} + 2}$ **135.** $\dfrac{1}{\sqrt{n^2 + 1} + n}$ **137.** 2

142. False   **143.** False

$2, -7$ **d.** 1   **e.** $x^2, 2x, -7$ **13. a.** $x^3 - 1$ **b.** 3   **c.** $1, -1$

.. 2 **e.** $2x^4, 3x^3, 4x^2, 5$ **17.** 3   **19.** 5   **21.** $5x^2 + 11x + 3$

$26x^2 - 29x + 10$ **29.** $y^2 + 3y + 2$ **31.** $4z^2 - 19z + 12$

39. $12d^2 + 4d - 8$ **41.** $r^3 + s^3$ **43.** $60c^3 - 49c^2 + 4$

$(x + 4)$ **53.** $-3x(5x + 4)$ **55.** $2xy(5x + 3 - 7y)$

63. $(6x + 1)(x + 4)$ **65.** $(17x + 4)(3x - 1)$ **67.** $(x^2 + 5)(x^2 + 1)$

.)   **75.** $(3x^2 - 1)(x^2 + 4)$ **77.** $(3x^3 - 4)(x^3 + 2)$ **79.** $(x - 3)(x + 3)$

$+ 3)$ **87.** $(a - 7)^2$ **89.** $(2x + 3)^2$ **91.** $(z^2 + 2w^2)^2$

$7.$ $(2 - x^2)(4 + 2x^2 + x^4)$ **99.** $(3x + 1)(x^2 + 2)$ **101.** $(x - 1)(ax + b)$

$- 1)(2x + 1)(4x^2 + 1)$ **109.** $b(3x + 4)(x - 1)(x + 1)$ **111.** $2b(6x + y)^2$

$1)$ **117.** not factorable over the integers   **119.** 8   **121.** 64

Prepare

127. $\dfrac{8}{5}$ **128.** $\dfrac{}{wy}$

131. $(x - 6)(x + 1)$ **132.** $(x - 4)(x^2 + 4x + 16)$

Exercise Set P.4, page 5

1. $\dfrac{x + 4}{3}$ **3.** $\dfrac{x - 3}{x - 2}$ **5.** $\dfrac{a^2 - 2a + 4}{a - 2}$ **7.** $-\dfrac{x + 8}{x + 2}$ **9.** $-\dfrac{4y^2 + 7}{y + 7}$ **11.** $-\dfrac{8}{a^3b}$ **13.** $\dfrac{10}{27q^2}$ **15.** $\dfrac{x(3x + 7)}{2x + 3}$ **17.** $\dfrac{x + 3}{2x + 3}$

19. $\dfrac{(2y + 3)(3y - 4)}{(2y - 3)(y + 1)}$ **21.** $\dfrac{1}{a - 8}$ **23.** $\dfrac{3p - 2}{r}$ **25.** $\dfrac{8x(x - 4)}{(x - 5)(x + 3)}$ **27.** $\dfrac{3y - 4}{y + 4}$ **29.** $\dfrac{7z(2z - 5)}{(2z - 3)(z - 5)}$ **31.** $\dfrac{-2x^2 + 14x - 3}{(x - 3)(x + 3)(x + 4)}$

33. $\dfrac{(2x - 1)(x + 5)}{x(x - 5)}$ **35.** $\dfrac{-q^2 + 12q + 5}{(q - 3)(q + 5)}$ **37.** $\dfrac{3x^2 - 7x - 13}{(x + 3)(x + 4)(x - 3)(x - 4)}$ **39.** $\dfrac{(x + 2)(3x - 1)}{x^2}$ **41.** $\dfrac{4x + 1}{x - 1}$ **43.** $\dfrac{x - 2y}{y(y - x)}$

45. $\dfrac{(5x + 9)(x + 3)}{(x + 2)(4x + 3)}$ **47.** $\dfrac{(b + 3)(b - 1)}{(b - 2)(b + 2)}$ **49.** $\dfrac{x - 1}{x}$ **51.** $2 - m^2$ **53.** $\dfrac{-x^2 + 5x + 1}{x^2}$ **55.** $\dfrac{-x - 7}{x^2 + 6x - 3}$ **57.** $\dfrac{2x - 3}{x + 3}$ **59.** $\dfrac{a + b}{ab(a - b)}$

61. $\dfrac{(b - a)(b + a)}{ab(a^2 + b^2)}$ **63. a.** ≈ 136.55 miles per hour   **b.** $\dfrac{2v_1v_2}{v_1 + v_2}$ **65.** 0.040   **67.** 9446 kilometers per second   **69.** $\dfrac{2x + 1}{x(x + 1)}$

71. $\dfrac{3x^2 - 4}{x(x - 2)(x + 2)}$ **73.** $\dfrac{x^2 + 9x + 25}{(x + 5)^2}$ **75.** $\dfrac{x(1 - 4xy)}{(1 - 2xy)(1 + 2xy)}$ **77.** $R\left[\dfrac{(1 + i)^n - 1}{i(1 + i)^n}\right]$

Prepare for Section P.5, page 59

79. $15x^2 - 22x + 8$ **80.** $25x^2 - 20x + 4$ **81.** $4\sqrt{6}$ **82.** $-54 + \sqrt{5}$ **83.** $\dfrac{17 + 8\sqrt{2}}{7}$ **84.** b

Exercise Set P.5, page 65

1. $9i$ **3.** $7i\sqrt{2}$ **5.** $4 + 9i$ **7.** $5 + 7i$ **9.** $8 - 3i\sqrt{2}$ **11.** $11 - 5i$ **13.** $-7 + 4i$ **15.** $8 - 5i$ **17.** -10 **19.** $-2 + 16i$ **21.** -40

23. -10 **25.** $19i$ **27.** $20 - 10i$ **29.** $22 - 29i$ **31.** 41 **33.** $12 - 5i$ **35.** $-114 + 42i\sqrt{2}$ **37.** $-6i$ **39.** $3 - 6i$ **41.** $\dfrac{7}{53} - \dfrac{2}{53}i$

43. $1 + i$ **45.** $\dfrac{15}{41} - \dfrac{29}{41}i$ **47.** $\dfrac{5}{13} + \dfrac{12}{13}i$ **49.** $2 + 5i$ **51.** $-16 - 30i$ **53.** $-11 - 2i$ **55.** $-i$ **57.** -1 **59.** $-i$ **61.** -1

63. $\dfrac{1}{2} + \dfrac{\sqrt{3}}{2}i$ **65.** $-\dfrac{3}{2} + \dfrac{\sqrt{3}}{2}i$ **67.** $\dfrac{1}{2} + \dfrac{1}{2}i$ **69.** $(x + 4i)(x - 4i)$ **71.** $(z + 5i)(z - 5i)$ **73.** $(2x + 9i)(2x - 9i)$ **79.** 0

Chapter P True/False Exercises, page 70

1. True   **2.** False; if $a = \dfrac{1}{2}$, then $\left(\dfrac{1}{2}\right)^2 = \dfrac{1}{4} < \dfrac{1}{2}$.   **3.** True   **4.** False; $\sqrt{2} + (-\sqrt{2}) = 0$, which is a rational number.

5. False. $\sqrt{-2}\sqrt{-6} = -4$ $(2 \oplus 4) \oplus 6 \neq 2 \oplus (4 \oplus 6)$.   **6.** False; $x > a$ is written as (a, ∞).   **7.** False; $\sqrt{(-2)^2} \neq -2$.   **8.** False. Let $a = 2$ and $b = 3$.   **9.** False. Let $a = 1$ and $b = 2$.   **10.** False. $\sqrt{-2} \cdot \sqrt{-8} = i\sqrt{2} \cdot i\sqrt{8} = i^2\sqrt{16} = -4$.

Chapter P Review Exercises, page 70

1. integer, rational number, real number, prime number [P.1]   **2.** irrational number, real number [P.1]   **3.** rational number, real number [P.1]
4. rational number, real number [P.1]   **5.** $\{1, 2, 3, 5, 7, 11\}$ [P.1]   **6.** $\{5\}$ [P.1]   **7.** Distributive property [P.1]   **8.** Commutative property of
addition [P.1]   **9.** Associative property of multiplication [P.1]   **10.** Closure property of addition [P.1]   **11.** Identity property of addition [P.1]
12. Identity property of multiplication [P.1]   **13.** Symmetric property of equality [P.1]   **14.** Transitive property of equality [P.1]
15. $(-4, 2]$ [P.1]   **16.** $(-\infty, -1] \cup (3, \infty)$ [P.1]
17. $-3 \le x < 2$ [P.1]   **18.** $x > -1$ [P.1]   **19.** 7 [P.1]   **20.** $\pi - 2$ [P.1]

21. $4 - \pi$ [P.1]   **22.** 11 [P.1]   **23.** 17 [P.1]   **24.** $\sqrt{5} + \sqrt{2}$ [P.1]   **25.** -36 [P.1]   **26.** $\dfrac{2}{27}$ [P.2]   **27.** $12x^8y^3$ [P.2]   **28.** $\dfrac{4a^2b^8}{9c^4}$ [P.2]

29. 5 [P.2]   **30.** -9 [P.2]   **31.** $x^{17/12}$ [P.2]   **32.** $4x^{1/2}$ [P.2]   **33.** $x^{3/4}y^2$ [P.2]   **34.** $x - y$ [P.2]   **35.** $4ab^3\sqrt{3b}$ [P.2]   **36.** $2a\sqrt{3ab}$ [P.2]

37. $6x\sqrt{2y}$ [P.2]   **38.** $3xy^2\sqrt{2xy}$ [P.2]   **39.** $\dfrac{3y\sqrt{15y}}{5}$ [P.2]   **40.** $-\dfrac{2\sqrt{10xyz}}{5z^2}$ [P.2]   **41.** $\dfrac{7\sqrt[3]{4x}}{2}$ [P.2]   **42.** $\dfrac{5\sqrt[3]{3y^2}}{3}$ [P.2]   **43.** $-3y^2\sqrt[3]{5x^2y}$ [P.2]

44. $-5y^2\sqrt[3]{2x}$ [P.2]   **45.** 6.2×10^5 [P.2]   **46.** 1.7×10^{-6} [P.2]   **47.** 35,000 [P.2]   **48.** 0.000000431 [P.2]   **49.** $-a^2 - 2a - 1$ [P.3]

50. $2b^2 + 8b - 8$ [P.3]   **51.** $6x^4 + 5x^3 - 13x^2 + 22x - 20$ [P.3]   **52.** $27y^3 - 135y^2 + 225y - 125$ [P.3]   **53.** $3(x + 5)^2$ [P.3]

54. $(5x - 3y)^2$ [P.3]   **55.** $4(5a^2 - b^2)$ [P.3]   **56.** $2(2a + 5)(4a^2 - 10a + 25)$ [P.3]   **57.** $\dfrac{3x - 2}{x + 4}$ [P.4]   **58.** $\dfrac{2x - 5}{4x^2 - 10x + 25}$ [P.4]

59. $\dfrac{2x + 3}{2x - 5}$ [P.4]   **60.** $\dfrac{2x + 1}{x + 3}$ [P.4]   **61.** $\dfrac{x(3x + 10)}{(x + 3)(x - 3)(x + 4)}$ [P.4]   **62.** $\dfrac{x(5x - 7)}{(x + 3)(x + 4)(2x - 1)}$ [P.4]   **63.** $\dfrac{2x - 9}{3x - 17}$ [P.4]   **64.** $\dfrac{x + 4}{5x + 8}$ [P.4]

65. $5 + 8i$ [P.5]   **66.** $2 - 3i\sqrt{2}$ [P.5]   **67.** $6 - i$ [P.5]   **68.** $-2 + 10i$ [P.5]   **69.** $8 + 6i$ [P.5]   **70.** $29 + 22i$ [P.5]   **71.** $8 + 6i$ [P.5]

72. i [P.5]   **73.** $-3 - 2i$ [P.5]   **74.** $-\dfrac{14}{25} - \dfrac{23}{25}i$ [P.5]

Chapter P Test, page 72

1. Distributive property [P.1]   **2.** $\{0, 1, 2, 3, 4, 5, 6, 7, 8, 9\}$ [P.1]   **3.** 7 [P.1]   **4.** $\dfrac{4}{9x^4y^2}$ [P.2]   **5.** $\dfrac{96bc^2}{a^5}$ [P.2]   **6.** 1.37×10^{-3} [P.2]   **7.** $\dfrac{x^{5/6}}{y^{9/4}}$ [P.2]

8. $7xy\sqrt[3]{3xy}$ [P.2]   **9.** $\dfrac{\sqrt[4]{8x}}{2}$ [P.2]   **10.** $\dfrac{3\sqrt{x} - 6}{x - 4}$ [P.2]   **11.** $x^3 - 2x^2 + 5xy - 2x^2y - 2y^2$ [P.3]   **12.** $(9x - 5)(9x + 5)$ [P.3]

13. $(7x - 1)(x + 5)$ [P.3]   **14.** $(a - 4b)(3x - 2)$ [P.3]   **15.** $2x(2x - y)(4x^2 + 2xy + y^2)$ [P.3]   **16.** $-\dfrac{x + 3}{x + 5}$ [P.4]   **17.** $\dfrac{(x - 6)(x + 1)}{(x + 3)(x - 2)(x - 3)}$ [P.4]

18. $\dfrac{x(x + 2)}{x - 3}$ [P.4]   **19.** $\dfrac{3a^2 - 3ab - 10a + 5b}{a(2a - b)}$ [P.4]   **20.** $\dfrac{x(2x - 1)}{2x + 1}$ [P.4]   **21.** $7 + 2i\sqrt{5}$ [P.5]   **22.** $2 + 2i$ [P.5]   **23.** $22 - 3i$ [P.5]

24. $\dfrac{11}{26} + \dfrac{23}{26}i$ [P.5]   **25.** i [P.5]

Exercise Set 1.1, page 82

1. 15   **3.** -4 **5.** $\dfrac{9}{2}$ **7.** $\dfrac{108}{23}$ **9.** $\dfrac{2}{9}$ **11.** 12   **13.** 16   **15.** 9   **17.** $\dfrac{1}{2}$ **19.** $\dfrac{22}{13}$ **21.** $\dfrac{95}{18}$ **23.** identity   **25.** conditional
equation   **27.** contradiction   **29.** identity   **31.** conditional equation   **33.** $-4, 4$ **35.** 7, 3   **37.** 8, -3 **39.** 2, -8 **41.** 20, -12
43. no solution   **45.** 12, -18 **47.** $\dfrac{a + b}{2}, \dfrac{a - b}{2}$ **49.** 2008   **51.** after 3 hours and after 5 hours 24 minutes   **53.** 72 square yards
55. 15 minutes   **57.** maximum 166 beats per minute, minimum 127 beats per minute   **61.** $\{x \,|\, x \ge -4\}$ **63.** $\{x \,|\, x \le -7\}$ **65.** $\left\{x \,\middle|\, x \ge -\dfrac{7}{2}\right\}$

Prepare for Section 1.2, page 84

67. $23\dfrac{1}{2}$ **68.** $\dfrac{4}{15}$ **69.** the distributive property   **70.** the associative property of multiplication   **71.** $\dfrac{11}{15}x$ **72.** $\dfrac{ab}{a + b}$

Exercise Set 1.2, page 92

1. $h = \dfrac{3V}{\pi r^2}$ **3.** $t = \dfrac{I}{Pr}$ **5.** $m_1 = \dfrac{Fd^2}{Gm_2}$ **7.** $d = \dfrac{a_n - a_1}{n - 1}$ **9.** $r = \dfrac{S - a_1}{S}$ **11.** 88.8   **13.** 9.5   **15.** 11.2   **17.** 100   **19.** 30 feet by 57 feet
21. 12 centimeters, 36 centimeters, 36 centimeters   **23.** 240 meters   **25.** 2 hours   **27.** 3 miles   **29.** 98   **31.** 850   **33.** $937.50

35. $7600 invested at 8%, $6400 invested at 6.5%   **37.** $3750   **39.** $18\dfrac{2}{11}$ grams   **41.** 64 liters   **43.** 1200 at $14 and 1800 at $25

45. $6\dfrac{2}{3}$ pounds of the $12 coffee and $13\dfrac{1}{3}$ pounds of the $9 coffee   **47.** 10 grams   **49.** 7.875 hours   **51.** $10.05 for book, $0.05 for bookmark   **53.** 6.25 feet   **55.** 40 pounds   **57.** 1384 feet   **59.** 84 years old

Prepare for Section 1.3, page 96

61. $(x + 6)(x - 7)$ **62.** $(2x + 3)(3x - 5)$ **63.** $3 + 4i$ **64.** 1   **65.** 1   **66.** 0

Exercise Set 1.3, page 107

1. $-3, 5$ **3.** $-\dfrac{1}{2}, 1$ **5.** $-24, \dfrac{3}{8}$ **7.** $0, \dfrac{7}{3}$ **9.** $2, 8$ **11.** ± 9 **13.** $\pm 2\sqrt{6}$ **15.** $\pm 2i$ **17.** $-1, 11$ **19.** $3 \pm 4i$ **21.** $-3 \pm 2\sqrt{2}$

23. $-3, 5$ **25.** $-2 \pm i$ **27.** $\dfrac{-3 \pm \sqrt{13}}{2}$ **29.** $\dfrac{-2 \pm \sqrt{6}}{2}$ **31.** $\dfrac{4 \pm \sqrt{13}}{3}$ **33.** $-3, 5$ **35.** $\dfrac{-1 \pm \sqrt{5}}{2}$ **37.** $\dfrac{-2 \pm \sqrt{2}}{2}$ **39.** $\dfrac{5}{6} \pm \dfrac{\sqrt{11}}{6}i$

41. $\dfrac{-3 \pm \sqrt{41}}{4}$ **43.** $-\dfrac{5}{6}, \dfrac{7}{4}$ **45.** $-2, \dfrac{4}{5}$ **47.** 81; two real solutions   **49.** -116; no real solutions   **51.** 0; one real solution   **53.** 2116; two real solutions   **55.** -111; no real solutions   **57.** 26.8 centimeters   **59.** width 43.2 inches; height 32.4 inches   **61.** 1996   **63.** 5800 or 11,000 racquets   **65.** 12 feet by 48 feet or 32 feet by 18 feet   **67.** 0.3 mile and 3.9 miles   **69.** 1.7 seconds   **71.** 1.8 seconds and 11.9 seconds   **73.** No   **75.** 9 people   **77.** 2006   **79. a.** 44.8 million pounds   **b.** 2007   **81. a.** $l = \left(\dfrac{1 + \sqrt{5}}{2}\right)w$ **b.** 163.4 feet

c. Answers will vary.   **83.** Yes   **85.** Yes

Prepare for Section 1.4, page 111

87. $x(x + 4)(x - 4)$ **88.** $x^2(x + 6)(x - 6)$ **89.** 4   **90.** 64   **91.** $x + 2\sqrt{x - 5} - 4$ **92.** $x - 4\sqrt{x + 3} + 7$

Exercise Set 1.4, page 119

1. $0, \pm 5$ **3.** $2, \pm 1$ **5.** $0, \pm 3$ **7.** $0, -5, 8$ **9.** $0, \pm 4$ **11.** $2, -1 \pm i\sqrt{3}$ **13.** 31   **15.** 2   **17.** no solution   **19.** no solution

21. $\dfrac{7}{2}$ **23.** -12 **25.** 1   **27.** 40   **29.** 3   **31.** 7   **33.** 7   **35.** 9   **37.** $\dfrac{5}{2}$ **39.** $1, -6$ **41.** $\pm\sqrt{7}, \pm\sqrt{2}$ **43.** $\pm 2, \pm\dfrac{\sqrt{6}}{2}$

45. $\sqrt[3]{2}, -\sqrt[3]{3}$ **47.** $1, 16$ **49.** $-\dfrac{1}{27}, 64$ **51.** $\pm\dfrac{\sqrt{15}}{3}$ **53.** ± 1 **55.** $\dfrac{256}{81}, 16$ **57.** $13\dfrac{1}{3}$ hours   **59.** 8 games   **61.** 9 words with three or more syllables   **63.** 3 inches   **65.** 10.5 millimeters   **67.** 87 feet   **69. a.** 8.93 inches   **b.** $5\sqrt{3}$ inches

71. $s = \left(\dfrac{-275 + 5\sqrt{3025 + 176T}}{2}\right)^2$

Prepare for Section 1.5, page 122

73. $\{x \mid x > 5\}$ **74.** 38   **75.** 2   **76.** $(2x + 3)(5x - 3)$ **77.** $\dfrac{7}{2}$ **78.** $\dfrac{5}{2}, 3$

Exercise Set 1.5, page 134

1. $\{x \mid x < 4\}$,

3. $\{x \mid x < -6\}$,

5. $\left\{x \mid x \geq -\dfrac{13}{8}\right\}$,

7. $\{x \mid x < 2\}$,

9. $\left\{x \mid -\dfrac{3}{4} < x \leq 4\right\}$,

11. $\left\{x \mid \dfrac{1}{3} \leq x \leq \dfrac{11}{3}\right\}$,

13. $\{x \mid x < -3 \text{ or } x \geq -1\}$,

15. $\{x \mid x < 1\}$,

17. $\left(-\infty, -\dfrac{3}{2}\right) \cup \left(\dfrac{5}{2}, \infty\right)$

19. $(-\infty, -8] \cup [2, \infty)$ **21.** $\left[-\dfrac{4}{3}, 8\right]$ **23.** $(-\infty, -4] \cup \left[\dfrac{28}{5}, \infty\right)$ **25.** $(-\infty, \infty)$ **27.** $\{4\}$ **29.** $(-\infty, -7) \cup (0, \infty)$ **31.** $[-4, 4]$

33. $(-5, -2)$ **35.** $(-\infty, -4] \cup [7, \infty)$ **37.** $(-4, 1)$ **39.** $\left[-\dfrac{29}{2}, -8\right)$ **41.** $\left[-4, -\dfrac{7}{2}\right)$ **43.** $(-\infty, -1) \cup (2, 4)$

45. $(-\infty, 5) \cup [12, \infty)$ **47.** $\left(-\dfrac{2}{3}, 0\right) \cup \left(\dfrac{5}{2}, \infty\right)$ **49.** $(-\infty, 5)$ **51.** if you write more than 57 checks a month **53.** $0 < h \le 26$ inches
55. at least 34 sales **57.** $20° \le C \le 40°$ **59.** $\{12, 14, 16\}, \{14, 16, 18\}$ **61.** 130.0 to 137.5 centimeters **63.** $(0, 210)$ **65.** at least
9791 books **67.** maximum radius 4.480 inches, minimum radius 4.432 inches **69.** $(-\infty, 3) \cup (3, 6) \cup (6, \infty)$ **71.** $(-3, \infty)$
73. $(-5, -1) \cup (1, 5)$ **75.** $(-7, -3] \cup [3, 7)$ **77.** $(a - \delta, a) \cup (a, a + \delta)$ **79.** more than 1 second but less than 3 seconds

Chapter 1 True/False Exercises, page 139

1. False; $(-3)^2 = 9$. **2.** False; one has solution set $\{3\}$, and the other has solution set $\{3, -4\}$. **3.** True **4.** True **5.** False; $100 > 1$ but
$\dfrac{1}{100} > \dfrac{1}{1}$. **6.** False; the discriminant is $b^2 - 4ac$. **7.** False; $\sqrt{1} + \sqrt{1} = 1 + 1 = 2$ but $1 + 1 = 2 \ne 2^2$. **8.** True **9.** False; $3x^2 - 48 = 0$
has roots of 4 and -4. **10.** True

Chapter 1 Review Exercises, page 140

1. $\dfrac{3}{2}$ [1.1] **2.** $\dfrac{11}{3}$ [1.1] **3.** $\dfrac{1}{2}$ [1.1] **4.** $\dfrac{11}{4}$ [1.1] **5.** $-\dfrac{38}{15}$ [1.4] **6.** $-\dfrac{1}{2}$ [1.4] **7.** $3, 2$ [1.3] **8.** $\dfrac{4}{3}, -\dfrac{3}{2}$ [1.3] **9.** $\dfrac{1 \pm \sqrt{13}}{6}$ [1.3]
10. $\dfrac{1}{2} \pm \dfrac{\sqrt{3}}{2} i$ [1.3] **11.** $0, \dfrac{5}{3}$ [1.4] **12.** $0, \pm 2$ [1.4] **13.** $\pm\dfrac{2\sqrt{3}}{3}, \pm\dfrac{\sqrt{10}}{2}$ [1.4] **14.** $\dfrac{4}{9}$ [1.4] **15.** $-5, 3$ [1.4] **16.** $6, -4$ [1.4]
17. 4 [1.4] **18.** 7 [1.4] **19.** -4 [1.4] **20.** 0 [1.4] **21.** $-2, -4$ [1.4] **22.** $\dfrac{11}{4}, \dfrac{9}{4}$ [1.4] **23.** $5, 1$ [1.1] **24.** $-1, -9$ [1.1] **25.** $2, -3$ [1.1]
26. $5, -\dfrac{1}{3}$ [1.1] **27.** $-2, -1$ [1.4] **28.** $1, -4, \dfrac{4}{3}$ [1.4] **29.** $(-\infty, 2]$ [1.5] **30.** $\left[\dfrac{6}{7}, \infty\right)$ [1.5] **31.** $[-5, 2]$ [1.5] **32.** $(-\infty, -1) \cup (3, \infty)$ [1.5]
33. $\left[\dfrac{145}{9}, 35\right]$ [1.5] **34.** $(86, 149)$ [1.5] **35.** $(-\infty, 0] \cup [3, 4]$ [1.5] **36.** $(-7, 0) \cup (3, \infty)$ [1.5] **37.** $(-\infty, -3) \cup (4, \infty)$ [1.5]
38. $(-\infty, -7) \cup [0, 5]$ [1.5] **39.** $\left(-\infty, \dfrac{5}{2}\right] \cup (3, \infty)$ [1.5] **40.** $\left[\dfrac{5}{2}, 5\right)$ [1.5] **41.** $\left(\dfrac{2}{3}, 2\right)$ [1.5] **42.** $(-\infty, 1] \cup [2, \infty)$ [1.5]
43. $(1, 2) \cup (2, 3)$ [1.5] **44.** $(a - b, a) \cup (a, a + b)$ [1.5] **45.** $h = \dfrac{V}{\pi r^2}$ [1.2] **46.** $t = \dfrac{A - P}{Pr}$ [1.2] **47.** $b_1 = \dfrac{2A - hb_2}{h}$ [1.2]
48. $w = \dfrac{P - 2l}{2}$ [1.2] **49.** $m = \dfrac{e}{c^2}$ [1.2] **50.** $m_1 = \dfrac{Fs^2}{Gm_2}$ [1.2] **51.** 80 [1.2] **52.** width = 12 feet by length = 15 feet [1.2]
53. 24 nautical miles [1.2] **54.** \$20.00 [1.2] **55.** \$1750 in the 4% account, \$3750 in the 6% account [1.2] **56.** Price of calculator is \$20.50.
Price of battery is \$0.50. [1.2] **57.** \$864 [1.2] **58.** length = 12 inches by width = 8 inches, or length = 8 inches by width = 12 inches [1.2]
59. 18 hours [1.4] **60.** 4024 adult tickets, 502 student tickets [1.2] **61.** ≈ 13 feet [1.2] **62.** $(12, 24)$ The revenue is greater than \$576 when
the price is between \$12 and \$24. [1.5] **63. a.** $|B - 218| > 48$ [1.5] **b.** $(0, 170) \cup (266, \infty)$ [1.5] **64.** more than \$425 but less than \$725 [1.5]
65. $0 < h \le 23.5$ inches [1.5] **66.** $[27, 82]$ [1.5] **67.** 9.39 to 9.55 inches [1.5] **68.** more than 0.6 mile but less than 3.6 miles from the city
center [1.5]

Chapter 1 Test, page 142

1. 3 [1.1] **2.** $-5, 11$ [1.1] **3.** $-\dfrac{1}{2}, \dfrac{8}{3}$ [1.3] **4.** $\dfrac{4 \pm \sqrt{14}}{2}$ [1.3] **5.** $\dfrac{5 \pm \sqrt{37}}{6}$ [1.3] **6.** discriminant: 1; two real solutions [1.3]
7. $x = \dfrac{c - cd}{a - c}, a \ne c$ [1.2] **8.** 3 [1.4] **9.** $\dfrac{8}{27}, -64$ [1.4] **10.** $-\dfrac{14}{3}$ [1.4] **11. a.** $\{x \mid x \le 8\}$ [1.5] **b.** $[-2, 5)$ [1.5] **12.** $[-4, -1) \cup [3, \infty)$ [1.5]
13. from $10\dfrac{7}{8}$ inches to $11\dfrac{7}{16}$ inches [1.5] **14.** 2 miles per hour [1.2] **15.** 2.25 liters [1.2] **16.** 15 hours [1.4] **17.** more than 100 miles [1.5]
18. more than 14.7 feet but less than 145.0 feet from a sideline [1.5] **19.** more than 1.25 miles but less than 10 miles from the city center [1.5]
20. 518 miles [1.4]

Cumulative Review Exercises, page 143

1. -11 [P.1] **2.** 1.7×10^{-4} [P.2] **3.** $8x^2 - 30x + 41$ [P.3] **4.** $(8x - 5)(x + 3)$ [P.3] **5.** $\dfrac{2x + 17}{x - 4}$ [P.4] **6.** $a^{11/12}$ [P.2] **7.** 29 [P.5] **8.** $\dfrac{10}{3}$ [1.1]
9. $\dfrac{2 \pm \sqrt{10}}{2}$ [1.3] **10.** $1, 5$ [1.1] **11.** 5 [1.4] **12.** $-6, 0, 6$ [1.4] **13.** $\pm\sqrt{3}, \pm\dfrac{\sqrt{10}}{2}$ [1.4] **14.** $\{x \mid x \le -1 \text{ or } x > 1\}$ [1.5]
15. $(-\infty, 4] \cup [8, \infty)$ [1.5] **16.** $\left\{x \,\middle|\, \dfrac{10}{7} \le x < \dfrac{3}{2}\right\}$ [1.5] **17.** length 58 feet, width 42 feet [1.2] **18.** 9475 to 24,275 printers [1.5]
19. 68 to 100 [1.5] **20.** between 14.3% and 23.1% [1.5]

Exercise Set 2.1, page 158

1.

3. a.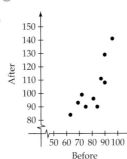

b. 23.4 beats per minute   **5.** $7\sqrt{5}$ **7.** $\sqrt{1261}$ **9.** $\sqrt{89}$ **11.** $\sqrt{38 - 12\sqrt{6}}$ **13.** $2\sqrt{a^2 + b^2}$ **15.** $-x\sqrt{10}$ **17.** $(12, 0), (-4, 0)$ **19.** $(3, 2)$ **21.** $(6, 4)$ **23.** $(-0.875, 3.91)$ **25.** 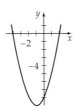   **27.**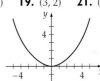

29. 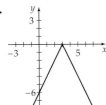   **31.**   **33.**   **35.** 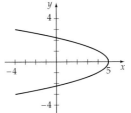   **37.**

39. $(6, 0), \left(0, \dfrac{12}{5}\right)$ 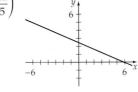   **41.** $(5, 0); \left(0, \sqrt{5}\right), \left(0, -\sqrt{5}\right)$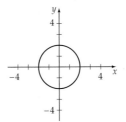

43. $(-4, 0); (0, 4), (0, -4)$ 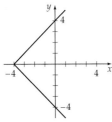   **45.** $(\pm 2, 0), (0, \pm 2)$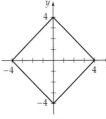

47. $(\pm 4, 0), (0, \pm 4)$ **49.** center $(0, 0)$, radius 6   **51.** center $(1, 3)$, radius 7   **53.** center $(-2, -5)$, radius 5

55. center $(8, 0)$, radius $\dfrac{1}{2}$ **57.** $(x - 4)^2 + (y - 1)^2 = 2^2$

59. $\left(x - \dfrac{1}{2}\right)^2 + \left(y - \dfrac{1}{4}\right)^2 = \left(\sqrt{5}\right)^2$ **61.** $(x - 0)^2 + (y - 0)^2 = 5^2$

63. $(x - 1)^2 + (y - 3)^2 = 5^2$ **65.** center $(3, 0)$, radius 2   **67.** center $(7, -4)$, radius 3

69. center $\left(-\dfrac{1}{2}, 0\right)$, radius 4   **71.** center $\left(\dfrac{1}{2}, -\dfrac{3}{2}\right)$, radius $\dfrac{5}{2}$

73. $(x + 1)^2 + (y - 7)^2 = 25$ **75.** $(x - 7)^2 + (y - 11)^2 = 121$ **77.**   **79.**   **81.**

83.   **85.**   **87.** $(13, 5)$ **89.** $(7, -6)$ **91.** $x^2 - 6x + y^2 - 8y = 0$ **93.** $9x^2 + 25y^2 = 225$
95. $(x + 3)^2 + (y - 3)^2 = 3^2$

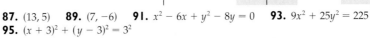

Prepare for Section 2.2, page 160

97. -4 **98.** $D = \{-3, -2, -1, 0, 2\}$; $R = \{1, 2, 4, 5\}$ **99.** $\sqrt{58}$ **100.** $x \geq 3$ **101.** $-2, 3$ **102.** 13

Exercise Set 2.2, page 174

1. a. 5   **b.** -4 **c.** -1 **d.** 1   **e.** $3k - 1$ **f.** $3k + 5$ **3. a.** $\sqrt{5}$ **b.** 3   **c.** 3   **d.** $\sqrt{21}$ **e.** $\sqrt{r^2 + 2r + 6}$ **f.** $\sqrt{c^2 + 5}$
5. a. $\dfrac{1}{2}$ **b.** $\dfrac{1}{2}$ **c.** $\dfrac{5}{3}$ **d.** 1   **e.** $\dfrac{1}{c^2 + 4}$ **f.** $\dfrac{1}{|2 + h|}$ **7. a.** 1   **b.** 1   **c.** -1 **d.** -1 **e.** 1   **f.** -1 **9. a.** -11 **b.** 6
c. $3c + 1$ **d.** $-k^2 - 2k + 10$ **11.** Yes   **13.** No   **15.** No   **17.** Yes   **19.** No   **21.** Yes   **23.** Yes   **25.** Yes   **27.** all real numbers
29. all real numbers   **31.** $\{x \mid x \neq -2\}$ **33.** $\{x \mid x \geq -7\}$ **35.** $\{x \mid -2 \leq x \leq 2\}$ **37.** $\{x \mid x > -4\}$ **39.**

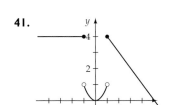

41.

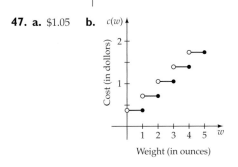

43.

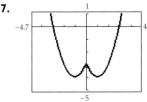

45.

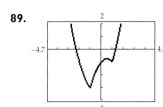

47. a. \$1.05   **b.**

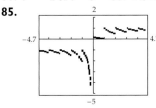

Weight (in ounces)

49. a, b, and **d.**   **51.** decreasing on $(-\infty, 0]$; increasing on $[0, \infty)$
53. increasing on $(-\infty, \infty)$ **55.** decreasing on $(-\infty, -3]$; increasing on $[-3, 0]$; decreasing on $[0, 3]$; increasing on $[3, \infty)$ **57.** constant on $(-\infty, 0]$; increasing on $[0, \infty)$
59. decreasing on $(-\infty, 0]$; constant on $[0, 1]$; increasing on $[1, \infty)$ **61.** g and F
63. a. $w = 25 - l$ **b.** $A = 25l - l^2$ **65.** $v(t) = 80{,}000 - 6500t, 0 \leq t \leq 10$
67. a. $C(x) = 2000 + 22.80x$ **b.** $R(x) = 37.00x$ **c.** $P(x) = 14.20x - 2000$
69. $h = 15 - 5r$ **71.** $d = \sqrt{(3t)^2 + 50^2}$ **73.** $d = \sqrt{(45 - 8t)^2 + (6t)^2}$

75. a. $L(x) = \left(\dfrac{1}{4\pi} + \dfrac{1}{16}\right)x^2 - \dfrac{5}{2}x + 25$ **b.** 25, 17.27, 14.09, 15.46, 21.37, 31.83
c. $[0, 20]$ **77. a.** $A(x) = \sqrt{900 + x^2} + \sqrt{400 + (40 - x)^2}$
b. 74.72, 67.68, 64.34, 64.79, 70   **c.** $[0, 40]$ **79.** 275, 375, 385, 390, 394

81. $c = -2$ or $c = 3$ **83.** 1 is not in the range of f.
85.

87.   **89.**

91. 4   **93.** 2   **95. a.** 36
b. 13   **c.** 12   **d.** 30
e. $13k - 2$ **f.** $8k - 11$
97. $4\sqrt{21}$ **99.** 1, -3

101.

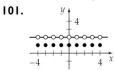

Prepare for Section 2.3, page 181

103. 7   **104.** -1 **105.** $-\dfrac{8}{5}$ **106.** $y = -2x + 9$ **107.** $y = \dfrac{3}{5}x - 3$ **108.** 2

Exercise Set 2.3, page 191

1. $-\dfrac{3}{2}$ **3.** $-\dfrac{1}{2}$ **5.** The line does not have slope.   **7.** 6   **9.** $\dfrac{9}{19}$ **11.** $\dfrac{f(3 + h) - f(3)}{h}$ **13.** $\dfrac{f(h) - f(0)}{h}$ **15.**

17. 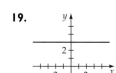 **19.**   **21.**   **23.** **25.** **27.** $y = x + 3$

29. $y = \dfrac{3}{4}x + \dfrac{1}{2}$

31. $y = (0)x + 4 = 4$

33. $y = -4x - 10$

35. $y = -\dfrac{3}{4}x + \dfrac{13}{4}$ **37.** $y = \dfrac{12}{5}x - \dfrac{29}{5}$ **39.** -2 **41.** $-\dfrac{1}{2}$ **43.** -4 **45.** 4 **47.** -20 **49.** $\dfrac{1}{3}$ **51.** $\dfrac{16}{3}$ **53.** $m = 2.875$. The

value of the slope indicates that the speed of sound in water increases 2.875 feet per second for a 1-degree increase in temperature.
55. a. $H(c) = 1.45c$ **b.** 26 miles per gallon   **57. a.** $N(t) = 2500t - 4,962,000$ **b.** 2008   **59. a.** $B(d) = 30d - 300$ **b.** The value of
the slope means that a 1-inch increase in the diameter of a log 32 feet long results in an increase of 30 board-feet of lumber that can be obtained
from the log.   **c.** 270 board-feet   **61.** line A, Michelle; line B, Amanda; line C, distance between Michelle and Amanda
63. a. $y = 1.842x - 18.947$ **b.** 147   **65.** $P(x) = 40.50x - 1782$, $x = 44$, the break-even point   **67.** $P(x) = 79x - 10,270$, $x = 130$, the break-
even point   **69. a.** \$275   **b.** \$283   **c.** \$355   **d.** \$8   **71. a.** $C(t) = 19,500.00 + 6.75t$ **b.** $R(t) = 55.00t$ **c.** $P(t) = 48.25t - 19,500.00$

d. approximately 405 days   **73.** $y = -\dfrac{3}{4}x + \dfrac{15}{4}$ **75.** $y = x + 1$ **77.** -5 ft   **79. a.** $Q = (3, 10)$, $m = 5$ **b.** $Q = (2.1, 5.41)$, $m = 4.1$

c. $Q = (2.01, 5.0401)$, $m = 4.01$ **d.** 4   **85.** $y = -2x + 11$ **87.** $5x + 3y = 15$ **89.** $3x + y = 17$ **93.** $\left(\dfrac{9}{2}, \dfrac{81}{4}\right)$

Prepare for Section 2.4, page 197

95. $(3x - 2)(x + 4)$ **96.** $x^2 - 8x + 16 = (x - 4)^2$ **97.** 26   **98.** $-\dfrac{1}{2}, 1$ **99.** $\dfrac{-3 \pm \sqrt{17}}{2}$ **100.** 1, 3

Exercise Set 2.4, page 206

1. d   **3.** b   **5.** g   **7.** c   **9.** $f(x) = (x + 2)^2 - 3$ **11.** $f(x) = (x - 4)^2 - 11$ **13.** $f(x) = \left(x - \left(-\dfrac{3}{2}\right)\right)^2 - \dfrac{5}{4}$

vertex: $(-2, -3)$ vertex: $(4, -11)$ vertex: $\left(-\dfrac{3}{2}, -\dfrac{5}{4}\right)$

axis of symmetry: $x = -2$ axis of symmetry: $x = 4$ axis of symmetry: $x = -\dfrac{3}{2}$

15. $f(x) = -(x - 2)^2 + 6$ **17.** $f(x) = -3\left(x - \dfrac{1}{2}\right)^2 + \dfrac{31}{4}$ **19.** vertex: $(5, -25)$, $f(x) = (x - 5)^2 - 25$
                                       21. vertex: $(0, -10)$, $f(x) = x^2 - 10$
  vertex: $(2, 6)$ vertex: $\left(\dfrac{1}{2}, \dfrac{31}{4}\right)$ **23.** vertex: $(3, 10)$, $f(x) = -(x - 3)^2 + 10$

  axis of symmetry: $x = 2$ axis of symmetry: $x = \dfrac{1}{2}$ **25.** vertex: $\left(\dfrac{3}{4}, \dfrac{47}{8}\right)$, $f(x) = 2\left(x - \dfrac{3}{4}\right)^2 + \dfrac{47}{8}$

                                      27. vertex: $\left(\dfrac{1}{8}, \dfrac{17}{16}\right)$, $f(x) = -4\left(x - \dfrac{1}{8}\right)^2 + \dfrac{17}{16}$

               29. $\{y \mid y \geq -2\}$, -1 and 3   **31.** $\left\{y \mid y \leq \dfrac{17}{8}\right\}$, 1 and $\dfrac{3}{2}$

               33. No, $3 \notin \left\{y \mid y \geq \dfrac{15}{4}\right\}$ **35.** -16, minimum   **37.** 11, maximum

               39. $-\dfrac{1}{8}$, minimum   **41.** -11, minimum   **43.** 35, maximum

45. a. 27 feet   **b.** $22\dfrac{5}{16}$ feet   **c.** 20.1 feet from the center   **47. a.** $w = \dfrac{600 - 2l}{3}$ **b.** $A = 200l - \dfrac{2}{3}l^2$ **c.** $w = 100$ feet, $l = 150$ feet
49. a. 12:43 P.M.   **b.** 91°F   **51.** 1993, 2500 homes   **53.** Yes   **55. a.** 41 miles per gallon   **b.** 34 miles per gallon   **57.** y-intercept $(0, 0)$;
x-intercepts $(0, 0)$ and $(-6, 0)$ **59.** y-intercept $(0, -6)$; no x-intercepts   **61.** 740 units yield a maximum revenue of \$109,520.   **63.** 85 units
yield a maximum profit of \$24.25.   **65.** $P(x) = -0.1x^2 + 50x - 1840$, break-even points: $x = 40$ and $x = 460$ **67. a.** $R(x) = -0.25x^2 + 30.00x$
b. $P(x) = -0.25x^2 + 27.50x - 180$ **c.** \$576.25   **d.** 55   **69. a.** $t = 4$ seconds   **b.** 256 feet   **c.** $t = 8$ seconds   **71.** 30 feet

73. $r = \dfrac{48}{4 + \pi} \approx 6.72$ feet, $h = r \approx 6.72$ feet   **77.** $f(x) = \dfrac{3}{4}x^2 - 3x + 4$ **79. a.** $w = 16 - x$ **b.** $A = 16x - x^2$ **81.** The discriminant is
$b^2 - 4(1)(-1) = b^2 + 4$, which is positive for all b.   **83.** increases the height of each point on the graph by c units   **85.** 4, 4

Prepare for Section 2.5, page 211

89. $x = -2$ **92.** 3, −1, −3, −3, −1   **93.** $(0, b)$ **94.** $(0, 0)$

Exercise Set 2.5, page 222

1. 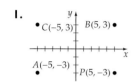   **3.** 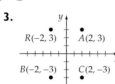   **5.** 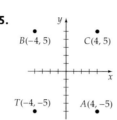   **7.** 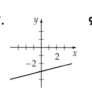   **9.** 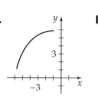   **11.**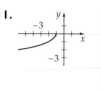

13. a. No **b.** Yes   **15. a.** No **b.** No   **17. a.** Yes **b.** Yes   **19. a.** Yes **b.** Yes   **21. a.** Yes **b.** Yes   **23.** No   **25.** Yes
27. Yes   **29.** Yes   **31.** 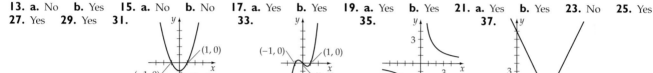   **33.**   **35.**   **37.**

39. 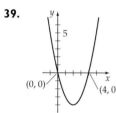   **41.**    **43.** even   **45.** odd   **47.** even   **49.** even   **51.** even   **53.** even   **55.** neither
57. a., b. 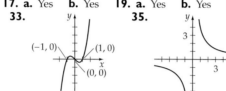   **59. a.** 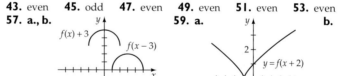   **b.**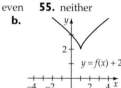

61. a. $(-5, 5), (-3, -2), (-2, 0)$ **b.** $(-2, 6), (0, -1), (1, 1)$ **63. a.** 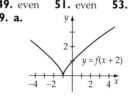   **b.**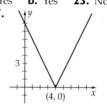

65. a. $(1, 3), (-2, -4)$ **b.** $(-1, -3), (2, 4)$ **67. a., b.**   **69.**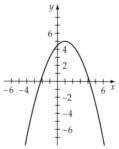

71. a. 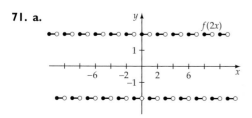   **b.**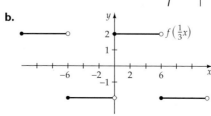

73. a.

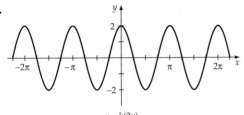

$y = h(2x)$

b.

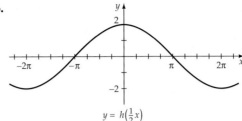

$y = h\left(\frac{1}{2}x\right)$

75.

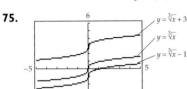

$y = \sqrt[3]{x} + 3$
$y = \sqrt[3]{x}$
$y = \sqrt[3]{x} - 1$

77.

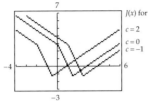

$J(x)$ for
$c = 2$
$c = 0$
$c = -1$

79.

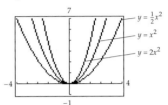

$y = \frac{1}{2}x^2$
$y = x^2$
$y = 2x^2$

81.

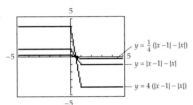

$y = \frac{1}{4}(|x-1| - |x|)$
$y = |x-1| - |x|$
$y = 4(|x-1| - |x|)$

83. a.

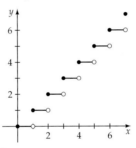

b.

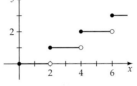

c.

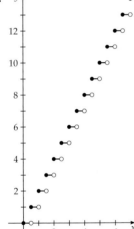

85. a. $f(x) = \dfrac{2}{(x+1)^2 + 1} + 1$ **b.** $f(x) = -\dfrac{2}{(x-2)^2 + 1}$

Prepare for Section 2.6, page 226

87. $x^2 + 1$ **88.** $6x^3 - 11x^2 + 7x - 6$ **89.** $18a^2 - 15a + 2$ **90.** $2h^2 + 3h$ **91.** all real numbers except $x = 1$ **92.** $[4, \infty)$

Exercise Set 2.6, page 235

1. $f(x) + g(x) = x^2 - x - 12$,   Domain is the set of all real numbers.
$f(x) - g(x) = x^2 - 3x - 18$,   Domain is the set of all real numbers.
$f(x) \cdot g(x) = x^3 + x^2 - 21x - 45$,   Domain is the set of all real numbers.
$\dfrac{f(x)}{g(x)} = x - 5$,   Domain $\{x \mid x \neq -3\}$

3. $f(x) + g(x) = 3x + 12$,   Domain is the set of all real numbers.
$f(x) - g(x) = x + 4$,   Domain is the set of all real numbers.
$f(x) \cdot g(x) = 2x^2 + 16x + 32$,   Domain is the set of all real numbers.
$\dfrac{f(x)}{g(x)} = 2$,   Domain $\{x \mid x \neq -4\}$

5. $f(x) + g(x) = x^3 - 2x^2 + 8x$, Domain is the set of all real numbers.
$f(x) - g(x) = x^3 - 2x^2 + 6x$, Domain is the set of all real numbers
$f(x) \cdot g(x) = x^4 - 2x^3 + 7x^2$, Domain is the set of all real numbers.
$\dfrac{f(x)}{g(x)} = x^2 - 2x + 7$, Domain $\{x \mid x \neq 0\}$

7. $f(x) + g(x) = 4x^2 + 7x - 12$, Domain is the set of all real numbers.
$f(x) - g(x) = x - 2$, Domain is the set of all real numbers.
$f(x) \cdot g(x) = 4x^4 + 14x^3 - 12x^2 - 41x + 35$, Domain is the set of all real numbers.
$\dfrac{f(x)}{g(x)} = 1 + \dfrac{x-2}{2x^2 + 3x - 5}$, Domain $\left\{x \mid x \neq 1, x \neq -\dfrac{5}{2}\right\}$

9. $f(x) + g(x) = \sqrt{x-3} + x$, Domain $\{x \mid x \geq 3\}$
$f(x) - g(x) = \sqrt{x-3} - x$, Domain $\{x \mid x \geq 3\}$
$f(x) \cdot g(x) = x\sqrt{x-3}$, Domain $\{x \mid x \geq 3\}$
$\dfrac{f(x)}{g(x)} = \dfrac{\sqrt{x-3}}{x}$, Domain $\{x \mid x \geq 3\}$

11. $f(x) + g(x) = \sqrt{4 - x^2} + 2 + x$, Domain $\{x \mid -2 \leq x \leq 2\}$
$f(x) - g(x) = \sqrt{4 - x^2} - 2 - x$, Domain $\{x \mid -2 \leq x \leq 2\}$
$f(x) \cdot g(x) = (\sqrt{4 - x^2})(2 + x)$, Domain $\{x \mid -2 \leq x \leq 2\}$
$\dfrac{f(x)}{g(x)} = \dfrac{\sqrt{4 - x^2}}{2 + x}$ Domain $\{x \mid -2 < x \leq 2\}$

13. 18   **15.** $-\dfrac{9}{4}$ **17.** 30   **19.** 12   **21.** 300   **23.** $-\dfrac{384}{125}$ **25.** $-\dfrac{5}{2}$ **27.** $-\dfrac{1}{4}$ **29.** 2   **31.** $2x + h$ **33.** $4x + 2h + 4$

35. $-8x - 4h$ **37.** $(g \circ f)(x) = 6x + 3$ **39.** $(g \circ f)(x) = x^2 + 4x + 1$ **41.** $(g \circ f)(x) = -5x^3 - 10x$ **43.** $(g \circ f)(x) = \dfrac{1 - 5x}{x + 1}$
$(f \circ g)(x) = 6x - 16$ $(f \circ g)(x) = x^2 + 8x + 11$ $(f \circ g)(x) = -125x^3 - 10x$
$(f \circ g)(x) = \dfrac{2}{3x - 4}$

45. $(g \circ f)(x) = \dfrac{\sqrt{1 - x^2}}{|x|}$ **47.** $(g \circ f)(x) = -\dfrac{2|5 - x|}{3}$ **49.** 66   **51.** 51   **53.** -4 **55.** 41   **57.** $-\dfrac{3848}{625}$ **59.** $6 + 2\sqrt{3}$
$(f \circ g)(x) = \dfrac{1}{x - 1}$ $(f \circ g)(x) = \dfrac{3|x|}{|5x + 2|}$

61. $16c^2 + 4c - 6$ **63.** $9k^4 + 36k^3 + 45k^2 + 18k - 4$ **65. a.** $A(t) = \pi(1.5t)^2$, $A(2) = 9\pi$ square feet ≈ 28.27 square feet   **b.** $V(t) = 2.25\pi t^3$, $V(3) = 60.75\pi$ cubic feet ≈ 190.85 cubic feet   **67. a.** $d(t) = \sqrt{(48 - t)^2 - 4^2}$ **b.** $s(35) = 13$ feet, $d(35) \approx 12.37$ feet   **69.** $(Y \circ F)(x)$ converts x inches to yards.   **71. a.** 99.8; this is identical to the slope of the line through $(0, C(0))$ and $(1, C(1))$.   **b.** 156.2   **c.** -49.7 **d.** -30.8
e. -16.4 **f.** 0

Chapter 2 True/False Exercises, page 241

1. False. Let $f(x) = x^2$. Then $f(3) = f(-3) = 9$, but $3 \neq -3$.   **2.** False. Consider $f(x) = x + 1$ and $g(x) = x^2 - 2$.   **3.** True   **4.** True
5. False. Let $f(x) = 3x$. $[f(x)]^2 = 9x^2$, whereas $f[f(x)] = f(3x) = 3(3x) = 9x$.   **6.** False. Let $f(x) = x^2$. Then $f(1) = 1$, $f(2) = 4$. Thus
$\dfrac{f(2)}{f(1)} = 4 \neq \dfrac{2}{1}$.   **7.** True   **8.** False. Let $f(x) = |x|$. Then $f(-1 + 3) = f(2) = 2$. $f(-1) + f(3) = 1 + 3 = 4$.   **9.** True   **10.** True   **11.** True
12. True   **13.** True

Chapter 2 Review Exercises, page 242

1. $\sqrt{181}$ [2.1]   **2.** $\sqrt{80} = 4\sqrt{5}$ [2.1]   **3.** $\left(-\dfrac{1}{2}, 10\right)$ [2.1]   **4.** $(2, -2)$ [2.1]   **5.** center $(3, -4)$, radius 9 [2.1]   **6.** center $(-5, -2)$,
radius 3 [2.1]   **7.** $(x - 2)^2 + (y + 3)^2 = 5^2$ [2.1]   **8.** $(x + 5)^2 + (y - 1)^2 = 8^2$, radius $= |-5 - (3)| = 8$ [2.1]   **9. a.** 2   **b.** 10   **c.** $3t^2 + 4t - 5$
d. $3x^2 + 6xh + 3h^2 + 4x + 4h - 5$ **e.** $9t^2 + 12t - 15$ **f.** $27t^2 + 12t - 5$ [2.2]   **10. a.** $\sqrt{55}$ **b.** $\sqrt{39}$ **c.** 0   **d.** $\sqrt{64 - x^2}$
e. $2\sqrt{64 - t^2}$ **f.** $2\sqrt{16 - t^2}$ [2.2]   **11. a.** 5   **b.** -11 **c.** $x^2 - 12x + 32$ **d.** $x^2 + 4x - 8$ [2.6]   **12. a.** 79   **b.** 56   **c.** $2x^2 - 4x + 9$
d. $2x^2 + 6$ [2.6]   **13.** $8x + 4h - 3$ [2.6]   **14.** $3x^2 + 3xh + h^2 - 1$ [2.6]   **15.** [2.2]   **16.** [2.2]

increasing on $[3, \infty)$
decreasing on $(-\infty, 3]$

f is increasing on $[0, \infty)$
f is decreasing on $(-\infty, 0]$

17. [2.2]

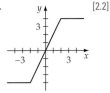

increasing on $[-2, 2]$
constant on $(-\infty, -2] \cup [2, \infty)$

18. [2.2]

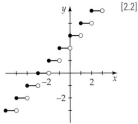

f is constant on..., $[-6, -5)$,
$[-5, -4), [-4, -3), [-3, -2),$
$[-2, -1), [-1, 0), [0, 1), \ldots$

19. [2.2]

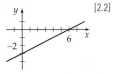

increasing on $(-\infty, \infty)$

20. [2.2]

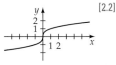

f is increasing on $(-\infty, \infty)$

21. Domain: $\{x \mid x \text{ is a real number}\}$ [2.2]   **22.** Domain: $\{x \mid x \le 6\}$ [2.2]   **23.** Domain: $\{x \mid -5 \le x \le 5\}$ [2.2]   **24.** Domain: $\{x \mid x \ne -3, x \ne 5\}$ [2.2]

25. $y = -2x + 1$ [2.3]   **26.** $y = \dfrac{11}{7}x$ [2.3]   **27.** $y = \dfrac{3}{4}x + \dfrac{19}{2}$ [2.3]   **28.** $y = \dfrac{5}{2}x + \dfrac{1}{2}$ [2.3]   **29.** $f(x) = (x + 3)^2 + 1$ [2.4]

30. $f(x) = 2(x + 1)^2 + 3$ [2.4]   **31.** $f(x) = -(x + 4)^2 + 19$ [2.4]   **32.** $f(x) = 4\left(x - \dfrac{3}{4}\right)^2 - \dfrac{5}{4}$ [2.4]   **33.** $f(x) = -3\left(x - \dfrac{2}{3}\right)^2 - \dfrac{11}{3}$ [2.4]

34. $f(x) = (x - 3)^2 + 0$ [2.4]   **35.** $(1, 8)$ [2.4]   **36.** $(0, -10)$ [2.4]   **37.** $(5, 161)$ [2.4]   **38.** $(-4, 30)$ [2.4]   **39.** $\dfrac{4\sqrt{5}}{5}$ [2.3]

40. a. $R = 13x$ **b.** $P = 12.5x - 1050$ **c.** $x = 84$ [2.3]   **41.** [2.5]

42. [2.5]

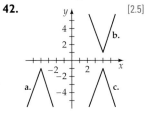

43. symmetric to the y-axis [2.5]   **44.** symmetric to the x-axis [2.5]   **45.** symmetric to the origin [2.5]   **46.** symmetric to the x-axis,
the y-axis, and the origin [2.5]   **47.** symmetric to the x-axis, the y-axis, and the origin [2.5]   **48.** symmetric to the origin [2.5]   **49.** symmetric
to the x-axis, the y-axis, and the origin [2.5]   **50.** symmetric to the origin [2.5]

51.

a. Domain is the set of all
real numbers.
Range: $\{y \mid y \le 4\}$
b. even [2.5]

52.

a. Domain is the set of all real
numbers.
Range is the set of all real
numbers.
b. g is neither even nor odd [2.5]

53.

a. Domain is the set of all
real numbers.
Range: $\{y \mid y \ge 4\}$
b. even [2.5]

54.

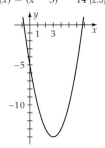

a. Domain: $\{x \mid -4 \le x \le 4\}$
Range: $\{y \mid 0 \le y \le 4\}$
b. even [2.5]

55.

a. Domain is the set of all
real numbers.
Range is the set of all real
numbers.
b. odd [2.5]

56.

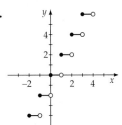

a. Domain: $\{x \mid x \text{ is a real number}\}$
Range: $\{y \mid y \text{ is an even integer}\}$
b. g is neither even nor odd [2.5]

57. $F(x) = (x + 2)^2 - 11$ [2.5]

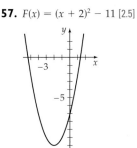

58. $A(x) = (x - 3)^2 - 14$ [2.5]

59. $P(x) = 3(x - 0)^2 - 4$ [2.5]   **60.** $G(x) = 2(x - 2)^2 - 5$ [2.5]   **61.** $W(x) = -4\left(x + \dfrac{3}{4}\right)^2 + \dfrac{33}{4}$ [2.5]   **62.** $T(x) = -2\left(x + \dfrac{5}{2}\right)^2 + \dfrac{25}{2}$

    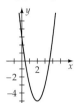      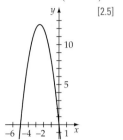

[2.5]

63. [2.5]   **64.** [2.5]   **65.** [2.5]

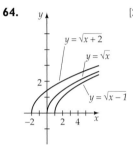   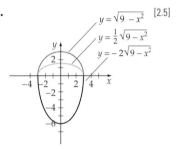

66. [2.2]   **67.** [2.2]   **68.** [2.2]

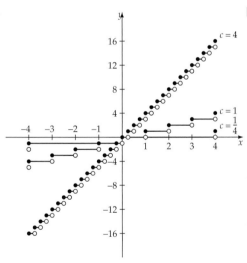   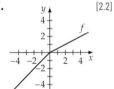   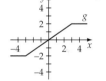

69. $f(x) + g(x) = x^2 + x - 6$, Domain is the set of all real numbers.
$f(x) - g(x) = x^2 - x - 12$, Domain is the set of all real numbers.
$f(x) \cdot g(x) = x^3 + 3x^2 - 9x - 27$, Domain is the set of all real numbers.
$\dfrac{f(x)}{g(x)} = x - 3$, Domain $\{x \mid x \neq -3\}$ [2.6]

70. $(f + g)(x) = x^3 + x^2 - 2x + 12$, Domain is the set of all real numbers.
$(f - g)(x) = x^3 - x^2 + 2x + 4$, Domain is the set of all real numbers.
$(fg)(x) = x^5 - 2x^4 + 4x^3 + 8x^2 - 16x + 32$, Domain is the set of all real numbers.
$\left(\dfrac{f}{g}\right)(x) = x + 2$, Domain is the set of all real numbers. [2.6]

71. 25, 25 [2.4]   **72.** −5 and 5 [2.4]   **73. a.** 18 feet per second   **b.** 15 feet per second   **c.** 13.5 feet per second   **d.** 12.03 feet per second
e. 12 feet per second [2.4]   **74. a.** 17 feet per second   **b.** 15 feet per second   **c.** 14 feet per second   **d.** 13.02 feet per second
e. 13 feet per second [2.4]

Chapter 2 Test, page 244

1. midpoint (1, 1); length $2\sqrt{13}$ [2.1]   **2.** $(-4, 0); \left(0, \sqrt{2}\right), \left(0, -\sqrt{2}\right)$ [2.1]   **3.** [2.1]

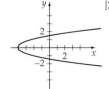   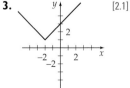

4. center $(2, -1)$; radius 3 [2.1]   **5.** domain $\{x \,|\, x \geq 4 \text{ or } x \leq -4\}$ [2.2]   **6.**

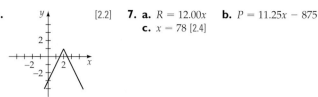

[2.2]   **7. a.** $R = 12.00x$ **b.** $P = 11.25x - 875$
c. $x = 78$ [2.4]

a. increasing on $(-\infty, 2]$
b. not constant on any interval
c. decreasing on $[2, \infty)$

8.

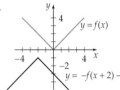

[2.5]   **9. a.** even   **b.** odd   **c.** neither [2.5]   **10.** $y = -\dfrac{2}{3}x + \dfrac{2}{3}$ [2.3]   **11.** -12, minimum [2.4]

12. $x^2 + x - 3; \dfrac{x^2 - 1}{x - 2}, x \neq 2$ [2.6]   **13.** $2x + h$ [2.6]   **14.** $4x^2 + 16x + 15$ [2.6]

15. a. 25 feet per second   **b.** 22.5 feet per second   **c.** 20.05 feet per second [2.6]

Cumulative Review Exercises, page 245

1. Commutative Property of Addition [P.1]   **2.** $\dfrac{6}{\pi}$, $\sqrt{2}$ [P.1]   **3.** $8x - 33$ [P.1]   **4.** $128x^5y^{10}$ [P.2]   **5.** $\dfrac{4}{3b^2}$ [P.2]   **6.** $6x^2 - 5x - 21$ [P.3]

7. $\dfrac{x + 9}{x + 3}$ [P.4]   **8.** $\dfrac{-2}{(2x - 1)(x - 1)}$ [P.4]   **9.** 0 [1.1]   **10.** $\dfrac{1 \pm \sqrt{5}}{2}$ [1.3]   **11.** $-\dfrac{7}{2}, 1$ [1.3]   **12.** $x = -\dfrac{2}{3}y + 5$ [1.1]   **13.** $\pm\sqrt{2}, \pm i$ [1.4]

14. $x > -4$ [1.5]   **15.** $\sqrt{17}$ [2.1]   **16.** -15 [2.2]   **17.** $y = -\dfrac{1}{2}x - 2$ [2.3]   **18.** 100 ounces [1.1]   **19.** Yes [2.4]   **20.** 0.04°F per minute [2.3]

Exercise Set 3.1, page 257

1. $5x^2 - 9x + 10 - \dfrac{10}{x + 3}$ **3.** $x^3 + 2x^2 - x + 1 + \dfrac{1}{x - 2}$ **5.** $x^2 + 4x + 10 + \dfrac{25}{x - 3}$ **7.** $x^3 + 7x^2 + 31x + 119 + \dfrac{475}{x - 4}$

9. $x^4 + 2x^3 + 2x - 1 - \dfrac{8}{x - 1}$ **11.** $4x^2 + 3x + 12 + \dfrac{17}{x - 2}$ **13.** $4x^2 - 4x + 2 + \dfrac{1}{x + 1}$ **15.** $x^4 + 4x^3 + 6x^2 + 24x + 101 + \dfrac{403}{x - 4}$

17. $x^4 + x^3 + x^2 + x + 1$ **19.** $8x^2 + 6$ **21.** $x^7 + 2x^6 + 5x^5 + 10x^4 + 21x^3 + 42x^2 + 85x + 170 + \dfrac{344}{x - 2}$

23. $x^5 - 3x^4 + 9x^3 - 27x^2 + 81x - 242 + \dfrac{716}{x + 3}$ **25.** 25   **27.** 45   **29.** -2230 **31.** -80 **33.** -187 **35.** Yes   **37.** No

39. Yes   **41.** Yes   **43.** No   **55.** $(x - 2)(x^2 + 3x + 7)$ **57.** $(x - 4)(x^3 + 3x^2 + 3x + 1)$ **59. a.** $19,968   **b.** $23,007   **61. a.** 336
b. 336; They are the same.   **63. a.** 100 cards   **b.** 610 cards   **65. a.** 400 people per square mile   **b.** 240 people per square mile
67. a. 304 cubic inches   **b.** 892 cubic inches   **69.** 13   **71.** Yes

Prepare for Section 3.2, page 259

73. 2   **74.** $\dfrac{9}{8}$ **75.** $[-1, \infty)$ **76.** $[1, \infty)$ **77.** $(x + 1)(x - 1)(x + 2)(x - 2)$ **78.** $\left(\dfrac{2}{3}, 0\right), \left(-\dfrac{1}{2}, 0\right)$

Exercise Set 3.2, page 271

1. up to the far left, up to the far right   **3.** down to the far left, up to the far right   **5.** down to the far left, down to the far right
7. down to the far left, up to the far right   **9.** $a < 0$ **11.** Vertex is $(-2, -5)$, minimum is -5.   **13.** Vertex is $(-4, 17)$, maximum is 17.
15. relative maximum $y \approx 5.0$ at   **17.** relative maximum $y \approx 31.0$ at   **19.** relative maximum $y \approx 2.0$ at $x \approx 1.0$,
$\quad\quad x \approx -2.1$, relative minimum $\quad\quad x \approx -2.0$, relative minimum $\quad\quad$ relative minima $y \approx -14.0$ at $x \approx -1.0$
$\quad\quad y \approx -16.9$ at $x \approx 1.4$ $\quad\quad y \approx -77.0$ at $x \approx 4.0$ $\quad\quad$ and $y \approx -14.0$ at $x \approx 3.0$

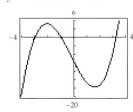

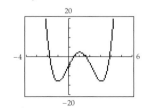

21. $-3, 0, 5$ **23.** $-3, -2, 2, 3$ **25.** $-2, -1, 0, 1, 2$ **33.** crosses the x-axis at $(-1, 0)$, $(1, 0)$, and $(3, 0)$ **35.** crosses the x-axis at $(7, 0)$;

intersects but does not cross at $(3, 0)$ **37.** crosses the x-axis at $(1, 0)$; intersects but does not cross at $\left(\dfrac{3}{2}, 0\right)$ **39.** crosses the x-axis at $(0, 0)$;
intersects but does not cross at $(3, 0)$

41.

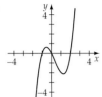

43.

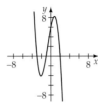

45.

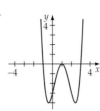

47. a. $V(x) = x(15 - 2x)(10 - 2x) = 4x^3 - 50x^2 + 150x$ **b.** 1.96 inches   **49.** 2.137 inches   **51.** \$464,000   **53. a.** 1918   **b.** 9.5 marriages per thousand population   **55. a.** 20.69 milligrams   **b.** 118 minutes   **57. a.** 3.24 inches   **b.** 4 feet from either end; 3.84 inches
c. 3.24 inches   **59.** between 3 and 4   **61.** $(5, 0)$ **63.** Shift the graph of $y = x^3$ horizontally 2 units to the right and vertically upward 1 unit.

Prepare for Section 3.3, page 275

65. $\dfrac{2}{3}, \dfrac{7}{2}$ **66.** $2x^2 - x + 6 - \dfrac{19}{x + 2}$ **67.** $3x^3 + 9x^2 + 6x + 15 + \dfrac{40}{x - 3}$ **68.** $1, 2, 3, 4, 6, 12$ **69.** $\pm 1, \pm 3, \pm 9, \pm 27$
70. $P(-x) = -4x^3 - 3x^2 + 2x + 5$

Exercise Set 3.3, page 286

1. 3 (multiplicity 2), -5 (multiplicity 1)   **3.** 0 (multiplicity 2), $-\dfrac{5}{3}$ (multiplicity 2)   **5.** 2 (multiplicity 1), -2 (multiplicity 1), -3 (multiplicity 2)

7. $\pm 1, \pm 2, \pm 4, \pm 8$ **9.** $\pm 1, \pm 2, \pm 3, \pm 4, \pm 6, \pm 12, \pm \dfrac{1}{2}, \pm \dfrac{3}{2}$ **11.** $\pm 1, \pm 2, \pm 4, \pm \dfrac{1}{2}, \pm \dfrac{1}{3}, \pm \dfrac{2}{3}, \pm \dfrac{4}{3}, \pm \dfrac{1}{6}$ **13.** $\pm 1, \pm 7, \pm \dfrac{1}{2}, \pm \dfrac{7}{2},$

$\pm \dfrac{1}{4}, \pm \dfrac{7}{4}$ **15.** $\pm 1, \pm 2, \pm 4, \pm 8, \pm 16, \pm 32$ **17.** upper bound 2, lower bound -5 **19.** upper bound 4, lower bound -4

21. upper bound 1, lower bound -4 **23.** upper bound 4, lower bound -2 **25.** upper bound 2, lower bound -1 **27.** one positive zero,
two or no negative zeros   **29.** two or no positive zeros, one negative zero   **31.** one positive zero, three or one negative zeros

33. three or one positive zeros, one negative zero   **35.** one positive zero, no negative zeros   **37.** $2, -1, -4$ **39.** $3, -4, \dfrac{1}{2}$

41. $\dfrac{1}{2}, -\dfrac{1}{3}, -2$ (multiplicity 2)   **43.** $\dfrac{1}{2}, 4, \sqrt{3}, -\sqrt{3}$ **45.** $6, 1 + \sqrt{5}, 1 - \sqrt{5}$ **47.** $5, \dfrac{1}{2}, 2 + \sqrt{3}, 2 - \sqrt{3}$

49. $1, -1, -2, -\dfrac{2}{3}, 3 + \sqrt{3}, 3 - \sqrt{3}$ **51.** $2, -1$ (multiplicity 2)   **53.** $0, -2, 1 + \sqrt{2}, 1 - \sqrt{2}$ **55.** -1 (multiplicity 3), 2

57. $-\dfrac{3}{2}, 1$ (multiplicity 2), 8   **59.** $n = 9$ inches   **61.** $x = 4$ inches   **63. a.** 26 pieces   **b.** 7 cuts   **65.** 7 rows   **67.** $x = 0.084$ inch

69. 1977 and 1986   **71.** 16.9 feet   **73. a.** 73 seconds   **b.** 93,000 digits   **75.** $B = 15$. The absolute value of each of the given zeros is less
than B.   **77.** $B = 11$. The absolute value of each of the zeros is less than B.

Prepare for Section 3.4, page 290

79. $3 + 2i$ **80.** $2 - i\sqrt{5}$ **81.** $x^3 - 8x^2 + 19x - 12$ **82.** $x^2 - 4x + 5$ **83.** $-3i, 3i$ **84.** $\dfrac{1}{2} - \dfrac{1}{2}i\sqrt{19}, \dfrac{1}{2} + \dfrac{1}{2}i\sqrt{19}$

Exercise Set 3.4, page 297

1. $2, -3, 2i, -2i$; $P(x) = (x - 2)(x + 3)(x - 2i)(x + 2i)$ **3.** $\dfrac{1}{2}, -3, 1 + 5i, 1 - 5i$; $P(x) = \left(x - \dfrac{1}{2}\right)(x + 3)(x - 1 - 5i)(x - 1 + 5i)$

5. 1 (multiplicity 3), $3 + 2i, 3 - 2i$; $P(x) = (x - 1)^3(x - 3 - 2i)(x - 3 + 2i)$

7. $-3, -\dfrac{1}{2}, 2 + i, 2 - i$; $P(x) = (x + 3)\left(x + \dfrac{1}{2}\right)(x - 2 - i)(x - 2 + i)$

9. $4, 2, \dfrac{1}{2} + \dfrac{3}{2}i, \dfrac{1}{2} - \dfrac{3}{2}i$; $P(x) = (x - 4)(x - 2)\left(x - \dfrac{1}{2} - \dfrac{3}{2}i\right)\left(x - \dfrac{1}{2} + \dfrac{3}{2}i\right)$ **11.** $1 - i, \dfrac{1}{2}$ **13.** $i, -3$ **15.** $2 + 3i, i, -i$

17. $1 - 3i, 1 + 2i, 1 - 2i$ **19.** $2i, 1$ (multiplicity 3) **21.** $5 - 2i, \dfrac{7}{2} + \dfrac{\sqrt{3}}{2}i, \dfrac{7}{2} - \dfrac{\sqrt{3}}{2}i$ **23.** $\dfrac{3}{2}, -\dfrac{1}{2} + \dfrac{\sqrt{7}}{2}i, -\dfrac{1}{2} - \dfrac{\sqrt{7}}{2}i$ **25.** $-\dfrac{2}{3}, \dfrac{3}{4}, \dfrac{5}{2}$

27. $-i, i, 2$ (multiplicity 2) **29.** -3 (multiplicity 2), 1 (multiplicity 2) **31.** $P(x) = x^3 - 3x^2 - 10x + 24$ **33.** $P(x) = x^3 - 3x^2 + 4x - 12$

35. $P(x) = x^4 - 10x^3 + 63x^2 - 214x + 290$ **37.** $P(x) = x^5 - 22x^4 + 212x^3 - 1012x^2 + 2251x - 1830$ **39.** $P(x) = 4x^3 - 19x^2 + 224x - 159$

41. $P(x) = x^3 + 13x + 116$ **43.** $P(x) = x^4 - 18x^3 + 131x^2 - 458x + 650$ **45.** $P(x) = x^5 - 4x^4 + 16x^3 - 18x^2 - 97x + 102$

47. $P(x) = 3x^3 - 12x^2 + 3x + 18$ **49.** $P(x) = -2x^4 + 4x^3 + 36x^2 - 140x + 150$ **51.** The Conjugate Pair Theorem does not apply because some of the coefficients of the polynomial are not real numbers.

Prepare for Section 3.5, page 298

53. $\dfrac{x - 3}{x - 5}$ **54.** $-\dfrac{3}{2}$ **55.** $\dfrac{1}{3}$ **56.** $x = 0, -3, \dfrac{5}{2}$ **57.** The degree of the numerator is 3. The degree of the denominator is 2.

58. $x + 4 + \dfrac{7x - 11}{x^2 - 2x}$

Exercise Set 3.5, page 311

1. $x = 0, x = -3$ **3.** $x = -\dfrac{1}{2}, x = \dfrac{4}{3}$ **5.** $y = 4$ **7.** $y = 30$

9. $x = -4, y = 0$ **11.** $x = 3, y = 0$ **13.** $x = 0, y = 0$ **15.** $x = -4, y = 1$ **17.** $x = 2, y = -1$

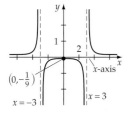 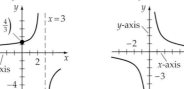 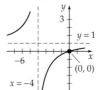 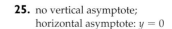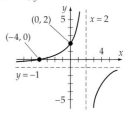

19. $x = 3, x = -3, y = 0$ **21.** $x = -3, x = 1, y = 0$ **23.** $x = -2, y = 1$ **25.** no vertical asymptote; horizontal asymptote: $y = 0$

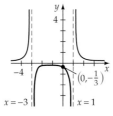 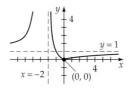 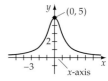

27. $x = 3, x = -3, y = 2$ **29.** $x = -1 + \sqrt{2}, x = -1 - \sqrt{2}, y = 1$ **31.** $y = 3x - 7$ **33.** $y = x$ **35.** $x = 0, y = x$

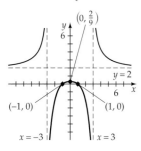 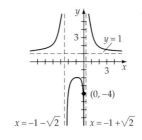 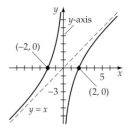

37. $x = -3, y = x - 6$ **39.** $x = 4, y = 2x + 13$ **41.** $x = -2, y = x - 3$ **43.** $x = 2, x = -2, y = x$

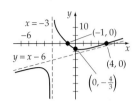

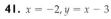

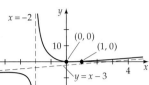

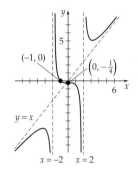

45.   **47.**   **49.**   **51.**

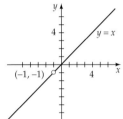

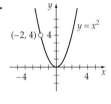

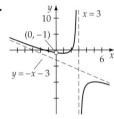

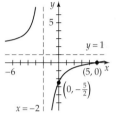

53. a. $76.43, $8.03, $1.19   **b.** $y = 0.43$. As the number of golf balls produced increases, the average cost per golf ball approaches $.43.

55. a. $1333.33   **b.** $8000   **c.**

57. a. $R(0) \approx 38.8\%, R(7) \approx 39.9\%, R(12) \approx 40.9\%$ **b.** $\approx 44.7\%$

59. a. 26,923, 68,293, 56,000   **b.** 2001   **c.** The population will approach 0.   **61. a.** 3.8 centimeters

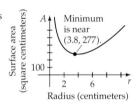

b. No   **c.** As the radius r increases without bound, the surface area approaches twice the area of a circle with radius r.   **63.** $(-2, 2)$

65. $(0, 1)$ and $(-4, 1)$ **67.** Answers will vary; one example is $\dfrac{x^2 + 1}{x^2 - x - 6}$.

Chapter 3 True/False Exercises, page 317

1. False; $P(x) = x - i$ has a zero of i, but it does not have a zero of $-i$.   **2.** False; Descartes' Rule of Signs indicates that $P(x) = x^3 - x^2 + x - 1$ has three or one positive zeros. In fact, P has only one positive zero.   **3.** True   **4.** True   **5.** False; $F(x) = \dfrac{x}{x^2 + 1}$ does not have a vertical asymptote.   **6.** False; $F(x) = \dfrac{(x - 2)^2}{(x - 3)(x - 2)} = \dfrac{x - 2}{x - 3}, x \neq 2$. The graph of F has a hole at $x = 2$.   **7.** True   **8.** True   **9.** True   **10.** True

11. True   **12.** False; $P(x) = x^2 + 1$ does not have a real zero.

Chapter 3 Review Exercises, page 318

1. $4x^2 + x + 8 + \dfrac{22}{x - 3}$ [3.1]   **2.** $5x^2 + 5x - 13 - \dfrac{11}{x - 1}$ [3.1]   **3.** $3x^2 - 6x + 7 - \dfrac{13}{x + 2}$ [3.1]   **4.** $2x^2 + 8x + 20$ [3.1]   **5.** $3x^2 + 5x - 11$ [3.1]

6. $x^3 + 2x^2 - 8x - 9$ [3.1]   **7.** 77 [3.1]   **8.** 22 [3.1]   **9.** 33 [3.1]   **10.** 558 [3.1]

The verifications in Exercises 11–14 make use of the concepts from Section 3.1.

15. [3.2]   **16.** [3.2]   **17.** [3.2]   **18.** [3.2]   **19.** [3.2]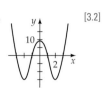

20. [3.2]   **21.** $\pm1, \pm2, \pm3, \pm6$ [3.3]   **22.** $\pm1, \pm2, \pm3, \pm5, \pm6, \pm10, \pm15, \pm30, \pm\dfrac{1}{2}, \pm\dfrac{3}{2}, \pm\dfrac{5}{2}, \pm\dfrac{15}{2}$ [3.3]

23. $\pm1, \pm2, \pm3, \pm4, \pm6, \pm12, \pm\dfrac{1}{3}, \pm\dfrac{2}{3}, \pm\dfrac{4}{3}, \pm\dfrac{1}{5}, \pm\dfrac{2}{5}, \pm\dfrac{3}{5}, \pm\dfrac{4}{5}, \pm\dfrac{6}{5}, \pm\dfrac{12}{5}, \pm\dfrac{1}{15}, \pm\dfrac{2}{15}, \pm\dfrac{4}{15}$ [3.3]   **24.** $\pm1, \pm2, \pm4, \pm8, \pm16, \pm32, \pm64$ [3.3]   **25.** ±1 [3.3]   **26.** $\pm1, \pm2, \pm\dfrac{1}{6}, \pm\dfrac{1}{3}, \pm\dfrac{1}{2}, \pm\dfrac{2}{3}$ [3.3]   **27.** no positive real zeros and three or one negative real zeros [3.3]

28. three or one positive real zeros, one negative real zero [3.3]   **29.** one positive real zero and one negative real zero [3.3]

30. five, three, or one positive real zeros, no negative real zeros [3.3]   **31.** $1, -2, -5$ [3.3]   **32.** 2, 5, 3 [3.3]   **33.** -2 (multiplicity 2), $-\dfrac{1}{2}, -\dfrac{4}{3}$ [3.3]   **34.** $-\dfrac{1}{2}, -3, i, -i$ [3.4]   **35.** 1 (multiplicity 4) [3.3]   **36.** $-\dfrac{1}{2}, 2 + 3i, 2 - 3i$ [3.4]   **37.** $-1, 3, 1 + 2i$ [3.4]   **38.** $-5, 2, 2 - i$ [3.4]

39. $P(x) = 2x^3 - 3x^2 - 23x + 12$ [3.4]   **40.** $P(x) = x^4 + x^3 - 5x^2 + x - 6$ [3.4]   **41.** $P(x) = x^4 - 3x^3 + 27x^2 - 75x + 50$ [3.4]

42. $P(x) = x^4 + 2x^3 + 6x^2 + 32x + 40$ [3.4]   **43.** vertical asymptote: $x = -2$, horizontal asymptote: $y = 3$ [3.5]

44. vertical asymptotes: $x = -3, x = 1$, horizontal asymptote: $y = 2$ [3.5]   **45.** vertical asymptote: $x = -1$, slant asymptote: $y = 2x + 3$ [3.5]

46. no vertical asymptote, horizontal asymptote: $y = 3$ [3.5]

47. [3.5]   **48.** [3.5]   **49.** [3.5]   **50.** [3.5]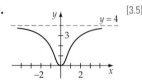

51. [3.5] 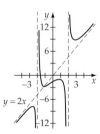   **52.** [3.5] 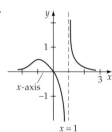   **53.** [3.5] 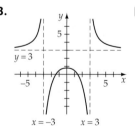   **54.** [3.5]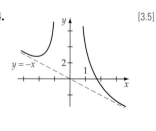

55. a. \$12.59, \$6.43   **b.** $y = 5.75$. As the number of skateboards produced increases, the average cost per skateboard approaches \$5.75. [3.5]
56. a. 15°F   **b.** 2.4°F   **c.** 0°F [3.5]   **57. a.** As the radius of the blood vessel approaches 0, the resistance gets larger.   **b.** As the radius of the blood vessel gets larger, the resistance approaches zero. [3.5]

Chapter 3 Test, page 319

1. $3x^2 - x + 6 - \dfrac{13}{x + 2}$ [3.1]   **2.** 43 [3.1]   **3.** The verification for Exercise 3 makes use of the concepts from Section 3.1.   **4.** up to the far left and down to the far right [3.2]   **5.** $0, \frac{2}{3}, -3$ [3.2]   **6.** $P(1) < 0, P(2) > 0$. Therefore, by the Zero Location Theorem, the continuous polynomial

function P has a zero between 1 and 2. [3.2]   **7.** 2 (multiplicity 2), -2 (multiplicity 2), $\frac{3}{2}$ (multiplicity 1), -1 (multiplicity 3) [3.3]

8. $\pm 1, \pm 3, \pm \frac{1}{2}, \pm \frac{3}{2}, \pm \frac{1}{3}, \pm \frac{1}{6}$ [3.3]   **9.** upper bound 4, lower bound -5 [3.3]   **10.** four, two, or zero positive zeros, no negative zero [3.3]

11. $\frac{1}{2}, 3, -2$ [3.3]   **12.** $2 - 3i, -\frac{2}{3}, -\frac{5}{2}$ [3.4]   **13.** 0, 1 (multiplicity 2), $2 + i, 2 - i$ [3.4]   **14.** $P(x) = x^4 - 5x^3 + 8x^2 - 6x$ [3.4]   **15.** vertical

asymptotes: $x = 3, x = 2$ [3.5]   **16.** horizontal asymptote: $y = \frac{3}{2}$ [3.5]   **17.** [3.5]   **18.** [3.5]

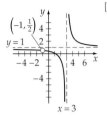

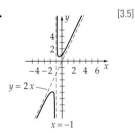

19. a. 5 words per minute, 16 words per minute, 25 words per minute   **b.** 70 words per minute [3.5]   **20.** 2.42 inches, 487.9 cubic inches [3.3]

Cumulative Review Exercises, page 321

1. $-1 + 2i$ [P.5]   **2.** $\frac{1 \pm \sqrt{5}}{2}$ [1.3]   **3.** 2, 10 [1.4]   **4.** $\{x \mid -8 \le x \le 14\}$ [1.5]   **5.** $\sqrt{281}$ [2.1]   **6.** Translate the graph of $y = x^2$ to the right

2 units and 4 units up. [2.5]   **7.** $2x + h - 2$ [2.6]   **8.** $32x^2 - 92x + 60$ [2.6]   **9.** $x^3 - x^2 + x + 11$ [2.6]   **10.** $4x^3 - 8x^2 + 14x - 32 + \frac{59}{x + 2}$

[3.1]   **11.** 141 [3.1]   **12.** The graph goes down. [3.2]   **13.** 0.3997 [3.2]   **14.** $\pm 1, \pm 2, \pm 4, \pm \frac{1}{3}, \pm \frac{2}{3}, \pm \frac{4}{3}$ [3.3]   **15.** zero positive real zeros,
three or one negative real zeros [3.3]   **16.** $-2, 1 + 2i, 1 - 2i$ [3.4]   **17.** $P(x) = x^3 - 4x^2 - 2x + 20$ [3.4]   **18.** $(x - 2)(x + 3i)(x - 3i)$ [3.4]
19. vertical asymptotes: $x = -3, x = 2$; horizontal asymptote: $y = 4$ [3.5]   **20.** $y = x + 4$ [3.5]

Exercise Set 4.1, page 334
1. 3   **3.** -3 **5.** 3   **7.** range   **9.** Yes   **11.** Yes   **13.** Yes

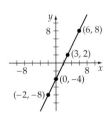

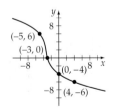

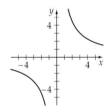

15. No

17. Yes   **19.** Yes   **21.** No   **23.** $\{(1, -3), (2, -2), (5, 1), (-7, 4)\}$ **25.** $\{(1, 0), (2, 1), (4, 2), (8, 3), (16, 4)\}$

27. $f^{-1}(x) = \frac{1}{2}x - 2$ **29.** $f^{-1}(x) = \frac{1}{3}x + \frac{7}{3}$ **31.** $f^{-1}(x) = -\frac{1}{2}x + \frac{5}{2}$ **33.** $f^{-1}(x) = \frac{x}{x - 2}, x \ne 2$ **35.** $f^{-1}(x) = \frac{x + 1}{1 - x}, x \ne 1$
37. $f^{-1}(x) = \sqrt{x - 1}, x \ge 1$ **39.** $f^{-1}(x) = x^2 + 2, x \ge 0$ **41.** $f^{-1}(x) = \sqrt{x + 4} - 2, x \ge -4$ **43.** $f^{-1}(x) = -\sqrt{x + 5} - 2, x \ge -5$
45. $V^{-1}(x) = \sqrt[3]{x}$. V^{-1} finds the length of a side of a cube given the volume.
47. Yes. Yes. A conversion function is a nonconstant linear function. All nonconstant linear functions have inverses that are also functions.

49. $s^{-1}(x) = \dfrac{1}{2}x - 12$ **51.** $E^{-1}(s) = 20s - 50{,}000$. From the monthly earnings s the executive can find $E^{-1}(s)$, the value of the software sold.

53. $44205833; f^{-1}(x) = \dfrac{1}{2}x + \dfrac{1}{2}, f^{-1}(44205833) = 22102917$

55. Because the function is increasing and 4 is between 2 and 5, c must be between 7 and 12.   **57.** between 2 and 5

59. between 3 and 7   **61.** $f^{-1}(x) = \dfrac{x-b}{a}, a \neq 0$ **63.** The reflection of f across the line given by $y = x$ yields f. Thus f is its own inverse.

65. Yes   **67.** No

Prepare for Section 4.2, page 337

69. 8   **70.** $\dfrac{1}{81}$ **71.** $\dfrac{17}{8}$ **72.** $\dfrac{40}{9}$ **73.** $\dfrac{1}{10}, 1, 10,$ and 100 **74.** $2, 1, \dfrac{1}{2},$ and $\dfrac{1}{4}$

Exercise Set 4.2, page 346

1. $f(0) = 1; f(4) = 81$ **3.** $g(-2) = \dfrac{1}{100}; g(3) = 1000$ **5.** $h(2) = \dfrac{9}{4}; h(-3) = \dfrac{8}{27}$ **7.** $j(-2) = 4; j(4) = \dfrac{1}{16}$ **9.** 9.19   **11.** 9.03   **13.** 9.74

15. a. $k(x)$ **b.** $g(x)$ **c.** $h(x)$ **d.** $f(x)$

17. 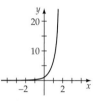   **19.**   **21.** 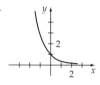   **23.**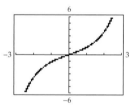

25. Shift the graph of f vertically upward 2 units.   **27.** Shift the graph of f horizontally to the right 2 units.
29. Reflect the graph of f across the y-axis.   **31.** Stretch the graph of f vertically away from the x-axis by a factor of 2.
33. Reflect the graph of f across the y-axis and then shift this graph vertically upward 2 units.

35. no horizontal asymptote   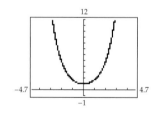   **37.** no horizontal asymptote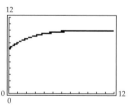

39. horizontal asymptote: $y = 0$ 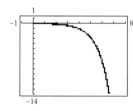   **41.** horizontal asymptote: $y = 10$

43. a. 122 million connections   **b.** 2006   **45. a.** 233 items per month; 59 items per month   **b.** The demand will approach 25 items per month.   **47. a.** 0.53   **b.** 0.89   **c.** 5.2 minutes   **d.** There is a 98% probability that at least one customer will arrive between 10:00 A.M. and 10:05.2 A.M.   **49. a.** 8.7%   **b.** 2.6%   **51. a.** 6400; 409,600   **b.** 11.6 hours   **53. a.** 515,000 people   **b.** 1997
55. a. 363 beneficiaries; 88,572 beneficiaries   **b.** 13 rounds   **57. a.** 141°F   **b.** after 28.3 minutes   **59. a.** 261.63 vibrations per second
b. No. The function $f(n)$ is not a linear function. Therefore, the graph of $f(n)$ does not increase at a constant rate.
63.   **65.** $(-\infty, \infty)$ **67.** $[0, \infty)$

Prepare for Section 4.3, page 351

68. 4   **69.** 3   **70.** 5   **71.** $f^{-1}(x) = \dfrac{3x}{2-x}$ **72.** $\{x|x \geq 2\}$ **73.** the set of all positive real numbers

Exercise Set 4.3, page 361

1. $10^1 = 10$ **3.** $8^2 = 64$ **5.** $7^0 = x$ **7.** $e^4 = x$ **9.** $e^0 = 1$ **11.** $\log_3 9 = 2$ **13.** $\log_4 \dfrac{1}{16} = -2$ **15.** $\log_b y = x$ **17.** $\ln y = x$

19. $\log 100 = 2$ **21.** 2   **23.** -5 **25.** 3   **27.** -2 **29.** -4 **31.**

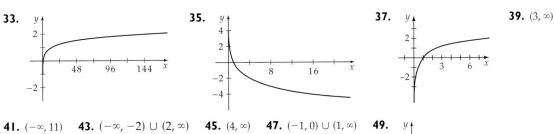

33.   **35.**   **37.**   **39.** $(3, \infty)$

41. $(-\infty, 11)$ **43.** $(-\infty, -2) \cup (2, \infty)$ **45.** $(4, \infty)$ **47.** $(-1, 0) \cup (1, \infty)$ **49.**

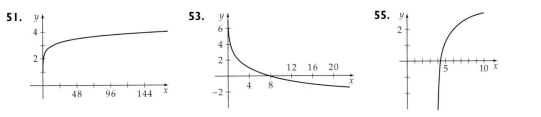

51.   **53.**   **55.**

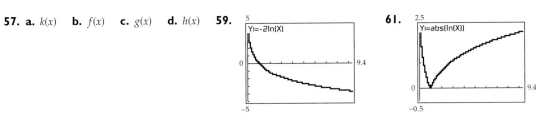

57. a. $k(x)$ **b.** $f(x)$ **c.** $g(x)$ **d.** $h(x)$ **59.**   **61.**

63.   **65.**   **67.**

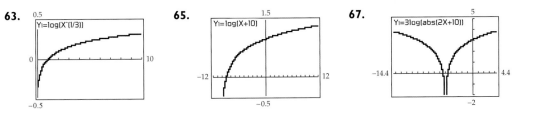

69. a. 2.0%   **b.** 45 months   **71. a.** 3298 units; 3418 units; 3490 units   **b.** 2750 units   **73.** 2.05 square meters
75. a. Answers will vary.   **b.** 96 digits   **c.** 3385 digits   **d.** 6,320,430 digits   **77.** f and g are inverse functions.
79. range of f: $\{y | -1 < y < 1\}$; range of g: all real numbers

Prepare for Section 4.4, page 364

81. ≈ 0.77815 for each expression   **82.** ≈ 0.98083 for each expression   **83.** ≈ 1.80618 for each expression
84. ≈ 3.21888 for each expression   **85.** ≈ 1.60944 for each expression   **86.** ≈ 0.90309 for each expression

Exercise Set 4.4, page 373

1. $\log_b x + \log_b y + \log_b z$ **3.** $\ln x - 4 \ln z$ **5.** $\frac{1}{2} \log_2 x - 3 \log_2 y$ **7.** $\frac{1}{2} \log_7 x + \frac{1}{2} \log_7 z - 2 \log_7 y$ **9.** $\log[x^2(x + 5)]$ **11.** $\ln(x + y)$
13. $\log\left[x^3 \cdot \sqrt[3]{y}(x + 1)\right]$ **15.** 1.5395   **17.** 0.8672   **19.** -0.6131 **21.** 0.6447   **23.**

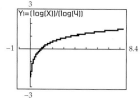

25.

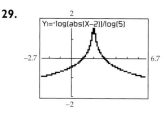

27.

29.

31. False; $\log 10 + \log 10 = 2$ but $\log(10 + 10) = \log 20 \neq 2$.   **33.** True   **35.** False; $\log 100 - \log 10 = 1$ but $\log(100 - 10) = \log 90 \neq 1$.
37. False; $\dfrac{\log 100}{\log 10} = \dfrac{2}{1} = 2$ but $\log 100 - \log 10 = 1$.   **39.** False; $(\log 10)^2 = 1$ but $2 \log 10 = 2$.   **41.** 2   **43.** 500^{501}
45. $1:870{,}551$; $1:757{,}858$; $1:659{,}754$; $1:574{,}349$; $1:500{,}000$ **47.** 10.4; base   **49.** 3.16×10^{-10} mole per liter
51. a. 82.0 decibels   **b.** 40.3 decibels   **c.** 115.0 decibels   **d.** 152.0 decibels   **53.** 10 times more intense   **55.** 5
57. $10^{6.5}I_0$ or about $3{,}162{,}277.7I_0$ **59.** 100 to 1   **61.** $10^{1.8}$ to 1 or about 63 to 1   **63.** 5.5   **65. a.** $M \approx 6$ **b.** $M \approx 4$ **c.** The results are close to the magnitudes produced by the amplitude-time-difference formula.

Prepare for Section 4.5, page 376

66. $\log_3 729 = 6$ **67.** $5^4 = 625$ **68.** $\log_a b = x + 2$ **69.** $x = \dfrac{4a}{7b + 2c}$ **70.** $x = \dfrac{3}{44}$ **71.** $x = \dfrac{100(A - 1)}{A + 1}$

Exercise Set 4.5, page 385

1. 6   **3.** $-\dfrac{3}{2}$ **5.** $-\dfrac{6}{5}$ **7.** 3   **9.** $\dfrac{\log 70}{\log 5}$ **11.** $-\dfrac{\log 120}{\log 3}$ **13.** $\dfrac{\log 315 - 3}{2}$ **15.** $\ln 10$ **17.** $\dfrac{\ln 2 - \ln 3}{\ln 6}$ **19.** $\dfrac{3 \log 2 - \log 5}{2 \log 2 + \log 5}$
21. 7   **23.** 4   **25.** $2 + 2\sqrt{2}$ **27.** $\dfrac{199}{95}$ **29.** -1 **31.** 3   **33.** 10^{10} **35.** 2   **37.** 5   **39.** $\log(20 + \sqrt{401})$ **41.** $\dfrac{1}{2} \log\left(\dfrac{3}{2}\right)$
43. $\ln(15 \pm 4\sqrt{14})$ **45.** $\ln(1 + \sqrt{65}) - \ln 8$ **47.** 1.61   **49.** 0.96   **51.** 2.20   **53.** -1.93 **55.** -1.34

57. a. 8500, 10,285   **b.** in 6 years   **59. a.** 60°F   **b.** 27 minutes   **61. a.**

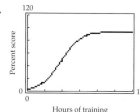

Percent score vs. Hours of training

  b. 48 hours
c. $P = 100$
d. As the number of hours of training increases, the test scores approach 100%.

63. a. 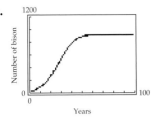 **b.** in 27 years or the year 2026 **c.** $B = 1000$ **d.** As the number of years increases, the bison population approaches but never reaches or exceeds 1000.

65. a. 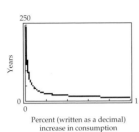 **b.** 78 years **c.** 1.9%

67. a. 116 feet per second **b.** $v = 150$ **c.** The velocity of the package approaches but never reaches or exceeds 150 feet per second.
69. a. 1.72 seconds **b.** $v = 100$ **c.** The object cannot fall faster than 100 feet per second.

71. a. 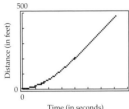 **b.** 2.6 seconds **73.** 138

75. The second step; because log 0.5 < 0, the inequality sign must be reversed. **77.** $x = \dfrac{y}{y - 1}$ **79.** $e^{0.336} \approx 1.4$

Prepare for Section 4.6, page 389
81. 1220.39 **82.** 824.96 **83.** −0.0495 **84.** 1340 **85.** 0.025 **86.** 12.8

Exercise Set 4.6, page 400

1. a. $9724.05 **b.** $11,256.80 **3. a.** $48,885.72 **b.** $49,282.20 **c.** $49,283.30 **5.** $24,730.82 **7.** 8.8 years **9.** $t = \dfrac{\ln 3}{r}$

11. 14 years **13. a.** 2200 bacteria **b.** 17,600 bacteria **15. a.** $N(t) \approx 22,600e^{0.01368t}$ **b.** 27,700 **17. a.** 10,755,000 **b.** 2042

19. a. 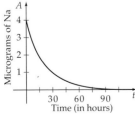 **b.** 3.18 micrograms **c.** ≈15.07 hours **d.** ≈30.14 hours **21.** ≈6601 years ago

23. ≈2378 years old **25. a.** 0.056 **b.** 42°F **c.** 54 minutes **27. a.** 211 hours **b.** 1386 hours **29.** 3.1 years

31. a. 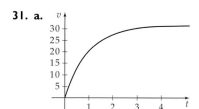   **b.** 0.98 second   **c.** $v = 32$ **d.** As time increases, the velocity approaches but never reaches or exceeds 32 feet per second.

33. a. 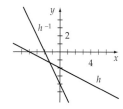   **b.** 2.5 seconds   **c.** ≈24.56 feet per second   **d.** The average speed of the object was approximately 24.56 feet per second during the period from $t = 1$ to $t = 2$ seconds.

35. a. 1900   **b.** 0.16   **c.** 200   **37. a.** 157,500   **b.** 0.04   **c.** 45,000   **39. a.** 2400   **b.** 0.12   **c.** 300

41. $P(t) \approx \dfrac{5500}{1 + 12.75e^{-0.37263t}}$ **43.** $P(t) \approx \dfrac{100}{1 + 4.55556e^{-0.22302t}}$ **45. a.** $158,000, $163,000   **b.** $625,000

47. a. $P(t) \approx \dfrac{1600}{1 + 4.12821e^{-0.06198t}}$ **b.** about 497 wolves   **49. a.** $P(t) \approx \dfrac{8500}{1 + 4.66667e^{-0.14761t}}$ **b.** 2010   **51.** 45 hours

53. a. 0.71 gram   **b.** 0.96 gram   **c.** 0.52 gram   **55. a.** 1.7%   **b.** 13.9%   **c.** 19.0%   **d.** 19.5%   **e.** 1.5%; $P \rightarrow 0$

57. 13,715,120,270 centuries

Chapter 4 True/False Exercises, page 409

1. False. $f(x) = x^2$ does not have an inverse function.   **2.** False. Let $f(x) = 2x$, $g(x) = 3x$. Then $f(g(0)) = 0$ and $g(f(0)) = 0$, but f and g are not inverse functions.   **3.** True   **4.** True   **5.** True   **6.** False; f is not defined for negative values of x, and thus $g(f(x))$ is undefined for negative values of x.   **7.** False; $h(x)$ is not an increasing function for $0 < b < 1$.   **8.** False; $j(x)$ is not an increasing function for $0 < b < 1$.
9. True   **10.** True   **11.** True   **12.** True   **13.** False; $\log x + \log y = \log(xy)$.   **14.** True   **15.** True   **16.** True

Chapter 4 Review Exercises, page 409

1. Yes [4.1]   **2.** Yes [4.1]   **3.** Yes [4.1]   **4.** No [4.1]

5. $f^{-1}(x) = \dfrac{x + 4}{3}$ [4.1]   **6.** $g^{-1}(x) = -\dfrac{1}{2}x + \dfrac{3}{2}$ [4.1]   **7.** $h^{-1}(x) = -2x - 4$ [4.1]   **8.** $k^{-1}(x) = k(x) = \dfrac{1}{x}$ [4.1]

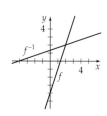   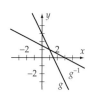   both k and k^{-1}

9. 2 [4.3]   **10.** 4 [4.3]   **11.** 3 [4.3]   **12.** π [4.3]   **13.** -2 [4.5]   **14.** 8 [4.5]   **15.** -3 [4.5]   **16.** -4 [4.5]
17. ± 1000 [4.5]   **18.** $\pm 10^{10}$ [4.5]   **19.** 7 [4.5]   **20.** ± 8 [4.5]

21.   [4.2]   **22.**   [4.2]   **23.** 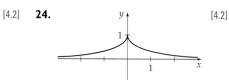   [4.2]   **24.**   [4.2]

25. [4.2] **26.** 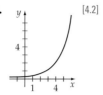 [4.2] **27.** [4.3] **28.** 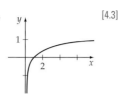 [4.3]

29. 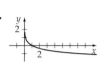 [4.3] **30.** 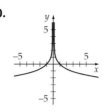 [4.3] **31.** 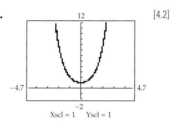 [4.2]

Xscl = 1   Yscl = 1

32. 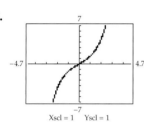 [4.2] **33.** $4^3 = 64$ [4.3] **34.** $\left(\dfrac{1}{2}\right)^{-3} = 8$ [4.3] **35.** $\left(\sqrt{2}\right)^4 = 4$ [4.3]

Xscl = 1   Yscl = 1

36. $e^0 = 1$ [4.3] **37.** $\log_5 125 = 3$ [4.3] **38.** $\log_2 1024 = 10$ [4.3] **39.** $\log_{10} 1 = 0$ [4.3] **40.** $\log_8 2\sqrt{2} = \dfrac{1}{2}$ [4.3]

41. $2 \log_b x + 3 \log_b y - \log_b z$ [4.4] **42.** $\dfrac{1}{2} \log_b x - 2 \log_b y - \log_b z$ [4.4] **43.** $\ln x + 3 \ln y$ [4.4] **44.** $\dfrac{1}{2} \ln x + \dfrac{1}{2} \ln y - 4 \ln z$ [4.4]

45. $\log\left(x^2 \sqrt[3]{x+1}\right)$ [4.4] **46.** $\log \dfrac{x^5}{(x+5)^2}$ [4.4] **47.** $\ln \dfrac{\sqrt{2xy}}{z^3}$ [4.4] **48.** $\ln \dfrac{xz}{y}$ [4.4] **49.** 2.86754 [4.4] **50.** 3.35776 [4.4]

51. -0.117233 [4.4] **52.** -0.578989 [4.4] **53.** $\dfrac{\ln 30}{\ln 4}$ [4.5] **54.** $\dfrac{\log 41}{\log 5} - 1$ [4.5] **55.** 4 [4.5] **56.** $\dfrac{1}{6}e$ [4.5] **57.** 4 [4.5] **58.** 15 [4.5]

59. $\dfrac{\ln 3}{2 \ln 4}$ [4.5] **60.** $\dfrac{\ln(8 \pm 3\sqrt{7})}{\ln 5}$ [4.5] **61.** 10^{1000} [4.5] **62.** $e^{(e^2)}$ [4.5] **63.** $1,000,005$ [4.5] **64.** $\dfrac{15 + \sqrt{265}}{2}$ [4.5] **65.** 81 [4.5]

66. $\pm\sqrt{5}$ [4.5] **67.** 4 [4.5] **68.** 5 [4.5] **69.** 7.7 [4.4] **70.** 5.0 [4.4] **71.** 3162 to 1 [4.4] **72.** 2.8 [4.4] **73.** 4.2 [4.4]

74. $\approx 3.98 \times 10^{-6}$ [4.4] **75. a.** $\$20,323.79$ **b.** $\$20,339.99$ [4.6] **76. a.** $\$25,646.69$ **b.** $\$25,647.32$ [4.6] **77.** $\$4,438.10$ [4.6]

78. a. 69.9% **b.** 6 days **c.** 19 days [4.6] **79.** $N(t) \approx e^{0.8047t}$ [4.6] **80.** $N(t) \approx 2e^{0.5682t}$ [4.6] **81.** $N(t) \approx 3.783e^{0.0558t}$ [4.6]

82. $N(t) \approx e^{-0.6931t}$ [4.6] **83. a.** $P(t) \approx 25,200e^{0.06155789t}$ **b.** $38,800$ [4.6] **84.** 340 years [4.6]

85. a. $P(t) \approx \dfrac{1400}{1 + \dfrac{17}{3}e^{-0.22458t}}$ **b.** 1070 [4.6] **86. a.** $21\dfrac{1}{3}$ **b.** $P(t) \to 128$ [4.6]

Chapter 4 Test, page 411

1. $f^{-1}(x) = \dfrac{1}{2}x + \dfrac{3}{2}$ [4.1] **2.** $f^{-1}(x) = \dfrac{8x}{4x - 1}$

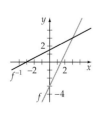

Domain f^{-1}: all real numbers except $\dfrac{1}{4}$

Range f^{-1}: all real numbers except 2 [4.1]

3. a. $b^c = 5x - 3$ [4.3] **b.** $\log_3 y = \dfrac{x}{2}$ [4.3] **4.** $2 \log_b z - 3 \log_b y - \dfrac{1}{2} \log_b x$ [4.4] **5.** $\log \dfrac{2x + 3}{(x - 2)^3}$ [4.4] **6.** 1.7925 [4.4]

7. [4.2]

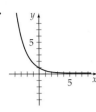

8. [4.3]

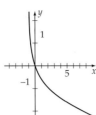

9. 1.9206 [4.5]   **10.** $\dfrac{5\ln 4}{\ln 28}$ [4.5]   **11.** 1 [4.5]   **12.** −3 [4.5]

13. a. $29,502.36   **b.** $29,539.62 [4.6]   **14.** 17.36 years [4.6]   **15. a.** 7.6   **b.** 63 to 1 [4.4]   **16. a.** $P(t) \approx 34,600e^{0.04667108t}$ **b.** 55,000 [4.6]

17. 690 years [4.6]   **18.** 32.2 minutes [4.5]   **19.** 53 feet per second [4.5]   **20. a.** $P(t) \approx \dfrac{1100}{1 + 5.875e^{-0.20429t}}$ **b.** ≈457 raccoons [4.6]

Cumulative Review Exercises, page 412

1. [2, 6] [1.1]   **2.** $\{x | 3 < x \le 6\}$ [1.5]   **3.** 7.8 [2.1]   **4.** 38.25 feet [2.4]   **5.** $4x^2 + 4x - 4$ [2.6]   **6.** $f^{-1}(x) = \dfrac{1}{3}x + \dfrac{5}{3}$ [4.1]

7. $A(W) = 2W^2$ [2.2]   **8.** 3 or 1 positive real zeros; 1 negative real zero [3.3]   **9.** $1, 4, -\sqrt{3}, \sqrt{3}$ [3.3]   **10.** $P(x) = x^3 - 4x^2 + 6x - 4$ [3.4]
11. vertical asymptote: $x = 4$, horizontal asymptote: $y = 3$ [3.5]   **12.** Domain: all real numbers; Range: $\{y | 0 < y \le 4\}$ [3.5]
13. decreasing function [4.2]   **14.** $4^y = x$ [4.3]   **15.** $\log_5 125 = 3$ [4.3]   **16.** 1.262 [4.4]   **17.** 7.1 [4.4]   **18.** 2.0149 [4.5]

19. 510 years old [4.6]   **20. a.** $P(x) \approx \dfrac{450}{1 + 1.8125e^{-0.13882x}}$ **b.** 310 wolves [4.6]

Exercise Set 5.1, page 425

1. $(2, -4)$ **3.** $\left(-\dfrac{6}{5}, \dfrac{27}{5}\right)$ **5.** $(3, 4)$ **7.** $(1, -1)$ **9.** $(3, -4)$ **11.** $(2, 5)$ **13.** $(-1, -1)$ **15.** $\left(\dfrac{62}{25}, \dfrac{34}{25}\right)$ **17.** no solution

19. $\left(c, -\dfrac{4}{3}c + 2\right)$ **21.** $(2, -4)$ **23.** $(0, 3)$ **25.** $\left(\dfrac{3}{5}c, c\right)$ **27.** $\left(-\dfrac{1}{2}, \dfrac{2}{3}\right)$ **29.** no solution   **31.** $(-6, 3)$ **33.** $\left(2, -\dfrac{3}{2}\right)$

35. $(2\sqrt{3}, 3)$ **37.** $\left(\dfrac{38}{17\pi}, \dfrac{3}{17}\right)$ **39.** $(\sqrt{2}, \sqrt{3})$ **41.** $125   **43.** plane: 120 mph, wind: 30 mph

45. boat: 25 mph, current: 5 mph   **47.** $12 per kilogram for iron, $16 per kilogram for lead

49. 8 grams of 40% gold, 12 grams of 60% gold   **51.** $\dfrac{9}{5}$ square units   **53.** 8   **55.** 42, 56, 70; 40, 42, 58; 42, 144, 150; 42, 440, 442

57. 90 people   **59.** (63,800, 12,760) The point indicates that a person with an adjusted gross income of $63,800 would pay the same tax using
either method.   **61.** $14,000 at 6%, $11,000 at 6.5%   **63.** $x = -\dfrac{58}{17}, y = \dfrac{52}{17}$ **65.** $x = -2, y = -1$ **67.** $x = \dfrac{153}{26}, y = \dfrac{151}{26}$

69. $x = 2 + 3i, y = 1 - 2i$ **71.** $x = 3 - 5i, y = 4i$

Prepare for Section 5.2, page 428

72. $y = \dfrac{2}{5}x - 3$ **73.** $z = -c + 13$ **74.** $\left(\dfrac{18}{5}, 4\right)$ **75.** $(2, -5)$ **76.** $(-2, -10)$ **77.** $(c, -4c + 9)$

Exercise Set 5.2, page 439

1. $(2, -1, 3)$ **3.** $(2, 0, -3)$ **5.** $(2, -3, 1)$ **7.** $(-5, 1, -1)$ **9.** $(3, -5, 0)$ **11.** $(0, 2, 3)$ **13.** $(5c - 25, 48 - 9c, c)$ **15.** $(3, -1, 0)$

17. no solution   **19.** $\left(\dfrac{1}{11}(50 - 11c), \dfrac{1}{11}(11c - 18), c\right)$ **21.** no solution   **23.** $\left(\dfrac{1}{29}(25 + 4c), \dfrac{1}{29}(55 - 26c), c\right)$ **25.** $(0, 0, 0)$

27. $\left(\dfrac{5}{14}c, \dfrac{4}{7}c, c\right)$ **29.** $(-11c, -6c, c)$ **31.** $(0, 0, 0)$ **33.** $y = 2x^2 - x - 3$ **35.** $x^2 + y^2 - 4x + 2y - 20 = 0$
37. center $(-7, -2)$, radius 13   **39.** 500 cars per hour   **41.** AC: 258 to 308, CD: 209 to 259, BD: 262 to 312   **43.** $d_1 = 9$ inches,
$d_2 = 3$ inches, $d_3 = 4$ inches   **45.** $(3, 5, 2, -3)$ **47.** $(1, -2, -1, 3)$ **49.** $(14a - 7b - 8, -6a + 2b + 5, a, b)$ **51.** $A = -\dfrac{13}{2}$

53. $A \neq -3, A \neq 1$ **55.** $A = -3$ **57.** $3x - 5y - 2z = -2$

Prepare for Section 5.3, page 442

58. $\{x \mid x \geq -3\}$ **59.** $-1 \pm \sqrt{3}$ **60.** $(1, -3)$ **61.**   **62.** 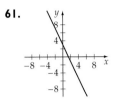   **63.**

Exercise Set 5.3, page 451

1.   **3.**   **5.**   **7.**   **9.**

11.   **13.** 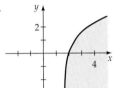   **15.**   **17.**   **19.**

21. no solution 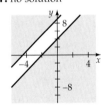   **23.**   **25.** 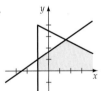   **27.**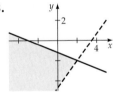

29.   **31.** 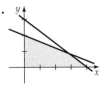   **33.** minimum at $(0, 8)$: 16   **35.** maximum at $(6, 5)$: 71   **37.** minimum at $\left(0, \dfrac{10}{3}\right)$: 20

39. maximum at $(0, 8)$: 56   **41.** minimum at $(2, 6)$: 18   **43.** maximum at $(3, 4)$: 25   **45.** 20 acres of wheat and 40 acres of barley

47. 0 starter sets and 18 pro sets   **49.**    **51.** 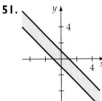   **53.**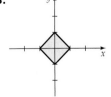

55.

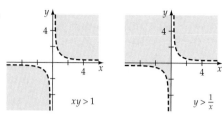

$xy > 1$ $y > \dfrac{1}{x}$

If x is a negative number, then the inequality is reversed when both sides of the inequality are divided by a negative number.

57. 24 ounces of group B and 0 ounces of group A yields a minimum cost of $2.40.

59. Two 4-cylinder engines and seven 6-cylinder engines yields a maximum profit of $2050.

Chapter 5 True/False Exercises, page 455

1. False; $\begin{cases} x + y = 1 \\ x + y = 3 \end{cases}$ has no solution.   **2.** True   **3.** False; a homogeneous system is one in which the constant term in each equation is zero.

4. True   **5.** False; $\begin{cases} x + y = 2 \\ x + 2y = 3 \end{cases}$ and $\begin{cases} 2x + 3y = 5 \\ 2x - 2y = 0 \end{cases}$ are two systems with the same solution but without common equations.   **6.** True

7. True   **8.** False; it is inconsistent.   **9.** False; $(-1, 1)$ satisfies the first inequality but not the second, and $(-2, -1)$ satisfies the second but not the first.

Chapter 5 Review Exercises, page 455

1. $\left(-\dfrac{18}{7}, -\dfrac{15}{28} \right)$ [5.1]   **2.** $\left(\dfrac{3}{2}, -3 \right)$ [5.1]   **3.** $(-3, -1)$ [5.1]   **4.** $(-4, 7)$ [5.1]   **5.** $(3, 1)$ [5.1]   **6.** $(-1, 1)$ [5.1]   **7.** $\left(\dfrac{1}{2}(5 - 3c), c \right)$ [5.1]

8. no solution [5.1]   **9.** $\left(\dfrac{1}{2}, 3, -1 \right)$ [5.2]   **10.** $\left(\dfrac{16}{3}, \dfrac{10}{27}, -\dfrac{29}{45} \right)$ [5.2]   **11.** $\left(\dfrac{1}{11}(7c - 3), \dfrac{1}{11}(16c - 43), c \right)$ [5.2]   **12.** $\left(\dfrac{74}{31}, -\dfrac{1}{31}, \dfrac{3}{31} \right)$ [5.2]

13. $\left(2, \dfrac{1}{2}(3c + 2), c \right)$ [5.2]   **14.** $\left(1, -\dfrac{2}{3}, \dfrac{1}{4} \right)$ [5.2]   **15.** $\left(\dfrac{14}{11}c, -\dfrac{2}{11}c, c \right)$ [5.2]   **16.** $(0, 0, 0)$ [5.2]   **17.** $\left(\dfrac{1}{2}(c + 1), \dfrac{1}{4}(3c - 1), c \right)$ [5.2]

18. $\left(\dfrac{65 - 11c}{16}, \dfrac{19 - c}{8}, c \right)$ [5.2]   **19.** [5.3]

20. [5.3]

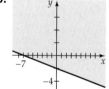

21. [5.3]

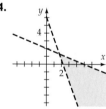

22. [5.3]

23. [5.3]

24. [5.3]

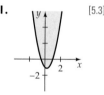

25. [5.3]

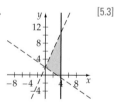

26. [5.3]

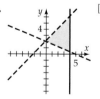

27. [5.3]

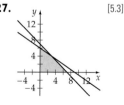

28. [5.3]

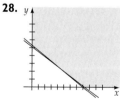

29. [5.3]

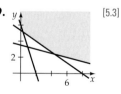

30. [5.3]   **31.** The maximum is 18 at (4, 5). [5.3]   **32.** The maximum is 44 at (6, 4). [5.3]

33. The minimum is 8 at (0, 8). [5.3]   **34.** The minimum is 20 at (10, 0). [5.3]   **35.** The minimum is 27 at (2, 5). [5.3]   **36.** The maximum is 43 at (3, 7). [5.3]   **37.** $y = \frac{11}{6}x^2 - \frac{5}{2}x + \frac{2}{3}$ [5.2]   **38.** $x^2 + y^2 - \frac{47}{11}x - \frac{21}{11}y + \frac{10}{11} = 0$ [5.2]   **39.** $z = -2x + 3y + 3$ [5.2]   **40.** $x = 15$ liters [5.1]

41. wind: 28 mph, plane: 143 mph [5.1]   **42.** 4 nickels, 3 dimes, 3 quarters; 1 nickel, 7 dimes, 2 quarters [5.2]   **43.** (0, 0, 0), (1, 1, 1), (1, −1, −1), (−1, −1, 1), (−1, 1, −1) [5.2]

Chapter 5 Test, page 457

1. (−3, 2) [5.1]   **2.** $\left(\frac{1}{2}(6 + c), c\right)$ [5.1]   **3.** $\left(\frac{173}{39}, \frac{29}{39}, -\frac{4}{3}\right)$ [5.2]   **4.** $\left(\frac{1}{4}(c + 3), \frac{1}{8}(7c + 1), c\right)$ [5.2]   **5.** $\left(\frac{1}{13}(c + 10), \frac{1}{13}(5c + 11), c\right)$ [5.2]

6. $\left(\frac{1}{14}c, -\frac{9}{14}c, c\right)$ [5.2]   **7.** [5.3]   **8.** [5.3]   **9.** 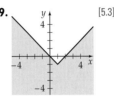 [5.3]

10. [5.3]   **11.** [5.3]   **12.** 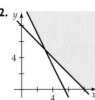 [5.3]   **13.** 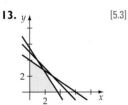 [5.3]

14. $\frac{680}{7}$ acres of oats and $\frac{400}{7}$ acres of barley [5.3]   **15.** $x^2 + y^2 - 2y - 24 = 0$ [5.2]

Cumulative Review Exercises, page 457

1. $-\frac{14}{27}$ [2.3]   **2.** $\{y \mid y \le -3\}$ [2.4]   **3.** 91 [P.1]   **4.** $\log_6[8x^3(x - 5)]$ [4.4]   **5.** −1, 2 [3.3]   **6.** $d_0 = \frac{Fd_1}{d_1 - F}$ [1.2]

7. $y = -\frac{1}{2}x$ [2.3]   **8.** even [2.5]   **9.** $\frac{300}{199}$ [4.5]   **10.** $3 + h$ [2.6]   **11.** 5 [2.2]   **12.** 30 [2.6]   **13.** 3 [4.3]

14. $-2x^2 + 3$ [2.6]   **15.** $x^3 + 2x^2 + 9x + 18$ [3.4]   **16.** $Q^{-1}(r) = \frac{r - 2}{2}$ [4.1]   **17.** $y = 2x + 1$ [3.5]   **18.** 81 [4.2]

19. 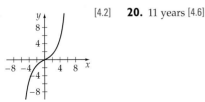 [4.2]   **20.** 11 years [4.6]

INDEX